T0181927

Computational Methods for Multiphase Flow

Predicting the behavior of multiphase flows is a problem of immense importance for both industrial and natural processes. Thanks to high-speed computers and advanced algorithms, the numerical simulation of such flows is rapidly gaining importance. Researchers and students alike need to have a one-stop account of the field. This book provides a comprehensive and self-contained graduate-level introduction to the computational modeling of multiphase flows. Each chapter is written by a recognized expert and contains extensive references to current research. The book is organized so that the chapters are essentially independent and it can be used for a range of advanced courses and the self-study of specific topics. In the first part a variety of different numerical methods for direct numerical simulations are described and illustrated with suitable examples. The second part is devoted to the numerical treatment of higher-level, averaged-equations models. No other book offers the simultaneous coverage of so many topics related to the computational modeling of multiphase flow. It will be welcomed by researchers and graduate students in engineering, physics, and applied mathematics.

Now in paperback, this edition incorporates authors' corrections since the first printing.

Computational Methods for Multiphase Flow

ANDREA PROSPERETTI
and
GRÉTAR TRYGGVASON

CAMBRIDGE
UNIVERSITY PRESS

CAMBRIDGE
UNIVERSITY PRESS

University Printing House, Cambridge CB2 8BS, United Kingdom

One Liberty Plaza, 20th Floor, New York, NY 10006, USA

477 Williamstown Road, Port Melbourne, VIC 3207, Australia

314-321, 3rd Floor, Plot 3, Splendor Forum, Jasola District Centre, New Delhi - 110025, India

103 Penang Road, #05-06/07, Visioncrest Commercial, Singapore 238467

Cambridge University Press is part of the University of Cambridge.

It furthers the University's mission by disseminating knowledge in the pursuit of
education, learning and research at the highest international levels of excellence.

www.cambridge.org
Information on this title: www.cambridge.org/9780521138611

© Cambridge University Press 2007, 2009

First published 2007
Reprinted in paperback with corrections 2009

A catalogue record for this publication is available from the British Library

ISBN 978-0-521-84764-3 Hardback
ISBN 978-0-521-13861-1 Paperback

Additional resources for this publication at www.cambridge.org/comp_mult_flow

Contents

v

Preface

Computation has made theory more relevant

This is a graduate-level textbook intended to serve as an introduction to computational approaches which have proven useful for problems arising in the broad area of multiphase flow. Each chapter contains references to the current literature and to recent developments on each specific topic, but the primary purpose of this work is to provide a solid basis on which to build both applications and research. For this reason, while the reader is expected to have had some exposure to graduate-level fluid mechanics and numerical methods, no extensive knowledge of these subjects is assumed. The treatment of each topic starts at a relatively elementary level and is developed so as to enable the reader to understand the current literature.

A large number of topics fall under the generic label of "computational multiphase flow," ranging from fully resolved simulations based on first principles to approaches employing some sort of coarse-graining and averaged equations. The book is ideally divided into two parts reflecting this distinction. The first part (Chapters 2–5) deals with methods for the solution of the Navier–Stokes equations by finite difference and finite element methods, while the second part (Chapters 9–11) deals with various reduced descriptions, from point-particle models to two-fluid formulations and averaged equations. The two parts are separated by three more specialized chapters on the lattice Boltzmann method (Chapter 6), the boundary integral method for Stokes flow (Chapter 7), and on averaging and the formulation of averaged equation (Chapter 8).

This is a multi-author volume, but we have made an effort to unify the notation and to include cross-referencing among the different chapters. Hopefully this feature avoids the need for a sequential reading of the chapters, possibly aside from some introductory material mostly presented in Chapter 1.

The objective of this work is to describe computational methods, rather than the physics of multiphase flow. With this aspect in mind, the primary criterion in the selection of specific examples has been their usefulness to illustrate the capabilities of an algorithm rather than the characteristics of particular flows. Selected figures are available in color via the book's website at `http://www.cambridge.org/comp_mult_flow`.

The original idea for this book was conceived when we chaired the Study Group on Computational Physics in connection with the Workshop on Scientific Issues in Multiphase Flow. The workshop, chaired by Prof. T.J. Hanratty, was sponsored by the U.S. Department of Energy and held on the campus of the University of Illinois at Urbana–Champaign on May 7–9 2002; a summary of the findings has been published in the *International Journal of Multiphase Flow*, Vol. 29, pp. 1041–1116 (2003). As we started to collect material and to receive input form our colleagues, it became clearer and clearer that multiphase flow computation has become an activity with a major impact in industry and research. While efforts in this area go back at least five decades, the great improvement in hardware and software of the last few years has provided a significant impulse which, if anything, can be expected to only gain momentum in the coming years.

Most multiphase flows inherently involve a multiplicity of both temporal and spatial scales. Phenomena at the scale of single bubbles, drops, solid particles, capillary waves, and pores determine the behavior of large chemical reactors, energy production systems, oil extraction, and the global climate itself. Our ability to see how the integration across all these scales comes about and what are its consequences is severely limited by this mind-boggling complexity. This is yet another area where computing offers a powerful tool for significant progress in our ability to understand and predict.

Basic understanding is achieved not only through the simulation of actual physical processes, but also with the aid of computational "experiments." Multiphase flows are notorious for the difficulties in setting up fully controlled physical experiments. However, computationally, it is possible, for example, to include or not include gravity, account for the effects of a well-characterized surfactant, and others. It is now possible to routinely compute the behavior of relatively simple systems, such as the breakup of jets and the shape of bubbles. The next few years are likely to result in an explosion of results for such relatively simple systems where computations will help us gain a very complete picture of the relevant physics over a large range of parameters. A strong impulse to these activities will be imparted by effective computational methods for multiscale problems, which are rapidly developing.

At a practical, industrial level, simulation must rely on an averaged description and closure models to account for the unresolved phenomena. The formulation of these closures will greatly benefit from the detailed simulation of the underlying microphysics. The situation is similar to single-phase turbulent flows where, in the last two decades, simulations have played a major role, e.g. in developing large-eddy models.

It is in the examination of very complex, very large-scale systems, where it is necessary to follow the evolution of an enormous range of scales for a long time, that the major challenges and opportunities lie. Such simulations, in which it is possible to get access to the complete data and to control accurately every aspect of the system, will not only revolutionize our predictive capability, but also open up new opportunities for controlling the behavior of such systems.

It is our firm belief that today we stand at the threshold of exciting developments in the understanding of multiphase flows for which computation will prove an essential element. All of us – authors and editors – sincerely hope that this book will contribute to further progress in this field.

Andrea Prosperetti
Gretar Tryggvason

Acknowledgments

The editors and the contributors to the present volume wish to acknowledge the help and support received by several individuals and organizations in connection with the preparation of this work.

- S. BALACHANDAR's research was supported by the ASCI Center for the Simulation of Advanced Rockets at the University of Illinois at Urbana-Champaign through the U.S. Department of Energy (subcontract number B341494).
- JERZY BLAWZDZIEWICZ would like to acknowledge the support provided by NSF CAREER grant CTS-0348175.
- HOWARD H. HU's research was supported by NSF grant CTS-9873236 and by DARPA through a grant to the University of Pennsylvania.
- M. YOUSUFF HUSSAINI would like to acknowledge NSF contract DMS 0108672, and the support and encouragement of Provost Lawrence G. Abele.
- LI-SHI LUO would like to acknowledge the support provided by NSF grant CTS-0500213.
- SREEKANTH PANNALA and SANKARAN SUNDARESAN would like thank Tom O'Brien, Madhava Syamlal and the MFIX team. The contribution has been partly authored by a contractor of the U.S. Government under Contract No. DE-AC05-00OR22725. Accordingly, the U.S. Government retains a non-exclusive, royalty-free license to publish or reproduce the published form of this contribution, or allow others to do so, for U.S. Government purposes.
- ANDREA PROSPERETTI expresses his gratitude to Drs. Anthony J. Baratta, Cesare Frepoli, Yao-Shin Hwang, Raad Issa, John H. Mahaffy, Randi Moe, Christopher J. Murray, Fadel Moukalled, Sylvain Pigny,

Iztok Tiselj, and Vaughn E. Whisker. His work was supported by NSF grant CTS-0210044 and by DOE grant DE-FG02-99ER14966.

- MARK SUSSMAN's contribution was supported in part by the National Science Foundation under contract DMS 0108672

- GRETAR TRYGGVASON would like to thank his graduate students and collaborators who have contributed to his work on multiphase flows. He would also like to acknowledge support by DOE grant DE-FG02-03ER46083, NSF grant CTS-0522581, as well as NASA projects NAG3-2535 and NNC05GA26G, during the preparation of this book.

- DUAN Z. ZHANG would like to acknowledge many important discussions and physical insights offered by Dr. F. H. Harlow. The Joint DoD/ DoE Munitions Technology Development Program provided the financial support for this work.

Contributors

S. (Bala) Balachandar Department of Mechanical and Aerospace Engineering, MAE-A Building, POB 116250, University of Florida, Gainesville, FL 32611, USA
s-bala@ufl.edu

Jerzy Blawzdziewicz Department of Mechanical Engineering, Yale University, P.O. Box 208286, New Haven, CT 06520-8286, USA
jerzy.blawzdziewicz@yale.edu

Shiyi Chen Department of Mechanical Engineering, Johns Hopkins University, 3400 N. Charles Street, Baltimore MD 21218, USA
syc@jhu.edu

Xiaoyi He Computational Modeling Center, Air Products and Chemicals, Inc., 7201 Hamilton Boulevard, Allentown, PA 18195-1501, USA
hex@airproducts.com

Howard H. Hu Department of Mechanical Engineering and Applied Mechanics, University of Pennsylvania, 229 Towne Building, 220 S. 33rd Street, Philadelphia, PA 19104-6315, USA
hhu@seas.upenn.edu

M. Yousuff Hussaini Department of Mathematics, Florida State University, Tallahassee, FL 32306, USA
myh@math.fsu.edu

Li-Shi Luo Department of Mathematics & Statistics, Old Dominion University, Norfolk, VA 23529, USA
lluo@odu.edu

Sreekanth Pannala Computational Mathematics Group, Computer Science & Mathematics Division, Oak Ridge National Laboratory, Oak Ridge, TN 37831, USA
pannalas@ornl.gov

Andrea Prosperetti Department of Mechanical Engineering, Johns
   Hopkins University, 3400 N. Charles Street, Baltimore MD 21218,
   USA; *and* Faculty of Applied Sciences and Burgers Centrum,
   University of Twente, 7500 AE Enschede, The Netherlands
   prosperetti@jhu.edu

Kyle D. Squires Mechanical and Aerospace Engineering Department,
   Arizona State University, Box 876106, Tempe, AZ 85287-6106, USA
   squires@asu.edu

Sankaran Sundaresan Department of Chemical Engineering, A-315
   Engineering Quadrangle, Princeton University, Princeton, NJ
   08544-5263, USA
   sundar@princeton.edu

Mark Sussman Department of Mathematics, Florida State University,
   Tallahassee, FL 32306, USA
   sussman@math.fsu.edu

Gretar Tryggvason Department of Mechanical Engineering, 100
   Institute Road, Worcester Polytechnic Institute, Worcester, MA
   01609-2280, USA
   gretar@wpi.edu

Duan Z. Zhang Theoretical Division, Fluid Dynamics Group, Los
   Alamos National Laboratory, Los Alamos, NM 87545, USA
   dzhang@lanl.gov

1

Introduction: A computational approach to multiphase flow

This book deals with multiphase flows, i.e. systems in which different fluid phases, or fluid and solid phases, are simultaneously present. The fluids may be different phases of the same substance, such as a liquid and its vapor, or different substances, such as a liquid and a permanent gas, or two liquids. In fluid–solid systems, the fluid may be a gas or a liquid, or gases, liquids, and solids may all coexist in the flow domain.

Without further specification, nearly all of fluid mechanics would be included in the previous paragraph. For example, a fluid flowing in a duct would be an instance of a fluid–solid system. The age-old problem of the fluid-dynamic force on a body (e.g. a leaf in the wind) would be another such instance, while the action of wind on ocean waves would be a situation involving a gas and a liquid.

In the sense in which the term is normally understood, however, *multiphase flow* denotes a subset of this very large class of problems. A precise definition is difficult to formulate as, often, whether a certain situation should be considered as a multiphase flow problem depends more on the point of view – or even the motivation – of the investigator than on its intrinsic nature. For example, wind waves would not fall under the purview of multiphase flow, even though some of the physical processes responsible for their behavior may be quite similar to those affecting gas–liquid stratified flows, e.g. in a pipe – a prime example of a multiphase system. The wall of a duct or a tree leaf may be considered as boundaries of the flow domain of interest, which would not qualify these as multiphase flow problems. However, the flow in a network of ducts, or wind blowing through a tree canopy, may be – and have been – studied as multiphase flow problems.

These examples point to a frequent feature of multiphase flow systems, namely the *complexity* arising from the mutual interaction of many subsystems. But – as a counterexample to the extent that it may be regarded as 'simple' – one may consider a single small bubble as an instance of multiphase flow, particularly if the study focuses on features that would be relevant to an assembly of such entities.

The interaction among many entities, such as bubbles, drops, or particles immersed in the fluid, is not the only source of the complexity usually exhibited by multiphase flow phenomena. There may be many other components as well, such as the very physics of the problem (e.g. the advancing of a solid–liquid–gas contact line, or the transition between different gas–liquid flow regimes), the simultaneous occurring of phenomena spanning widely different scales (e.g. oil recovery, where the flow at the single pore level impacts the behavior of the entire reservoir), the presence of a disturbed interface (e.g. surface waves on a falling film, or large, highly deformable drops or bubbles), turbulence, and others.

This complexity strongly limits the usefulness of purely analytical methods. For example, even for the flow around bodies with a simple shape such as spheres, most analytical results are limited to very small or very large Reynolds numbers. The more common and interesting situation of intermediate Reynolds numbers can hardly be studied by these means. When two or more bodies interact, or the ambient flow is not simple, the power of analytical methods is reduced further.

In a laboratory, it may even be difficult to set up a multiphase flow experiment with the necessary degree of control: the breakup of a drop in a turbulent flow or a precise characterization of the bubble or drop size distribution may be examples of such situations. Furthermore, many of the experimental techniques developed for single-phase flow encounter severe difficulties in their extension to multiphase systems. For example, even at volume fractions of a few percent, a bubbly flow may be nearly opaque to optical radiation so that visualization becomes problematic. The clustering of suspended particles in a turbulent flow depends on small-scale details which it may be very difficult to resolve. Little information about atomization can be gained by local probes, while adequate seeding for visualization may be impossible.

In this situation, numerical simulation becomes an essential tool for the investigation of multiphase flow. In a limited number of cases, computation can solve actual practical problems which lend themselves to direct numerical simulation (e.g. the flow in microfluidic devices), or for which sufficiently reliable mathematical models exist. But, more frequently, computation is the only available tool to investigate crucial physical aspects of the situation of interest, for example the role of gravity, or surface tension, which can be set to arbitrary values unattainable with physical experimentation.

Furthermore, the complexity of multiphase flows often requires reduced descriptions, for example by means of averaged equations, and the formulation of such reduced models can greatly benefit from the insight provided by computational results.

The last decade has seen the development of powerful computational capabilities which have marked a turning point in multiphase flow research. In the chapters that follow, we will give an overview of many of these developments on which future progress will undoubtedly be built.

1.1 Some typical multiphase flows

Having given up on the idea of providing a definition, we may illustrate the scope of multiphase flow phenomena by means of some typical examples. Here we encounter an embarrassment of riches. In technology, electric power generation, sprays (e.g. in internal combustion engines), pipelines, catalytic oil cracking, the aeration of water bodies, fluidized beds, and distillation columns are all legitimate examples. As a matter of fact, it is estimated that over half of anything which is produced in a modern industrial society depends to some extent on a multiphase flow process. In Nature, one may cite sandstorms, sediment transport, the "white water" produced by breaking waves, geysers, volcanic eruptions, acquifiers, clouds, and rain. The number of items in these lists can easily be made arbitrarily large, but it may be more useful to consider with a minimum of detail a few representative situations.

A typical example of a multiphase flow of major industrial interest is a *fluidized bed* (see Section 10.4). Conceptually, this device consists of a vertical vessel containing a bed of particles, which may range in size from tens of microns to centimeters. A fluid (a liquid or, more frequently, a gas) is pumped through the porous bottom of the vessel and through the bed. As the flow velocity is increased, initially one observes an increasing pressure drop across the bed. However, when the pressure drop reaches a value close to the weight of the bed per unit area, the particles become suspended in the fluid stream and the bed is said to be fluidized. These systems are useful as they promote an intimate contact between the particles and the fluid which facilitates, e.g., the combustion of material with a low caloric content (such as low-grade coal, or even domestic garbage), the *in situ* absorption of the pollutants deriving from the combustion (e.g. limestone particles absorbing SO_2), the action of a catalyst (e.g. in oil cracking), and others. In order for the bed to fulfill these functions, it is desirable that it remain homogeneous, which is exceedingly difficult to obtain. Indeed, under most conditions, one observes large volumes of fluid, called bubbles, which contain a much smaller concentration of particles than the average, and which rise through the bed venting at its surface. In the regime commonly called "channeling," these particle-free fluid structures span the entire height of the bed. It is evident that both bubbling and channeling reduce the effectiveness of the system as they cause a large fraction of the fluid to leave the bed contacting only a limited number of particles. The transition from the state of uniform fluidization to the bubbling regime is thought to be the result of an instability which is still incompletely understood after several decades of study. The resulting uncertainty hampers both design tasks, such as scale-up, and performance, by requiring operation with conservative safety margins. Several different types of fluidized beds exist. Figure 1.1 shows a diagram of a *circulating* fluidized bed, so called because the particles are ejected from the top of the riser and then returned to the bed. The figure illustrates the wide variety of situations encountered in this system: the dense particle flow in the standpipe, the fast and dilute flow in the riser, the balance between centrifugal and gravitational forces in the cyclones, and wall effects.

It is evident that a system of this complexity is way beyond the reach of direct numerical simulation. Indeed, the mathematical models in use rely on averaged equations which, however, still suffer from several problems as will be explained in Chapters 8 and 10. Attempts to improve these equations must rely on a good understanding of the flow through assemblies of particles or, at the very least, of the flow around a particle suspended in a fluid stream, possibly spatially non-uniform and temporally varying. Furthermore, interactions with the walls are important. These considerations are a powerful motivation for the development of numerical methods for the detailed simulation of particle–fluid flow. Some methods suitable for this purpose are described in Chapters 4 and 5 of this book.

An important natural phenomenon involving fluid–particle interactions is *sediment transport* in rivers, coastal areas, and others. A significant difference with the case of fluidized beds is that, in this case, gravity tends to act orthogonally to the mean flow. This circumstance greatly affects the balance of forces on the particles, increasing the importance of lift. This component of the hydrodynamic force on bodies of a general shape is still insufficiently understood and, again, the computational methods described in Chapters 2–5 are an effective tool for its investigation.

A *bubble column* is the gas–liquid analog of a fluidized bed. The bubbles are introduced at the bottom of a liquid-filled column with the purpose of increasing the interfacial area available for a gas–liquid chemical reaction,

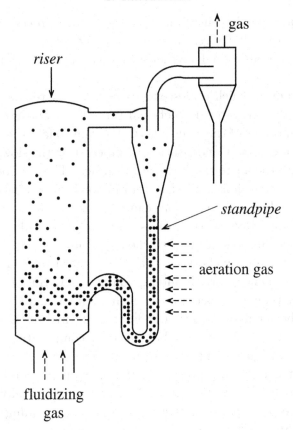

riser

gas

standpipe

aeration gas

fluidizing
gas

Fig. 1.1. This figure shows schematically one of several different configurations of a circulating fluidized bed loop used in engineering practice. The particles flow downward through the aerated "standpipe," and enter the bottom of a fast fluidized bed "riser." The particles are centrifugally separated from the gas in a train of "cyclones." In this diagram, the particles separated in the primary cyclone are returned to the standpipe while the fate of the particles removed from the secondary cyclone is not shown.

of aerating the liquid, or even to lift the liquid upward in lieu of a pump. Spatial inhomogeneities arise in systems of this type as well, and their effect can be magnified by the occurrence of coalescence which may produce very large gas bubbles occupying nearly the entire cross-section of the column and separated by so-called liquid "slugs." The transition from a bubbly to a slug-flow regime is a typical phenomenon of gas–liquid flows, of great practical importance but still poorly understood. Here, in addition to understanding how the bubbles arrange themselves in space, it is necessary to model the forces which cause coalescence and the coalescence process itself. These are evidently major challenges in free-surface flows: Chapters 10 and 11 describe some computational methods capable of shedding light on such phenomena.

Another system in which coalescence plays a major role is in *clouds* and rain formation. Small water droplets fall very slowly and are easy prey to the convective motions of the atmosphere. For rain to fall, the drops need to grow to a sufficient size. Condensation is impeded by the slowness of vapor diffusion through the air to reach the drop surface. The only possible explanation of the observed short time scale for rain formation is the occurrence of coalescence. Simple random collisions caused by turbulence are very unlikely in dilute conditions. Rather, the process must rely on a subtler influence of turbulence which can be studied with the aid of an approximation in which the finite size of the droplets is (partially) disregarded. This approach to the study of turbulence–particle interaction is a powerful one described in Chapter 9. This is another example in which a critical ingredient to improve modeling is a better understanding of fluctuating hydrodynamic forces on particle assemblies which can only be gained by computational means.

Other important gas–liquid flows occur in *pipelines*. Here free gas may exist because it is originally present at the inlet, as in many oil pipelines, but it may also be due to the ex-solution of gases originally dissolved in the liquid as the pressure along the pipeline falls. Depending on the liquid and gas flow rates and on the slope of the pipeline, one may observe a whole variety of flow regimes such as bubbly, stratified, wavy, slug, annular, and others. Each one of them reacts differently to an imposed pressure gradient. For example, in a stratified flow, a given pressure drop would produce a much larger flow rate of the gas phase than of the liquid phase, unlike a bubbly or slug-flow regime. In slug flow, solid surfaces such as pumps and tube walls are often subjected to large fluctuating forces which may cause dangerous vibration and fatigue. It is therefore of great practical importance to be able to predict which flow regime would occur in a given situation, the operational limits to remain in the desired regime, and how the system would react to transients such as start-ups and shut-downs. The experimental effort devoted to this subject has been very considerable, but progress has proven to be frustratingly slow and elusive. The computational methods described in Chapters 3, 10, and 11 are promising tools for a better understanding of these problems.

Even remaining at the level of the momentum coupling between the phases, all of the examples described so far are challenging enough that a complete understanding is not yet available. When energy coupling becomes important, such as in combustion and boiling, the difficulties increase and, with them, the prospect of progress by computational means. *Boiling* is the premier process by which electric power is generated world-wide, and is considered to be a vital means of heat removal in the computers of the future and human activities in space. Yet, this is another instance of those processes which have been very reluctant to yield their secrets in spite of nearly a century of experimental and theoretical work. Vital questions such as nucleation site density, bubble–bubble interaction, and critical heat flux are still for the most part unanswered. For space applications, understanding the role of gravity is an absolute prerequisite but microgravity experimentation is costly and fraught with difficulties. Once again, computation is a most attractive proposition. In this book, space constraints prevent us from getting very far into the treatment of nonadiabatic multiphase flow. A very brief treatment of energy coupling in the context of averaged equations is presented in Chapter 11.

1.2 A guided tour

The book can be divided into two parts, arranged in order of increasing complexity of the systems for which the methods described can be used. The first part, consisting of Chapters 2–7, describes methods suitable for the detailed solution of the Navier–Stokes equations for typical situations of interest in multiphase flow. Chapter 8 introduces the concept of averaged equations, and methods for their solution take up the second part of the book, Chapters 9 to 11.

In Chapter 2 we introduce the idea of direct numerical simulation of multiphase flows, discussing the motivation behind such simulations and what to expect from the results. We also give a brief overview of the various numerical methods used for such simulations and present in some detail elementary techniques for the solution of the Navier–Stokes equations. In Chapter 3, numerical methods for fluid–fluid simulations are discussed. The methods presented all rely on the use of a fixed Cartesian grid to solve the fluid equations, but the phase boundary is tracked in different ways, using either marker functions or connected marker particles. Computation of flows over stationary solid particles is discussed in Chapter 4. We first give an overview of methods based on the use of fixed Cartesian grids, along similar lines as the methods presented in Chapter 3, and then move on to methods based on body-fitted grids. While less versatile, these latter methods are capable of producing very accurate results for relatively high Reynolds number, thus providing essentially exact solutions that form the basis for the modeling of forces on single particles. Simulations of more complex solid-particle flows are introduced in Chapter 5, where several versions of finite element arbitrary Lagrangian–Eulerian methods, based on unstructured tetrahedron grids that adapt to the particles as they move, are used to simulate several moving solid particles. One of the important applications of simulations of this type may be in formulating closures of the averaged quantities necessary for the modeling of multiphase flows in average terms. Chapter 6 introduces the lattice Boltzman method for multiphase flows and in Chapter 7 we discuss boundary integral methods for Stokes flows of two immiscible fluids or solid particles in a viscous fluid. While restricted to a somewhat special class of flows, boundary integral methods can reduce the computational effort significantly and yield very accurate results.

Chapters 8–11 constitute the second part of the book and deal with situations for which the direct solution of the Navier–Stokes equations would require excessive computational resources and the use of reduced descriptions becomes necessary. The basis for these descriptions is some form of averaging applied to the exact microscopic laws and, accordingly, the first chapter of this group outlines the averaging procedure and illustrates how the various reduced descriptions in the literature and in the later chapters are rooted in it. A useful approximate treatment of disperse flows – primarily particles suspended in a gas – is based on the use of point-particle models, which are considered in Chapter 9. In these models, the fluid momentum equation is augmented by point forces which represent the effect of the particles, while the particle trajectories are calculated in a Lagrangian fashion by adopting simple parameterizations of the fluid-dynamic forces. The fluid component of the model, therefore, looks very much like the ordinary Navier–Stokes equations, and it can be treated by the same methods developed for single-phase computational fluid dynamics. At present, this is the only well-developed reduced-description approach capable of incorporating the direct numerical simulation of turbulence, and efforts are currently under way to apply to it the ideas and methods of large-eddy simulation.

The point-particle model is only valid when the particle concentration is so low that particle–particle interactions can be neglected, and the particles are smaller than the smallest flow length scale, e.g. in turbulent flow, the Kolmogorv scale. Therefore, while useful, the range of applicability of the approach is rather limited. The following two chapters deal with models based on a different philosophy of broader applicability, that of *interpenetrating continua*. In the underlying conceptual picture it is supposed that the various phases are simultaneously present in each volume element in proportions which vary with time and position. Each phase is described by a continuity, momentum, and energy equation, all of which contain terms describing the exchange of mass, momentum, and energy among the phases. Numerically, models of this type pose special challenges due to the nearly omnipresent instabilities of the equations, the constraint that the volume fractions occupied by each phase necessarily lie between 0 and 1, and many others.

In principle, the interpenetrating-continua modeling approach is very broadly applicable to a large variety of situations. A model suitable for one application, for example stratified flow in a pipeline, differs from that applicable to a different one, for example, pneumatic transport, mostly in the way in which the interphase interaction terms are specified. It turns out that, for computational purposes, most of these specific models share a very similar structure. A case in point is the vast majority of multiphase flow models adopted in commercial codes. Two broad classes of numerical methods are available. In the first one, referred to as the *segregated* approach and described in Chapter 10, the various balance equations are solved sequentially in an iterative fashion starting from an equation for the pressure. The general idea is derived from the well-known $SIMPLE$ method of single-phase computational fluid mechanics. The other class of methods, described in Chapter 11, adopts a more coupled approach to the solution of the equations and is suitable for faster transients with stronger interactions among the phases.

1.3 Governing equations and boundary conditions

In view of the prominent role played by the incompressible single-phase Navier–Stokes equations throughout this book, it is useful to summarize them here. It is assumed that the reader has a background in fluid mechanics and, therefore, no attempt at a derivation or an in-depth discussion will be made. Our main purpose is to set down the notation used in later chapters and to remind the reader of some fundamental dimensionless quantities which will be frequently encountered.

If $\rho(\mathbf{x}, t)$ and $\mathbf{u}(\mathbf{x}, t)$ denote the fluid density and velocity fields at position $\mathbf{x}$ and time t, the equation of continuity is

$$\frac{\partial \rho}{\partial t} + \boldsymbol{\nabla} \cdot (\rho \mathbf{u}) = 0. \tag{1.1}$$

For incompressible flows this equation reduces to

$$\boldsymbol{\nabla} \cdot \mathbf{u} = 0. \tag{1.2}$$

This latter equation embodies the fact that each fluid particle conserves its volume as it moves in the flow.

In conservation form, the momentum equation is

$$\frac{\partial}{\partial t}(\rho \mathbf{u}) + \boldsymbol{\nabla} \cdot (\rho \mathbf{uu}) = \boldsymbol{\nabla} \cdot \boldsymbol{\sigma} + \rho \mathbf{f}, \tag{1.3}$$

in which $\mathbf{f}$ is an external force per unit volume acting on the fluid. Very often, the force $\mathbf{f}$ will be the acceleration of gravity $\mathbf{g}$. However, as in Chapter 9, one may think of very small suspended particles as exerting point forces which can also be described by the field $\mathbf{f}$. The stress tensor $\boldsymbol{\sigma}$ may be decomposed into a pressure p and viscous part $\boldsymbol{\tau}$:

$$\boldsymbol{\sigma} = -p\hat{\boldsymbol{I}} + \boldsymbol{\tau}, \tag{1.4}$$

in which $\hat{\boldsymbol{I}}$ is the identity two-tensor. In most of the applications that follow, we will be dealing with Newtonian fluids, for which the viscous part of the stress tensor is given by

$$\boldsymbol{\tau} = 2\mu \mathbf{e}, \qquad \mathbf{e} = \frac{1}{2}\left(\boldsymbol{\nabla}\mathbf{u} + \boldsymbol{\nabla}\mathbf{u}^{\mathrm{T}}\right), \tag{1.5}$$

in which μ is the coefficient of (dynamic) viscosity, $\mathbf{e}$ the rate-of-strain tensor, and the superscript T denotes the transpose; in component form:

$$e_{ij} = \frac{1}{2}\left(\frac{\partial u_i}{\partial x_j} + \frac{\partial u_j}{\partial x_i}\right), \tag{1.6}$$

in which $\mathbf{x} = (x_1, x_2, x_3)$. With (1.5), (1.3) takes the familiar form of the Navier–Stokes momentum equation for a Newtonian, constant-properties fluid:

$$\frac{\partial \mathbf{u}}{\partial t} + \boldsymbol{\nabla} \cdot (\mathbf{uu}) = -\frac{1}{\rho}\boldsymbol{\nabla}p + \nu\boldsymbol{\nabla}^2\mathbf{u} + \mathbf{f}, \tag{1.7}$$

in which $\nu = \mu/\rho$ is the kinematic viscosity. Because of (1.2), this equation may be written in non-conservation form as

$$\frac{\partial \mathbf{u}}{\partial t} + (\mathbf{u} \cdot \boldsymbol{\nabla})\,\mathbf{u} = -\frac{1}{\rho}\boldsymbol{\nabla}p + \nu\boldsymbol{\nabla}^2\mathbf{u} + \mathbf{f}, \tag{1.8}$$

where the notation implies that the i-th component of the second term is given by

$$[(\mathbf{u} \cdot \boldsymbol{\nabla})\,\mathbf{u}]_i = \sum_{j=1}^{3} u_j \frac{\partial u_i}{\partial x_j}. \tag{1.9}$$

When the force field $\mathbf{f}$ admits a potential $\mathcal{U}$, $\mathbf{f} = -\boldsymbol{\nabla}\mathcal{U}$, one may introduce the *reduced* or *modified pressure*, i.e. the pressure in excess of the hydrostatic contribution,

$$p^{\mathrm{r}} = p + \rho\,\mathcal{U} \tag{1.10}$$

in terms of which (1.8) becomes

$$\frac{\partial \mathbf{u}}{\partial t} + (\mathbf{u} \cdot \boldsymbol{\nabla})\,\mathbf{u} = -\frac{1}{\rho}\boldsymbol{\nabla} p^{\mathrm{r}} + \nu \nabla^2 \mathbf{u}. \tag{1.11}$$

In particular, for the gravitational force, $\mathcal{U} = -\rho \mathbf{g} \cdot \mathbf{x}$.

We have already noted at the beginning of this chapter that multiphase flows are often characterized by the presence of interfaces. When there is a mass flux $\dot{m}$ across (part of) the boundary S separating two phases 1 and 2 as, for example, in the presence of phase change at a liquid–vapor interface, conservation of mass requires that

$$\dot{m} \equiv \rho_2\,(\mathbf{u}_2 - \mathbf{w}) \cdot \mathbf{n} = \rho_1\,(\mathbf{u}_1 - \mathbf{w}) \cdot \mathbf{n} \tag{1.12}$$

where $\mathbf{n}$ is the unit normal and $\mathbf{w} \cdot \mathbf{n}$ the normal velocity of the interface itself. An expression for this quantity is readily found if the interface is represented as

$$S(\mathbf{x}, t) = 0. \tag{1.13}$$

Indeed, at time $t + dt$, we will have $S(\mathbf{x} + \mathbf{w}dt, t + dt) = 0$ from which, after a Taylor series expansion,

$$\frac{\partial S}{\partial t} + \mathbf{w} \cdot \boldsymbol{\nabla} S = 0 \qquad \text{on} \qquad S = 0. \tag{1.14}$$

But the unit normal, directed from the region where $S < 0$ to that where $S > 0$, is given by

$$\mathbf{n} = \frac{\boldsymbol{\nabla} S}{|\boldsymbol{\nabla} S|}, \tag{1.15}$$

so that

$$\mathbf{n} \cdot \mathbf{w} = -\frac{1}{|\boldsymbol{\nabla} S|}\frac{\partial S}{\partial t}. \tag{1.16}$$

If $S = 0$ denotes an impermeable surface, as in the case of a solid wall, $\dot{m} = 0$ so that $\mathbf{n} \cdot \mathbf{u} = \mathbf{n} \cdot \mathbf{w}$. In this case, by (1.12), (1.16) becomes the so-called *kinematic boundary condition*:

$$\frac{\partial S}{\partial t} + \mathbf{u} \cdot \boldsymbol{\nabla} S = 0 \qquad \text{on} \qquad S = 0. \tag{1.17}$$

At solid surfaces, for viscous flow, one usually imposes the *no-slip* condition, which requires the tangential velocity of the fluid to match that of the boundary:

$$\mathbf{n} \times (\mathbf{u} - \mathbf{w}) = 0 \qquad \text{on} \qquad S = 0. \tag{1.18}$$

(It is well known that there are situations, such as contact line motion, where this relation does not reflect the correct physics. Several more or less *ad hoc* models to treat these cases exist, but a "standard" one has yet to emerge.) Upon combining (1.14) and (1.18) one easily finds, for an impermeable surface,

$$\mathbf{u} = \mathbf{w} \qquad \text{on} \qquad S = 0. \tag{1.19}$$

The tangential velocity of a fluid interface can only be unambiguosly defined when the interface points carry some attribute other than their geometric location in space, such as the concentration of a surfactant[1]. For a purely geometric interface, the tangential velocity is meaningless as a mapping of the interface on itself cannot have physical consequences. For example, in the case of an expanding sphere such as a bubble, a rotation around the fixed center cannot have quantitative effects. In the case of two fluids separated by a purely geometric interface, the velocity field of each fluid must individually satisfy (1.17) but, rather than (1.18), the proper condition is one of continuity of the tangential velocity:

$$\mathbf{n} \times (\mathbf{u}_1 - \mathbf{u}_2) = 0. \tag{1.20}$$

It is interesting to note that, while both (1.18) and (1.20) are essentially phenomenological relations, in the case of inviscid fluids with a constant surface tension (1.20) is actually a consequence of the conservation of tangential momentum provided $\dot{m} \neq 0$. When $\dot{m} = 0$, the combination of (1.17) for each fluid and (1.20) renders the entire velocity continuous across the interface:

$$\mathbf{u}_1 = \mathbf{u}_2 \qquad \text{on} \qquad S = 0. \tag{1.21}$$

When the interface separates a liquid from a gas or a vapor, the dynamical effects of the latter can often be modeled in terms of pressure alone, neglecting viscosity. In this case, only the normal condition (1.17) applies, but not the tangential condition (1.20).

For solid boundaries with a prescribed velocity, the condition (1.19), possibly augmented by suitable conditions at infinity and at the initial instant, is sufficient to find a well-defined solution to the Navier–Stokes equations (1.2)

---

[1] In spite of its simplicity, the interface model described here is often adequate for many applications. Much more sophisticated models exist as described, for example, in Edwards *et al.* (1991).

and (1.7) or (1.8). For a free surface, a further condition is required to determine the motion of the surface itself. This condition arises from a momentum balance across the interface which stipulates that the jump in the surface tractions $\mathbf{t} = \boldsymbol{\sigma} \cdot \mathbf{n}$, combined with the momentum fluxes, be balanced by the action of surface tension:

$$(\boldsymbol{\sigma}_2 - \boldsymbol{\sigma}_1) \cdot \mathbf{n} - \dot{m}(\mathbf{u}_2 - \mathbf{u}_1) = -\boldsymbol{\nabla} \cdot [(\boldsymbol{I} - \mathbf{nn})\,\gamma] = -(\boldsymbol{I} - \mathbf{nn}) \cdot \boldsymbol{\nabla}\gamma + \gamma\kappa\mathbf{n}, \tag{1.22}$$

where γ is the surface tension coefficient and

$$\kappa = \boldsymbol{\nabla} \cdot \mathbf{n}, \tag{1.23}$$

the local mean curvature of the surface. It will be recognized that $\boldsymbol{I} - \mathbf{nn}$ is the projector on the plane tangent to the interface. The signs in Eq. (1.22) are correct provided S is defined so that $S > 0$ in fluid 2 and $S < 0$ in fluid 1. In practice, it is more convenient to decompose this condition into its normal and tangential parts. The former is

$$-p_2 + p_1 + \mathbf{n} \cdot (\boldsymbol{\tau}_2 - \boldsymbol{\tau}_1) \cdot \mathbf{n} - \dot{m}(\mathbf{u}_2 - \mathbf{u}_1) \cdot \mathbf{n} = \gamma\kappa \tag{1.24}$$

while the tangential component is, by (1.20),

$$\mathbf{n} \times (\boldsymbol{\tau}_2 - \boldsymbol{\tau}_1) \cdot \mathbf{n} = -\mathbf{n} \times \boldsymbol{\nabla}\gamma. \tag{1.25}$$

If, in place of p, the reduced pressure p^{r} defined in (1.10) is used, the right-hand sides of (1.22) and (1.24) acquire an additional contribution necessary to cancel the difference between the potentials $\mathcal{U}$ in the two fluids; for example, (1.24) becomes

$$-p_2^{\mathrm{r}} + p_1^{\mathrm{r}} + \mathbf{n} \cdot (\boldsymbol{\tau}_2 - \boldsymbol{\tau}_1) \cdot \mathbf{n} = \gamma\kappa + \rho_1\mathcal{U}_1 - \rho_2\mathcal{U}_2. \tag{1.26}$$

Let us now consider a rigid body of mass m_{b}, inertia tensor $\mathbf{J}_{\mathrm{b}}$, volume $\mathcal{V}_{\mathrm{b}}$ and surface S_{b} immersed in the fluid. According to the laws of dynamics, the motion of such a body is governed by an equation specifying the rate of change of the linear momentum

$$\frac{d}{dt}(m_{\mathrm{b}}\mathbf{v}) = \mathbf{F}^{\mathrm{h}} + \mathbf{F}^{\mathrm{e}} + m_{\mathrm{b}}\mathbf{g}, \tag{1.27}$$

and of the angular momentum

$$\frac{d}{dt}(\mathbf{J}_{\mathrm{b}} \cdot \boldsymbol{\Omega}) = \mathbf{L}^{\mathrm{h}} + \mathbf{L}^{\mathrm{e}}. \tag{1.28}$$

Here $\mathbf{v}$ is the velocity of the body center of mass, $\boldsymbol{\Omega}$ the angular velocity about the center of mass, and $\mathbf{F}$ and $\mathbf{L}$ denote forces and couples, respectively; the superscripts "h" and "e" distinguish between forces and couples of hydrodynamic and other, external, origin. The former are given by

$$\mathbf{F}^h = \int_{S_b} \boldsymbol{\sigma} \cdot \mathbf{n} \, dS_b, \qquad \mathbf{L}^h = \int_{S_b} \mathbf{x} \times [\boldsymbol{\sigma} \cdot \mathbf{n}] \, dS_b, \qquad (1.29)$$

where $\mathbf{x}$ is measured from the center of mass and the unit normal $\mathbf{n}$ is directed outward from the body. When the fluid stress in $\mathbf{F}^h$ is expressed in terms of the ordinary pressure p, the buoyancy force arises as part of the hydrodynamic force. Sometimes it may be more useful to express the fluid stress in terms of the reduced pressure p^r defined in (1.10). In the case of gravity, $\mathcal{U} = -\rho \mathbf{g} \cdot \mathbf{x}$ and (1.27) takes the form

$$\frac{d}{dt} (m_b \mathbf{v}) = \mathbf{F}_r^h + \mathbf{F}^e + (m_b - \rho \mathcal{V}_b) \, \mathbf{g}. \qquad (1.30)$$

The position $\mathbf{X}$ of the center of mass and the orientation of the body (for example, the three Euler angles), $\boldsymbol{\Theta}$, depend on time according to the kinematic relations

$$\frac{d\mathbf{X}}{dt} = \mathbf{v}, \qquad \frac{d\boldsymbol{\Theta}}{dt} = \boldsymbol{\Omega}, \qquad (1.31)$$

respectively.

1.4 Some dimensionless groups

The use of dimensional analysis and dimensionless groups is well-established in ordinary fluid dynamics and it is no less useful in multiphase flow. Each problem will have one or more characteristic length scales such as particle size, duct diameter, and others. The spatial scale of each problem can therefore be represented by a characteristic length L and, possibly, dimensionless ratios of the other scales to L. A similar role may be played by an intrinsic time scale τ due, for example, to an imposed time dependence of the flow or a force oscillating with a prescribed frequency, and by a velocity scale U. We introduce dimensionless variables $\mathbf{x}_*$, t_*, and $\mathbf{u}_*$ by writing

$$\mathbf{x} = L\mathbf{x}_*, \qquad t = \tau t_*, \qquad \mathbf{u} = U\mathbf{u}_*. \qquad (1.32)$$

Furthermore, we let

$$\nabla p = \frac{\Delta P}{L} \nabla_* p_*, \qquad \mathbf{f} = f \mathbf{f}_* \qquad (1.33)$$

where ∇_* denotes the gradient operator with respect to the dimensionless coordinate $\mathbf{x}_*$, ΔP is an appropriate pressure-difference scale, and f

a representative value of **f**. Then the continuity equation remains formally unaltered,

$$\nabla_* \cdot \mathbf{u}_* = 0, \tag{1.34}$$

while the momentum equation (1.8) becomes

$$\frac{1}{Sl}\frac{\partial \mathbf{u}_*}{\partial t_*} + (\mathbf{u}_* \cdot \nabla_*)\,\mathbf{u}_* = -\frac{\Delta p}{\rho U^2}\nabla_* p_* + \frac{1}{Re}\nabla_*^2 \mathbf{u}_* + \frac{fL}{U^2}\mathbf{f}_*. \tag{1.35}$$

Here we have introduced the *Strouhal number Sl*, defined by

$$Sl = \frac{U\tau}{L}, \tag{1.36}$$

which expresses the ratio of the intrinsic time scale τ to the convective time scale L/U. When no external or imposed time scale is present, $\tau = L/U$ and $Sl = 1$. The *Reynolds number Re* is defined by

$$Re = \frac{\rho LU}{\mu} = \frac{LU}{\nu}, \tag{1.37}$$

and, in addition to its usual meaning of the ratio of inertial to viscous forces, can be interpreted as the ratio of the viscous diffusion time L^2/ν to the convective time scale L/U. When the force **f** is gravity, $f = g = |\mathbf{g}|$ and the group

$$Fr = \frac{U^2}{gL} \tag{1.38}$$

is known as the *Froude number.*

The appropriate pressure-difference scale depends on the situation. When fluid inertia is important, pressure differences scale proportionally to ρU^2 so that we may take $\Delta P = \rho U^2$ to find

$$\frac{1}{Sl}\frac{\partial \mathbf{u}_*}{\partial t_*} + (\mathbf{u}_* \cdot \nabla_*)\,\mathbf{u}_* = -\nabla_* p_* + \frac{1}{Re}\nabla_*^2 \mathbf{u}_* + \frac{fL}{U^2}\mathbf{f}_*. \tag{1.39}$$

Frequently $Sl = 1$ and this equation becomes

$$\frac{\partial \mathbf{u}_*}{\partial t_*} + (\mathbf{u}_* \cdot \nabla_*)\,\mathbf{u}_* = -\nabla_* p_* + \frac{1}{Re}\nabla_*^2 \mathbf{u}_* + \frac{fL}{U^2}\mathbf{f}_*. \tag{1.40}$$

On the other hand, when the flow is dominated by viscosity, the proper pressure scale is $\Delta P = \mu U/L$ and the equation becomes

$$\frac{1}{Sl}\frac{\partial \mathbf{u}_*}{\partial t_*} + (\mathbf{u}_* \cdot \nabla_*)\,\mathbf{u}_* = -\frac{1}{Re}\nabla_* p_* + \frac{1}{Re}\nabla_*^2 \mathbf{u}_* + \frac{fL}{U^2}\mathbf{f}_*. \tag{1.41}$$

A special situation arises when $Re \ll 1$ and $Re/Sl = (L^2/\nu)/\tau \ll 1$ as, then, the left-hand side of this equation is negligible; in dimensional form, what remains is

$$-\nabla p + \mu \nabla^2 \mathbf{u} + \rho \mathbf{f} = 0, \tag{1.42}$$

which, together with (1.34), are known as the *Stokes equations*.

Additional dimensionless groups arise from the boundary conditions. In the case of inertia-dominated pressure scaling, the normal stress condition (1.24) leads to

$$-p_{*2} + p_{*1} + \frac{1}{Re} \mathbf{n} \cdot (\boldsymbol{\tau}_{*2} - \boldsymbol{\tau}_{*1}) \cdot \mathbf{n} = \frac{1}{We} \kappa_* \tag{1.43}$$

where $\kappa_* = L\kappa$ and the *Weber number*, expressing the ratio of inertial and surface-tension-induced pressures, is defined by

$$We = \frac{\rho L U^2}{\gamma}. \tag{1.44}$$

In some cases, the characteristic velocity is governed by buoyancy, which leads to the estimate $U \sim \sqrt{(|\rho - \rho'|/\rho)gL}$. A typical case is the rise of large gas bubbles (density ρ') in a free liquid or in a liquid-filled tube. In these cases, equation (1.44) becomes

$$Eo = Bo = \frac{|\rho - \rho'|gL^2}{\gamma}, \tag{1.45}$$

a combination known as the *Eötvös number* or *Bond number*. When $\rho' \ll \rho$, Eo is simply written as

$$Eo = \frac{\rho g L^2}{\gamma}. \tag{1.46}$$

The *Morton number*, defined by

$$Mo = \frac{g\mu^4}{\rho\gamma^3}, \tag{1.47}$$

is often useful as, for fixed g, it only depends on the fluid properties. If the Reynolds number is expressed in terms of the characteristic velocity $\sqrt{gL}$, one immediately verifies that $Mo = (Eo^3/Re^4)$. The Reynolds number constructed with the velocity $\sqrt{(|\rho - \rho'|/\rho)gL}$ is the *Galilei number*

$$Ga = \frac{\sqrt{g\rho|\rho - \rho'|L^3}}{\mu}. \tag{1.48}$$

In the opposite case of viscosity-dominated pressure scaling, the normal-stress condition (1.24) becomes

$$-p_{*2} + p_{*1} + \mathbf{n} \cdot (\boldsymbol{\tau}_{*2} - \boldsymbol{\tau}_{*1}) \cdot \mathbf{n} = \frac{1}{Ca} \kappa_* \tag{1.49}$$

where the *capillary number*, expressing the ratio of viscous to capillary stresses, is defined by

$$Ca = \frac{\mu U}{\gamma}. \tag{1.50}$$

For small-scale phenomena dominated by surface tension and viscosity, the characteristic time due to the flow, L/U, is of the order of $\sqrt{\rho L^3/\gamma}$, while the intrinsic time scale is the diffusion time L^2/ν. In this case the inverse of the Strouhal number (1.36) is known as the *Ohnesorge number*

$$Oh = \frac{\mu}{\sqrt{\rho \gamma L}}. \tag{1.51}$$

An important dimensionless parameter governing the dynamics of a particle in a flow is the *Stokes number* defined as the ratio of the characteristic time of the particle response to the flow to that of the flow itself:

$$St = \frac{\tau_b}{\tau}. \tag{1.52}$$

This ratio can be estimated as follows. Let U_r denote the characteristic particle–fluid relative velocity and A its projected area on a plane normal to the relative velocity. When inertia is important, the order of magnitude of the hydrodynamic force $|\mathbf{F}^h|$ may be estimated in terms of a *drag coefficient* C_d defined by

$$C_d = \frac{F^h}{\frac{1}{2}\rho A U_r^2}, \tag{1.53}$$

In problems where the scale of the relative velocity is determined by a balance between the hydrodynamic and gravity forces, U_r may be estimated as

$$U_r \sim \sqrt{\frac{1}{C_d}\left|\frac{\rho_b}{\rho} - 1\right| L g} \tag{1.54}$$

where ρ_b is the density of the body and $L = \mathcal{V}_b/A$ is a characteristic body length defined in terms of the body volume $\mathcal{V}_b$. The characteristic relaxation time of the body velocity in the flow, τ_b, may be determined by balancing the left-hand side of the body momentum equation, $\rho_b \mathcal{V}_b U_r/\tau_b$, with the hydrodynamic force to find

$$\tau_b \sim \frac{L}{C_d U_r}\frac{\rho_b}{\rho} \sim \frac{\rho_b}{\rho}\sqrt{\frac{L}{C_d |\rho_b/\rho - 1| g}}. \tag{1.55}$$

When the Reynolds number of the relative motion, $Re_b = LU_r/\nu$, is small,

$C_d \simeq 1/Re_b$ and we have

$$U_r \sim \left| \frac{\rho_b}{\rho} - 1 \right| \frac{L^2 g}{\nu}, \qquad \text{from which} \qquad \tau_b \sim \frac{\rho_b}{\rho} \frac{L^2}{\nu} \qquad (1.56)$$

so that

$$St \sim \frac{\rho_b}{\rho} \frac{L^2}{\nu \tau}. \qquad (1.57)$$

In particular, for a sphere of radius a, $\tau_b = 2\rho_b a^2/(9\rho\nu)$ and one finds

$$St = \frac{2\rho_b a^2}{9\mu\tau}. \qquad (1.58)$$

2

Direct numerical simulations of finite Reynolds number flows

In this chapter and the following three, we discuss numerical methods that have been used for direct numerical simulations of multiphase flow. Although direct numerical simulations, or DNS, mean slightly different things to different people, we shall use the term to refer to computations of complex unsteady flows where all continuum length and time scales are fully resolved. Thus, there are no modeling issues beyond the continuum hypothesis. The flow within each phase and the interactions between different phases at the interface between them are found by solving the governing conservation equations, using grids that are finer and time steps that are shorter than any physical length and time scale in the problem.

The detailed flow field produced by direct numerical simulations allows us to explore the mechanisms governing multiphase flows and to extract information not available in any other way. For a single bubble, drop, or particle, we can obtain integrated quantities such as lift and drag and explore how they are affected by free stream turbulence, the presence of walls, and the unsteadiness of the flow. In these situations it is possible to take advantage of the relatively simple geometry to obtain extremely accurate solutions over a wide range of operating conditions. The interactions of a few bubbles, drops, or particles is a more challenging computation, but can be carried out using relatively modest computational resources. Such simulations yield information about, for example, how bubbles collide or whether a pair of buoyant particles, rising freely through a quiescent liquid, orient themselves in a preferred way. Computations of one particle can be used to obtain information pertinent to modeling of dilute multiphase flows, and studies of a few particles allow us to assess the importance of rare collisions. It is, however, the possibility of conducting DNS of thousands of freely interacting particles that holds the greatest promise. Such simulations can yield data for not only the collective lift and drag of dense systems, but also about how the particles are distributed and what impact the formation of structures and clusters has on the overall behavior of the flow. Most industrial size systems, such as fluidized bed reactors or bubble columns, will remain out of reach of direct numerical simulations for the foreseeable future (and even if they were possible, DNS is unlikely to be used for routine design). However, the size of systems that *can* be studied is growing rapidly. It is realistic today to conduct DNS of fully three-dimensional systems resolved by several hundred grid points in each spatial direction. If we assume that a single bubble can be adequately resolved by 25 grid points (sufficient for clean bubbles at relatively modest Reynolds numbers), that the bubbles are, on the average, one bubble diameter apart (a void fraction of slightly over 6%), and that we have a uniform grid with 1000^3 grid points, then we would be able to accommodate 8000 bubbles. High Reynolds numbers and solid particles or drops generally require higher resolution. Furthermore, the number of bubbles that we can simulate on a given grid obviously depends strongly on the void fraction. It is clear, however, that DNS has opened up completely new possibilities in the studies of multiphase flows which we have only started to explore.

In addition to relying on explosive growth in available computer power, progress in DNS of multiphase flows has also been made possible by the development of numerical methods. Advecting the phase boundary poses unique challenges and we will give a brief overview of such methods below, followed by a more detailed description in the next few chapters. In most cases, however, it is also necessary to solve the governing equations for the fluid flow. For body-fitted and unstructured grids, these are exactly the same as for flows without moving interfaces. For the "one-fluid" approach introduced in Chapter 3, we need to deal with density and viscosity fields that change abruptly across the interface and singular forces at the interface, but otherwise the computations are the same as for single-phase flow. Methods developed for single-phase flows can therefore generally be used to solve the fluid equations. After we briefly review the different ways of computing multiphase flows, we will therefore outline in this chapter a relatively simple method to compute single-phase flows using a regular structured grid.

2.1 Overview

Many methods have been developed for direct numerical simulations of multiphase flows. The oldest approach is to use one stationary, structured grid for the whole computational domain and to identify the different fluids by markers or a marker function. The equations expressing conservation of mass, momentum and energy hold, of course, for any fluid, even when density and viscosity change abruptly and the main challenge in this approach is to accurately advect the phase boundary and to compute terms concentrated at the interface, such as surface tension. In the marker-and-cell (MAC) method of Harlow and collaborators at Los Alamos (Harlow and Welch, 1965) each fluid is represented by marker points distributed over the region that it occupies. Although the MAC method was used to produce some spectacular results, the distributed marker particles were not particularly good at representing fluid interfaces. The Los Alamos group thus replaced the markers by a marker function that is a constant in each fluid and is advected by a scheme specifically written for a function that changes abruptly from one cell to the next. In one dimension this is particularly straightforward and one simply has to ensure that each cell fills completely before the marker function is advected into the next cell. Extended to two and three dimensions, this approach results in the volume-of-fluid (VOF) method.

The use of a single structured grid leads to relatively simple as well as efficient methods, but early difficulties experienced with the volume-of-fluid method have given rise to several other methods to advect a marker function. These include the level-set method, originally introduced by Osher and Sethian (1988) but first used for multiphase flow simulations by Sussman, Smereka, and Osher (1994), the CIP method of Yabe and collaborators (Takewaki, Nishiguchi and Yabe, 1985; Takewaki and Yabe, 1987), and the phase field method used by Jacqmin (1997). Instead of advecting a marker function and inferring the location of the interface from its gradient, it is also possible to mark the interface using points moving with the flow and reconstruct a marker function from the interface location. Surface markers have been used extensively for boundary integral methods for potential flows and Stokes flows, but their first use in Navier–Stokes computations was by Daly (1969a,b) who used them to calculate surface tension effects with the MAC method. The use of marker points was further advanced by the introduction of the immersed boundary method by Peskin (1977), who used connected marker points to follow the motion of elastic boundaries immersed in homogeneous fluids, and by Unverdi and Tryggvason (1992) who used connected marker points to advect the boundary between two different fluids and to compute surface tension from the geometry of the interface. Methods based on using a single structured grid, identifying the interface either by a marker function or connected marker points, are discussed in some detail in Chapter 3 of this book.

The attraction of methods based on the use the "one-fluid" formulation on stationary grids is their simplicity and efficiency. Since the interface is, however, represented on the grid as a rapid change in the material properties, their formal accuracy is generally limited to first order. Furthermore, the difficulty that the early implementations of the "one-fluid" approach experienced, inspired several attempts to develop methods where the grid lines were aligned with the interface. These attempts fall, loosely, into three categories. Body-fitted grids, where a structured grid is deformed in such a way that the interface always coincides with a grid line; unstructured grids where the fluid is resolved by elements or control volumes that move with it in such a way that the interface coincides with the edge of an element; and what has most recently become known as sharp interface methods, where a regular structured grid is used but something special is done at the interface to allow it to stay sharp.

Body-fitted grids that conform to the phase boundaries greatly simplify the specification of the interaction of the phases across the interface. Furthermore, numerical methods on curvilinear grids are well developed and a high level of accuracy can be maintained both within the different phases and along their interfaces. Such grids were introduced by Hirt, Cook, and Butler (1970) for free surface flows, but their use by Ryskin and Leal (1983, 1984) to find the steady state shape of axisymmetric buoyant bubbles brought their utility to the attention of the wider fluid dynamics community. Although body-fitted curvilinear grids hold enormous potential for obtaining accurate solutions for relatively simple systems such as one or two spherical particles, generally their use is prohibitively complex as the number of particles increases. These methods are briefly discussed in Chapter 4. Unstructured grids, consisting usually of triangular (in two-dimensions) and tetrahedral (in three-dimensions) shaped elements offer extreme flexibility, both because it is possible to align grid lines to complex boundaries and also because it is possible to use different resolution in different parts of the computational domain. Early applications include simulations of the breakup of drops by Fritts, Fyre, and Oran (1983) but more recently unstructured moving grids have been used for simulations of multiphase particulate systems, as discussed in Chapter 5. Since body-fitted grids are usually limited to relatively simple geometries and methods based on unstructured grids are complex to implement and computationally expensive, several authors have sought to combine the advantages of the single-fluid approach and methods based on a more accurate representation of the interface. This approach was pioneered by Glimm and collaborators many years ago (Glimm, 1982; Glimm and McBryan, 1985) but has recently re-emerged in methods that can be referred to collectively as "sharp interface" methods. In these methods the fluid domain is resolved by a structured grid, but the interface treatment is improved by, for example, introducing special difference formulas that incorporate the jump across the interface (Lee and LeVeque, 2003), using "ghost points" across the interface (Fedkiw *et al.*, 1999), or restructuring the control volumes next to the interface so that the face of the control volume is aligned with the interface (Udaykumar *et al.*, 1997). While promising, for the most part these methods have yet to prove that they introduce fundamentally new capabilities and that the extra complication justifies the increased accuracy. We will briefly discuss "sharp interface methods" for simulations of the motion of fluid interfaces in Chapter 3 and in slightly more detail for fluid–solid interactions in Chapter 4.

2.2 Integrating the Navier–Stokes equations in time

For a large class of multiphase flow problems, including most of the systems discussed in this book, the flow speeds are relatively low and it is appropriate to treat the flow as incompressible. The unique role played by the pressure for incompressible flows, where it is not a thermodynamic variable, but takes on whatever value is needed to enforce a divergence-free velocity field, requires us to pay careful attention to the order in which the equations are solved. There is, in particular, no explicit equation for the pressure and therefore such an equation has to be found as a part of the solution process. The standard way to integrate the Navier–Stokes equations is by the so called "projection method," introduced by Chorin (1968) and Yanenko (1971). In this approach, the velocity is first advanced without accounting for the pressure, resulting in a field that is in general not divergence-free. The pressure necessary to make the velocity field divergence-free is then found and the velocity field corrected by adding the pressure gradient.

We shall first work out the details for a simple first-order explicit time integration scheme and then see how it can be modified to generate a higher order scheme. To integrate equations (1.2) and (1.7) (or 1.8) in time, we write

$$\frac{\mathbf{u}^{n+1} - \mathbf{u}^n}{\Delta t} + \mathbf{A}_\mathrm{h}(\mathbf{u}^n) = -\frac{1}{\rho}\nabla_\mathrm{h} p + \nu \mathbf{D}_\mathrm{h}(\mathbf{u}^n) + \mathbf{f}_\mathrm{b}^n \qquad (2.1)$$

$$\nabla_\mathrm{h} \cdot \mathbf{u}^{n+1} = 0. \qquad (2.2)$$

The superscript n denotes the variable at the beginning of a time step of length Δt and $n + 1$ denotes the new value at the end of the step. $\mathbf{A}_\mathrm{h}$ is a numerical approximation to the advection term, $\mathbf{D}_\mathrm{h}$ is a numerical approximation to the diffusion term, and $\mathbf{f}_\mathrm{b}$ is a numerical approximation to any other force acting on the fluid. ∇_h means a numerical approximation to the divergence or the gradient operator.

In the projection method the momentum equation is split into two parts by introducing a temporary velocity $\mathbf{u}^*$ such that $\mathbf{u}^{n+1} - \mathbf{u}^n = \mathbf{u}^{n+1} - \mathbf{u}^* + \mathbf{u}^* - \mathbf{u}^n$. The first part is a predictor step, where the temporary velocity field is found by ignoring the effect of the pressure:

$$\frac{\mathbf{u}^* - \mathbf{u}^n}{\Delta t} = -\mathbf{A}_h(\mathbf{u}^n) + \nu \mathbf{D}_h(\mathbf{u}^n) + \mathbf{f}_b^n. \tag{2.3}$$

In the second step – the projection step – the pressure gradient is added to yield the final velocity at the new time step:

$$\frac{\mathbf{u}^{n+1} - \mathbf{u}^*}{\Delta t} = -\frac{1}{\rho}\nabla_h p^{n+1}. \tag{2.4}$$

Adding the two equations yields exactly equation (2.1).

To find the pressure, we use equation (2.2) to eliminate $\mathbf{u}^{n+1}$ from equation (2.4), resulting in Poisson's equation:

$$\frac{1}{\rho}\nabla_h^2 p^{n+1} = \frac{1}{\Delta t}\nabla_h \cdot \mathbf{u}^* \tag{2.5}$$

since the density ρ is constant. Once the pressure has been found, equation (2.4) is used to find the projected velocity at time step $n+1$. We note that we do not assume that $\nabla_h \cdot \mathbf{u}^n = 0$. Usually, the velocity field at time step n is not exactly divergence-free but we strive to make the divergence of the new velocity field, at $n+1$, zero.

As the algorithm described above is completely explicit, it is subject to relatively stringent time-step limitations. If we use standard centered second-order approximations for the spatial derivatives, as done below, stability analysis considering only the viscous terms requires the step size Δt to be bounded by

$$\Delta t < C_\nu \frac{h^2}{\nu} \tag{2.6}$$

where $C_\nu = 1/4$ and $1/6$ for two- and three-dimensional flows, respectively, and h is the grid spacing. The advection scheme is unstable by itself, but it is stabilized by viscosity if the step size is limited by

$$\Delta t < \frac{2\nu}{q^2}, \tag{2.7}$$

where $q^2 = \mathbf{u} \cdot \mathbf{u}$. More sophisticated methods for the advection terms, which are stable in the absence of viscosity and can therefore also be used to integrate the Euler equations in time, are generally subject to the Courant–Friedrichs–Lewy (CFL) condition[1]. For one-dimensional flow,

$$\Delta t < \frac{h}{(|u|)}. \tag{2.8}$$

Many advection schemes are implemented by splitting, where the flow is sequentially advected in each coordinate direction. In these cases the one-dimensional CFL condition applies separately to each step. For fully multidimensional schemes, however, the stability analysis results in further reduction of the size of the time step. General discussions of the stability of different schemes and the resulting maximum time step can be found in standard textbooks, such as Hirsch (1988), Wesseling (2001), or Ferziger and Perić (2002). In an unsteady flow, the CFL condition on the time step is usually not very severe, since accuracy requires the time step to be sufficiently small to resolve all relevant time scales. The limitation due to the viscous diffusion, equation (2.6), can be more stringent, particularly for slow flow, and the viscous terms are frequently treated implicitly, as discussed below. For problems where additional physics must be accounted for, other stability restrictions may apply. When surface tension is important, it is generally found, for example, that it is necessary to limit the time step in such a way that a capillary wave travels less than a grid space in one time step.

The simple explicit forward-in-time algorithm described above is only first-order accurate. For most problems it is desirable to employ at least a second-order accurate time integration method. In such methods the nonlinear advection terms can usually be treated explicitly, but the viscous terms are often handled implicitly, for both accuracy and stability. If we use a second-order Adams–Bashforth scheme for the advection terms and a second-order Crank–Nicholson scheme for the viscous term, the predictor step is (Wesseling, 2001)

$$\frac{\mathbf{u}^* - \mathbf{u}^n}{\Delta t} = -\frac{3}{2}\mathbf{A}_\mathrm{h}(\mathbf{u}^n) + \frac{1}{2}\mathbf{A}_\mathrm{h}(\mathbf{u}^{n-1}) + \frac{\nu}{2}\Big(\mathbf{D}_\mathrm{h}(\mathbf{u}^n) + \mathbf{D}_\mathrm{h}(\mathbf{u}^*)\Big), \tag{2.9}$$

and the correction step is

$$\frac{\mathbf{u}^{n+1} - \mathbf{u}^*}{\Delta t} = -\nabla_\mathrm{h}\phi^{n+1}. \tag{2.10}$$

---

[1] Lewy is often spelled Levy. This is incorrect. Hans Lewy (1904–1988) was a well-known mathematician.

Here, ϕ is not exactly equal to the pressure, since the viscous term is not computed at the new time level but at the intermediate step. It is easily seen that

$$-\nabla\phi^{n+1} = -\nabla p^{n+1} + \frac{\nu}{2}\left(\mathbf{D}_{\mathrm{h}}(\mathbf{u}^{n+1}) - \mathbf{D}_{\mathrm{h}}(\mathbf{u}^*)\right). \qquad (2.11)$$

The intermediate velocity $\mathbf{u}^*$ does not satisfy the divergence-free condition, and a Poisson equation for the pseudo-pressure is obtained as before from Eq. (2.10) as

$$\nabla^2_{\mathrm{h}}\phi^{n+1} = \frac{\nabla_{\mathrm{h}} \cdot \mathbf{u}^*}{\Delta t}. \qquad (2.12)$$

This multidimensional Poisson's equation must be solved before the final velocity, $\mathbf{u}^{n+1}$, can be obtained. Since the viscous terms are treated implicitly, we must rearrange equation (2.9) to yield a Helmholtz equation for the intermediate velocity $\mathbf{u}^*$

$$\mathbf{D}_{\mathrm{h}}(\mathbf{u}^*) - \frac{2\nu}{\Delta t}\mathbf{u}^* = \frac{3\Delta t}{2}\mathbf{A}_{\mathrm{h}}(\mathbf{u}^n) - \frac{\Delta t}{2}\mathbf{A}_{\mathrm{h}}(\mathbf{u}^{n-1}) - \mathbf{D}_{\mathrm{h}}(\mathbf{u}^n) - \frac{2\nu}{\Delta t}\mathbf{u}^n = \mathrm{RHS}$$
$$(2.13)$$

which needs to be solved with the appropriate boundary conditions. Considerable effort has been devoted to the solution of Poisson's and Helmholtz's equations and several packages are available (see www.mgnet.org, for example), particularly for rectangular grids. For curvilinear body-fitted grids the solution of Helmholtz's equation can sometimes be simplified by using implicit time advancement of the viscous term selectively only along certain directions. The viscous term in the other directions can be treated explicitly along with the nonlinear terms. Usually, the wall-normal viscous term is treated implicitly, while the tangential viscous terms can be treated explicitly (Mittal, 1999; Bagchi and Balachandar, 2003b). The resulting viscous time-step limitation, arising only from the tangential contributions to (2.13), is usually not very stringent.

The above two-step formulation of the time-splitting scheme is not unique; another variant is presented in Chapter 9. We refer the reader to standard textbooks, such as Ferziger and Perić (2002) and Wesseling (2001) for further discussions.

2.3 Spatial discretization

Just as there are many possible time integration schemes, the spatial discretization of the Navier–Stokes equations – where continuous variables are replaced by discrete representation of the fields and derivatives are approximated by relations between the discrete values – can be accomplished in many ways. Here we use the finite-volume method and discretize the governing equations by dividing the computational domain into small control volumes of finite size. In the finite-volume method we work with the average velocity in each control volume, and approximate each term in equations (2.3) and (2.4) by its average value over the control volume. To derive numerical approximations to the advection and the viscous terms, we first find the averages over each control volume:

$$\mathbf{A}(\mathbf{u}^n) = \frac{1}{\Delta V} \int_V \nabla \cdot (\mathbf{u}^n \mathbf{u}^n) \, dv = \frac{1}{\Delta V} \oint_S \mathbf{u}^n (\mathbf{u}^n \cdot \mathbf{n}) \, ds \qquad (2.14)$$

and

$$\mathbf{D}(\mathbf{u}^n) = \frac{1}{\Delta V} \int_V \nabla^2 \mathbf{u}^n \, dv. \qquad (2.15)$$

Here, ΔV is the volume of the control volume, S is the surface of the control volume, and we have used the divergence theorem to convert the volume integral of the advection term to a surface integral. It is also possible to rewrite the viscous term as a surface integral, but for constant-viscosity fluids it is generally simpler to work with the volume integral. Numerical approximations to these terms, $\mathbf{A}_h$ and $\mathbf{D}_h$, are found by evaluating the integrals numerically.

Similarly, a numerical approximation to the continuity equation (2.2), $\nabla_h \cdot \mathbf{u}^{n+1} = 0$, is found by first integrating over the control volume and then rewriting the integral as a surface integral:

$$\frac{1}{\Delta V} \int_V \nabla \cdot \mathbf{u}^{n+1} \, dv = \frac{1}{\Delta V} \oint_S \mathbf{u}^{n+1} \cdot \mathbf{n} \, ds. \qquad (2.16)$$

The surface integral is then evaluated numerically. While we started here with the differential form of the governing equations, averaging over each cell, the discrete approximations could just as well have been obtained by starting with the conservation principles in integral form.

To carry out the actual computations, we must specify the control volumes to be used, and how the surface and volume integrals are approximated. Here, we will take the simplest approach and use square or cubic control volumes, defined by a regular array of grid points, separated by a distance h. For simplicity, we assume a two-dimensional flow as the extension to three dimensions involves no new concepts. We start by picking a control volume around a point where the pressure is stored and identify it by the integer pair (i, j). The control volumes to the left and the right are given by $(i - 1, j)$ and $(i + 1, j)$, respectively. Similarly, $(i, j - 1)$ and $(i, j + 1)$ refer to the control volumes below and above. The locations of the edges are identified by half-indices $(i \pm 1/2, j)$ and $(i, j \pm 1/2)$. The pressure

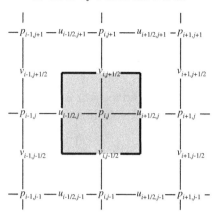

Fig. 2.1. The notation used for a standard staggered MAC mesh. The pressure control volume, centered at the (i, j) node, is outlined.

control volume, centered at (i, j), is shown by a thick line in Fig. 2.1. To derive a discrete approximation for the incompressibility condition, we evaluate equation (2.16) for the pressure control volume centered at (i, j). The integrals along the edges of the control volume are approximated by the midpoint rule, using the normal velocities at the edges. The normal velocities on the right and the left boundaries are $u_{i+1/2,j}$ and $u_{i-1/2,j}$, respectively. Similarly, $v_{i,j+1/2}$ and $v_{i,j-1/2}$ are the normal velocities on the top and the bottom boundary. The discrete approximation to the incompressibility condition is therefore:

$$u_{i+1/2,j}^{n+1} - u_{i-1/2,j}^{n+1} + v_{i,j+1/2}^{n+1} - v_{i,j-1/2}^{n+1} = 0 \qquad (2.17)$$

since the grid spacing is the same in both directions.

The velocities needed in equation (2.17) are the normal velocities to the boundary of the control volume. Although methods have been developed to allow us to use co-located grids (discussed below and in Section 10.3.4), where the velocities are stored at the same points as the pressures, here we proceed in a slightly different way and to define new control volumes, one for the u velocity, centered at $(i+1/2, j)$ and another one for the v velocity, centered at $(i, j+1/2)$. Such *staggered* grids result in a very robust numerical method that is – in spite of the complex looking indexing – relatively easily implemented. The control volume for $u_{i+1/2,j}$ is shown in the left frame of Fig. 2.2 and the control volume for $v_{i,j+1/2}$ in the right frame.

Continuing with the first-order method described above, the discrete forms of equations (2.3) and (2.4) for the u velocity in a control volume centered at

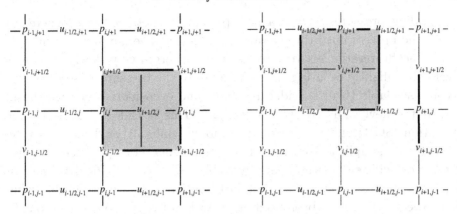

Fig. 2.2. The control volumes for the velocities on a staggered grid. The u-velocity control volume, centered at $(i + 1/2, j)$, is shown on the left and the v-velocity control volume, centered at $(i, j + 1/2)$, is shown on the right.

$(i + 1/2, j)$ and the v velocity in a control volume centered at $(i, j + 1/2)$ are:

$$u^*_{i+1/2,j} = u^n_{i+1/2,j} + \Delta t\left(-(A_x)^n_{i+1/2,j} + \nu(D_x)^n_{i+1/2,j} + (f_x)_{i+1/2,j}\right)$$

$$v^*_{i,j+1/2} = v^n_{i,j+1/2} + \Delta t(-(A_y)^n_{i,j+1/2} + \nu(D_y)^n_{i,j+1/2} + (f_y)_{i,j+1/2}\Big) \quad (2.18)$$

for the predictor step, and

$$u^{n+1}_{i+1/2,j} = u^*_{i+1/2,j} - \frac{1}{\rho}\frac{\Delta t}{h}(p^{n+1}_{i+1,j} - p^{n+1}_{i,j})$$

$$v^{n+1}_{i,j+1/2} = v^*_{i,j+1/2} - \frac{1}{\rho}\frac{\Delta t}{h}(p^{n+1}_{i,j+1} - p^{n+1}_{i,j}) \quad (2.19)$$

for the projection step.

To find an equation for the pressure, we substitute the expression for the correction velocities at the edges of the pressure control volume, equations (2.19), into the continuity equation (2.17). When the density is constant, we get:

$$\frac{p^{n+1}_{i+1,j} + p^{n+1}_{i-1,j} + p^{n+1}_{i,j+1} + p^{n+1}_{i,j-1} - 4p^{n+1}_{i,j}}{h^2}$$

$$= \frac{\rho}{\Delta t}\left(\frac{u^*_{i+1/2,j} - u^*_{i-1/2,j} + v^*_{i,j+1/2} - v^*_{i,j-1/2}}{h}\right). \quad (2.20)$$

The pressure Poisson equation, (2.20), can be solved by a wide variety of methods. The simplest one is iteration, where we isolate $p^{n+1}_{i,j}$ on the left-hand side and compute it by using already estimated values for the surrounding pressures. Once a new pressure is obtained everywhere, we repeat the process until the pressure does not change any more. This iteration (Jacobi iteration) is very robust but converges very slowly. It can be accelerated slightly by using the new values of the pressure as soon as they become available (Gauss–Seidel iteration) and even more by extrapolating toward the new value at every iteration. This is called successive over-relaxation (SOR) and was widely used for a while although now its value is mostly its simplicity during code development. For "production runs", much more efficient methods are available. For constant-density flows in regular geometries, solvers based on fast Fourier transforms or cyclic reduction (Sweet, 1974) have been used extensively for over two decades, but for more complex problems, such as variable-density or nonsimple domains or boundary conditions, newer methods such as multigrid iterative methods are needed (Wesseling, 2004). Solving the pressure equation is generally the most time-consuming part of any simulation involving incompressible flows.

Now that we have determined the layout of the control volumes, we can write explicit formulas for the advection and diffusion terms. The simplest approach is to use the midpoint rule to approximate the integral over each edge in equation (2.14) and to use a linear interpolation for the velocities at those points where they are not defined. This results in:

$$
\begin{aligned}
(A_x)_{i+1/2,j}^n = \frac{1}{h}\Bigg\{ &\left(\frac{u_{i+3/2,j}^n + u_{i+1/2,j}^n}{2}\right)^2 - \left(\frac{u_{i+1/2,j}^n + u_{i-1/2,j}^n}{2}\right)^2 \\
&+ \left(\frac{u_{i+1/2,j+1}^n + u_{i+1/2,j}^n}{2}\right)\left(\frac{v_{i+1,j+1/2}^n + v_{i,j+1/2}^n}{2}\right) \\
&- \left(\frac{u_{i+1/2,j}^n + u_{i+1/2,j-1}^n}{2}\right)\left(\frac{v_{i+1,j-1/2}^n + v_{i,j-1/2}^n}{2}\right) \Bigg\}
\end{aligned}
$$

$$
\begin{aligned}
(A_y)_{i,j+1/2}^n = \frac{1}{h}\Bigg\{ &\left(\frac{u_{i+1/2,j}^n + u_{i+1/2,j+1}^n}{2}\right)\left(\frac{v_{i,j+1/2}^n + v_{i+1,j+1/2}^n}{2}\right) \\
&- \left(\frac{u_{i-1/2,j+1}^n + u_{i-1/2,j}^n}{2}\right)\left(\frac{v_{i,j+1/2}^n + v_{i-1,j+1/2}^n}{2}\right) \\
&+ \left(\frac{v_{i,j+3/2}^n + v_{i,j+1/2}^n}{2}\right)^2 - \left(\frac{v_{i,j+1/2}^n + v_{i,j-1/2}^n}{2}\right)^2 \Bigg\}.
\end{aligned}
$$

The viscous term, equation (2.15), is approximated by the Laplacian at the center of the control volume, found using centered differences and by assuming that the average velocities coincide with the velocity at the center of the control volume:

$$(D_x)_{i+1/2,j}^n = \frac{u_{i+3/2,j}^n + u_{i-1/2,j}^n + u_{i+1/2,j+1}^n + u_{i+1/2,j-1}^n - 4u_{i+1/2,j}^n}{h^2}$$

$$(D_y)_{i,j+1/2}^n = \frac{v_{i+1,j+1/2}^n + v_{i-1,j+1/2}^n + v_{i,j+3/2}^n + v_{i,j-1/2}^n - 4v_{i,j+1/2}^n}{h^2}. \quad (2.21)$$

The staggered grid used above results in a very robust method, where pressure and velocities are tightly coupled. The grid does, however, require fairly elaborate bookkeeping of where each variable is and, for body-fitted or unstructured grids, the staggered grid arrangement can be cumbersome. Considerable effort has therefore been devoted to the development of methods using *co-located* grids, where all the variables are stored at the same spatial locations. The simplest approach is probably the one due to Rhie and Chow (1983), where we derive the pressure equation using the predicted velocities interpolated to the edges of the basic control volume. Thus, after we find the temporary velocities at the pressure points, $u_{i,j}^*$ and $v_{i,j}^*$, we interpolate to find

$$u_{i+1/2,j}^* = \frac{1}{2}(u_{i+1,j}^* + u_{i,j}^*)$$

$$v_{i,j+1/2}^* = \frac{1}{2}(v_{i,j+1}^* + v_{i,j}^*). \quad (2.22)$$

We then "pretend" that we are working with staggered grids and that the edge velocities at the new time step are given by

$$u_{i+1/2,j}^{n+1} = u_{i+1/2,j}^* - \frac{\Delta t}{\rho h}(p_{i+1,j}^{n+1} - p_{i,j}^{n+1})$$

$$v_{i,j+1/2}^{n+1} = v_{i,j+1/2}^* - \frac{\Delta t}{\rho h}(p_{i,j+1}^{n+1} - p_{i,j}^{n+1}), \quad (2.23)$$

which is identical to equation (2.19), except that the temporary edge velocities are found by interpolation (equation 2.22).

The continuity equation for the pressure control volume is still given by equation (2.17), and substitution of the velocities given by equation (2.23) yields exactly equation (2.20), but with the velocities on the right-hand side now given by equation (2.22), rather than equations (2.19). Although we enforce incompressibility using the corrected edge velocities so that the pressure gradient is estimated using pressure values next to each other, the new cell-centered velocities are found using the average pressure at the edges. Thus,

$$u_{i,j}^{n+1} = u_{i,j}^* - \frac{\Delta t}{\rho h}(p_{i+1,j}^{n+1} - p_{i-1,j}^{n+1})$$

$$v_{i,j}^{n+1} = v_{i,j}^* - \frac{\Delta t}{\rho h}(p_{i,j+1}^{n+1} - p_{i,j-1}^{n+1}). \tag{2.24}$$

The Rhie–Chow method has been successfully implemented by a number of authors and applied to a range of problems. For regular structured grids it offers no advantage over the staggered grid, but for more complex grid layouts and for cut-cell methods (discussed in Chapter 4) it is easier to implement. A different approach for the solution of the fluid equations on co-located grids is described in Section 10.3.4.

The method described above works well for moderate Reynolds number flows and short enough computational times, but for serious computational studies, particularly at high Reynolds numbers, it usually needs to be refined in various ways. In early computations the centered difference scheme for the advection terms was sometimes replaced by the upwind scheme. In this approach, the momentum is advected through the left boundary of the $(i+1/2, j)$ control volume in Fig. 2.2 using $u_{i-1/2,j}$ if the flow is from the left to right and using $u_{i+1/2,j}$ if the flow is from the right to left. While this resulted in much improved stability properties, the large numerical diffusion of the scheme and the low accuracy has all but eliminated it from current usage. By using more sophisticated time integration schemes, such as the Adams–Bashforth method described above or Runge–Kutta schemes, it is possible to produce stable schemes based on centered differences for the advection terms that are subject to the time-step limit given by (2.8), rather than equation (2.7). However, for situations where the velocity changes rapidly over a grid cell this approach can produce unphysical oscillations. These oscillations do not always render the results unusable, and the problem only shows up in regions of high gradients. However, as the Reynolds number becomes higher the problem becomes more serious. To overcome these problems, many authors have resorted to the use of higher order upwind methods. These methods are almost as accurate as centered difference schemes in regions of fully resolved smooth flows and much more robust in regions where the solution changes rapidly and the resolution is marginal. The best-known approach is the quadratic upstream interpolation for convective kinematics (QUICK) method of Leonard (1979), where values at the cell edges are interpolated by upstream biased third-order polynomials. Thus, for example, the u velocity at (i, j) in Fig. 2.2 is found from

$$u_{i,j} = \begin{cases} (1/8)\big(3u_{i+1/2,j} + 6u_{i-1/2,j} - u_{i-3/2,j}\big), & \text{if } u_{i,j} > 0; \\ (1/8)\big(3u_{i-1/2,j} + 6u_{i+1/2,j} - u_{i+3/2,j}\big), & \text{if } u_{i,j} < 0. \end{cases} \qquad (2.25)$$

While QUICK and its variants are not completely free of "wiggles" for steep enough gradients, in practice they are much more robust than the centered difference scheme and much more accurate than first-order upwind. Other authors have used the second-order ENO (essentially nonoscillatory) scheme described in Chapter 3 (for the advection of the level set function) to find the edge velocities.

2.4 Boundary conditions

At solid walls the boundary conditions for the Navier–Stokes equations are well defined. As we saw in Section 1.3, the velocity is simply equal to the wall velocity. When staggered grids are used to resolve a rectangular geometry, the grid is arranged in such a way that the boundaries coincide with the location of the normal velocities. In the discrete version of the equations, the relative normal velocity is then simply zero on the side of the control volume that coincides with the wall. Since the location of the tangent velocity component is, however, half a grid space away from the wall, imposing the tangent wall velocity is slightly more complicated. Usually, this is done by the introduction of "ghost points" on the other side of the wall, half a grid space away from the wall. The tangent velocity at this point is specified in such a way that linear interpolation gives the correct wall velocity. Thus, if the wall velocity is u_{b} and the velocity at the first point inside the domain (half a cell from the wall) is $u_{1/2}$, the velocity at the ghost point is $u_{-1/2} = 2u_{\mathrm{b}} - u_{1/2}$. Figure 2.3 shows a ghost point near a solid boundary.

While the enforcement of the tangential velocity boundary conditions is perhaps a little kludgy, it is in the implementation of the boundary conditions for the pressure where the true elegance of the staggered grid manifests itself best. As seen before, the pressure equation is derived by substituting the discrete equations for the correction velocities into the discrete continuum equation. For cells next to the boundary the normal velocity at the wall is known and there is no need to substitute the correction velocity for the boundary edge. The pressure equation for the cell will therefore only contain pressures for nodes inside the domain. This has sometimes led to declarations to the effect that on staggered grids no boundary conditions are needed for the pressure. This is, of course, not correct. The discrete

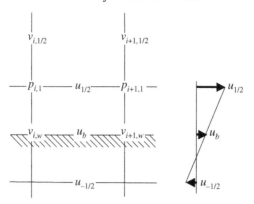

Fig. 2.3. A cell next to a solid wall. To impose the tangent velocity u_w, a "ghost" point is introduced half a grid cell outside the boundary.

pressure equation for cells next to the boundary is different than for interior nodes since the boundary conditions are incorporated during the derivation of the equation. For the pressure cell in Fig. 2.4 the inflow on the left is given $(u_{b,j})$, so that we only substitute equations (2.19) for the right, top, and bottom boundaries, yielding:

$$\frac{p_{i+1,j}^{n+1} + p_{i,j+1}^{n+1} + p_{i,j-1}^{n+1} - 3p_{i,j}^{n+1}}{h^2}$$
$$= \frac{\rho}{\Delta t} \left\{ \frac{u_{i+1/2,j}^* - u_{b,j} + v_{i,j+1/2}^* - v_{i,j-1/2}^*}{h} \right\} \qquad (2.26)$$

which does not include a pressure to the left of the control volume. Similar equations are used for the other boundaries. At corner nodes, the velocity is known at two edges of the control volume. This expression can also be derived by substituting equation (2.19), written for $u_{i-1/2,j}^*$, into equation (2.20) and using that $u_{i-1/2,j}^{n+1} = u_{b,j}$.

For co-located grids, it is necessary to specify boundary conditions for the intermediate velocity in terms of the desired boundary conditions to be satisfied by the velocity at the end of the time step. For the simple case where a Dirichlet boundary condition on velocity, $\mathbf{u}_{n+1} = \mathbf{u}_b$, is specified, the boundary condition for the intermediate velocity can be expressed as

$$\mathbf{u}^* \cdot \mathbf{n} = \mathbf{u}_b \cdot \mathbf{n} \qquad \text{and} \qquad \mathbf{u}^* \cdot \mathbf{t} = \mathbf{u}_b \cdot \mathbf{t} + \Delta t (2\nabla \phi^n - \nabla \phi^{n-1}) \cdot \mathbf{t}, \qquad (2.27)$$

where $\mathbf{n}$ is the unit vector normal to the surface on which the boundary condition is applied and $\mathbf{t}$ represents the unit vector(s) parallel to the surface. From equation (2.10) the corresponding boundary condition for ϕ^{n+1} is the homogeneous Neumann condition: $\nabla \phi^{n+1} \cdot \mathbf{n} = 0$.

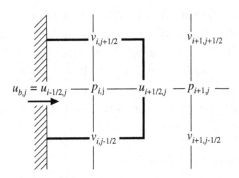

Fig. 2.4. A cell next to an inflow boundary. The normal velocity $u_{b,j}$ is given.

While the discrete boundary conditions for solid walls are easily obtained, the situation is quite different for inflow and outflow boundaries. Generally the challenges are to obtain as uniform (or at least well-defined) inflow as possible and to design outflow boundary conditions that have as little effect on the upstream flow as possible. Thus, the numerical problem is essentially the same as the one faced by the experimentalist and just as the experimentalist installs screens and flow straighteners, the computationalist must use *ad hoc* means to achieve the same effect.

The inclusion of a well-defined inflow, such as when a flow from a long pipe enters a chamber, is relatively easy, particularly in computations using staggered grids. Usually the normal and the tangential velocities are specified and once the normal velocity is given, the pressure equation can be derived in the same way as for rigid walls. Although a given velocity is easily implemented numerically, the inflow must be carefully selected and placed sufficiently far away from the region of interest in order to avoid imposing artificial constraints on the upstream propagation of flow disturbances. In many cases, however, such as for flow over a body, the upstream boundary condition does not represent a well-defined inflow, but rather a place in the flow where the modeler has decided that the influence of the body is sufficiently small so that the disturbance can be ignored there. For multiphase flow such situations arise frequently in studies of the flow over a single particle and we shall defer the discussion of how these are handled to Chapter 4. Although the specification of the inflow velocity conditions is by far the most common approach to simulations of internal flow, many practical situations call for the specification of the overall pressure drop. For discussions of how to handle such problems we refer the reader to standard references such as Wesseling (2001).

The outflow is a much more difficult problem. Ideally, we want the domain to be sufficiently long so that the outflow boundary has essentially no effect on the flow region of interest. Practical considerations, however, often force us to use relatively short computational domains. In reality the flow continues beyond the boundary and we accommodate this fact by assuming that the flow is relatively smooth. Thus, the trick is to do as little as possible to disturb the flow. Below we list a few of the proposals that have been put forward in the literature to allow the flow to exit the computational domain as gently as possible:

Convective (e.g. Kim and Choi, 2002):

$$\frac{\partial \mathbf{u}^*}{\partial t} + c\frac{\partial \mathbf{u}^*}{\partial n} = 0 \tag{2.28}$$

Parabolic (e.g. Magnaudet, Rivero, and Fabre, 1995):

$$\frac{\partial^2 u_n^*}{\partial n^2} = 0, \quad \frac{\partial u_\tau^*}{\partial n} = 0, \quad \frac{\partial^2 p}{\partial n \partial \tau} = 0 \tag{2.29}$$

Zero gradient (e.g. Kim, Elghobashi, & Sirignano, 1998):

$$\frac{\partial \mathbf{u}^*}{\partial n} = 0 \tag{2.30}$$

Zero second gradient (e.g. Shirayama, 1992):

$$\frac{\partial^2 \mathbf{u}^*}{\partial n^2} = 0 \tag{2.31}$$

In the above, n and τ indicate directions normal and tangential to the outer boundary and c is a properly selected advection velocity. Many other boundary conditions have been proposed in the literature to account for outflows in a physically plausible, yet computationally efficient way. The outflow boundary conditions are generally imposed for the velocity and an equation for the pressure is then derived in the same way as for solid boundaries. While the straightforward treatment of pressure for solid walls on staggered grids carries over to both in and outflow boundaries, co-located grids generally require us to come up with explicit approximations for the pressure at in and outflow boundaries. A more extensive discussion of outflow boundary conditions can be found in Section 6.5 in Wesseling (2001) and Johansson (1993), for example. Spectral simulations, owing to their global nature, place a more stringent nonreflection requirement at the outflow boundary. A buffer domain or viscous sponge technique is often used to implement a nonreflecting outflow boundary condition as discussed in Chapter 4.

3

Immersed boundary methods for fluid interfaces

Nearly half a century of computational fluid dynamics has shown that it is very hard to beat uniform structured grids in terms of ease of implementation and computational efficiency. It is therefore not surprising that a large fraction of the most popular methods for finite Reynolds number multiphase flows today are methods where the governing equations are solved on such grids. The possibility of writing one set of governing equations for the whole flow field, frequently referred to as the "one-fluid" formulation, has been known since the beginning of large-scale computational studies of multiphase flows. It was, in particular, used by researchers at the Los Alamos National Laboratory in the early 1960s for the marker-and-cell (MAC) method, which permitted the first successful simulation of the finite Reynolds number motion of free surfaces and fluid interfaces. This approach was based on using marker particles distributed uniformly in each fluid to identify the different fluids. The material properties were reconstructed from the marker particles and sometimes separate surface markers were also introduced to facilitate the computation of the surface tension. While the historical importance of the MAC method for multiphase flow simulations cannot be overstated, it is now obsolete. In current usage, the term "MAC method" usually refers to a projection method using a staggered grid.

When the governing equations are solved on a fixed grid, the different fluids must be identified by a marker function that is advected by the flow. Several methods have been developed for that purpose. The volume-of-fluid (VOF) method is the oldest and, after many improvements and innovations, continues to be widely used. Other marker function methods include the level-set method, the phase-field method, and the constrained interpolated propagation (CIP) method[1]. Instead of advecting the marker function

---

[1] While the initials CIP have stayed constant since the introduction of the method the actual name of the method has evolved. CIP initially stood for *cubic interpolated pseudoparticle*

directly, the boundary between the different fluids can be tracked using marker points, and the marker function then reconstructed from the location of the interface. Methods using marker points are generally referred to as "front-tracking" methods, and have been developed by a number of investigators. In addition to the difference in the way the phase boundary is advected, the inclusion of surface tension is fundamentally different in front-tracking and marker-function methods.

In this chapter we will discuss solution techniques based on the "one-fluid" formulation. In essentially all cases the Navier–Stokes equations are solved on a fixed grid using a projection method. Once the density and viscosity fields have been determined, the various methods are the same, although different implementations generally rely on slightly different flow solvers. The advection of the fluid interface, or – equivalently – the density and viscosity fields, is what distinguishes one method from another. We first present numerical solution techniques for the Navier–Stokes equations for flows with discontinuous density and viscosity fields, leaving unspecified how these fields are updated. Once we have discussed the integration of the momentum equations, we will describe in detail how the phase boundary is advected using the different methods. But first we must introduce the "one-fluid" formulation of the governing equations.

3.1 The "one-fluid" approach

When multiphase flow is simulated by solving a single set of equations for the whole flow field, it is necessary to account for differences in the material properties of the different fluids and to add appropriate interface terms for interfacial phenomena, such as surface tension. Since these terms are concentrated at the boundary between the different fluids, they are represented by delta (δ) functions. When the equations are discretized, the δ-functions must be approximated along with the rest of the equations. The material properties and the flow field are, in general, also discontinuous across the interface and all variables must therefore be interpreted in terms of generalized functions.

The various fluids can be identified by a step (Heaviside) function H which is 1 where a particular fluid is located and 0 elsewhere. The interface itself is marked by a nonzero value of the gradient of the step function. To find

---

then for *cubic interpolated propagation* and most recently for *constrained interpolation profile*.

the gradient it is most convenient to express H in terms of the integral over multidimensional δ-functions. For a two-dimensional field

$$H(x,y) = \int_A \delta(x-x')\delta(y-y')\, da', \qquad (3.1)$$

where the integral is over an area A bounded by a contour S. H is obviously 1 if the point (x,y) is enclosed by S and 0 otherwise. To find the gradient of H, first note that since the gradient is with respect to the unprimed variables, the gradient operator can be put under the integral sign. Since the gradient of the δ-function is antisymmetric with respect to the primed and unprimed variables, the gradient with respect to the unprimed variables can be replaced by the gradient with respect to the primed variables. The resulting area (or volume, in three dimensions) integral can be transformed into a line (surface) integral by a variation of the divergence theorem for gradients. Symbolically:

$$\nabla H = \int_A \nabla \left[\delta(x-x')\delta(y-y')\right] da' \qquad (3.2)$$

$$= -\int_A \nabla' \left[\delta(x-x')\delta(y-y')\right] da'$$

$$= -\oint_S \delta(x-x')\delta(y-y')\mathbf{n}'\, ds'.$$

Here the prime on the gradient symbol denotes the gradient with respect to the primed variables and $\mathbf{n}'$ is the outward unit normal vector to the interface. Although we have used that S is a closed contour, the contribution of most of the integral is zero. We can therefore replace the integral by one over a part of the contour and drop the circle on the integral:

$$\nabla H = -\int_S \delta(x-x')\delta(y-y')\mathbf{n}'\, ds'. \qquad (3.3)$$

By introducing local coordinates tangent (s) and normal (n) to the front, we can write

$$\delta(x-x')\delta(y-y') = \delta(s)\delta(n), \qquad (3.4)$$

and evaluate the integral as

$$-\int_S \delta(x-x')\delta(y-y')\mathbf{n}'\, ds' = -\int_S \delta(s')\delta(n')\mathbf{n}'\, ds' = -\delta(n)\mathbf{n}. \qquad (3.5)$$

This allows us to use a one-dimensional delta function of the normal variable, instead of the two-dimensional one in equation (3.2). Although we have assumed a two-dimensional flow in the discussion above, the same arguments apply to three-dimensional space.

If the density of each phase is assumed to be constant, the density at each point in the domain can be represented by the constant densities and the Heaviside function:

$$\rho(x, y) = \rho_1 H(x, y) + \rho_0[1 - H(x, y)]. \tag{3.6}$$

Here, ρ_1 is the density of the fluid in which $H = 1$ and ρ_0 is the density where $H = 0$. The gradient of the density is given by

$$\nabla\rho = \rho_1\nabla H - \rho_0\nabla H = (\rho_1 - \rho_0)\nabla H \tag{3.7}$$
$$= \Delta\rho \int \delta(x - x')\delta(y - y')\mathbf{n}' \, ds' = \Delta\rho\delta(n)\mathbf{n},$$

where $\Delta\rho = \rho_0 - \rho_1$. Similar equations can be derived for other material properties.

The Navier–Stokes equations, as derived in Chapter 1, allow for arbitrary changes in the material properties of the fluids. While the differential form does not, strictly speaking, allow discontinuous material properties, numerical methods based on the finite-volume method are equivalent to working with the integral form of the governing equations, where no smoothness assumption is made. Furthermore, if we work with generalized functions, then we can use the equations in their original form. This is the approach that we take here. The equations as written in Chapter 1 do not, however, account for interface effects such as surface tension. Surface tension acts only at the interface and we can add this force to the Navier–Stokes equations as a singular interface term by using a δ-function,

$$\rho\frac{\partial\mathbf{u}}{\partial t} + \rho\nabla\cdot\mathbf{u}\mathbf{u} = -\nabla p + \rho\mathbf{f} + \nabla\cdot\mu(\nabla\mathbf{u} + \nabla^T\mathbf{u}) + \gamma\kappa\delta(n)\mathbf{n}. \tag{3.8}$$

Here, κ is the curvature for two-dimensional flows and twice the mean curvature in three-dimensions, and $\mathbf{n}$ is a properly oriented unit vector normal to the front. n is a normal coordinate to the interface, with $n = 0$ at the interface. As in Chapter 1, $\mathbf{u}$ is the velocity, p is the pressure, and $\mathbf{f}$ is a body force. With the singular term added, this equation is valid for the whole flow field, including flows with interfaces across which ρ, and the viscosity field, μ, change discontinuously. This fact justifies the "one-fluid" denomination.

For incompressible flows, the mass conservation equation is the same as for a single-phase flow

$$\nabla\cdot\mathbf{u} = 0, \tag{3.9}$$

showing that volume is conserved. For a single-phase flow where the density is constant, there is no need to follow the motion of individual fluid particles. If the density varies from one particle to another, but remains constant for each particle as it moves (as it must do for an incompressible flow), it is necessary to follow the motion of each fluid particle. This can be done by integrating the equation

$$\frac{D\rho}{Dt} = 0. \tag{3.10}$$

For multiphase flows with well-defined interfaces, where the density of each phase is a constant, we only need to find H and then construct the density directly from H as discussed above. The same arguments hold for the viscosity and other properties of the fluid.

The "one-fluid" equations are an exact rewrite of the Navier–Stokes equations for the fluid in each phase and the interface boundary conditions. The governing equations as listed above assume that the only complication in multifluid flows is the presence of a moving phase boundary with a constant surface tension. While these equations can be used to describe many problems of practical interest, additional complications quickly emerge. The presence of a surfactant or contaminants at the interface between the fluids is perhaps the most common one, but other effects such as phase changes, are common too. We will not address these more complex systems in this book.

There is, however, one issue that needs to be addressed here. While the Navier–Stokes equations, with the appropriate interface conditions, govern the evolution of a system of two fluids separated by a sharp interface, the topology of the interface can change by processes that are not included in the continuum description. Topology changes are common in multiphase flow, such as when drops or bubbles break up or coalesce. These changes can be divided into two broad classes: films that rupture and threads that break. If a large drop approaches another drop or a flat surface, the fluid in between must be "squeezed" out before the drops are sufficiently close so that the film between them becomes unstable to attractive forces and ruptures. A long, thin cylinder of one fluid will, on the other hand, generally break by Rayleigh instability where one part of the cylinder becomes sufficiently thin so that surface tension "pinches" it in two. The exact mechanisms of how threads snap and films break are still being actively investigated. There are, however, good reasons to believe that threads can become infinitely thin in a finite time and that their breaking is "almost" described by the Navier–Stokes equations (Eggers, 1995). Films, on the other hand, are generally believed to rupture due to short-range attractive forces, once they are a few hundred angstroms thick (Edwards, Brenner, and Wasan, 1991). At the moment, most numerical simulations of topological changes treat the process in a very *ad hoc* manner and simply fuse interfaces together when they come close enough and allow threads to snap when they are thin enough. "Enough" in this context is generally about a grid spacing and it should be clear that this approach can, when the exact time of rupture is important, as it sometimes is for thin films, lead to results that do not converge under grid refinement. The pinching of threads, on the other hand, appears to be much less sensitive to the exact numerical treatment and results that are essentially independent of the grid can be obtained. How to incorporate more exact models for the rupture of films into simulations of multiphase flows is a challenging problem that is, as of this writing, mostly unsolved. Current efforts on multiscale computing, however, hold great promise.

The one-fluid formulation of the Navier–Stokes equations allows multiphase flow to be treated in more or less the same way as homogeneous flows and any standard algorithm based on fixed grids can, in principle, be used to integrate the discrete Navier–Stokes equations in time. The main difference between a Navier–Stokes solver for multiphase flows and the simple method outlined in Chapter 2 is that we must allow for variable density and viscosity and add the surface tension as a body force. For the simple first-order method of Chapter 2, the equation for the predicted velocity (equation 2.3) is modified by explicitly identifying at which time the density is computed and by adding the surface tension:

$$\mathbf{u}^* = \mathbf{u}^n + \Delta t\left(-\mathbf{A}_h(\mathbf{u}^n) + \frac{1}{\rho^n}\mathbf{D}_h(\mathbf{u}^n) + \frac{1}{\rho^n}\mathbf{f}_b^n + \frac{1}{\rho^n}\mathbf{F}_\gamma^n\right). \qquad (3.11)$$

The exact calculation of the surface tension term $\mathbf{F}_\gamma$ generally depends on how the marker function is advected, which is discussed below. The Poisson equation for the pressure is the same, except that the density is no longer a constant and must be explicitly included under the divergence operator

$$\nabla_h \frac{1}{\rho^n} \cdot \nabla_h p = \frac{1}{\Delta t}\nabla_h \cdot \tilde{\mathbf{u}}. \qquad (3.12)$$

This simple-looking change has rather profound implications for the solution of the pressure equation, since highly developed methods for separable equations cannot be used. Considerable progress has, however, been made in the development of efficient methods for elliptic equations with variable coefficients like equation (3.12). We refer the reader to Wesseling (2004) for a discussion and additional references. The projected velocity at time step $n+1$ is found, as before, from equation (2.4) from Chapter 2, but with the density computed at time n:

$$\mathbf{u}^{n+1} = \tilde{\mathbf{u}} - \frac{\Delta t}{\rho^n}\nabla_{\mathrm{h}}P. \tag{3.13}$$

Several variations of this algorithm are possible. Here we have used the momentum equations in the non-conservative form. If the conservative form is used, the density must be available at time $n+1$ so the density must be advected at the beginning of the time step. The conservative form can, however, lead to unrealistically large velocities around the interface, so the non-conservative form is generally recommended. For serious simulations, higher order time integration is usually used. The spatial discretization of the advection terms is the same as discussed in Chapter 2, but we must use the full deformation tensor for the viscous terms, since the viscosity is generally not constant. Although the viscous fluxes are usually computed by simply using the linearly interpolated value of viscosity for the boundaries of the control volumes, the resulting approximation does not ensure the proper continuity of the viscous fluxes when the viscosity changes rapidly. Many authors have found that working with the inverse of the viscosities improves the results (Patankar, 1980; Ferziger, 2003).

3.2 Advecting a marker function

When a fluid interface is captured on a fixed grid, a marker function f is introduced such that $f = 1$ in one fluid and $f = 0$ in the other fluid. The marker function is advected by the fluid and once the fluid velocity is known, f can, in principle, be updated by integrating

$$\frac{\partial f}{\partial t} + \mathbf{u} \cdot \nabla f = 0 \tag{3.14}$$

in time. Here, f is essentially the Heaviside function H introduced by equation (3.1), but in numerical implementations f can have a smooth transition zone between one value and the other. We will, therefore, use f to denote a marker function introduced for computational reasons. Integrating equation (3.14) for discontinuous data is, in spite of its apparent simplicity, one of the hard problems in computational fluid dynamics.

The importance of understanding the difficulties in advecting a discontinuous marker function justifies devoting space to examining it in some detail. To simplify the discussion, we focus on the one-dimensional advection equation, with constant advection velocity, $U > 0$:

$$\frac{\partial f}{\partial t} + U\frac{\partial f}{\partial x} = 0. \tag{3.15}$$

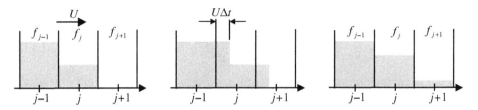

Fig. 3.1. The advection of a scalar with a constant velocity U, using the first-order upwind method. The function f is assumed to be a constant in each cell, equal to the cell average f_j^n (first frame). After the function has been moved a distance $U\Delta t$ with the velocity (middle frame), the function is averaged in each cell, giving a new cell average f_j^{n+1}.

We introduce the flux $F = Uf$ and write

$$\frac{\partial f}{\partial t} + \frac{\partial F}{\partial x} = 0. \tag{3.16}$$

To discretize this equation we use a spatial grid with evenly spaced grid points denoted by $j-1$, j, $j+1$ and so on. The grid is sketched in Fig. 3.1. The grid points represent the centers of cells or control volumes whose boundaries are located half-way between them. The boundaries of cell j are denoted by $j \pm 1/2$ and the average value of f in cell j is defined by

$$f_j = \frac{1}{h} \int_{j-1/2}^{j+1/2} f(x)\,dx, \tag{3.17}$$

where h is the grid spacing. Integrating equation (3.16) in space and time yields

$$f_j^{n+1} - f_j^n = \int_t^{t+\Delta t} F_{j+1/2}\,dt - \int_t^{t+\Delta t} F_{j-1/2}\,dt. \tag{3.18}$$

It is important to appreciate that this is an exact result. The physical meaning of equation (3.18) is that the average value of f_j increases by the difference in the flux of f in and out of the control volume. Thus, the quality of the results depends solely on how we compute F and its time integral at each cell boundary.

Since the velocity U is constant, the f profile is advected by simply shifting it to the right and the time integral of the flux across the boundary can be computed if the exact shape of f is known. Since we only store the average value f_j, we have to make some assumptions about the distribution of f in each cell. If we take f to be a constant in each cell, equal to the average f_j, then the time integral of the flux can be computed exactly as

$$\int_t^{t+\Delta t} F_{j+1/2} \, dt = \Delta t U f_j. \tag{3.19}$$

The average value of f_j is then updated by:

$$f_j^{n+1} = f_j^n - \frac{\Delta t U}{h}(f_j^n - f_{j-1}^n). \tag{3.20}$$

Although we have introduced this method for advecting the solution by geometric considerations, it is also easily derived by approximating equation (3.15) using a first-order *upwind* (or one-sided, using the value of f at grid points j and $j-1$) finite difference approximation for the spatial derivative and a first-order forward approximation for the time derivative.

We can think of the advection of f as a two-step process. First the solution is simply shifted by $U\Delta t$ to the right. Then it is averaged to give a new uniform distribution of f in each cell. These steps in updating f_j are shown in Fig. 3.1. The initial distribution of f is shown on the left-hand side. The solution moves to the right with velocity U and the shaded area in the middle frame shows the exact solution after time Δt. However, only the average value is stored in each cell and the shaded area in the frame on the right represents the distribution after the averaging. This distribution is then used to take the next time step. The main problem with this approach is that by assuming that f in each cell is equal to the average value and by using the average value to compute the fluxes, f starts flowing out of cell j long before it is full. Or, in other words, cell $j+1$ acquires f before cell j is full. This leads to the very rapid *artificial diffusion* that the first-order advection scheme is so notorious for.

The poor performance of the method described above is entirely due to the assumption that f was uniform in each cell. The advection itself, after all, was exact. This suggests that we should try to use a more sophisticated representation for the distribution of f in each cell. If we assume that f is distributed linearly, it is still relatively easy to compute the advection exactly. There are, however, several ways to select the slopes of the line describing how f is distributed. If we assume that the slope in cell j is given by $s_j = (f_{j+1} - f_j)/h$, we obtain the centered second-order Lax–Wendroff scheme where the average is updated by:

$$f_j^{n+1} = f_j^n - \frac{\Delta t U}{2h}(f_{j+1}^n - f_{j-1}^n) + \frac{\Delta t^2 U^2}{2h^2}(f_{j+1}^n - 2f_j^n + f_{j-1}^n). \tag{3.21}$$

The performance of the two approaches outlined above is shown in Fig. 3.2 where the results of advecting a discontinuous function by both the low-order

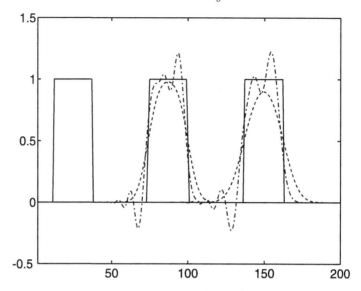

Fig. 3.2. The advection of a "blob" of f using a first-order upwind method (dashed line) and the second-order Lax–Wendroff method (dash-dotted line) in a constant velocity field $U = 1$. The exact solution is shown by a solid line. The domain has length 4 and has been resolved by 201 grid points. The time step is $\Delta t = 0.5h$. The initial conditions are shown on the left, and the solution is then shown after 125 and 250 time steps.

upwind scheme (equation 3.20) and the Lax–Wendroff scheme (equation 3.21) are presented. The initial data is the rectangle on the left. The solutions are then shown after 125 and 250 steps. Results for the upwind scheme are shown by the dashed line and the results for the higher order scheme are shown by the dash-dot line. The exact solution is the initial condition translated without a change of shape as shown by the solid line. Obviously, neither method performs very well. For the low-order scheme the discontinuity is rapidly smeared so that the sharp interface disappears, and the high-order scheme produces unacceptable wiggles or oscillations. The numerical solutions also deteriorate in time, whereas the exact solution propagates unchanged. Selecting the slopes in each cell differently results in different high-order schemes. While there are differences in their behavior, they all show oscillations near the discontinuity.

The higher order methods can be improved considerably by the use of nonlinear filters that prevent the formation of oscillations. Such "monotone" schemes have been developed to a very high degree of sophistication for gas dynamics applications where they have been used to capture a shock in essentially one grid cell. These high-order advection schemes – designed to keep shocks sharp and monotonic – are intended to deal with general

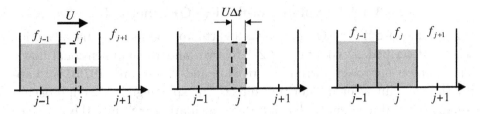

Fig. 3.3. VOF "reconstruction" for the advection of a scalar f in one dimension. Given the average value of f in each cell, f_j^n, and which neighboring cell is full and which is empty, the exact location of the interface in a half-full cell can be determined (left frame). The interface is then moved a distance $U\Delta t$ (middle frame) and a new cell averaged value f_j^{n+1} found by averaging (right frame).

hyperbolic systems such as the Euler equations. Further aspects of these schemes are discussed in Chapter 10, Section 10.2. Our equation, as well as the initial data, permit a simpler approach. The velocity field is given (or computed separately) so the equation is linear and f_j is simply 1 or 0 in most of the domain. In a partially full cell, where the value of f_j is somewhere between 0 and 1 we must have $f = 1$ in a part of the cell and $f = 0$ in the remainder. Thus, given the cell average f_j and information about whether it is the $j + 1$ or the $j - 1$ cell that is full, we can reconstruct f in cell j exactly. The advection of f_j, using this reconstruction, is shown in Fig. 3.3 where we have assumed that the full cell is on the left. On the left side of cell j, $f = 1$ and on the right hand side, $f = 0$ (dashed line in the first frame). This distribution is then advected a distance $U\Delta t$ (middle frame) and the advected distribution used to find the new average. In this example the interface remained in cell j, and no f flowed into cell $j + 1$. If the j cell is nearly full, however, then obviously some f must flow into the next cell. To account for both cases, we write the time integral of the fluxes as

$$\int_t^{t+\Delta t} F_{j+1/2} \, dt = \begin{cases} 0, & \Delta t \le (1 - f_j)h/U, \\ h - (f_j + U\Delta t), & \Delta t > (1 - f_j)h/U. \end{cases} \tag{3.22}$$

Thus, given f_j, we can compute the fluxes exactly. The advection of f_j, using this approach, is shown schematically in Fig. 3.3 and the exact solution in Fig. 3.2 (the solid line) was computed using this algorithm.

The ease by which we can advect f_j exactly in one dimenension is misleading. For two- and three-dimensional flow fields the advection becomes much more complex and the difficulties encountered in higher dimensions have stimulated the development of several methods to advect a discontinuous marker function. We will survey a few of the more common and/or successful ones below.

3.3 The volume-of-fluid (VOF) method

The volume-of-fluid or VOF method is an attempt to generalize the advection scheme described by equation (3.22) to two- and three-dimensional flows. The simplest approach, introduced by Noh and Woodward (1976), is to apply the one-dimensional algorithm directly by splitting. In this approach – usually referred to as simple line interface calculation or SLIC – the interface separating the $f = 1$ part from the $f = 0$ part of the cell is approximated by a straight line parallel to the y-axis for advection in the x-direction and parallel to the x-axis for y-advection (for two-dimensional flow). Hirt and Nichols (1981) proposed a slightly different method where the interface was still approximated by straight lines parallel to the coordinate axis, but the same orientation was used for both the x- and the y-directions. To determine whether the interface should be horizontal or vertical, Hirt and Nichols found the normal to the interface, using values in the neighboring cells and selected the orientation of the interface depending on whether the normal was more closely aligned with the x- or the y-axis. Although perhaps more appealing than the original SLIC method, tests by Rudman (1997) suggest that the Hirt–Nichols method is not significantly more accurate. In addition to distorting the interfaces, both methods generate a considerable amount of "floatsam" and "jetsam" where pieces of the interface break away in an unphysical way.

Although the method of Hirt and Nichols perhaps did not improve significantly on the SLIC approach, it nevertheless suggested that the key to improving the behavior of the advection scheme was the *reconstruction* of the interface in each cell, using the values of the marker function in the neighboring cells. In the method introduced by Youngs (1982), the interface was approximated by a straight-line segment in each cell, but the line could be oriented arbitrarily with respect to the coordinate axis. The orientation of the line is determined by the normal to the interface, which is found by considering the average value of f in both the cell under consideration, as well as neighboring cells. The results of the advection depend on the accuracy of the interface reconstruction and finding the normal accurately; therefore these operations are critical for piecewise linear interface calculation (PLIC) methods. Several methods have been proposed. Rudman (1997) recommends using

$$n_{i,j}^x = \frac{1}{h}(f_{i+1,j+1} + 2f_{i+1,j} + f_{i+1,j-1} - f_{i-1,j+1} - 2f_{i-1,j} - f_{i-1,j-1}) \quad (3.23)$$

$$n_{i,j}^y = \frac{1}{h}(f_{i+1,j+1} + 2f_{i,j+1} + f_{i-1,j+1} - f_{i+1,j-1} - 2f_{i,j-1} - f_{i-1,j-1}) \quad (3.24)$$

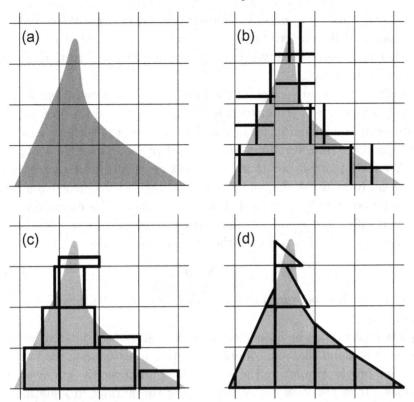

Fig. 3.4. VOF reconstruction of the solution for advection in two-dimensions. (a) The original interface. (b) The original SLIC reconstruction where the interface is always perpendicular to the advection direction. (c) The Hirt–Nichols reconstruction where the interface is parallel to one grid line. (d) PLIC reconstruction using straight lines with "optimal" orientation. Figure adapted from Rudman (1997).

introduced by Kothe, Mjolsness, and Torrey (1991). Once the normal and the average value of f in each cell, $f_{i,j}$, is known, the exact location of the interface can be determined. In two-dimensions the line segment can cross any of two adjacent or opposite cell faces, so there are four basic interface configurations. In three-dimensions there is a considerably larger number of possible configurations, adding to the complexity of the method. Figure 3.4, adapted from Rudman (1997), shows the main difference between the interface reconstruction using SLIC, the Hirt–Nichols VOF method, and PLIC. While the linear reconstruction captures slowly varying sections of the interface very well, it usually does less well for segments that are changing rapidly. Notice that the line segments are not continuous across cell boundaries.

Once the interface in each cell has been constructed, the fluxes from one cell to another are computed by geometric considerations. While fully multidimensional VOF schemes have been developed, generally it is found that split schemes, where the advection is first done in one direction and then the other, work better.

In addition to the advection of the interface, the computation of surface tension has been greatly improved in recent versions of the VOF method. In many early implementations of the VOF method the surface tension was simply ignored. However, in a widely cited paper, Brackbill, Kothe, and Zemach (1992) pointed out that the curvature of an interface represented by a marker function can be computed by taking the divergence of a normal field computed in a region extended off the interface. This extension usually exists naturally, particularly if the marker function is smoothed slightly. Once the normal field $\mathbf{n}$ is found, the mean curvature is given by

$$\kappa = \nabla \cdot \mathbf{n}. \tag{3.25}$$

This approach has become known as the continuous surface force (CSF) method.

Equation (3.25) can be implemented in several slightly different ways and generally it is found that smoothing the marker function to spread the transition zone before computing the normal field improves the curvature computations. It is also found that computing the normal in equation (3.8) using the same discretization used for the pressure gradient helps.

While the reconstruction of the interface in three dimensions is generally a fairly complex task, VOF remains one of the most widely used interface tracking method and impressive results have been obtained using the most advanced implementations. Many commercial CFD codes now include the option of simulating free-surface or multiphase flows using the VOF method. A review of the VOF method can be found in Scardovelli and Zaleski (1999).

3.4 Front tracking using marker points

Instead of updating the value of the marker function by finding the fluxes in and out of each cell, we can mark the boundary between the two fluids using connected marker points that are advected with the flow and then reconstruct the marker function from the location of the front. Marker points have been used extensively for boundary integral simulations of potential flows and Stokes flows, but the use of connected marker points to identify a boundary between two fluids governed by the full Navier–Stokes equations appears to originate with Daly (1969a,b), who used the points to compute surface tension. The advection of the material properties was,

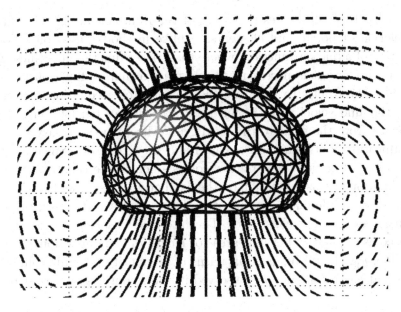

Fig. 3.5. Explicit tracking of a three-dimensional fluid interface. The interface is represented by an unstructured triangulated grid that moves with the fluid. As the interface stretches, points and elements are added and deleted.

however, done by the original MAC method where unconnected markers were distributed throughout the flow region. Further progress was made by Peskin (1977) who used marker points to represent elastic one-dimensional fibers in homogeneous flows and by Glimm and collaborators (Glimm, 1982, Glimm and McBryan, 1985) who tracked shocks in compressible flows by connected marker points. The first use of marker points to capture immiscible fluid interfaces in a flow governed by the full Navier–Stokes equations appears to be Unverdi and Tryggvason (1992). The method of Unverdi and Tryggvason and its derivatives have now been used to examine a large range of multiphase flows. A recent review of the method and its applications can be found in Tryggvason *et al.* (2001).

We describe the Unverdi and Tryggvason method in some detail here. Figure 3.5 shows a rising buoyant bubble whose surface is tracked by a triangular surface grid while the fluid velocity is found by solving the conservation equations on a fixed grid. The data structure forming the front consists of marker points whose coordinates are stored and triangular elements that connect the marker points. The elements also carry information about adjacent elements and material properties, in those cases where they vary over the front. Both the points and the elements are stored in a linked list to facilitate restructuring of the front, including the addition and deletion of points and elements. Unlike methods based on following the motion of a marker function, where the marker function identifies where each fluid is, here it is necessary to construct a marker function from the location of the front. To integrate the Navier–Stokes equations in time to update the fluid velocity on the fixed grid, the surface tension must also be transferred from the front to the fixed grid. Thus, it is necessary to set up communications between the front grid and the fixed grid. Since the front represents a δ-function, the transfer corresponds to the construction of an approximation to this δ-function on the fixed grid. This "smoothing" can be done in several different ways, but it is always necessary to ensure that the quantity transferred is conserved. The interface variables, ψ_f, are usually expressed as a quantity per unit area (or length in two dimensions), but the grid value, ψ_g, should be given in terms of a quantity per unit volume. To ensure that the total value is conserved in the smoothing, we must require that:

$$\int_{\Delta s} \psi_f(s)\, ds = \int_{\Delta v} \psi_g(\mathbf{x})\, dv. \tag{3.26}$$

The discrete version of this condition is

$$\psi_{ijk} = \sum_f \psi_f w_{ijk}^f \frac{\Delta s_f}{h^3} \tag{3.27}$$

for a three-dimensional smoothing. Here ψ_f is a discrete approximation to the front value and ψ_{ijk} is an approximation to the grid value. Δs_f is the area of the front element. The weights, w_{ijk}^f, determine what fraction of the front value each grid points gets. They must satisfy

$$\sum_{ijk} w_{ijk}^f = 1, \tag{3.28}$$

but can be selected in several different ways. The number of grid points used in the smoothing depends on the particular weighting function. Since the weights have finite support, there is a relatively small number of front elements that contribute to the value at each point of the fixed grid. Therefore, even though the summation in equation (3.27) is, in principle, over all of the front points, in practice it is sufficient to sum only over those points whose "range" includes the given grid point.

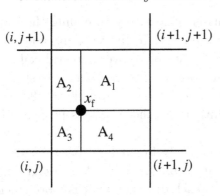

Fig. 3.6. Area weighting for smoothing and interpolation. For distribution of a front property at x_p, grid point (i,j) gets the fraction determined by A_1, grid point $(i+1,j)$ the fraction determined by A_2, and so on. When interpolating, the process is reversed and the grid values are added up at the front point, weighted by the area fractions.

We usually write the weighting functions as products of one-dimensional functions. In three dimensions, for example, the weight for the grid point (i,j,k) for smoothing a quantity from point $\mathbf{x}_f = (x_f, y_f, z_f)$ is written as

$$w^f_{ijk}(\mathbf{x}_f) = d(x_f - ih)\, d(y_f - jh)\, d(z_f - kh). \tag{3.29}$$

For two-dimensional interpolation, the third term is set to unity. $d(r)$ can be constructed in different ways. The simplest interpolation is the area (volume) weighting:

$$d(r) = \begin{cases} (h-r)/h, & 0 < r < h, \\ (h+r)/h, & -h < r < 0, \\ 0, & |r| \geq h . \end{cases} \tag{3.30}$$

Figure 3.6 gives a simple geometric interpretation of the weights. The weight for grid point (i,j) is the area fraction A_1, the weight for $(i+1,j)$ is A_2, and so on. A smoother weighting function, distributing the front quantities to a larger number of grid points, was introduced by Peskin (1977) who suggested:

$$d(r) = \begin{cases} (1/4h)(1 + \cos(\pi r/2h)), & |r| < 2h, \\ 0, & |r| \geq 2h. \end{cases} \tag{3.31}$$

Other weighting functions can be found in the literature, see Lai and Peskin (2000), for example.

On the fixed grid the front is approximated by a transition zone from one fluid to the other and it is, in principle, desirable that this zone be as thin as possible. This is achieved by spreading the front quantities to the grid points closest to the front. In the VOF method the transition between one fluid and the other takes place over one grid cell and for front tracking the area weighting function results in a similarly sharp transition. A very narrow transition zone does, however, usually result in increased small-scale anisotropy of the solution and the functions proposed by Peskin provide a smoother transition.

In many cases, particularly for high surface tension, the smoother functions improve the results considerably. Area weighting, however, involves only two grid points in each direction so it is more efficient in three dimensions where it requires values from eight grid points versus 27 for Peskin's interpolation functions. It also allows for simpler treatment of boundaries.

Since the fluid velocities are computed on the fixed grid and the front moves with the fluid, the velocity of the interface points must be found by interpolating from the fixed grid. The interpolation starts by identifying the grid points closest to the front point. The grid value is then interpolated by

$$\psi_{\mathrm{f}} = \sum_{ijk} w^{\mathrm{f}}_{ijk} \psi_{ijk}, \tag{3.32}$$

where the summation is over the points on the fixed grid that are close to the front point and ψ stands for one of the velocity components. It is generally desirable that the interpolated front value be bounded by the grid values and that the front value be the same as the grid value if a front point coincides with a grid point. Although it is not necessary to do so, the same weighting functions used to smooth front values onto the fixed grid are usually used to interpolate values to the front from the fixed grid. Once the velocity of each front point has been determined, its new position can be found by integration. If a simple first-order explicit Euler integration is used, the new location of the front points is given by

$$\mathbf{x}^{n+1}_{\mathrm{f}} = \mathbf{x}^{n}_{\mathrm{f}} + \mathbf{v}^{n}_{\mathrm{f}} \Delta t, \tag{3.33}$$

where $\mathbf{x}^{n}_{\mathrm{f}}$ is the front position at the beginning of the time step, $\mathbf{v}_{\mathrm{f}}$ is the front velocity, and Δt is the time step.

When the interface is tracked using marker particles, the fluid properties, such as the density, are not advected directly. Thus it is necessary to update these properties (or a marker function that is used to determine these properties) at every time step. There are several ways to do this. The simplest methods is, of course, to loop over the interface elements and set the value of the property on the fixed grid as a function of the shortest normal distance from the interface. Since the interface is usually restricted to move less than the size of one fixed grid mesh, this update can be limited to the grid points in the immediate neighborhood of the interface. This straightforward approach fails, however, when two interfaces are very close to each other, or when an interface folds back on itself, such that two front segments are located between the same two fixed grid points. In these cases the proper value on the fixed grid depends on which interface segment is being considered. To overcome this problem, Unverdi and Tryggvason (1992) use the fact that the gradient of the marker function is a δ-function located at the interface. When this δ-function is smoothed onto a grid, the gradients for interfaces that are very close simply cancel. Once the grid gradient field has been constructed, the marker function must be recovered. Again, this can be done in several ways. The simplest approach is to integrate the grid gradient along grid lines from a point where the marker function is known. Since this can lead to slight grid dependence, depending upon which direction the integration is carried out, the recovery is often done by first taking the numerical divergence of the discrete grid gradient and then using this field as a source term for a Poisson equation

$$\nabla_h^2 f = \nabla_h \cdot (\nabla f)_f. \tag{3.34}$$

Here $(\nabla f)_f$ is the gradient of the marker function as constructed from the front. This equation can be solved by any standard method, but since the interface generally moves by less than a grid space, it is usually simplest to solve it by an iteration confined to a narrow band of points around the interface.

For interfaces identified by connected marker points, the computation of surface tension is, at least in principle, relatively straightforward. In most cases it is the total force on a small section of the front that is needed. In two dimensions we are generally working with a line element connecting two points and in three dimensions it is the force on a surface element connecting three points that is needed. Thus, the challenge is to find

$$\delta \mathbf{F}_\gamma = \int_{\Delta s} \gamma \kappa \mathbf{n} \, ds. \tag{3.35}$$

For a two-dimensional flow we use the definition of the curvature of a plane curve, $\kappa \mathbf{n} = \partial \mathbf{t}/\partial s$, to write equation (3.35) as

$$\delta \mathbf{F}_\gamma = \int_{\Delta s} \gamma \kappa \mathbf{n} \, ds = \gamma \int_{\Delta s} \frac{\partial \mathbf{t}}{\partial s} \, ds = \gamma (\mathbf{t}_2 - \mathbf{t}_1). \tag{3.36}$$

Therefore, instead of having to find the curvature, it is only necessary to find the tangents of the endpoints of each element. The simplest approach is to fit a parabola to the interface points and differentiate to obtain the tangent vectors. For higher accuracy, a polynomial is fitted through more points, then differentiated to give the tangent vector.

For three-dimensional surfaces we use the fact that the mean curvature can be written as

$$\kappa \mathbf{n} = (\mathbf{n} \times \nabla) \times \mathbf{n}. \tag{3.37}$$

The force on a surface element is therefore,

$$\delta \mathbf{F}_\gamma = \gamma \int_{\delta A} \kappa \mathbf{n} \, da = \gamma \int_{\delta A} (\mathbf{n} \times \nabla) \times \mathbf{n} \, da = \gamma \oint_L \mathbf{m} \, dl, \tag{3.38}$$

where we have used the Stokes theorem to convert the area integral into a line integral along the edges of the element. Here, $\mathbf{m} = \mathbf{t} \times \mathbf{n}$, where $\mathbf{t}$ is a vector tangent to the edge of the element and $\mathbf{n}$ is a normal vector to the surface. The cross-product is a vector that is in the surface and is normal to the edge of the element. The surface tension coefficient times this vector gives the "pull" on the edge and the net "pull" is obtained by integrating around the edges. If the element is flat, the net force is zero, but if the element is curved, the net force is normal to it when the surface tension coefficient is constant. As in two dimensions, this formulation ensures that the net force on a closed surface is zero, as long as the force on the common edge of two elements is computed in the same way. Once the force on each element is known, it can be smoothed onto the fixed grid using equation (3.27).

One of the major differences between methods based on advecting a marker function and methods based on tracking the interface using marker particles is their treatment of interfaces that come close together. Physically, thin filaments of one fluid embedded in another fluid can snap and thin films can rupture. When a marker-function is advected directly, the interface is simply the place where the marker-function changes from one value to the other. If two interfaces come together, they therefore fuse together when the cells between them fill up (or empty out). Thus, topological changes (snapping of filaments and breaking of films) take place automatically. Interfaces marked by connected marker particles, on the other hand, do not reconnect unless something special is done. Which behavior is preferred? It depends. Multiphase flows obviously do undergo topological changes and in many cases marker-function methods reproduce such processes in a physically realistic way. In other cases, the topological change, particularly the rupturing of films, depends sensitively on how rapidly the thin film drains, and fusing the interfaces together once the film thickness is comparable to the grid size may not be the right thing to do. It can, in particular, lead to solutions that do not converge under grid refinement, since the time of rupture depends directly on the grid size. In these cases, using connected marker particles gives the user more control over when the rupture takes place. Front tracking does not, however, introduce any new physics and the accurate modeling of thin films and the incorporation of appropriate models to capture their rupture remains a difficulty that is still not solved. We should note that while the "natural" behavior for the marker-function method is always to fuse interfaces together and for the front-tracking method never to do so, each method can be modified to behave like the other.

3.5 The level-set method

The level-set method, introduced by Osher and Sethian (1988) and further developed by Sussman, Smereka, and Osher (1994) for multiphase flow simulations, has emerged as the main alternative to the volume-of-fluid method for the direct advection of a marker function. As with VOF and other marker-function advection methods, level-set methods make no assumption about the connectivity of the interface. For example, if the interface separating liquid and gas should undergo a topological transition (where a drop breaks into two or more smaller drops or two drops coalesce), or form a sharp corner or cusp, there is no user intervention or extra coding necessary in order to continue the computation. Besides being robust, level-set methods allow one to accurately represent interfacial quantities such as the interfacial normal and curvature. Unlike interfaces in the VOF method, where the transition from one fluid to the next takes place over one grid cell, the level-set function transitions smoothly across the liquid–gas interface; therefore standard discretizations of the normal and the curvature using the level-set function can be as accurate as needed (e.g. second-order accurate or higher). In this section, we describe the equations and their accompanying discretizations for advecting and "reinitializing" the level-set function. We also describe the discretization details associated with coupling the level-set representation with a "one-fluid" multiphase Navier–Stokes solver.

In the level-set method, the interface separating two phases is represented using the level-set function ϕ. For a gas–liquid system, we can define ϕ to be positive in the liquid and negative in the gas. In other words,

$$\phi(\mathbf{x}, t) = \begin{cases} +d, & \mathbf{x} \text{ in the liquid,} \\ -d, & \mathbf{x} \text{ in the gas,} \end{cases}$$

where d represents the normal distance to the interface at time t.

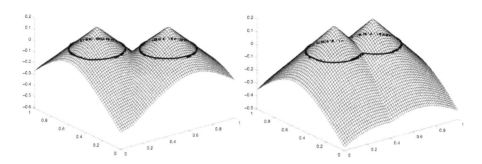

Fig. 3.7. Level-set representation of an interface. The interface consists of two
distinct circles on the left, but on the right the circles are closer and form one
interface. The level-set function is constructed as a distance function.

Figure 3.7 shows the level-set function ϕ initialized as a distance function.
On the left the two contours represent two distinct circular fluid blobs, but
on the right the fluid blobs have merged into one.

The level-set function ϕ is advected by

$$\frac{\partial \phi}{\partial t} + \mathbf{u} \cdot \nabla \phi = 0, \tag{3.39}$$

where $\mathbf{u}$ is the fluid velocity. The level-set equation is derived using the fact
that the level-set function should be constant along particle paths. In other
words,

$$\frac{d\phi(\mathbf{x}(t), t)}{dt} = 0. \tag{3.40}$$

Equation (3.40), together with $d\mathbf{x}(t)/dt \equiv \mathbf{u}$, implies that

$$\frac{d\mathbf{x}(t)}{dt} \cdot \nabla \phi(\mathbf{x}, t) + \frac{\partial \phi(\mathbf{x}, t)}{\partial t} = 0,$$

which leads to equation (3.39).

In principle, the level-set function can be advected by any sufficiently accurate scheme for hyperbolic equations. In practice, however, the following
simple second-order Runge–Kutta/ENO (Shu and Osher, 1989) numerical
discretization of equation (3.39) has often been used. We first rewrite equa-
tion (3.39) as $\phi_t = L\phi$. Assuming that $\phi_{i,j}^n$ and $\mathbf{u}_{i,j}^n$ are valid discrete values
defined at $t = t^n$, $x = x_i$, and $y = y_j$, we advance the solution to $t = t^{n+1}$
by first finding a predicted value for the level-set function

$$\phi_{i,j}^* = \phi_{i,j}^n + \Delta t L \phi^n \tag{3.41}$$

and then correcting the solution by

$$\phi_{i,j}^{n+1} = \phi_{i,j}^n + \frac{\Delta t}{2}(L\phi^n + L\phi^*). \tag{3.42}$$

The operator $L\phi$ is discretized by

$$L\phi = -u_{i,j}\frac{\phi_{i+1/2,j} - \phi_{i-1/2,j}}{h} - v_{i,j}\frac{\phi_{i,j+1/2} - \phi_{i,j-1/2}}{h}, \tag{3.43}$$

where the value of ϕ at the cell boundaries is found from

$$\phi_{i+1/2,j} = \begin{cases} \phi_{i,j} + \frac{1}{2}M(D_x^+\phi_{i,j}, D_x^-\phi_{i,j}), & \frac{1}{2}(u_{i+1,j} + u_{i,j}) > 0 \\ \phi_{i+1,j} - \frac{1}{2}M(D_x^+\phi_{i+1,j}, D_x^-\phi_{i+1,j}), & \frac{1}{2}(u_{i+1,j} + u_{i,j}) < 0. \end{cases} \tag{3.44}$$

M is a switch defined by

$$M(a,b) = \begin{cases} a, & |a| < |b| \\ b, & |b| \le |a| \end{cases} \tag{3.45}$$

and the differences are defined as

$$D_x^+\phi_{i,j} = \phi_{i+1,j} - \phi_{i,j} \qquad D_x^-\phi_{i,j} = \phi_{i,j} - \phi_{i-1,j}. \tag{3.46}$$

The equation for $\phi_{i,j+1/2}$ is similar.

While the $\phi = 0$ contour is accurately advected in this way, the level-set function off the interface will, in general, not remain a distance function. Since preserving the level-set function as a distance function (at least in the vicinity of the interface) is important for the coupling with the fluid equations in simulations of multiphase flows, it is necessary to *reinitialize* the level-set function. Without reinitialization, the magnitude of the gradient of the level-set function, $|\nabla\phi|$, can become very large or very small near the zero level-set of ϕ. The large gradients in the level-set function lead to an overall loss of accuracy of equation (3.39) and in variables (e.g. velocity) that may depend on ϕ. In the reinitialization process ϕ is converted into a new level-set function ϕ_d in which ϕ and ϕ_d share the same zero level-set, and ϕ_d satisfies

$$|\nabla\phi_d| = 1 \tag{3.47}$$

for values (x,y) within M cells of the zero level-set,

$$|\phi_d| < Mh.$$

A level-set function that satisfies equation (3.47) is called a distance function because $\phi_d(x,y)$ is the signed normal distance to the zero level-set of ϕ_d. For example, the level-set functions,

$$\phi(x,y) = x^2 + y^2 - 1$$

and

$$\phi_d(x, y) = \sqrt{x^2 + y^2} - 1$$

share the same zero level-set, but ϕ_d is a distance function and ϕ is not.

Sussman, Smereka, and Osher (1994) introduced an iterative approach to reinitialize ϕ. The advantage of an iterative approach is that if the level-set function ϕ is already close to a distance function, then only a few iterations are necessary to turn ϕ into the valid distance function ϕ_d. The reinitialization step is achieved by solving the following partial differential equation,

$$\frac{\partial \phi_d}{\partial \tau} = \text{sgn}(\phi)(1 - |\nabla \phi_d|), \tag{3.48}$$

with initial conditions,

$$\phi_d(\mathbf{x}, 0) = \phi(\mathbf{x}),$$

where

$$\text{sgn}(\phi) = \begin{cases} -1, & \phi < 0 \\ 0, & \phi = 0 \\ 1, & \phi > 0 \end{cases}$$

and τ is an artificial time. The steady solutions of equation (3.48) are distance functions. Furthermore, since $\text{sgn}(0) = 0$, then $\phi_d(\mathbf{x}, \tau)$ has the same zero level-set as $\phi(\mathbf{x})$. A nice feature of this reinitialization procedure is that the level-set function is reinitialized near the front first. To see this we rewrite equation (3.48) as,

$$\frac{\partial \phi_d}{\partial \tau} + \mathbf{w} \cdot \nabla \phi_d = \text{sgn}(\phi_d), \tag{3.49}$$

where

$$\mathbf{w} = \text{sgn}(\phi) \frac{\nabla \phi_d}{|\nabla \phi_d|}.$$

It is evident that equation (3.49) is a nonlinear hyperbolic equation with the characteristic velocities pointing *outwards* from the interface in the direction of the normal. This means that ϕ_d will be reinitialized to $|\nabla \phi_d| = 1$ near the interface first. Since we only need the level-set function to be a distance function near the interface, it is not necessary to solve equation (3.49) to steady state. We may use a fixed number of iterations in order to ensure the distance function property near the interface. For example, if the iteration step size is $\Delta \tau = h/2$, and the interface is spread over a thickness $2Mh$, then we can stop the iteration process after $2M$ time steps.

The same simple second-order Runge–Kutta/ENO numerical discretization used to advance the level-set function (equation 3.39) in time can be used for the reinitialization, where the new distance function, ϕ_d, satisfies equation (3.47) for $|\phi_d| < Mh$ (Sussman, Smereka, and Osher, 1994). We first rewrite equation (3.49) as $\partial \phi_d / \partial \tau = L\phi_d$. We solve equation (3.49) for $\tau = 0, \ldots, Mh$, assuming that $\Delta \tau = h/2$; therefore we take $2M$ fictitious time steps. $(\phi_d)_{i,j}^n$ are discrete values defined at $\tau = \tau^n$, $x = x_i$ and $y = y_j$. As before, we first find a predicted value by an Euler step $(\phi_d)_{i,j}^* = (\phi_d)_{i,j}^n + \Delta \tau L\phi_d^n$ and then correct the solution using the midpoint rule, $(\phi_d)_{i,j}^{n+1} = (\phi_d)_{i,j}^n + (\Delta \tau / 2)(L\phi_d^n + L\phi_d^*)$. For the discretization of $L\phi$ we have

$$L\phi = \mathrm{sgn}_{Mh}(\phi)\left(1 - \sqrt{\left(\frac{\tilde{D}_x}{h}\right)^2 + \left(\frac{\tilde{D}_y}{h}\right)^2}\right), \tag{3.50}$$

where

$$\tilde{D}_x = \begin{cases} \tilde{D}_x^+, & \mathrm{sgn}(\phi)D_x^+(\phi_d)_{i,j} < 0 \text{ and} \\ & \mathrm{sgn}(\phi)D_x^-(\phi_d)_{i,j} < -\mathrm{sgn}(\phi)D_x^+(\phi_d)_{i,j} \\ \tilde{D}_x^- & \mathrm{sgn}(\phi)D_x^-(\phi_d)_{i,j} > 0 \text{ and} \\ & \mathrm{sgn}(\phi)D_x^+(\phi_d)_{i,j} > -\mathrm{sgn}(\phi)D_x^-(\phi_d)_{i,j} \\ \frac{1}{2}(\tilde{D}_x^+ + \tilde{D}_x^-) & \text{otherwise} \end{cases} \tag{3.51}$$

and the modified differences are given by

$$\begin{aligned} \tilde{D}_x^+ &= D_x^+(\phi_d)_{i,j} - \frac{1}{2}M(D_x^+D_x^-(\phi_d)_{i,j}, D_x^+D_x^-(\phi_d)_{i+1,j}) \\ \tilde{D}_x^- &= D_x^-(\phi_d)_{i,j} + \frac{1}{2}M(D_x^+D_x^-(\phi_d)_{i,j}, D_x^+D_x^-(\phi_d)_{i-1,j}). \end{aligned} \tag{3.52}$$

Here, $M(a,b)$ is defined by equation (3.45) and $D_x^+\phi_{i,j}$ and $D_x^-\phi_{i,j}$ by equations (3.46). The smoothed sign function in equation (3.50) is defined by

$$\mathrm{sgn}_{Mh}(\phi) = \begin{cases} 1 & \phi \geq Mh \\ -1 & \phi \leq -Mh \\ (\phi/Mh) - \frac{1}{\pi}\sin(\pi\phi/Mh) & \text{otherwise} \end{cases} \tag{3.53}$$

To couple the level-set method with a "one-fluid" multiphase flow algorithm, we need to use the level-set function to determine the density ρ, viscosity μ, and surface tension term $\gamma\kappa\delta(n)\mathbf{n}$. From the level-set function, we generate a marker function f which is, in effect, a smoothed Heaviside function,

$$f(\phi) = \begin{cases} 0, & \text{if } \phi < -Mh \\ \frac{1}{2}(1 + \frac{\phi}{Mh} + \frac{1}{\pi}\sin(\pi\frac{\phi}{Mh})), & \text{if } |\phi| \leq Mh \\ 1 & \text{if } \phi > Mh. \end{cases} \tag{3.54}$$

The density ρ and viscosity μ are given by,

$$\rho(\phi) = \rho_1 f(\phi) + \rho_0(1 - f(\phi))$$

and

$$\mu(\phi) = \mu_1 f(\phi) + \mu_0(1 - f(\phi)).$$

The surface tension term is given by

$$\gamma \kappa \delta(n) \mathbf{n} = \gamma \kappa(\phi) \frac{df(\phi)}{d\phi} \nabla(\phi), \tag{3.55}$$

where the curvature is given by

$$\kappa(\phi) = \nabla \cdot \frac{\nabla \phi}{|\nabla \phi|}. \tag{3.56}$$

Since the level-set function is naturally a smooth function, the discretization of the curvature, equation (3.56), can be done using standard central differences:

$$2h\kappa(\phi) \approx \left(\frac{\phi_x}{|\nabla \phi|}\right)_{i+1/2,j+1/2} + \left(\frac{\phi_x}{|\nabla \phi|}\right)_{i+1/2,j-1/2}$$

$$- \left(\frac{\phi_x}{|\nabla \phi|}\right)_{i-1/2,j+1/2} - \left(\frac{\phi_x}{|\nabla \phi|}\right)_{i-1/2,j-1/2}$$

$$+ \left(\frac{\phi_y}{|\nabla \phi|}\right)_{i+1/2,j+1/2} - \left(\frac{\phi_y}{|\nabla \phi|}\right)_{i+1/2,j-1/2}$$

$$+ \left(\frac{\phi_y}{|\nabla \phi|}\right)_{i-1/2,j+1/2} - \left(\frac{\phi_y}{|\nabla \phi|}\right)_{i-1/2,j-1/2}$$

where

$$(\phi_x)_{i+1/2,j+1/2} \approx (\phi_{i+1,j+1} + \phi_{i+1,j} - \phi_{i,j+1} - \phi_{i,j})/(2h)$$

and

$$(\phi_y)_{i+1/2,j+1/2} \approx (\phi_{i+1,j+1} - \phi_{i+1,j} + \phi_{i,j+1} - \phi_{i,j})/(2h).$$

We note that an appropriately smoothed Heaviside function (equation 3.54) is critical and without it the solutions may not converge under grid refinement; we recommend using an interfacial thickness of six cells ($M = 3$). In order for f(ϕ) to be a valid "smoothed" Heaviside function (equation 3.54), ϕ must be reinitialized every time step so that ϕ is a distance function within Mh of the zero level-set.

    Early implementations of the level set method generally did not have good mass conservation properties. Later improvements have enabled the practioner to have considerably more control over the error in mass conservation. The simplest approach, used by Herrmann (2008), for example, is to use a finer grid for the advection of the level set function than for solving the fluid dynamics equations. The increased accuracy reduces the mass loss and since the grid only has to be fine around the interface, the fine grid advection can be implemented in ways that do not significantly increase the computational cost. Other authors have introduced conservative level set methods (Olsson, Kreiss and Zahedi, 2007), used marker particles to help accurately track the interface (Enright et al, 2002), or coupled the level set method with a VOF method to advect the level set function in a conservative way (Sussman, 2003).

We point out that other "field" approaches, e.g. phase-field methods, or VOF methods, can also treat coalescence and breakup without any special treatment. The level-set method is, however, significantly simpler than advanced VOF methods and extends to three-dimensional flow in a straightforward way. For more detailed discussions of the use of level-set functions to capture fluid interfaces, including more accurate/efficient discretizations, we refer the reader to Osher and Fedkiew (2003), Sethian (1999), Sussman and Fatemi (1999), Chang *et al.*, (1996), and Smereka (1996). To emphasize the ease of implementation/flexibility of the level-set approach, we point out that the level-set method can seamlessly be coupled with finite element methods or spectral element methods; see, e.g. Sussman and Hussaini (2003).

3.6 Other methods to advect the marker function

In addition to the volume-of-fluid, the level-set, and the front-tracking methods discussed above, several other methods have been proposed to simulate multiphase flows, using the "one-fluid" formulation of the governing equations. In some cases the difference is only in the way the marker function, or the fluid interface, is advected, but in other cases the differences are in how the governing fluid equations are treated.

In the CIP (constrained interpolated propagation) method, equation (3.14) is supplemented by equations for the derivatives of f, obtained by differentiating equation (3.14). Generally the equation for the derivative will be the same as equation (3.14), except that the right-hand side is not zero. However, the derivatives are first advected in the same way as f, and a correction for the effect of the nonzero right-hand side then added. To do the advection, a cubic polynomial is fitted to f and its derivatives and the solution profile obtained is translated by $U\Delta t$ (in one dimension) to give the new nodal values of f and its derivatives. Although a fully multidimensional version of the method has been developed, splitting, where each coordinate direction is done separately, works well. The original method generally shows slight oscillations near a sharp interface but these can be reduced by the use of rational polynomials (the RCIP method). For a recent review of the method and the various extensions, as well as a few applications, see Yabe, Xiao, and Utsumi (2001).

While the CIP scheme is simply another method to solve equation (3.14), the phase-field method is based on modifying the governing equations by incorporating some of the physical effects that are believed to govern the structure of a thin interface. Although the smoothed region between the different fluids is described in a thermodynamically consistent way, in actual implementation the thickness of the transition is much larger than it is in real systems and it is not clear whether keeping the correct thermodynamics in an artificially thick interface has any advantages over methods that model the behavior of the transition zone in other ways. The phase function, which identifies the different fluids, is updated by nonlinear advection–diffusion known as the Cahn–Hilliard equation. The diffusion terms smear an interface that is becoming thin due to straining, but an antidiffusive part prevents the interface from becoming too thick if the interface is being compressed. The Navier–Stokes equations are also modified by adding a term that results in surface tension in the interface zone. The key to the modification is the introduction of a properly selected free-energy function, ensuring that the thickness of the interface remains of the same order as the grid spacing. The phase-field approach has found widespread use in simulation of solidification, but its use for fluid dynamic simulations is relatively limited. Jacqmin (1999) analyzed the method in some detail and showed examples of computations of a two-dimensional Rayleigh–Taylor instability in the Boussinesq limit. Other applications of the phase-field method to simulations of the motion of a fluid interface include Jamet et al. (2001) who discussed using it for flows with phase change, Verschueren, van de Vosse, and Meijer (2001) who examined thermocapillary flow instabilities in a Hele–Shaw cell, and Jacqmin (2000) who studied contact-line dynamics of a diffuse fluid interface.

Although a comprehensive comparison of the accuracy of the phase-field method for simulations of the motion of a fluid interface has not been done, the results that have been produced suggest that the method is comparable to other methods that use fixed grids and the one-fluid formulation of the governing equations. The smoothing of the transition zone and the use of front capturing (rather than explicit tracking) is likely to make the phase-field method similar to the level-set approach.

In the CIP and the phase-field method, like VOF, level-set, and front-tracking methods, the fluid interface is treated just like the rest of the fluid domain when the Navier–Stokes equations are solved. This requires replacing the sharp interface with a smooth transition zone where the transition from one fluid to the other takes place over a few grid cells. This is done both for the material properties (density and viscosity) as well as for surface tension (in the VOF method the interface is confined to one grid cell but for the surface tension calculation additional smoothing is usually necessary).

The success of the various methods based on the one-fluid formulation has demonstrated beyond a doubt the power of this approach. It has, however, also become clear that the smoothing of the interface over several grid cells limits the obtainable accuracy and often makes it difficult to obtain detailed information about the solution at the interface. Several methods have recently been proposed to improve the treatment of the interface while retaining most of the advantages of the "one-fluid" treatment. Such "sharp interface" methods originated with the work of Glimm *et al.* (1982, 1985, 1986), but more recent contributions include the "ghost fluid" method of Fedkiw *et al.* (1999) who assign fictitious values to grid points on the other side of a fluid discontinuity, the immersed interface method of Lee and LeVeque (2003), where the finite difference approximations are modified near the interface to account for the discontinuity, and the method of Udaykumar, Mittal, and Shyy (1999), where the grid cells near the interface are distorted in such a way that their faces coincide with the interface. We will discuss some of these methods in slightly greater detail in Chapter 4, in the context of simulations of flows over solid particles. None of the "sharp interface" methods have yet been used as extensively as the ones we have described above and the jury is still out as to whether the added complexity is necessary for many of the physical problems that are currently being simulated. The development of methods for multiphase flows is, however, currently an active area and it seems likely that new and improved methods will continue to emerge.

3.7 Computational examples

The observation that no matter how promising an algorithm may be theoretically, its utility can only be assessed by actual tests, is so fundamental to algorithm development that it probably deserves the status of an axiom or a law in scientific computing. In this section we will first discuss tests aimed at evaluating the performance of the various methods outlined above and then show a few examples of applications to real physical systems.

3.7.1 Validation problems

The two main challenges in following accurately the motion of the interface between two immiscible fluids are the accurate advection of the interface and the computation of the surface tension. To address the performance of the various advection schemes that have been developed, several authors have examined the passive advection of a patch of one fluid by a given velocity

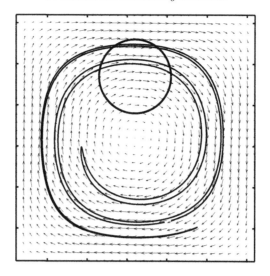

Fig. 3.8. The advection test case introduced by Rider and Kothe (1995). The circular blob near the top is deformed in a vortical flow field, until time $T = 16$ and then rotated back into its original position. The domain is given by $[0,1] \times [0,1]$. The undeformed blob is circular, with radius 0.15 and is initially located at $(0.5, 0.75)$.

field. Two tests have emerged as the most popular ones. In the first one, originally introduced by Zalesak (1979), a circular blob, with a rectangular notch almost cutting through the circle, is rotated with a fluid velocity corresponding to a solid body rotation. The circle should stay intact as it rotates, but generally it deteriorates and the degree of deterioration after a given number of rotations is used to distinguish between the different advection schemes. While this test assesses the ability of the advection scheme to preserve the sharp corners in the initial conditions, it does not test how well the scheme performs for shear flow. To evaluate the performance of various methods in such flows, another test case, introduced by Rider and Kothe (1995), has become essentially *the* standard advection test. Here, an initially circular blob of one fluid is advected in a vortex given by

$$ U(x,y) = \cos \pi x \times \sin \pi y, \qquad V(x,y) = -\sin \pi x \times \cos \pi y \qquad (3.57) $$

in a $[0,1] \times [0,1]$ domain. The radius of the blob is 0.15 and it is initially located off the vortex center so it is deformed by the flow. The computations are generally run until the blob has deformed significantly and then the flow is reversed until the blob should have returned to its original shape.

Figure 3.8 shows the boundaries of the initial circular blob, the boundary after time $T = 16$, and the boundary after the velocity field has been reversed. The computations are done using connected marker particles, the velocity is given on a 32^2 grid, and area weighting is used to interpolate the velocities. The time integration is done using a second-order Runge–Kutta scheme with $\Delta t = 0.005$. Here, a large number of marker particles were used, so the circle returns to essentially exactly its original position when the sign of the velocity field is reversed.

The vortical flow shown in Fig. 3.8 was used by Rider and Kothe (1995) and Rudman (1997) to study the performance of several advection schemes. The methods examined included various implementations of the VOF method (the original SLIC method, the Hirt–Nichols method, and a PLIC method), the first-order upwind method, the piecewise parabolic method of Colella and Woodward (1984), and level-sets. Rudman (1997) used the original Youngs' version of the PLIC method but Rider and Kothe (1995) used a modified version of a least-square reconstruction technique introduced by Pilliod and Puckett (1997). Both authors found that the best results were obtained by the PLIC method. Rider and Kothe (1995) found that the level-set method ranked next and then SLIC. Rudman (1997) found that the Hirt and Nichols version of the method offered little improvement over the original SLIC. Rider and Kothe (1995) also tested a method based on uniformly distributed marker particles used to represent the blob. While more expensive than the other methods tested, the markers generally performed better than the marker functions. Neither of these authors used connected marker particles, but as Fig. 3.8 shows, even on a relatively coarse grid this method produces essentially the exact solution.

In addition to advection, the accurate computation of surface tension poses a challenge to methods designed for simulations of multiphase flows. To test the performance of surface tension calculations three tests have emerged as the most common ones. The first is simply a static test where the pressure inside a circular or a spherical drop is compared with the predictions of Laplace's law. Every method in current use passes this test for moderate values of the surface tension coefficient. The second test is a comparison of the oscillation frequency of a circular or spherical drop (see Lamb, 1932, for example) with analytical results for small-amplitude oscillations. In addition to the oscillation frequency, the viscous decay of the drop is sometimes also compared to theoretical predictions. The third test is essentially the same as the first one, but we look at the velocity field, which should be zero for a circular or spherical drop. For high surface tension, all methods based on the one-fluid formulation on fixed grids generally suffer from artificial currents induced by a mismatch between the resolution of the pressure and the surface tension terms. In early implementations of surface tension in VOF methods using the continuous surface force (CSF)

(Brackbill, Kothe, and Zemach, 1992; Lafaurie *et al.*, 1994) these parasitic currents posed a serious limitation on the range of problems that could be solved. While the front-tracking method of Unverdi and Tryggvason (1992) also shows parasitic currents at high surface tension, these are generally smaller. The origin of these currents is now reasonably well understood and remedies are emerging. The key seems to be that the normal vector to the interface should be found as the gradient of the indicator function and discretized in exactly the same way as the pressure gradient. If the curvature is constant this results in no parasitic currents. If the curvature is not constant it must be found from information about the geometry of the interface. In the PROST-VOF method of Renardy and Renardy (2002) the curvature is found by fitting a parabolic surface to the interface, resulting in a significant reduction of the parasitic currents. For phase-field methods, Jamet, Torres, and Brackbill (2002) have developed an energy-conserving discretization of the surface tension terms and shown that the parasitic currents are greatly reduced by its use. A similar remedy should work for level-set methods. For front-tracking methods, Shin *et al.* (2005) introduced a hybrid technique where the normal vector is found by differentiating the reconstructed marker function but a grid curvature is found from the front. This approach seems to essentially eliminate any spurious currents.

In addition to the problems with high surface tension, other difficulties emerge as the various governing parameters take on larger values. It is, in particular, generally true that more care has to be taken when the ratios of the material properties of the different fluids become large. It is therefore important to test each method under those circumstances, and although different authors have used different tests, no *standard* tests have emerged.

High density ratios cause two problems. At high Reynolds numbers, the appearance of high and irregular velocities near the interface can sometimes destroy the solution when the fully conservative form of the advection terms is used. These high velocities can be traced to an incompatibility in the advection of mass and momentum and can be overcome simply by using the nonconservative form of the advection terms (see Esmaeeli and Tryggvason, 2005, for a discussion). This is slightly counterintuitive since the conservative form is usually preferred for computations of shocks in compressible flows. The other problem is the solution of the elliptic equation for the pressure. Iterative solvers generally take a long time to converge, but – as long as the density is positive – a simple iterative technique like SOR will always converge with the proper selection of the over-relaxation constant. We should note, however, that for a density ratio of 1000, say, a very minor error – relative to the density in the heavier liquid – can lead to negative density values. More efficient multigrid solvers may, on the other hand, fail to converge. While some progress has been made, the development of efficient pressure solvers for high-density ratios remains high on the wish-list of nearly everybody engaged in the use of one-fluid methods for multiphase flow simulations. In many applications, relatively modest values of the density and viscosity ratios can be used as approximations for larger differences, without influencing the solution too much. This is particularly true when the dynamics of the low viscosity/density fluid is largely controlled by what the high viscosity/density fluid is doing. In other cases, such as for wind-generated breakup and droplet suspensions, it is more important to use the correct properties of both fluids.

For large viscosity ratios there are generally inaccuracies around the interface due to the smoothing of the transition zone. As the error is generally "pushed" to the less viscous side, it tends to be small when the flow in the less viscous liquid has relatively minor impact on the global behavior. This is usually the case for bubbles, and for short-time evolution of drops, such as for drop collisions and splashing. For viscous drops suspended in a less viscous liquid, the errors in the viscous terms are more serious and the convergence rate is worse. It has been known for quite some time, in the context of VOF methods, that taking the harmonic mean of the viscosities at points where the viscosity is not defined, instead of the arithmetic average, ameliorated the discontinuity of the viscous stresses (Patankar, 1980). Ferziger (2003) expanded this approach to interfaces smoothed over several grid points.

3.7.2 Applications to physical problems

The "one-fluid" approach has been used to examine multifluid problems since the beginning of computational fluid mechanics. Using the marker-and-cell (MAC) method, Harlow and Welch (1965) studied gravity currents by looking at the motion of fluids caused by the sudden release of a dam and Harlow and Welch (1966) examined the nonlinear evolution of the Rayleigh–Taylor instability where a heavy fluid falls into a lighter one. Both the breaking dam and the Rayleigh–Taylor problems have since become standard test problems for multifluid codes. The MAC method, modified in several ways, has been used to study a number of other free-surface and interface problems, including splashing of drops on a liquid layer, collision of drops, free surface waves and cavitation. While many of these results were in excellent agreement with experiments, limited computer power restricted the available grid size and confined most of the studies to two-dimensional (or axisymmetric) systems.

During the last decade a large number of computations have been carried out for a large number of systems and it would be impractical to provide a complete review of these simulations and what has been learned from the results. We will, therefore, simply show a few examples of recent results, mostly for bubbly flows, demonstrating the capabilities of current methods.

Homogeneous bubbly flow with many buoyant bubbles rising together in an initially quiescent fluid is perhaps one of the simplest examples of dispersed flows. Such flows can be simulated using periodic domains where the bubbles in each period interact freely, but the configuration is repeated infinitely many times in each coordinate direction. In the simplest case there is only one bubble per period so the configuration of the bubbles does not change as they rise. While such regular arrays are unlikely to be seen in an experiment, they provide a useful reference configuration for a freely evolving array. As the number of bubbles in each period is increased, the regular array becomes unstable and the bubbles generally rise unsteadily, repeatedly undergoing close interactions with other bubbles. The behavior is, however, statistically steady and the average motion (averaged over long enough time) does not change. While the number of bubbles clearly influences the average motion for a small enough number of bubbles per period, the hope is that once the size of the system is large enough, information obtained by averaging over each period will be representative of a truly homogeneous bubbly flow.

Tryggvason and collaborators have examined the motion of nearly spherical bubbles in periodic domains in a number of papers, using improved versions of the front-tracking method originally described by Unverdi and Tryggvason (1992). Esmaeeli and Tryggvason (1998) examined a case where the average rise Reynolds number of the bubbles remained relatively small (1–2) and Esmaeeli and Tryggvason (1999) looked at another case where the Reynolds number was 20–30. Bunner and Tryggvason (2002a,b) simulated a much larger number of three-dimensional bubbles using a parallel version of the method used by Esmaeeli and Tryggvason. Their largest simulations followed the motion of 216 three-dimensional buoyant bubbles per periodic domain for a relatively long time. Simulations of freely evolving arrays were compared with regular arrays and it was found that while freely evolving bubbles at low Reynolds numbers rise faster than a regular array (in agreement with Stokes flow results), at higher Reynolds numbers the trend is reversed and the freely moving bubbles rise slower. The time averages of the two-dimensional simulations were generally well converged, but exhibited a dependency on the size of the system. This dependency was stronger for the low Reynolds number case than the moderate Reynolds number one.

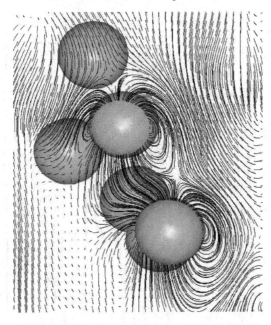

Fig. 3.9. A close-up of the flow field around a few bubbles, from a simulation of the buoyant rise of 91 bubbles in a periodic domain. Here $Eo = 1$ and $M = 1.234 \times 10^{-6}$ and the void fraction is 6%. The simulation was done using a 192^3 grid. From Bunner (2000).

The effect of deformability was studied by Bunner and Tryggvason (2003) who found that relatively modest deformability could lead to a streaming state where bubbles gathered in a "stream" or a "chimney." For extensive discussion of the results, and the insight generated by these studies, we refer the reader to the original papers. Figures 3.9 and 3.10 show two examples from the simulations of Bunner and Tryggvason. In Fig. 3.9 a close-up of the flow field around a few nearly spherical bubbles is shown. The streamlines are plotted in a plane cutting through a few of the bubbles and it is clear that the flow field is well resolved. In Fig. 3.10, one frame from a simulation of 27 deformable bubbles, while the bubbles are still relatively uniformly distributed, is shown. In addition to the bubbles and the velocity vectors in a plane cutting through the domain, the vorticity is shown by gray-scale contours. The vorticity is highest (light gray) in the wake of the bubbles, as expected.

While bubbles in initially quiescent liquid in fully periodic domains are a natural starting point for direct numerical simulation of bubbly flows, many of the important questions that need to be addressed for bubbly flows involve walls. Figure 3.11, from Lu, Fernandez, and Tryggvason (2005), shows one frame from a simulation of 16 bubbles in the so-called "minimum

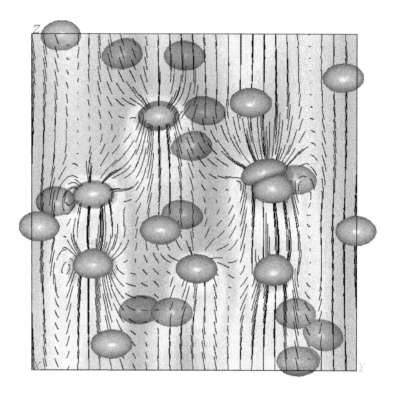

Fig. 3.10. One frame from a simulation of 27 deformable bubbles in a periodic domain. Here $Eo = 5$ and $M = 1.543 \times 10^{-4}$ and the void fraction is 2%. The simulation was done using a 128^3 grid. See also http://www.cambridge.org/comp_mult_flow. From Bunner (2000).

turbulent channel" of Jimenez and Moin (1991). In addition to the bubbles, isocontours of spanwise vorticity are shown, with different color indicating positive and negative vorticity. The contours on the bottom wall show the local wall shear. The goal of this simulation is to cast some light on the mechanisms underlying drag reduction due to the injection of bubbles near the wall and to provide data that may be useful for the modeling of such flows. Experimental studies (see Merkle and Deutsch, 1990; Kato *et al.*, 1995, and Kodama *et al.*, 2003, for a review) show that the injection of a relatively small number of bubbles into a turbulent boundary layer can result in a significant drag reduction. The flow rate in this simulation is kept constant so the total wall drag changes as the flow evolves. At the time shown, total wall shear has been reduced by about 15%. While drag reduction is usually found for deformable bubbles, nearly spherical bubbles

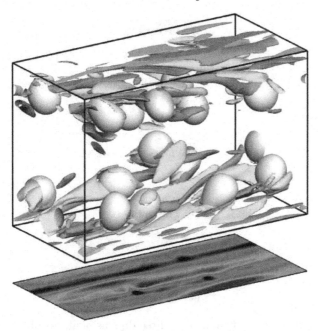

Fig. 3.11. One frame from a simulation of 16 bubbles in a turbulent channel flow. The wall Reynolds number is $Re^+ = 135$ and the initial flow, computed using a spectral code, is fully turbulent. The computations were done using a grid of $256 \times 128 \times 192$ grid points, uniformly spaced in the streamwise and the spanwise direction, but unevenly spaced in the wall normal direction. See also http://www.cambridge.org/comp_mult_flow. From Lu, Fernandez, and Tryggvason (2005).

generally lead to drag increase since they come closer to the wall and are slowed down by the viscous sublayer. It is unlikely that the subtle effect of bubble deformation on the wall drag found here could have been uncovered in any way other than by direct numerical simulations.

The results in Figs 3.9–3.11 were calculated using marker points to track the interface between the air in the bubbles and the ambient liquid. The remaining examples in this chapter have been computed using the level-set method. In Fig. 3.12 we show the evolution of a buoyant bubble where an air bubble rises in silicone oil (density ratio 642:1, viscosity ratio 27:1) computed using a 64^3 grid. According to the chart in Bhaga and Weber (1981), the bubble should rise unsteadily and "wobble." Although the computation is done using fully three-dimensional bubbles, the initial conditions are axisymmetric and the bubble therefore does not show the sidewise motion observed experimentally. The bubble does, however, oscillate. Two computations showing a fully three-dimensional motion are shown in Fig. 3.13 where the

Fig. 3.12. Numerical computation of a three-dimensional "wobbly" bubble. Results compare well with predicted behavior (Clift, Grace, and Weber, 1978). Parameters correspond to air inside the bubble and silicone oil surrounding the bubble. Density ratio and viscosity ratio are 642:1 and 27:1, respectively. See also http://www.cambridge.org/comp_mult_flow. From Ohta *et al.* (2004).

bubble on the left is rising along a spiral path, but the bubble on the right wobbles as it rises.

In Fig. 3.14 we show the evolution of a fully three-dimensional bubble collapse. Initially, a high-pressure spherical adiabatic bubble, with radius 1/2, is placed close to a solid wall in a quiescent liquid. Here, the bubble center is initially located a dimensionless distance of 3/2 from the wall. The overall size of the computational domain is $16 \times 16 \times 8$. For a similar simulation using an axisymmetric geometry, see Sussman (2003). As the bubble collapses, a thin jet, directed into the bubble, forms on the side of the bubble away from the wall and eventually penetrates through the bubble and impinges the solid wall. To capture this jet accurately, local patches of finer grids are used where needed. The overall coarse grid resolution is $64 \times 64 \times 32$ and three levels of adaptivity are used, giving an effective resolution of $512 \times 512 \times 256$ grid points. The emergence of localized small features is common in multiphase flow simulations and the use of adaptive mesh refinement (AMR) greatly increases the range of problems that can be solved in a reasonable amount of time. Other examples of the use of AMR in multifluid simulations include Agresar *et al.* (1998), Roma, Peskin,

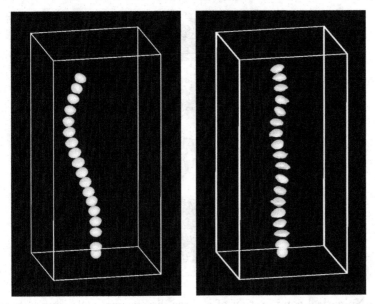

Fig. 3.13. The unsteady rise of a single bubble. Left frame: $Re = 700$, $Eo = 1.9$; Right frame: $Re = 1000$, $Eo = 7.5$. In both cases $M = 1.9 \times 10^{-10}$. Each bubble is shown at several times. From Ohta $et\ al.$ (2004).

and Berger (1999), and Sussman $et\ al.$ (1999). Simulations following an interface using a level-set on an adaptive unstructured mesh can be found in Zheng $et\ al.$ (2005).

Our final example is a three-dimensional computation from Aleinov, Puckett, and Sussman (1999) of the ejection of ink by a microscale jetting devices, using a level-set method with adaptive grid refinement. Figure 3.15 shows three snapshots of the evolution after a drop has formed and as the thin filament trailing the drop breaks up into satellite drops due to surface tension. The density ratio in this computation corresponds to that for air/water. By adjusting parameters such as inlet pressure, geometry, etc. one gets different numbers/sizes of satellite drops which in turn affect the quality of the printing, once the drops land on the paper.

The level-set method has recently been used by several researchers to study various multiphase flow problems. These include Tran and Udaykumar (2004) (high-speed solid–fluid interaction), Carlson, Mucha, and Turk (2004) (low-speed solid-fluid interaction), Dhir (2001) (boiling), and Goktekin, Bargteil, and O'Brien (2004) (viscoelastic multiphase flows). Other applications can be found in the book by Osher and Fedkiw (2003).

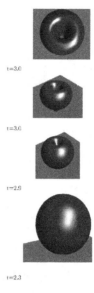

Fig. 3.14. Numerical computation of a three-dimensional collapsing bubble in the presence of a rigid boundary. Three levels of grid adaptivity are used, giving an effective fine grid resolution of $512 \times 512 \times 256$ grid points. See also http://www.cambridge.org/comp_mult_flow. From Sussman (2005).

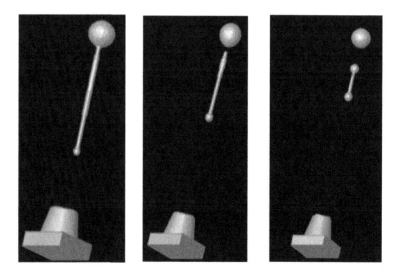

Fig. 3.15. Computation of jetting of ink using a fully three-dimensional code with adaptive grid refinement. Fluid is ejected from the solid block at the bottom. The jet is shown at three times, as it forms one large drop and several satellite drops. From Aleinov, Puckett, and Sussman (1999).

3.8 Conclusion

In this chapter we have discussed how to compute the motion of a fluid interface using fixed grids. We introduced the "one-fluid" formulation of the governing equations and discussed in some detail the various strategies that have been developed to update the marker function identifying the different fluids. Although the original idea of the "one-fluid" approach goes back to the very beginning of computational fluid dynamics, major progress has been made in the last decade. It is, in particular, safe to say that given enough computational resources we can now routinely solve a large range of problems that involve the motion of two fluids. It is possible, for example, to simulate accurately the evolution of finite Reynolds number disperse flows of several hundred bubbles and drops for sufficiently long time that reliable values can be obtained for various statistical quantities.

The "one-fluid" approach outlined in this chapter is likely to continue to remain an important strategy to handle multifluid and multiphase problems. This approach is relatively simple and enormously powerful. While the different methods of updating the marker function may at first appear very different – and there certainly are differences in the complexity and accuracy of these methods – their similarities are probably more important. There is, in particular, a considerable amount of migration of ideas developed in the context of one method to other methods. In the near future we are likely to see further convergence of the different methodologies, as well as the development of hybrid methods combining the best aspects of the different techniques.

4

Structured grid methods for solid particles

In this chapter we discuss numerical methods based on structured grids for direct numerical simulations of multiphase flows involving solid particles. We will focus on numerical approaches that are designed to solve the governing equations in the fluid and the interaction between the phases at their interfaces.

In methods employing structured grids, there are two distinct possibilities for handling the geometric complexities imposed by the phase boundaries. The first approach is to precisely define the phase boundaries and use a body-fitted grid in one or both phases, as necessary. The curvilinear grid that conforms to the phase boundaries greatly simplifies the specification of interaction processes that occur across the interface. Furthermore, numerical methods on curvilinear grids are well developed and a desired level of accuracy can be maintained both within the different phases and along their interfaces. The main difficulty with this approach is grid generation; for example, for the case of flow in a pipe with several hundred suspended particles, obtaining an appropriate structured body-fitted grid around all the particles is a nontrival task. There are several structured grid generation packages that are readily available. Nevertheless, the computational cost of grid generation increases rapidly as geometric complexity increases. This can be severely restrictive, particularly in cases where the phase boundaries are in time-dependent motion. The cost of grid generation can then overwhelm the cost of computing the flow.

The alternative to a body-fitted grid that is increasingly becoming popular are Cartesian grid methods. Here, irrespective of the complex nature of the internal and external boundaries between the phases, only a regular Cartesian mesh is employed and the governing equations are solved on this simple mesh. The interfaces between the different phases are treated as embedded boundaries within the regular Cartesian mesh. Although the general approach draws heavily on the "one-fluid" methodology described in Chapter 3, several aspects must be treated differently. The two key issues to consider are: (1) how the interface is defined and advected on the Cartesian mesh, and (2) how the interfacial interaction is accounted for in the bulk motion. Based on the above two issues a further classification of the Cartesian grid methods is possible. In the sharp-interface Cartesian grid approach the sharpness of the interface between the phases is recognized both in its advection through the Cartesian mesh and also in accounting for its influence on the bulk motion (see Leveque and Li, 1994; Pember *et al.*, 1995; Almgren *et al.*, 1997; Ye *et al.*, 1999; Calhoun and LeVeque, 2000; Udaykumar *et al.*, 2001). In the computational cells cut through by the interface, the governing equations are solved by carefully computing the various local fluxes (or gradients). The above references have shown that by carefully modifying the computational stencil at the interface cells a formal second-order accuracy can be maintained.

In the immersed boundary technique a sharp interface between the phases is defined as a surface in three dimensions (curve in two dimensions) and advected over time accurately. The influence of the interface on the adjacent phases is enforced by suitable source terms in the governing equations and in this process the interface is typically diffused over a few mesh cells. Different variants of feedback and direct forcing have been proposed for accurate implementation of the interface effect (Peskin, 1977; Goldstein, Handler, and Sirovich, 1993; Saiki and Biringen, 1994; Fadlun *et al.*, 2000; Kim, Kim, and Choi, 2001). Thus the jump conditions across the interface are enforced only in an integral sense. The advantage of the immersed boundary technique over the sharp-interface approach is that the former is simpler to implement in a code. Furthermore, as the interface moves across the fixed Cartesian grid, the sharp interface approach requires recomputation of the discrete operators close to the interface. Accomplishing this in three dimensions in an efficient manner can be a challenge.

In this chapter we will first discuss the sharp-interface Cartesian method in some detail. The particular interpolation scheme to be presented in this context will be that developed by Ye *et al.* (1999) and Udaykumar *et al.* (2001). Other implementations of the sharp-interface method are similar in spirit and therefore will only be mentioned briefly. We will then discuss the immersed boundary technique. Here we will discuss the feedback forcing approach suggested by Goldstein, Handler, and Sirovich (1993), the direct forcing introduced by Fadlun *et al.* (2000) and the discrete-time mass and momentum forcing developed by Kim, Kim, and Choi (2001). The body-fitted structured mesh will then be briefly introduced in the context of a finite-volume implementation. Finally we will discuss a pseudospectral method on a structured body-fitted grid for the spectrally accurate simulation of complex flows around an isolated sphere.

4.1 Sharp-interface Cartesian grid method

In this section we will discuss the sharp-interface Cartesian grid approach that is appropriate for multiphase flow simulations. Consider the simplest case of a fluid flow around a complex distribution of rigid solid particles. For this initial discussion we also consider the solid particles to be stationary, so that the interface between the fluid and solid phases is stationary. A regular Cartesian grid will be chosen, and the grid may be clustered in order to achieve enhanced resolution of the flow in the neighborhood of the embedded particles. Other than this requirement, the choice of the grid can be quite independent of the actual shape, size, or location of the particles. The sharp-interface method to be described below is based on the recent paper by Ye *et al.* (1999) to which the reader is referred for details.

4.1.1 Fluid–solid interface treatment

The fluid–solid boundaries are represented using marker points which are connected by piecewise linear or quadratic curves. These boundaries overlay the background Cartesian mesh and here we will describe how to treat the Cartesian cells that are cut through by the interfacial boundaries. When cut cells are used to incorporate a complex boundary, it is generally easier to work with co-located grids where the pressure and the velocities are all given at the same location.

The first step is to go through a *reshaping procedure*, where the shape of the cells adjacent to the boundaries is redefined by changing the cell faces. As outlined by Ye *et al.* (1999), the cells cut by the interface whose center lies within the fluid are reshaped by discarding the part of the cell that lies within the solid. On the other hand, if the center of the cut cell lies within the solid, then it is not treated independently, but it is attached to a neighboring cell to form a larger cell. Thus the resulting control volumes are of trapezoidal shape as shown in Fig. 4.1. The details of this reshaping procedure can be found in Udaykumar *et al.* (1997) and Udaykumar, Mittal, & Shyy (1999).

The next issue is to define fluxes and gradients across the faces of the trapezoidal cell in terms of interface information available as cell-centered values in the neighborhood. Particular attention must be paid to maintain

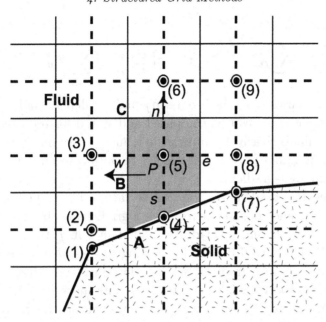

Fig. 4.1. Schematic of the fluid–solid interface cutting through the underlying Cartesian mesh. The figure shows the reshaping of the shaded cell located adjacent to the interface. Also marked are the nine points that form the stencil in two dimensions, which are required to compute the necessary fluxes and gradients for the cell marked P. Three of the nine points are interface points and the other five are cell centers. This figure is copied from Ye *et al.* (1999).

second-order accuracy. A multidimensional polynomial interpolating through the neighboring cell centers and interface points is constructed, which then is used to calculate appropriate fluxes and gradient information. For example, for calculating the fluxes through the left face of the trapezoidal cell (marked AC in Fig. 4.1) the following interpolation function is used

$$u(x, y) = c_1 xy^2 + c_2 y^2 + c_3 xy + c_4 y + c_5 x + c_6, \qquad (4.1)$$

where the coefficients $c_1 - c_6$ are obtained by fitting equation (4.1) to the values of u at the six points marked 1 to 6 in Fig. 4.1. These six points are carefully chosen to surround the left face and they include four cell centers and two points along the interface. The function value and the gradient at the midpoint of the face (marked B) can be evaluated from (4.1) by substituting $x = x_B$ and $y = y_B$. The result can be rearranged to express the velocity and its x-gradient as a weighted sum of the velocity at the six points:

$$u_w = \sum_{j=1}^{6} \alpha_j u_j \quad \text{and} \quad \left(\frac{\partial u}{\partial x}\right)_w = \sum_{j=1}^{6} \beta_j u_j. \tag{4.2}$$

For fixed boundaries, the coefficients α_j and β_j depend only on the cell geometry and therefore can be computed and stored for repeated use. The above polynomial is clearly asymmetric in x and y; however, it can be shown that it is the minimal polynomial needed to obtain second-order accuracy for the x-flux and x-gradient.

   The wall normal gradient on the slanted face of the trapezoid is decomposed into its x- and y-components. The y-gradient is obtained to second-order accuracy by fitting a quadratic polynomial in y through the three points marked 4, 5, and 6 in Fig. 4.1. Owing to the orientation of the cell, the calculation of the x-gradient is a bit more involved. Here, for second-order accuracy, a polynomial fitted to the points marked 1, 2, 3, 7, 8, and 9 must be constructed for computing the x-gradient at the midpoint of the slanted face. The fluxes and the gradients on the other faces can be defined similarly (for details, see Ye *et al.*, 1999). When all the different terms are inserted into the finite-volume version of the governing equations, the resulting compact stencil involves nine points for the trapezoidal cell which are shown in Fig. 4.1. Three of the nine points are boundary points while the other six are cell centers. Note that the nine-point stencil is larger than the regular five-point stencil used for the interior cells (see Fig. 4.1). The intersection of the fluid–solid interface with the Cartesian mesh will result in trapezoidal cells of other types as well and the stencil for such cells can be similarly formed.

4.1.2 Other issues

The above finite-volume discretization of the equation for $\mathbf{u}^*$ (2.9), results in a linear system of the following form for the x-component of the velocity:

$$\sum_{k=1}^{M} \chi_P^k u_k = R_P \tag{4.3}$$

where P denotes the cell in which the control volume momentum balance is applied and it runs over all the cells. The coefficients χ correspond to the weights for the M nodes of the stencil around the point P. For the interior cells the stencil is of width $M = 5$ and for the boundary cells $M = 9$. The banded linear operator on the left-hand side corresponds to the finite-volume discretization of the Helmholtz operator and the right-hand side is the net effect of the explicit terms in (2.9). A similar linear system results from the finite-volume discretization of the Poisson equation (2.12). These linear systems are generally so large that they cannot be solved directly. Several options exist for their iterative solution. BICGSTAB has been recommended for the solution of the Helmholtz equation for the intermediate velocity and multigrid methods are preferred for the Poisson equation (Udaykumar *et al.*, 2001). GMRES is an attractive alternative for the iterative solution of the above large linear system (Saad, 1996).

The above discretization has been explained for a two-dimensional geometry. Extension of the formalism to three-dimensional geometries is in principle straightforward, although a three-dimensional implementation can be complicated. Udaykumar *et al.* (2001) have presented an extension of the above formalism to account for a moving fluid–solid interface over a fixed Cartesian grid. An interesting aspect of the moving interface is that new cells can appear in the fluid domain, with no prior velocity or pressure information in them, since they were previously part of the solid. Strategies for handling such situations have been discussed by Udaykumar *et al.* (2001).

Other approaches have been proposed for handling complex interfaces maintaining their sharpness within a Cartesian discretization. In particular, the approach proposed by Calhoun and LeVeque (2000) is quite interesting. A capacity function is introduced to account for the fact that some of the Cartesian cells are only partly in the fluid region. This allows the partial cells to be treated in essentially the same way as the other interior full cells. An explicit wave-propagation method, implemented in the CLAWPACK software package (see `http://www.amath.washington.edu/rjl/clawpack/cartesian`) is used for the advection term, while the diffusion term is treated with an implicit finite-volume algorithm. An overall second-order accuracy was demonstrated in Calhoun and LeVeque (2000).

Another approach to treating the interface between a solid and a fluid, and also between two different fluids, in a sharp fashion, is the ghost fluid method (Fedkiw *et al.*, 1999; Liu, Fedkiw, and Kang, 2000). The attractiveness of this approach is that it accounts for the jump in material properties, normal gradient of tangential velocities and pressure across the interface in a sharp manner, without modifying the symmetric structure of the linear operators involved in the solution of the intermediate velocity and pressure. The influence of the jump conditions appears only on the right-hand side. Thus, this approach is easier to implement and extends to more complex situations. However, this approach as of now has only been formulated to first-order accuracy.

4.2 Immersed boundary technique

The immersed boundary technique has been proposed and developed for over three decades by Peskin and his group at the Courant Institute (see, for example, Peskin, 1977; McQueen and Peskin, 1989). The idea is to employ a regular Cartesian grid, but apply additional appropriately distributed momentum and mass sources within the domain in order to satisfy the requisite interfacial conditions at the interface between the phases. Here again we will limit our attention to the simple case of a fluid flow around rigid solid particles.

The fluid–solid interface is defined as a surface, S, in three dimensions and a body force, $\mathbf{f}$, is introduced into the Navier–Stokes equations as

$$\frac{\partial \mathbf{u}}{\partial t} + \mathbf{u} \cdot \nabla \mathbf{u} = -\nabla p + \frac{1}{Re}\nabla^2 \mathbf{u} + \mathbf{f} \; . \tag{4.4}$$

The forcing is chosen so that it enforces the desired velocity boundary condition on the interface, S. Ideally we desire the forcing to be identically zero in the entire fluid region, so that the standard Navier–Stokes equation is faithfully solved there. The forcing must be localized along the interface and perhaps extended into the solid. Clearly there is no explicit restriction on the shape of the interface or the boundary condition to be satisfied on it. Irrespective of these details the flow can be solved on a Cartesian grid that is suitably chosen to provide enhanced resolution where needed.

It is clear that there can be different choices for the force field that can accomplish the stated goals. In fact, there are different variants of the immersed boundary technique which differ primarily in the manner in which the force field is computed. The popular forcing techniques in use are feedback forcing (Goldstein, Handler, and Sirovich, 1993), direct forcing (Fadlun et al., 2000) and discrete mass and momentum forcing (Kim, Kim, and Choi, 2001). Here we will discuss these forcing techniques in some detail. We will adopt the second-order time-splitting method presented in Chapter 2 (equations 2.9–2.13) and consider the forcing to be applied during the advection-diffusion stage (equation 2.9). Thus the forcing will be applied to enforce the appropriate interface condition on the intermediate velocity, $\mathbf{u}^*$, which in turn will make the final divergence-free velocity satisfy the specified interface condition. The pressure correction step (2.10 and 2.12) remains unaffected by the added forcing.

4.2.1 Feedback forcing

Goldstein, Handler, and Sirovich (1993) proposed a forcing of the following form to enforce a Dirichlet velocity condition $\mathbf{u} = \mathbf{V}$ on the interface

$$\mathbf{f}(\mathbf{x}_s, t) = \alpha \int_0^t \left[\mathbf{u}^*(\mathbf{x}_s, t') - \mathbf{V}(\mathbf{x}_s, t') \right] dt' + \beta \left[\mathbf{u}^*(\mathbf{x}_s, t) - \mathbf{V}(\mathbf{x}_s, t) \right]. \quad (4.5)$$

The above forcing attempts to drive $\mathbf{u}^*$ towards $\mathbf{V}$ as a damped oscillator. The constants α and β determine the frequency and damping of the oscillator as $(1/2\pi)\sqrt{|\alpha|}$ and $-\beta/(2\sqrt{|\alpha|})$. We desire the frequency of damping to be much larger than all relevant inverse time scales of fluid motion and the damping coefficient to be adequately large, in order for the velocity condition at the interface to be satisfied rapidly. This requires the constants α and β to be chosen as large positive constants.

Goldstein *et al.* (1993) applied the feedback forcing in the context of a pseudospectral numerical method and difficulties with localizing the forcing within the global nature of the method were addressed. Saiki and Biringen (1996) showed that the feedback forcing is far more effective in the context of finite difference schemes.

When the forcing is applied only to control the interfacial velocity condition, the resulting flow field will include a nonphysical velocity distribution in the region covered by the solid. Goldstein *et al.* (1993) used this solid-phase velocity to smooth the discontinuity in velocity derivative that might otherwise arise at the fluid–solid interface. However, in their application, Saiki and Biringen (1996) observed the above approach to sometime lead to an overall nonphysical solution. This erroneous behavior was rectified by imposing an extended forcing that will enforce the solid-phase velocity conditions not only at the interface, but also inside the solid region.

One of the primary difficulties with the feedback forcing has been the severe restriction it imposes on the computational time step. For an explicit treatment of the forcing using the Adams–Bashforth scheme the time step limitation has been shown to be (Goldstein *et al.*, 1993)

$$\Delta t \leq \frac{-\beta - \sqrt{(\beta^2 - 2\alpha k)}}{\alpha}, \quad (4.6)$$

where k is an $O(1)$ flow-dependent constant. The above limitation generally results in time steps that are more than two orders of magnitude smaller than required by the standard CFL condition. Such very small time steps arise from the large values of α and β chosen for (4.5) and the associated very small time scale of the oscillator. It was shown by Fadlun *et al.* (2000) that if an implicit scheme is used for the forcing instead of the explicit scheme, considerably larger time steps can be employed. Nevertheless, even with the implicit treatment, the time step that satisfies stability requirements is still small compared to the CFL condition.

4.2.2 Direct forcing

Mohd–Yusof (1997) developed a very simple direct procedure for forcing the required velocity condition at the interface, which was later adapted and tested by Fadlun *et al.* (2000). In this approach the forcing is determined directly without solving an equation of the form (4.5).

Fadlun *et al.* (2000) implemented the direct forcing in the context of a fully staggered second-order accurate finite difference formulation on a regular Cartesian mesh. At the velocity nodes, away from the interface, a discretized version of the Helmholtz equation (2.13) is solved. In two dimensions the discretized Helmoltz operator results in a five-point stencil that includes the left, right, bottom and top neighbors and in three dimensions the stencil extends to seven points. There is no forcing at the internal points which are more than one grid away from the interface. At the grid points adjacent to the interface on the fluid side, the discretized equation to be solved is determined by requiring that the interpolated velocity at the interface satisfy the appropriate condition. For example, in Fig. 4.2, the velocity at node (i) will be determined such that when interpolated along with the velocity at node $(i+1)$, the required interfacial velocity will be $\mathbf{V}(\mathbf{x}_s)$ (see equation 2.27). A simple linear fit can be used to obtain

$$\alpha \mathbf{u}_i^* + \beta \mathbf{u}_{i+1}^* = V(\mathbf{x}_s), \qquad (4.7)$$

where the coefficients depend only on the geometric details of the grid spacing and the intersection of the interface with the Cartesian grid. At the grid point marked i the above equation is solved instead of the discretized version of the Helmoltz equation (2.13). In effect, if an appropriate momentum forcing were to be added to the right-hand side of (2.13), then the resulting equation would reduce to (4.7). Thus there is a momentum forcing implicit in the above formalism.

It is clear from Fig. 4.2 that there are other options for enforcing the velocity boundary condition along the interface and equation (4.7) is just a particular choice. The local forcing and the resulting fluid flow are likely to be dependent on the particular manner in which the interpolation is carried out. Nevertheless, Fadlun *et al.* (2000) have applied the above formalism to a variety of problems (vortex ring formation, flow around a sphere, and flow inside an internal combustion chamber) and demonstrated

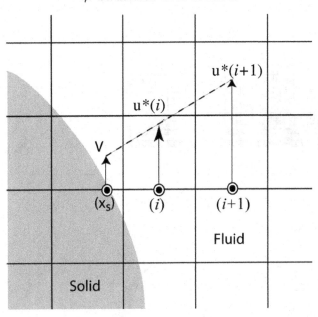

Fig. 4.2. Schematic of the fluid–solid interface cutting through the underlying Cartesian mesh. The figure shows linear interpolation between the interface and the two neighboring velocity points. This figure is copied from Fadlun *et al.* (2000).

second-order convergence. One of the primary advantages of the direct forcing over feedback forcing is that the time-step limitation is not so severe. In fact, in the above applications it was demonstrated that time steps as large as the CFL limit can be taken without any numerical instability.

As a demonstration of the flexibility of this approach, here we show results obtained in Fadlun *et al.* (2000) for the case of a uniform flow over a stationary solid sphere. The Reynolds number of the flow was varied from 100 to 1000. By enforcing a symmetry condition about the sphere midplane only a steady axisymmetric solution was sought. Figure 4.3(a) shows the computational domain and a background nonuniform Cartesian grid. Figure 4.3(b) shows the pressure contours obtained at a Reynolds number of 200 with a two-dimensional grid of 129 × 257 points, respectively, in the radial and axial directions. The results of Fornberg (1988), who used a highly accurate spectral simulation, will be used as the reference. The reference steady drag coefficients, $C_{d,ref}$, is compared with the immersed boundary results for varying grid resolution. Figure 4.3(c) shows a plot of the deviation in C_d from the reference value for increasing grid resolution. From the log-log plot it is clear that convergence is better than second-order.

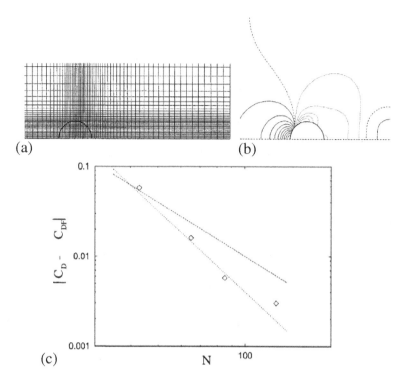

Fig. 4.3. Simulation of axisymmetric flow over a sphere using direct forcing (reproduced from Fadlun *et al.*, 2000, with permission from Elsevier). (a) Sample nonuniform Cartesian grid employed with resolution of 129×257 points along the radial and axial directions respectively; (b) pressure contours at Re=200; (c) log-log plot of error in the computer drag coefficient from the reference value of Fornberg (1988) vs grid resolution. The dash and dotted lines are of slope -2 and -3, respectively.

4.2.3 Discrete-time mass and momentum forcing

Kim, Kim, and Choi (2001) pointed out that the direct forcing employed by Fadlun *et al.* (2000) applies the forcing inside the fluid, i.e. even outside the solid region. This is evident from the fact that the momentum equation for the *i*-th node in Fig. 4.2 is modified to enforce equation (4.7). Kim *et al.* (2001) proposed a modification of the direct forcing strategy that strictly applies the momentum forcing only on the fluid–solid interface and inside the solid region. Thus the Navier–Stokes equations are faithfully solved in the entire region occupied by the fluid. Their reformulation, however, requires that a mass source also be included along the interface. In other words, the continuity equation is modified to

$$\nabla \cdot \mathbf{u} = q, \tag{4.8}$$

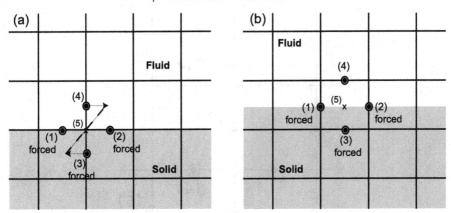

Fig. 4.4. Schematic of the discrete-time momentum forcing for two different scenarios. (a) Here the interface is exactly along the wall normal velocity points; (b) here the interface is along the tangential velocity points. In both frames the interpolation is between points marked (3) and (4), such that the appropriate component of intermediate velocity is satisfied at the point marked (5).

where q is strictly zero inside the fluid, but can take nonzero values along the interface and possibly in the solid region.

Here, following Kim *et al.* (2001), we discuss the second-order accurate finite difference implementation for a staggered grid. Consider the simple situation shown in Fig. 4.4, where the interface between the fluid and the solid is such that some of the velocity points are right on the interface, while others are not. In frame (a), wall normal velocity points marked (1) and (2) lie on the interface and at these points the intermediate velocity condition as given in (2.27) is enforced directly. For the tangential component of velocity at the point marked (4), the tangential component of the discrete Helmholtz equation (2.13) is solved. At the point marked (3), which is just inside the solid, the tangential component of the intermediate velocity is chosen such that when interpolated with point (4), the tangential interface velocity as given in (2.27) is also satisfied (at the point marked 5). In frame (b) a different scenario is shown, where the tangential velocity points marked (1) and (2) are on the interface.

In Fig. 4.4 the interface was conveniently considered to be parallel to the grid lines. But in general the situation is likely to be more complex. Consider a fluid–solid interface cutting through the Cartesian grid as shown in Fig. 4.5. In the context of the staggered grid scheme, points marked (1), (3), and (5) are inside the fluid and here the x-component of the discretized Helmoltz equation (2.13) is solved. Points marked (2), (4), and (6) are also inside the fluid and here the y-component of the discretized

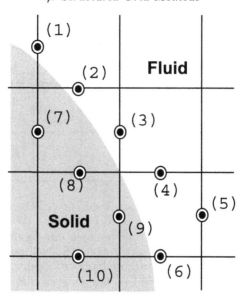

Fig. 4.5. Schematic of the discrete-time momentum forcing where the interface cuts through the Cartesian grid in a complex manner. Generated based on Kim *et al.* (2001).

Helmoltz equation is solved. The other four points are inside the solid. The x- or y-components of velocity at these points are determined so that with an interpolation along with the points inside the fluid domain the velocity boundary condition (2.27) at a set of suitably chosen points along the interface is satisfied. As pointed out by Kim *et al.* (2001), depending on the situation, a linear or bilinear interpolation will be required. The interpolation schemes and how the point on the interface, where the boundary condition is satisfied, is chosen are given in detail in Kim *et al.* (2001). Note that this procedure ensures that the governing momentum equations are accurately solved inside the fluid region and the momentum forcing is limited to the interface and the interior of the solid.

It is clear from Fig. 4.4(b) that the incompressibility condition needs to be modified in the cells intersected by the interface. In this figure the x-component of velocity is zero at (1) and (2) due to the no-slip condition, while the y-component of velocity at points (3) and (4) is of the same magnitude but directed in the opposite direction in order to interpolate to no-penetration at the surface (point marked 5). Thus there is a net flow into (or out of) this entire cell, which must be compensated. In other words, as suggested by Kim *et al.* (2001), the Poisson equation for pseudopressure (2.12) must be modified to

$$\nabla^2 \phi^{n+1} = \frac{\nabla \cdot \mathbf{u}^\star - q}{\Delta t}, \tag{4.9}$$

where q is the local mass source/sink.

Following the advection–diffusion step for the intermediate velocity, the computation of the mass source proceeds as follows. Consider the simple situation displayed in Fig. 4.6 in two dimensions. Here we desire the final velocity to satisfy the divergence free conditions within the triangular fluid region bounded by the corners BCD. Assuming a stationary interface, this can be expressed in terms of the discrete velocities as

$$u_{i+1}\Delta y + v_{j+1}\Delta x = 0. \tag{4.10}$$

For the entire rectangle containing the body and the fluid (marked by the corners ABCD) a discretization of (4.9) yields

$$u_{i+1}\Delta y + v_{j+1}\Delta x = u_i \Delta y + v_j \Delta x + q\Delta x \Delta y. \tag{4.11}$$

The above two equations can be combined to yield the following expression for the mass source/sink

$$q = -\frac{u_i \Delta y + v_j \Delta x}{\Delta x \Delta y}. \tag{4.12}$$

Note that in the above procedure, since continuity is satisfied exactly for the fluid region, the mass source/sink pertains entirely to the solid region. Furthermore, in equation (4.12), q is defined in terms of the final velocity at the end of the time step, since it is this velocity that we require to satisfy (4.10). However, the final $(n+1)$-th time-step velocity is not available before solving the Poisson equation for pseudopressure. Therefore, in equation (4.12), the velocity is replaced with the intermediate velocity, $\mathbf{u}^*$, which evaluates q to $O(\Delta t^2)$ accuracy. This q is then used in the Poisson equation (2.12) and finally a pressure correction is applied to get the final velocity of the time step. Figure 4.6 and the above discussion cover the special case where the interface passes through the cell corners. As indicated by Kim *et al.* (2001), the more general case can be considered in a similar manner to obtain the appropriate q.

An interesting aspect of the implementation in Kim *et al.* (2001) is that the immersed boundary technique is applied on a cylindrical rather than Cartesian grid. The authors have thus demonstrated the generality of the immersed boundary technique. Second-order accuracy of the scheme was established in three different test cases: decaying vortices, flow over a circular cylinder and flow over a sphere. A comparison of results with those

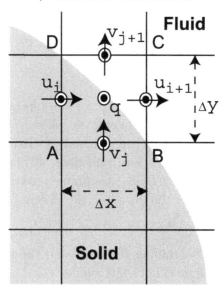

Fig. 4.6. Schematic of the discrete-time mass forcing where the interface cuts through the Cartesian grid in a complex manner. Generated based on Kim *et al.* (2001).

of Fadlun *et al.* (2000) clearly showed that in this approach, by carefully forcing only along the interface and in the interior of the solid, and by taking into account the mass source/sink, a significant improvement in accuracy is achieved. For example, for the case of flow over a sphere at $Re = 100$, the highly accurate simulations of Fornberg (1988) yields $C_{d,\mathrm{Ref}} = 1.085$, which compares very well with the results of both Kim *et al.* (2001) ($C_d = 1.087$) and Fadlun *et al.* (2000) ($C_d = 1.0794$). However there is an improvement in accuracy with mass and momentum forcing. Kim *et al.* (2001) have carried out their immersed boundary simulations in three dimensions and obtained accurate results for the shedding of vortices behind a sphere above a Reynolds number of about 280.

4.2.4 Using the exact near-body velocity

While using forcing spread around the interface leads to a reasonable representation of the flow around a solid body, the techniques described above where the velocity at the points next to the body are found by interpolation, lead to considerable improvement. An even higher accuracy could be attained if the velocity next to the body could be set exactly. Prosperetti and collaborators (Takagi *et al.* 2003; Zhang and Prosperetti 2003; Prosperetti and Zhang 2005) have shown that this is, indeed, possible, at least for bodies with a simple shape. Since the relative velocity near the surface of a rigid body vanishes, they suggest that the Navier–Stokes equations in the body rest frame can be linearized. A suitable variable transformation turns these linearized equations into the Stokes equations the solution of which can be expressed analytically as a series expansion (Lamb 1932; Kim and Karrila 1991). By matching this solution to the velocity at grid points next to the surface, they can determine the coefficients in the series solution. As the Reynolds number increases, the region adjacent to the body where inertia is negligible becomes smaller and smaller but, for any finite Reynolds number, this region will be finite. The method can then be applied provided that the grid is sufficiently fine to have nodes in this region. Prosperetti and Zhang (2005) demonstrate that even for $O(100)$ Reynolds numbers, where the flow separates, this approach yields very accurate results with a relatively modest resolution. In addition to examining flows over stationary spheres, Prosperetti *et al.* have used their method to simulate the motion of several freely moving spheres in turbulent flow. A similar approach based on a different series expansion was used by Kalthoff, Schwarzer, and Herrmann (1997).

4.3 Body-fitted grids

In this section we will address techniques where the gridding is dependent on the details of the interface. The interface is first accurately defined with a two-dimensional structured discretization, which is then extended into the fluid domain with a structured three-dimensional mesh. We will first briefly discuss the finite-volume technique in the context of a body-fitted curvilinear structured mesh. The finite-volume technique provides great flexibility in terms of the geometric complexity that it can handle. However, it is generally second-order accurate or less. We will then consider the highly accurate fully spectral method, whose application requires a structured body-fitted grid. Of course, the fully spectral method will be limited in terms of the geometric complexity that it can handle. However, it is well suited for problems where one is interested in the complexity of the flow (such as turbulence) as opposed to geometric complexity.

4.3.1 Finite-volume technique

Consider the simple case of a solid sphere in a channel flow. A structured mesh that covers the fluid region outside the sphere and extends over the entire Cartesian geometry of the channel is required. There are different approaches to construct such a structured grid: algebraic methods, methods based on solving partial differential equations, variational approach, etc.

For complex problems involving a complicated fluid–soild interface, one might need to resort to a block structured approach. Here the overall computational domain is divided into blocks and within each block a three-dimensional structured mesh is defined and the adjacent blocks communicate at the block interface. The simplest implementation is one where the discretization within the blocks is conforming, that is, at the block boundaries the adjacent blocks share common nodal points. A more complex implementation involves nonconforming blocks. This requires interpolation between the different meshes at the block boundaries. The nonconforming blocks provide greater flexibility in handling more complex geometries. Nevertheless within each block the grid can be considered to provide a local curvilinear coordinate.

Within each block the physical domain is mapped onto a computational domain, which is usually defined to be a unit square in two dimensions and a unit cube in three dimensions. A transformation from the physical coordinates (x, y) to the computational coordinates (ξ, η) is defined:

$$\xi = \xi(x, y) \qquad \text{and} \qquad \eta = \eta(x, y). \tag{4.13}$$

The above transformation can be inverted to give

$$x = x(\xi, \eta) \qquad \text{and} \qquad y = y(\xi, \eta). \tag{4.14}$$

From the above coordinate transformations, the transformation of gradients can also be calculated by the chain rule as

$$\begin{bmatrix} \frac{\partial}{\partial x} \\ \frac{\partial}{\partial y} \end{bmatrix} = \begin{bmatrix} \xi_x & \eta_x \\ \xi_y & \eta_y \end{bmatrix} \begin{bmatrix} \frac{\partial}{\partial \xi} \\ \frac{\partial}{\partial \eta} \end{bmatrix} \qquad \text{and} \qquad \begin{bmatrix} \frac{\partial}{\partial \xi} \\ \frac{\partial}{\partial \eta} \end{bmatrix} = \begin{bmatrix} x_\xi & y_\xi \\ x_\eta & y_\eta \end{bmatrix} \begin{bmatrix} \frac{\partial}{\partial x} \\ \frac{\partial}{\partial y} \end{bmatrix}, \tag{4.15}$$

where subscripts denote differentiation. Here the forward and backward transformation matrices are inverse of each other:

$$\begin{bmatrix} \xi_x & \eta_x \\ \xi_y & \eta_y \end{bmatrix} = \begin{bmatrix} x_\xi & y_\xi \\ x_\eta & y_\eta \end{bmatrix}^{-1}. \tag{4.16}$$

The above transformation allows for the governing equations to be written in terms of the computational coordinates. The convective and diffusive fluxes through the four bounding surfaces of each computational cell (six surfaces in three dimensions), as well as the requisite gradients, can be computed in terms of differences between adjacent cell volumes. These standard steps were addressed in previous chapters and additional details can be found in textbooks on finite-volume or finite difference methods (see, e.g. Anderson,

Tannehill, and Pletcher, 1984; Wesseling, 2001, and others) and therefore will not be covered here in any detail.

The key point to note is that the fluid–solid interface does not require any special treatment, since it is naturally one of the surfaces bounding the boundary cells. From the specified boundary condition, fluxes through this surface and required gradients can be computed. Thus the governing equations can be solved similarly in all the cells, without any special treatment. Examples of the application of such methods for flows over a bubble or a spherical particle are given by Dandy and Dwyer (1990), Magnaudet, Rivero, and Fabre (1995), Johnson and Patel (1999), Kim, Elghobashi, and Sirignano (1993), Kurose and Komori (1999), Kim and Choi (2002), Mei and Adrian (1992), and Shirayama (1992), and Yang, Prosperetti, and Takagi (2003).

4.3.2 Spectral methods

Spectral methods are global in nature and provide the highest possible accuracy since as the grid resolution is increased, the error decreases exponentially rather than algebraically. This high accuracy, however, comes at a cost. Due to their global nature, spectral methods are limited to simple geometries, because all the boundaries of the computational domain, including fluid–solid interfaces and inflow and outflow surfaces, must conform to the chosen coordinate system. In problems where it can be applied, however, the spectral method provides the best possible option in terms of accuracy and computational efficiency. For instance, a good application of the use of the method in multiphase flow has been in the investigation of particle–turbulence interaction (Bagchi and Balachandar, 2003a,b). Here the geometric complexity was simplified to a single sphere subjected to an isotropic turbulent cross-flow. By considering such a simplified geometric situation all the computational resources have been devoted to accurately addressing the details of turbulence and its interaction with the sphere.

An important development in adapting the spectral methodology to complex geometries is the development of spectral element methods (Karniadakis, Israeli, and Orszag, 1991). Here the computational domain is divided into elements and within each element a spectral representation is used with an orthogonal polynomial basis. Across the interface between the elements continuity of the function and its derivatives is imposed. Again as an example, in the context of flow over a spherical particle, spectral (Chang and Maxey, 1994; Mittal, 1999; Elcrat, Fornberg, and Miller, 2001; Bagchi and Balachandar, 2002c), spectral element (Fischer, Leaf, and Restrepo, 2002), and mixed spectral element/Fourier expansion (Tomboulides and Orszag, 2000) have been employed.

In closing this introduction to spectral methods, it must be pointed out that one of the earliest attempts to use the immersed boundary technique (Goldstein, Handler, and Sirovich, 1993) was in the context of spectral methods. The immersed boundary method, as described above in Section 4.2, gives greater geometric flexibility. However, it has been shown that the imposition of a sharp interface through the immersed boundary technique is not fully consistent with the global nature of the spectral approach (Saiki and Biringen, 1994). This incompatibility shows up in the solution as oscillations. Therefore interfaces have to be sufficiently numerically diffused to suppress the so-called Gibbs phenomena.

In the following discussion we will present the full spectral method for the solution of complex flow over a single rigid spherical particle. We will consider the case of a freely translating and rotating particle, where the particle motion is in response to hydrodynamic and external forces and torque. We will formulate the problem in a frame of reference translating with the particle so that in the computational domain the particle is always at the same location, thus greatly simplifying the computations. We will also continue our considerations of inflow and outflow specification for spectral methods.

4.3.2.1 Governing equations

We consider a sphere moving with a time-dependent velocity $\mathbf{v}(\tau)$ and angular velocity $\mathbf{\Omega}(\tau)$ and start by formulating the governing equations for the translating noninertial frame of reference. The undisturbed ambient flow (that would exist in the absence of the particle) $\mathbf{U}(\mathbf{X}, \tau)$ is taken to be both time and space varying. We will consider interactions of the particle with ambient flows such as shear flows, straining flows, vortical flows, and turbulent flows. The presence of the particle and its motion introduces a disturbance field $\mathbf{U}'(\mathbf{x}, \tau)$ and the resultant flow is given by $\mathbf{U} + \mathbf{U}'$. The governing equations (in dimensional terms) can be written in terms of the perturbation field as

$$\nabla \cdot \mathbf{U}' = 0, \tag{4.17}$$

$$\frac{\partial \mathbf{U}'}{\partial \tau} + \mathbf{U}' \cdot \nabla \mathbf{U}' + \mathbf{U}' \cdot \nabla \mathbf{U} + \mathbf{U} \cdot \nabla \mathbf{U}' = -\frac{1}{\rho} \nabla p' + \nu \nabla^2 \mathbf{U}', \tag{4.18}$$

where the total pressure is given by the sum $P + p'$, and the undisturbed ambient pressure P is related to $\mathbf{U}$ by

$$-\frac{1}{\rho}\nabla P = \frac{\partial \mathbf{U}}{\partial \tau} + \mathbf{U} \cdot \nabla \mathbf{U} - \nu \nabla^2 \mathbf{U}. \tag{4.19}$$

For simple ambient flows such as a linear shear, straining, or vortical flows, the corresponding ambient pressure field P can be readily obtained. At large distances from the particle the disturbance field decays to zero, while on the surface of the particle, the no-slip and no-penetration conditions require

$$\mathbf{U}'(\mathbf{X}, \tau) = \mathbf{v}(\tau) + \mathbf{\Omega}(\tau) \times (\mathbf{X} - \mathbf{X}_p) - \mathbf{U}(\mathbf{X}). \tag{4.20}$$

Here, $\mathbf{X}_p$ is the current location of the center of the particle.

To proceed, we attach a reference frame to the moving particle. For the case of a spherical particle, it is sufficient to translate the frame with the moving particle, ignoring rotation. In the noninertial reference frame $(\mathbf{x} = \mathbf{X} - \mathbf{X}_p, t)$ that moves with the particle, the perturbation velocity is

$$\mathbf{u}(\mathbf{x}, t) = \mathbf{U}'(\mathbf{X}, \tau) - \mathbf{v}(\tau), \tag{4.21}$$

and the governing equations can be rewritten as

$$\nabla \cdot \mathbf{u} = 0, \tag{4.22}$$

and

$$\frac{\partial \mathbf{u}}{\partial t} + \frac{d\mathbf{v}}{dt} + \mathbf{u} \cdot \nabla \mathbf{u} + \mathbf{U} \cdot \nabla \mathbf{u} + \mathbf{u} \cdot \nabla \mathbf{U} + \mathbf{v} \cdot \nabla \mathbf{U}$$
$$= -\frac{1}{\rho}\nabla p' + \nu \nabla^2 \mathbf{u}. \tag{4.23}$$

The boundary conditions for $\mathbf{u}$ are as follows. Far away from the particle the velocity approaches the undisturbed ambient flow $\mathbf{u} = -\mathbf{v}$; on the surface of the particle, which is now defined by $|\mathbf{x}| = R$: $\mathbf{u} = \mathbf{\Omega}(\tau) \times \mathbf{x} - \mathbf{U}(\mathbf{X})$. The term $d\mathbf{v}/dt$ on the left-hand side of the above equation accounts for the nonrotating, noninertial frame. The advantage of using a moving frame of reference is that in this frame the geometry of the computational domain remains fixed.

In addition to the equations for the fluid flow in the region outside the particle, we need to translate and rotate the particle using its equations of motion, from Section 1.3 in Chapter 1:

$$m_p \frac{d\mathbf{v}}{dt} = \mathbf{F} + \mathbf{F}_{\text{ext}} \tag{4.24}$$

$$\frac{d\mathbf{X}_p}{dt} = \mathbf{v}, \tag{4.25}$$

$$I \frac{d\mathbf{\Omega}}{dt} = \mathbf{T} + \mathbf{T}_{\text{ext}}. \tag{4.26}$$

Here, m_p is the mass of the particle and I is its moment of inertia $I = (2/5)m_p R^2$. The net hydrodynamic force, $\mathbf{F}$, and torque, $\mathbf{T}$, acting on the particle can be obtained by integrating the pressure and the normal and tangential stresses on its surface:

$$\mathbf{F} = \int_S \left[-(p' + P + \tau_{rr}) \, \mathbf{e}_r + \tau_{r\theta} \, \mathbf{e}_\theta + \tau_{r\phi} \, \mathbf{e}_\phi \right] dS \tag{4.27}$$

$$\mathbf{T} = R \int_S \mathbf{e}_r \times (\tau_{r\theta} \mathbf{e}_\theta + \tau_{r\phi} \mathbf{e}_\phi) \, dS. \tag{4.28}$$

The integrals are over the surface of a sphere of radius R and the stress components, τ_{rr}, $\tau_{r\theta}$, and $\tau_{r\phi}$ are computed on the surface from the velocity field $\mathbf{u}$. In the above, $\mathbf{e}_r$, $\mathbf{e}_\theta$, and $\mathbf{e}_\phi$ are the unit vectors in a spherical coordinate system whose origin is at the center of the particle. $\mathbf{F}_{\text{ext}}$ and $\mathbf{T}_{\text{ext}}$ are the externally applied force and torque, respectively.

The governing equations (4.23) can be nondimensionalized with the diameter of the particle $d_p = 2R$ as the length scale and the relative velocity $|\mathbf{u}_r| = |\mathbf{U}(\mathbf{X}_p, \tau) - \mathbf{v}(\tau)|$ (assumed constant for simplicity) as the velocity scale. The resulting nondimensional equations are unchanged, except that the kinematic viscosity is replaced by the inverse of the Reynolds number $Re = d_p|\mathbf{u}_r|/\nu$. The nondimensional force and torque coefficients, are defined as follows:

$$C_F = \frac{|\mathbf{F}|}{\frac{1}{2} \rho |\mathbf{u}_r|^2 \pi (d_p/2)^2} \tag{4.29}$$

$$C_T = \frac{|\mathbf{T}|}{\frac{1}{2} \rho |\mathbf{u}_r|^2 \pi (d_p/2)^3}. \tag{4.30}$$

4.3.2.2 Temporal discretization

The time-splitting technique can be applied to the above governing equations just as efficiently as to the original Navier–Stokes equations. In the evaluation of the intermediate velocity, the nonlinear term in equations (2.9) and (2.13) is now given by

$$\mathbf{A} = \frac{d\mathbf{v}}{dt} + \mathbf{u} \cdot \nabla \mathbf{u} + \mathbf{U} \cdot \nabla \mathbf{u} + \mathbf{u} \cdot \nabla \mathbf{U} + \mathbf{v} \cdot \nabla \mathbf{U}. \tag{4.31}$$

The entire nonlinear term is treated with an explicit time advancement scheme. If desired, the $d\mathbf{v}/dt$ and the $\mathbf{v} \cdot \nabla \mathbf{U}$ terms can be treated implicitly. After the Helmholtz equations for the intermediate velocity, the Poisson equation for the pseudopressure, ϕ, is solved.

In the evaluation of equation (4.27) the pressure and the surface stresses must be evaluated at the same time level. The equations of particle motion and rotation (equations 4.24–4.26) must also be integrated in time with a consistent scheme. A simple approach is to use an explicit scheme, which, however may be unstable under certain conditions. In this connection see Section 5.2 in Chapter 5.

4.3.2.3 Computational domain

Although we usually seek to find the flow around a particle in an unbounded domain, the computational domain is generally truncated at a finite distance away from the particle where the far-field boundary conditions are imposed (see Fig. 4.7a). The placement of the outer boundary is dictated by a few competing requirements: the computational domain must be small enough so that the flow can be adequately resolved with as few grid points as possible, thus reducing the computational cost; the inflow and free-stream sections of the outer boundary must be sufficiently far from the sphere so that the far-field boundary condition can be imposed with high accuracy; and the outflow section of the outer boundary must also be sufficiently far downstream of the sphere so that any unavoidable approximation of the outflow boundary condition is prevented from affecting the flow upstream. Many simulations of flow over a sphere, especially those using spectral methods, have been done using an outer domain that is a sphere or an ellipsoid of radius about 10–15 times the particle diameter (Kurose and Komori, 1999; Johnson and Patel, 1999; Mittal, 1999; Bagchi and Balachandar, 2002c). This large spherical outer domain places the inflow and outflow boundaries sufficiently far away from the particle. The particle also causes very little blockage, since in the plane normal to the flow direction the particle covers only about 0.1–0.25% of the total cross-sectional area.

The influence of the placement of the outer boundary was investigated by Bagchi & Balachandar (2002c) who compared the results from a simulation using one domain with an outer radius of 15 times the particle diameter to another with a larger outer radius of 30 times the particle diameter. The resolution was maintained the same in both cases by proportionately increasing the number of grid points in the larger domain. The drag and lift coefficients obtained from the two different domains for a linear shear flow past a spherical particle at $Re = 100$ were $1.1183, -0.0123$, and $1.1181, -0.0123$,

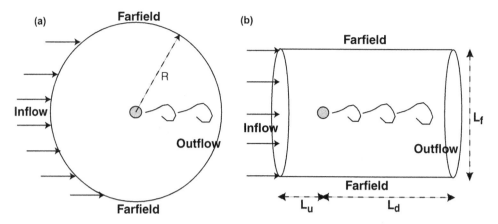

Fig. 4.7. Schematic of the computational domain. Frame (a) shows a spherical outer domain, which includes inflow, outflow, and far-field boundaries. Frame (b) shows an alternative where the outer computational domain is a cylinder. With this option the placement of upstream, downstream, and far-field boundaries can be independently controlled.

respectively, clearly demonstrating the domain independence of the results. By selecting a spherical outer boundary, a spherical coordinate system can be adopted, making the use of spectral methods possible.

An alternate computational domain is a large cylinder (Legendre and Magnaudet, 1998; Tomboulides and Orszag, 2000; Kim and Choi, 2002), surrounding the particle (see Fig. 4.7b). The advantage of this choice is that the distance from the particle to the inflow, L_u, outflow, L_d, and far-field boundaries, $L_f/2$, can be independently controlled. The geometric transition from the inner spherical boundary to the cylinder-like outer boundary can be easily handled with an unstructured mesh (Tomboulides and Orszag, 2000) or with a structured mesh using generalized curvilinear coordinates (Magnaudet, Rivero and Fabre, 1995; Legendre and Magnaudet, 1998).

4.3.2.4 Boundary conditions

Appropriate specification of the boundary conditions, especially at the outflow section of the outer boundary, plays an important role in the overall accuracy of the numerical solution. Only carefully derived boundary conditions will produce stable and consistent results. For the time-splitting technique, boundary conditions are required for the intermediate velocity, $\mathbf{u}^*$, and the pseudopressure, ϕ^{n+1}, in the solution of the Helmholtz (2.13) and

the Poisson (2.12) equations. A homogeneous Neumann boundary condition
for the pressure

$$\frac{\partial \phi^{n+1}}{\partial n} = 0 \qquad (4.32)$$

is generally applied both at the outer boundary and on the surface of the
particle. Here, $\partial/\partial n$ indicates the gradient normal to the surface. Note that
the total pressure does not satisfy a Neumann condition, since $\partial P/\partial n$ may
not be zero on the boundary, depending on the nature of the imposed ambient flow. The forces arising from the ambient flow are therefore accounted
for exactly and the Neumann condition for the perturbation pressure only
influences the viscous terms (Bagchi and Balachandar, 2002c, 2003b).

On the surface of the particle, no-slip and no-penetration conditions are
imposed: $\mathbf{u}^{n+1} = \mathbf{\Omega} \times \mathbf{x} - \mathbf{U}$. In anticipation of the pressure-correction step,
the appropriate boundary condition for the intermediate velocity is then

$$\mathbf{u}^* = \mathbf{\Omega} \times \mathbf{x} - \mathbf{U} + \Delta t (2(\nabla \phi')^n - (\nabla \phi')^{n-1}). \qquad (4.33)$$

This condition, combined with the homogeneous Neumann boundary condition for the pressure (equation 4.32), guarantees zero penetration through
the surface. The no-slip condition is satisfied to order $O(\Delta t^3)$, since the
pressure gradient terms on the right-hand side of equation (4.33) only approximate $(\nabla p')^{n+1}$. The nondimensional time step Δt is typically about
two orders of magnitude smaller than unity and thus the slip error arising
from the time splitting scheme is usually quite small and can be neglected.

At the inflow section of the outer boundary, a Dirichlet boundary condition specifying the undisturbed ambient flow is enforced. For the intermediate velocity this translates to the following boundary condition:

$$\mathbf{u}^* = -\mathbf{v} + \Delta t (2(\nabla p')^n - (\nabla p')^{n-1}). \qquad (4.34)$$

The Helmholtz equation (2.13) requires a boundary condition everywhere,
including the outflow section of the outer boundary. Several methods proposed in the literature were discussed in Chapter 2. Spectral simulations,
however, owing to their global nature, demand a more stringent enforcement
of the nonreflection conditions at the outflow boundary. A buffer domain or
a viscous sponge technique is often used there (Streett and Macaraeg, 1989;
Karniadakis and Triantafyllou, 1992; Mittal and Balachandar, 1996). The
idea is to parabolize the governing equations smoothly by multiplying the
diffusion term in (2.13) by a filter function:

$$f(r, \theta)\nabla^2 \mathbf{u}^* - \frac{2}{\nu \Delta t}\mathbf{u}^* = \text{RHS}. \qquad (4.35)$$

The filter function is defined in such a way that $f \to 1$ for most of the computational domain, but as the outflow boundary is approached, $f \to 0$ smoothly. Thus the diffusion term remains unaltered nearly everywhere except at the outflow boundary where the equation for $\mathbf{u}^*$ is parabolized. The method therefore does not require any outflow boundary condition and equation (4.35) can be solved to obtain the velocity at the boundary. Instead of filtering the entire viscous term one could filter just the radial component of the viscous term:

$$f(r,\theta)\nabla_r^2 \mathbf{u}^* + \nabla_{\theta,\phi}^2 \mathbf{u}^* - \frac{2}{\nu \Delta t}\mathbf{u}^* = \text{RHS}, \qquad (4.36)$$

where ∇_r^2 is the radial component of the Laplacian and the θ and ϕ components are together represented by $\nabla_{\theta,\phi}^2$. In this case a two-dimensional Helmholtz equation in θ and ϕ must be solved at the outer boundary, along with the inflow condition, in order to find the outflow velocity. Even for a time-dependent wake, if the mean wake location is axisymmetric, the filter function can be taken to be axisymmetric, i.e. dependent only on r and θ. Bagchi and Balachandar (2002c) have, for example, used the following wake filter function:

$$f(r,\theta) = 1 - \exp\left[-\nu_1 \left|\frac{r-R_0}{1/2-R_0}\right|^{\gamma_1}\right] \exp\left[-\nu_2 \left|\frac{\theta}{\pi}\right|^{\gamma_2}\right]. \qquad (4.37)$$

Large values for ν_1 and ν_2, of the order of 40, and values for γ_1 and γ_2 of the order of 4 were used to localize the filtered region close to the outflow outer boundary.

4.3.2.5 Spectral discretization

Here we will first describe a spatial discretization that has been used in several pseudospectral simulations of the flow around a sphere (Chang and Maxey, 1994; Mittal, 1999; Bagchi and Balachandar, 2002c). For a general discussion of spectral methods, the reader is referred to the books by Gottlieb and Orszag (1977), Canuto *et al.* (1988), Boyd (1989), and Fornberg (1996). The domain is taken to be a sphere with a large nondimensional radius, R_0, concentric with the particle. This facilitates the choice of spherical coordinate (r, θ, ϕ) system (Fig. 4.8a) over the domain

$$1/2 \le r \le R_0, \quad 0 \le \theta \le \pi, \quad \text{and} \quad 0 \le \phi \le 2\pi.$$

A Chebyshev expansion in the inhomogeneous radial direction is appropriate. The grid points along this direction are given by the Gauss–Lobatto

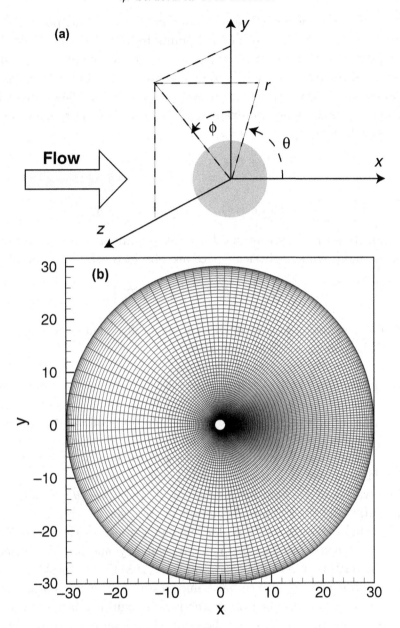

Fig. 4.8. Computational grid. (a) Schematic of the spherical coordinate and (b) a projection on the plane $\phi = 0$ of the computational grid.

collocation points, which are first defined on $[-1, 1]$ by

$$\xi_i = -\cos\left[\frac{\pi(i-1)}{N_r - 1}\right], \tag{4.38}$$

for $i = 1, 2, \ldots, N_r$, where N_r is the number of radial grid points. A grid mapping is then used to map the grid points from $[-1, 1]$ to $[1/2, R_0]$. The mapping can also be used to cluster points close to the sphere, in order to provide enhanced resolution of the complex flow features in that region. Typically an algebraic grid stretching is used (Canuto *et al.*, 1988). Bagchi and Balachandar (2002c) have, for example, used the following radial stretching in their simulations

$$\hat{\xi} = C_0 + C_1 \xi - C_0 \xi^2 + C_3 \xi^3, \tag{4.39}$$

$$C_1 = \frac{1}{2}(-\gamma_1 + 2C_0 + 3), \quad C_3 = \frac{1}{2}(\gamma_1 - 2C_0 - 1). \tag{4.40}$$

The parameters C_0 and γ_1 are used to vary the amount of stretching. The computational points in physical space are obtained using the mapping

$$r_i = \frac{1}{2}\hat{\xi}_i \left(\frac{1}{2} - R_0\right) + \frac{1}{2}\left(\frac{1}{2} + R_0\right). \tag{4.41}$$

A Fourier expansion is used in the azimuthal direction since ϕ is periodic over 2π. The collocation points in ϕ are

$$\phi_k = \frac{2\pi(k-1)}{N_\phi} \tag{4.42}$$

for $k = 1, 2, \ldots, N_\phi$, where N_ϕ is the number of grid points in ϕ. The natural range for the variable θ is between 0 and π. Therefore, in order to use a Fourier collocation in θ, it is necessary to add some symmetry restrictions. We should note that a scalar, the radial component of a vector, and the radial derivative of a scalar, are continuous at the poles ($\theta = 0$ and π), but the tangential and the azimuthal components of a vector, and the tangential and azimuthal derivatives of a scalar, change sign across the poles. This is the so-called 'parity' problem in spherical geometry, and has been discussed by Merilees (1973), Orszag (1974), and Yee (1981). The problem of pole parity does not arise if surface harmonics are used (Elcrat, Fornberg, and Miller, 2001). However, spectral methods using surface harmonics require $O(N)$ operations per mode, while those based on Fourier series require only $O(\log N)$ operations. For the study described here, a suitable Fourier expansion in θ is derived formally by following Shariff's (1993) approach. We start by considering a typical term in the expansion:

$$\begin{Bmatrix} c \\ u_r \\ u_\theta \end{Bmatrix} = \begin{Bmatrix} \alpha \\ \beta \\ \gamma \end{Bmatrix} r^p \exp(im\theta)\exp(ik\phi), \tag{4.43}$$

where c represents a scalar. The method requires that a scalar and the Cartesian components of a vector each independently be analytic at the poles. Such a requirement results in the following expansions:

$$c = \begin{cases} \sum \alpha_{pmk} T_p(r)\cos(m\theta)\exp(ik\phi) & \text{even} \quad k \\ \sum \alpha_{pmk} T_p(r)\sin(m\theta)\exp(ik\phi) & \text{odd} \quad k, \end{cases} \tag{4.44}$$

$$u_r = \begin{cases} \sum \beta_{pmk} T_p(r)\cos(m\theta)\exp(ik\phi) & \text{even} \quad k \\ \sum \beta_{pmk} T_p(r)\sin(m\theta)\exp(ik\phi) & \text{odd} \quad k, \end{cases} \tag{4.45}$$

and

$$u_\theta = \begin{cases} \sum \gamma_{pmk} T_p(r)\sin(m\theta)\exp(ik\phi) & \text{even} \quad k \\ \sum \gamma_{pmk} T_p(r)\cos(m\theta)\exp(ik\phi) & \text{odd} \quad k. \end{cases} \tag{4.46}$$

Here, T_p represents the p-th Chebyshev polynomial, $i = \sqrt{-1}$, and m and k are the wavenumbers in the θ- and ϕ-directions. α, β, and γ are the coefficients in the expansions and are functions of p, m, and k. The expansion for u_ϕ follows that of u_θ.

The collocation points in θ are normally equispaced:

$$\hat{\theta}_j = \frac{\pi}{N_\theta}\left[j - \frac{1}{2}\right], \tag{4.47}$$

for $j = 1, 2, ..., N_\theta$, where N_θ is the number of grid points in θ. However, for flow over a sphere the complexity, and therefore the resolution requirement, is far greater in the wake on the leeward side than on the windward side. A grid stretching that will cluster points in the wake region will make better use of the available grid points. One requirement, however, is that the grid stretching preserve the periodic nature of the θ-direction. Such a grid stretching has been suggested by Augenbaum (1989) and was employed in the recent simulation of Bagchi and Balachandar (2002). Here,

$$\theta_j = \tan^{-1}\left[\frac{\sin(\hat{\theta}_j)(1 - \hbar^2)}{\cos(\hat{\theta}_j)(1 + \hbar^2) - 2\hbar}\right], \tag{4.48}$$

where $\hbar$ is the parameter that controls the degree of stretching. A value of $\hbar = -0.35$ has been shown to provide sufficient resolution in the wake (Bagchi and Balachandar, 2002c). A projection on the plane $\phi = 0$ of a typical mesh with enhanced resolution in the wake through grid stretching is shown in Fig. 4.8(b). Such enhanced resolution has been widely employed.

Due to the topology of the grid, the azimuthal resolution is spatially nonuniform. The resolution is much finer near the poles compared to the equator. Furthermore, the azimuthal grid spacing increases linearly with increasing radial distance from the particle. The viscous time-step restriction, due to the smallest grid spacing, is avoided by the implicit treatment of the radial and azimuthal diffusion terms. However, the time-step size is still restricted by the convective stability (CFL) condition. Ideally, it is desirable to have high spatial resolution only in regions where the local flow structure demands it. A simple strategy to remove higher ϕ resolution in regions where it is not needed is to filter the high-frequency components. Such a filter necessarily has to be a function of both r and θ.

There are some constraints that must be satisfied by the pole filter. It must be sufficiently smooth in all its variables to preserve spectral convergence. The analytic nature of the scalar and vector fields requires that in the limit $\theta \to 0$ and π, only the azimuthal modes $k = 0$ and $k = \pm 1$ exist in the expansions (4.44)–(4.46). Physically, $k = 0$ is the axisymmetric mode and it does not contribute to the pole stability constraint. The $k = \pm 1$ modes are the most unstable ones and lead to bifurcation in the flow (Natarajan and Acrivos, 1993). Hence, these modes must be retained over the entire computational domain. From the CFL criterion, it can be inferred that as long as the ϕ-spectra of the velocity field decay faster than k^{-2}, the time step is dictated by the $k = \pm 1$ mode. Based on this observation, the following pole filter can be introduced

$$f_\phi = 1 - \exp\left[-\lambda_1 Y^{\lambda_2}\right], \tag{4.49}$$

where $Y = r \sin \theta$. Here λ_1 and λ_2 are functions of k and are determined by the conditions

$$f_\phi(k) = \frac{1}{k} \quad \text{at} \quad Y = Y_{\min} \qquad \text{and} \qquad f_\phi = 0.9 \quad \text{at} \quad Y = kY_{\min}. \tag{4.50}$$

$Y_{\min}$ is the value of Y at the grid point closest to the pole of the particle (note that in equation 4.47 the θ-discretization has been chosen to avoid the $\theta = 0$ and π points). Thus, the filter function attempts to achieve at least k^{-2} decay at the point closest to the pole. However, it approaches unity exponentially so that at a distance $kY_{\min}$ from the pole, f_ϕ approaches 0.9. Thus, the filtering is localized to a very small region near the poles of the particle. The size of the filtered region slowly increases with the azimuthal mode number. The filter is applied on the intermediate velocity field $\mathbf{u}^\star$ at the end of the advection–diffusion step.

4.3.2.6 Sample results

As an example of the use of a spectral method, we present here the results from an investigation of the interaction of isotropic turbulence with a single rigid spherical particle. The isotropic turbulent field was obtained from a separate computation by Langford (2000) using a 256^3 grid for a $(2\pi)^3$ box. The turbulent field is periodic in all three directions and hence can be extended to an arbitrarily large volume. If the Kolmogorov length and velocity scales η and v_k are chosen as the reference length and velocity scales, then some of the important measures of the isotropic turbulence are: root mean square of the turbulent velocity fluctuation $u_{\mathrm{rms}}/v_k = 6.5$, box size $L/\eta = 757.0$, and Taylor microscale $\lambda/\eta = 25.2$. The isotropic flow field is, however, characterized by a single parameter, the microscale Reynolds number $Re_\lambda = 164$.

A rigid spherical particle was placed in a uniform cross-flow, V, with an instantaneous realization of the isotropic turbulence field superimposed. Figure 4.9 shows the isotropic turbulence in a plane cutting through the domain. The small black solid dot is a sphere of diameter $d/\eta = 10$ drawn to scale. The larger circle represents the outer boundary of the computational domain. Figure 4.9 shows the relative size of the different length scales. As the box of isotropic turbulence is advected past the particle, the turbulent field is interpolated onto the inflow section of the outer boundary of the computational domain, giving the inflow conditions.

Here we employ a Fourier–Chebyshev spherical grid with grid clustering in the wake as described above and shown in Fig. 4.8. The grid resolution is chosen to satisfy two criteria: first, the size of the largest grid spacing in the spherical domain is less than that of the grid used to simulate the isotropic turbulence, in order to guarantee resolution of the freestream turbulence. Second, the grid size is adequate to resolve the thin shear layers and the wake structures generated by the particle. A typical grid used in the simulations has 141 points in the radial direction, 160 in the θ-direction and 128 in the ϕ-direction.

An important outcome of this study is a direct measurement of the effect of free-stream turbulence on the particle drag. The mean drag coefficient for a sphere, as a function of Re, obtained from a variety of experimental sources, is shown in Fig. 4.10. Also plotted in the figure, for reference, is the standard drag correlation for a stationary or steadily moving sphere in a steady uniform ambient flow, equation (8.71). The scatter in the experimental data clearly illustrates the degree of disagreement about the effect of turbulence. For example, in the moderate Reynolds number regime, the

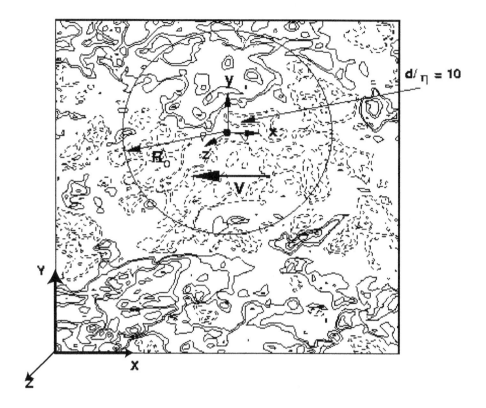

Fig. 4.9. Schematic of the particle flow configuration. Drawn to scale, a particle of $d/\eta = 10$ is shown here. The larger circle surrounding the particle represents the outer boundary of the spectral computational domain attached to the particle. The outer box represents the $(2\pi)^3$ box in which the isotropic turbulence was generated. From Bagchi and Balachandar (2003a).

measurements of Uhlherr and Sinclair (1970), Zarin and Nichols (1971), and Brucato, Grisafi, and Montante (1998) indicate a substantial increase in the drag coefficient in a turbulent flow. The numerical study by Mohd-Yusof (1996) also illustrated a drag increase of nearly 40% in a free-stream turbulence of 20% intensity. On the other hand, the results of Rudolff and Bachalo (1988) tend to suggest a reduction in the drag coefficient due to ambient turbulence. In contrast, Warnica, Renksizbulut and Strong (1994) suggest that the drag on a spherical liquid drop is not significantly different from the standard drag. The experiments of Wu and Faeth (1994a,b) also suggest little influence of turbulence on the mean drag.

The results of the present simulation are also shown in the figure. The scatter in the experimental data can be explained in terms of two basic

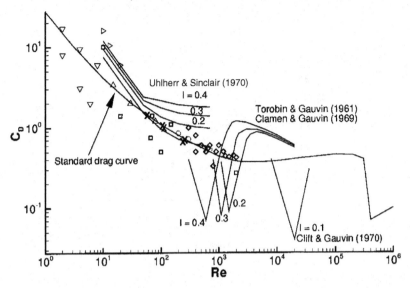

Fig. 4.10. A summary of the results on the effect of turbulence on the drag coefficient. × Present results; □ Gore and Crowe (1990); ◇ Sankagiri and Ruff (1997); ○ Zarin and Nichols (1971); △ Warnica *et al.* (1994); ∇ Rudolf and Bachalo (1988); ▷ Brucato *et al.* (1998). The standard drag curve is obtained using the Schiller–Neumann formula, equation 8.71 (see also Clift, Grace, and Weber, 1978). The parameter *I* is the ratio of the rms velocity of the free-stream turbulence to the mean relative velocity between the particle and the fluid. From Bagchi and Balachandar (2003a).

mechanisms that are invariably present in such experiments. First, it is now recognized that particles settling in turbulence prefer downwash regions to upwash regions and this inertial bias increases the settling velocity and thereby decrease the interpreted mean drag coefficient. Second, at finite Re the relation between drag and relative velocity is nonlinear. For a particle in an unsteady flow the increase in the drag coefficient for a given reduction in flow velocity is larger than the decrease if the velocity is increased by the same amount. A drag coefficient based on the average settling velocity is therefore likely to be larger than for a particle moving with the same velocity in a quiescent fluid. In the simulations discussed above the trajectory bias is clearly avoided and by carefully defining the mean drag coefficient the nonlinear effect is also eliminated. Thus the simulation results clearly demonstrate that free-stream turbulence has very little systematic inherent effect on the mean drag. Detailed information on fluctuating drag and lift forces and their modeling was also obtained (see Bagchi and Balachandar, 2003a).

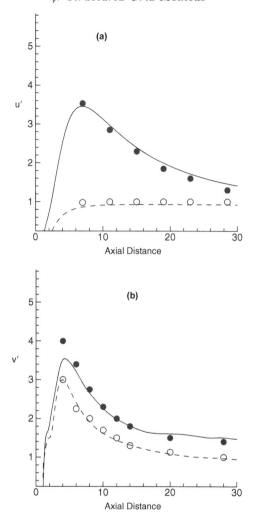

Fig. 4.11. Comparison of the DNS results with the experimental results of Wu and Faeth (1994a,b). Root-mean-square of streamwise (solid lines and filled symbols) and cross-stream (broken lines and open symbols) fluctuations along the axis of the sphere normalized by the freestream RMS are shown. Lines are the DNS results and symbols are experimental result. (a) $d/\eta = 1.5$, $I = 0.1$, $Re_p = 107$; (b) $d/\eta = 9.6$, $I = 0.1$, $Re_p = 610$. From Bagchi and Balachandar (2004).

The results from the simulations described above can also be used to investigate the effect of the particle on free-stream turbulence. Such information can provide valuable information about turbulence modulation due to the dispersed phase. Figure 4.11 compares the rms velocity fluctuation from the simulations with the experimental results of Wu and Faeth (1994a,b). Both the streamwise and the transverse velocity fluctuation statistics are shown for two different particle Reynolds numbers of $Re = 114$

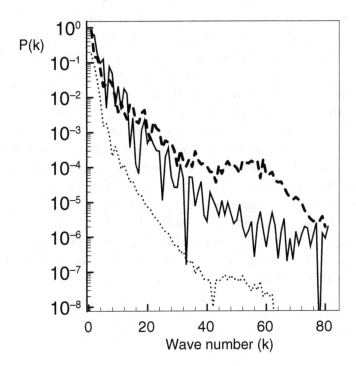

Fig. 4.12. Instantaneous velocity spectra along all three coordinate directions for the case of $d/\eta = 9.59$, $l = 0.1$, $Re_p = 610$. Solid line: radial spectra, dash line: θ-spectra, and dotted line: ϕ-spectra. From Bagchi and Balachandar (2003a).

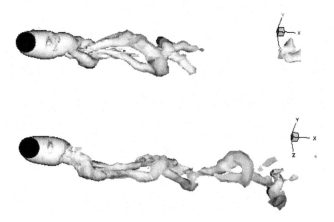

Fig. 4.13. 3D vortex topology in a free-stream turbulent flow at two different time instances. $d/\eta = 9.59$, $l = 0.1$, $Re_p = 610$. From Bagchi and Balachandar (2004).

and $Re = 610$. In the first case the particle diameter is $d/\eta = 1.5$, while in the second case $d/\eta = 9.59$ (Bagchi and Balachandar, 2004). It is clear that the results from the simulations are in good agreement with the experimental result, suggesting adequate resolution of the free-stream turbulence and the wake. For spectral methods the question of spatial resolution can be easily addressed by plotting the spectra along all three directions. For example, Fig. 4.12 shows the velocity spectra measured at a point within the separated shear layer slightly downstream of the sphere. A decay of six-to-nine orders of magnitude is observed in the radial, tangential, and azimuthal spectra. Similar investigations of the spectral decay at other critical points within the flow suggest adequate resolution over the entire domain. The effect of free-stream turbulence on the wake dynamics was investigated in detail and reported in Bagchi and Balachandar (2004). As an example, Fig. 4.13 shows the three-dimensional structure of the wake at $Re = 610$ for a sphere of $d/\eta = 9.59$ with a free-stream turbulence whose rms fluctuations are 10% of the relative velocity. At this Reynolds number the wake is unsteady, even in a uniform free-stream, but the presence of free-stream turbulence greatly enhances the shedding and the wake turbulence substantially increases.

5

Finite element methods for particulate flows

In Chapter 4 we described finite difference methods for the direct numerical simulations of fluid–solid systems in which the local flow field around moving particles is resolved numerically without modeling. Here we describe methods of the finite element type for the same purpose.

The first method is termed the ALE (arbitrary Lagrangian–Eulerian) particle mover. The ALE particle mover uses a technique based on a combined formulation of the fluid and particle momentum equations, together with an arbitrary Lagrangian–Eulerian moving unstructured finite element mesh technique to deal with the movement of the particles. It was first developed by Hu *et al.* (1992, 1996, 2001). The method has been used to solve particle motions in both Newtonian and viscoelastic fluids in two- and three-dimensional flow geometries. It also handles particles of different sizes, shapes, and materials.

The second method is based on a stabilized space-time finite element technique to solve problems involving moving boundaries and interfaces. In this method, the temporal coordinate is also discretized using finite elements. The deformation of the spatial domain with time is reflected simply in the deformation of the mesh in the temporal coordinate. Using this technique, Johnson and Tezduyar (1999) were able to simulate the sedimentation of 1000 spheres in a Newtonian fluid at a Reynolds number of 10.

The third numerical method is termed the DLM (distributed Lagrange multiplier) particle mover. The basic idea of the DLM particle mover is to extend the problem from a time-dependent geometrically complex domain to a stationary fictitious domain, which is larger but simpler. On this fictitious domain, the constraints of the rigid-body motion of the particles are enforced using a distributed Lagrange multiplier. The DLM particle mover was recently introduced by Glowinski *et al.* (1999), and has been extended

113

to handle viscoelastic fluids. It has been used to simulate the sedimentation and fluidization of over 1000 spheres in Newtonian fluids.

5.1 Governing equations

In a system of solid rigid particles suspended in a fluid, the motion of the fluid and that of the particles are fully coupled. The motion of the particles is determined by the hydrodynamic forces and moments exerted on them by the surrounding fluid. In turn, the fluid motion is strongly influenced by the particle motion and, as in the case of sedimentation, even entirely driven by the particle motion. In most applications, the Reynolds number of the flow based on the particle size is usually not small, and therefore the inertia of both the fluid and the particles has to be accounted for in the model. We start by recapitulating in this section the governing equations.

Let us consider the motion of N rigid solid objects (particles) in an incompressible fluid. Denote by $\Omega_0(t)$ the domain occupied by the fluid at a given time instant $t \in [0, T]$, and $\Omega_\alpha(t)$ the domain occupied by the α-th particle $(\alpha = 1, 2, \ldots, N)$, with $P(t) = \cup \Omega_\alpha(t)$. The boundaries of $\Omega_0(t)$ and $\Omega_\alpha(t)$ are denoted by $\partial\Omega_0(t)$ and $\partial\Omega_\alpha(t)$, respectively. The governing equations for the fluid motion in $\Omega_0(t)$ are the conservation of mass, equation (1.1)

$$\nabla \cdot \mathbf{u} = 0, \tag{5.1}$$

and momentum, equation (1.2).

$$\rho\frac{d\mathbf{u}}{dt} = \rho\frac{\partial\mathbf{u}}{\partial t} + \rho\left(\mathbf{u} \cdot \nabla\right)\mathbf{u} = \rho\mathbf{f} + \nabla \cdot \boldsymbol{\sigma}. \tag{5.2}$$

The stress tensor is decomposed into a pressure and a viscous part. For a Newtonian fluid, the latter has the simple form given in (1.4). For a viscoelastic fluid, the stress tensor may be expressed as

$$\boldsymbol{\sigma} = -p\boldsymbol{I} + 2\mu_2\mathbf{e}[\mathbf{u}] + \boldsymbol{\tau}_{\mathrm{p}}, \tag{5.3}$$

where $\boldsymbol{\tau}_{\mathrm{p}}$ is the "polymer contribution" to the stress and may be governed by a constitutive equation like an Oldroyd-B fluid model (see, e.g. Bird *et al.*, 1987):

$$\lambda\left(\frac{d\boldsymbol{\tau}_{\mathrm{p}}}{dt} - (\nabla\mathbf{u})^{\mathrm{T}} \cdot \boldsymbol{\tau}_{\mathrm{p}} - \boldsymbol{\tau}_{\mathrm{p}} \cdot \nabla\mathbf{u}\right) + \boldsymbol{\tau}_{\mathrm{p}} = 2\mu_1\mathbf{e}[\mathbf{u}]. \tag{5.4}$$

The parameter λ is the fluid relaxation time and $\mu = \mu_1 + \mu_2$ is the fluid viscosity. The Newtonian fluid can be considered as a special case with $\mu_2 = \mu$ and $\boldsymbol{\tau}_{\mathrm{p}} = 0$. In general, the viscosity and relaxation time of the fluid are functions of the local shear rate of the flow; for example, viscosity laws like Bird–Carreau, power-law, Bingham, and Herschel–Bulkley may apply (see e.g. Bird *et al.*, 1987).

Rigid particles satisfy Newton's second law for the translational motion, equation (1.25)

$$m_\alpha \frac{d\mathbf{v}_\alpha}{dt} = \mathbf{F}_\alpha^e + \mathbf{F}_\alpha^h = \mathbf{F}_\alpha^e + \int_{\partial\Omega_\alpha(t)} \boldsymbol{\sigma} \cdot \mathbf{n}\, dS, \tag{5.5}$$

where $\mathbf{F}^e$ is a constant driving force such as the particle weight. The Euler equation (1.26) for the angular velocity, which we rewrite omitting the non-hydrodynamic couple is,

$$\frac{d(J_\alpha \boldsymbol{\Omega}_\alpha)}{dt} = \mathbf{L}_\alpha^h = \int_{\partial\Omega_\alpha(t)} (\mathbf{x} - \mathbf{X}_\alpha) \times (\boldsymbol{\sigma} \cdot \mathbf{n})\, dS. \tag{5.6}$$

In these equations, the unit normal $\mathbf{n}$ points out of the particle. The position $\mathbf{X}_\alpha$ of the center of mass and the orientation $\boldsymbol{\Theta}_\alpha$ of the α-th particle are updated according to equations (1.29).

The boundary of the fluid domain, $\partial\Omega_0(t)$, can be decomposed into three nonoverlapping sections, $\partial\Omega_u$, $\partial\Omega_\sigma$, and $\partial\Omega_\alpha$, each one with a different type of boundary condition:

$$\mathbf{u} = \mathbf{u}_g, \quad \mathbf{x} \in \partial\Omega_u, \tag{5.7}$$

$$\boldsymbol{\sigma} \cdot \mathbf{n} = 0, \quad \mathbf{x} \in \partial\Omega_\sigma, \tag{5.8}$$

$$\mathbf{u} = \mathbf{v}_\alpha + \boldsymbol{\Omega}_\alpha \times (\mathbf{x} - \mathbf{X}_\alpha), \quad \mathbf{x} \in \partial\Omega_\alpha, \tag{5.9}$$

where $\mathbf{u}_g$ is a prescribed fluid velocity. Expression (5.9) represents the no-slip condition on the particle surface. For viscoelastic materials, conditions on the elastic stress are normally imposed on the inflow boundary $\partial\Omega_{in}$,

$$\boldsymbol{\tau}_p = \boldsymbol{\tau}_{in}, \quad \mathbf{x} \in \partial\Omega_{in}, \tag{5.10}$$

in which $\boldsymbol{\tau}_{in}$ is the prescribed stress.

At $t = 0$, the initial conditions for the particle variables are

$$\mathbf{X}_\alpha(0) = \mathbf{X}_\alpha^0, \ \boldsymbol{\Theta}_\alpha(0) = \boldsymbol{\Theta}_\alpha^0, \ \mathbf{v}_\alpha(0) = \mathbf{v}_\alpha^0, \ \boldsymbol{\Omega}_\alpha(0) = \boldsymbol{\Omega}_\alpha^0, \ \alpha = 1, 2, \ldots, N, \tag{5.11}$$

and, for the velocity and stress fields,

$$\mathbf{u} = \mathbf{u}_0, \qquad \boldsymbol{\tau} = \boldsymbol{\tau}_0, \tag{5.12}$$

where the initial velocity, $\mathbf{u}_0$, should be divergence-free.

5.2 Fully explicit scheme and its stability

A simple approach to numerically simulate the fluid–particle motion is to decouple the motion of the fluid and solid at each time step. A fully explicit scheme would be the following:

Scheme 1. Fully explicit scheme

Initialization: $t^0 = 0$ and $n = 0$ (n is the index for time step)

Initialize $\mathbf{u}(\mathbf{x}, 0)$, and $\mathbf{v}_\alpha(0)$, $\boldsymbol{\Omega}_\alpha(0)$, $\mathbf{X}_\alpha(0)$, $\boldsymbol{\Theta}_\alpha(0)$ for
$\alpha = 1, 2, \ldots, N$.

Do $n = 0, 1, 2, \ldots, M$ (over time steps)

(i) Select time step Δt^{n+1}: $t^{n+1} = t^n + \Delta t^{n+1}$.

(ii) Update particle positions and orientations $\mathbf{X}_\alpha(t^{n+1})$, $\boldsymbol{\Theta}_\alpha(t^{n+1})$, using the particle velocity $\mathbf{v}_\alpha(t^n), \boldsymbol{\Omega}_\alpha(t^n)$.

(iii) Discretize the flow domain around the new particle positions.

(iv) Solve the flow field $\mathbf{u}(\mathbf{x}, t^{n+1})$ and $p(\mathbf{x}, t^{n+1})$ by a conventional numerical method, using the particle velocities $\mathbf{v}_\alpha(t^n), \boldsymbol{\Omega}_\alpha(t^n)$ as boundary conditions.

(v) Calculate the hydrodynamic forces $\mathbf{F}_\alpha^{\mathrm{h}}(t^{n+1})$ and moments $\mathbf{L}_\alpha^{\mathrm{h}}(t^{n+1})$ acting on the particles, using the updated flow field.

(vi) Update the particle velocities $\mathbf{v}_\alpha(t^{n+1})$, $\boldsymbol{\Omega}_\alpha(t^{n+1})$, using the computed forces and moments.

End-Do

This scheme is simple and easy to implement but, unfortunately, it can be unstable under certain circumstances (see Hu *et al.*, 1992). To illustrate this point, consider the initial stage of the motion of a particle accelerating from rest in an infinite expanse of quiescent fluid. In the early stages of the motion, the particle velocity is very small, and the dominant hydrodynamic force on the particle is the virtual (or added) mass force caused by the acceleration of the fluid surrounding the particle. The translational motion of the particle then takes the form

$$m\frac{d\mathbf{v}}{dt} = \mathbf{F}^{\mathrm{e}} + \mathbf{F}^{\mathrm{h}} \approx \mathbf{F}^{\mathrm{e}} - m_{\mathrm{v}}\frac{d\mathbf{v}}{dt} \tag{5.13}$$

where m_{v} is the virtual mass. By using the fully explicit scheme described above, the hydrodynamic force on the particle is calculated based on the particle velocity at the previous time step. A backward Euler finite difference discretization of (5.13) with an equal time increment Δt leads to

$$m\frac{\mathbf{v}(t^{n+1}) - \mathbf{v}(t^n)}{\Delta t} = \mathbf{F}^{\mathrm{e}} - m_{\mathrm{v}}\frac{\mathbf{v}(t^n) - \mathbf{v}(t^{n-1})}{\Delta t}. \tag{5.14}$$

This finite difference equation can be solved with the result

$$\mathbf{v}(t^{n+1}) - \mathbf{v}(t^n) = \frac{2\mathbf{F}^e}{m + m_v}\Delta t \left[1 - \left(-\frac{m_v}{m}\right)^n\right] + \left(-\frac{m_v}{m}\right)^n \left[\mathbf{v}(t^1) - \mathbf{v}(t^0)\right].$$

(5.15)

Obviously, whenever the added mass is larger than the mass of the particle, $m_v \geq m$, the particle velocity oscillates with increasingly large amplitude as the computation proceeds. Therefore, this scheme is not stable. This result is general, and does not depend on the backward Euler finite difference discretization of (5.13). The actual value of the virtual mass, m_v, depends on the particle shape and flow geometry. For a spherical particle in an infinite medium, the virtual mass equals one-half the mass of the fluid displaced by the particle. However, if the same sphere is moving through a tightly fitting tube, the value of the virtual mass is much higher, since the sphere needs to accelerate more fluid both ahead and behind itself. Therefore, the condition of $m_v \geq m$ could be encountered, for example, in the motion of light particles in a fluid, or of heavy particles in confined geometries. To avoid the stability problem of the fully explicit scheme, a coupled scheme for solving the flow field and particle velocities at a given time step is needed. For example, in an implicit scheme, the solution of the flow field and the forces and moments acting on the particles could be determined iteratively with the velocities of the particles at the same time instant. However, it is more efficient to treat the fluid and the solid particles as a single system and generate a combined formulation for this system.

5.3 Combined fluid–solid formulation

In fluid–particle systems, due to the complex, irregular nature of the domain occupied by the fluid, finite element techniques are particularly powerful in the discretization of the governing fluid equations. In order to apply the finite element method, we first seek a weak formulation that incorporates both the fluid and particle momentum equations (5.2), (5.5), and (5.6).

Introduce a function space $\mathcal{V}$,

$$\mathcal{V} = \left\{ \begin{array}{l} \mathbf{U} \equiv (\mathbf{u}, \mathbf{v}_1, \ldots, \mathbf{v}_N, \mathbf{\Omega}_1, \ldots, \mathbf{\Omega}_N) \mid \mathbf{u} \in \mathcal{H}^1(\Omega_0)^3, \mathbf{v}_\alpha \in R^3, \mathbf{\Omega}_\alpha \in R^3; \\ \mathbf{u} = \mathbf{v}_\alpha + \mathbf{\Omega}_\alpha \times (\mathbf{x} - \mathbf{X}_\alpha), \ \mathbf{x} \in \partial\Omega_\alpha(t); \mathbf{u} = \mathbf{u}_g, \ \mathbf{x} \in \partial\Omega_u \end{array} \right\}.$$

(5.16)

where $\mathcal{H}^1(\Omega_0)^3$ is the functional space which consists of all functions that have square-integrable first order derivatives for the three-dimensional velocity field in the fluid, and R^3 stands for each space of the particle velocities (three translational and three angular velocity components per particle). The space $\mathcal{V}$ is a natural space for the velocity of the fluid–solid mixture.

The space for the pressure is chosen as the square-integrable space $L^2(\Omega_0)$ with a zero value at a fixed point in the domain, and is denoted by Q:

$$Q = \left\{ p \mid p \in L^2(\Omega_0); \ p(\mathbf{x}_0) = 0 \right\}. \tag{5.17}$$

Similarly, the space for the elastic polymer stress tensor is selected to be $L^2(\Omega_0)^6$, which represents six independent components, and is denoted by $\mathcal{T}$,

$$\mathcal{T} = \left\{ \boldsymbol{\tau} \mid \boldsymbol{\tau} \in L^2(\Omega_0)^6; \ \boldsymbol{\tau} = \boldsymbol{\tau}_{\text{in}}, \ \mathbf{x} \in (\partial\Omega)_{\text{in}} \right\}. \tag{5.18}$$

To derive the weak formulation of the combined fluid and particle equations of motion, we consider a test function, the variation of $\mathbf{U}$,

$$\tilde{\mathbf{U}} = \left(\tilde{\mathbf{u}}, \tilde{\mathbf{v}}_1, \ldots, \tilde{\mathbf{v}}_N, \tilde{\boldsymbol{\Omega}}_1, \ldots, \tilde{\boldsymbol{\Omega}}_N \right) \in \mathcal{V}_0. \tag{5.19}$$

belonging to the variational space $\mathcal{V}_0$ which is the same as $\mathcal{V}$, except that $\tilde{\mathbf{u}} = 0$ on $\partial\Omega_u$. We shall define the variational space $\mathcal{T}_0$ to be the same as $\mathcal{T}$, except that $\boldsymbol{\tau} = \mathbf{0}$ on $\partial\Omega_{\text{in}}$. Multiplying the fluid momentum equation (5.2) by the test function for the fluid velocity, $\tilde{\mathbf{u}}$, and integrating over the fluid domain at a time instant t, we have

$$\int_{\Omega_0} \rho \left(\frac{d\mathbf{u}}{dt} - \mathbf{f} \right) \cdot \tilde{\mathbf{u}} \, dV + \int_{\Omega_0} (\boldsymbol{\sigma} : \nabla\tilde{\mathbf{u}}) \, dV + \sum_{\alpha=1}^{N} \int_{\partial\Omega_\alpha} (\boldsymbol{\sigma} \cdot \mathbf{n}) \cdot \tilde{\mathbf{u}} \, dS = 0. \tag{5.20}$$

On the particle surface, the test function for the fluid velocity will be replaced by the test functions for the particle velocity according to the no-slip condition (5.9), or

$$\int_{\partial\Omega_\alpha} (\boldsymbol{\sigma} \cdot \mathbf{n}) \cdot \tilde{\mathbf{u}} \, dS = \tilde{\mathbf{v}}_\alpha \cdot \int_{\partial\Omega_\alpha} (\boldsymbol{\sigma} \cdot \mathbf{n}) \, dS + \tilde{\boldsymbol{\Omega}}_\alpha \cdot \int_{\partial\Omega_\alpha} (\mathbf{x} - \mathbf{X}_\alpha) \times (\boldsymbol{\sigma} \cdot \mathbf{n}) \, dS. \tag{5.21}$$

Furthermore, using the equations of motion for the particles (5.5) and (5.6), we obtain

$$\int_{\partial\Omega_\alpha} (\boldsymbol{\sigma} \cdot \mathbf{n}) \cdot \tilde{\mathbf{u}} dS = \tilde{\mathbf{v}}_\alpha \cdot \left(m_\alpha \frac{d\mathbf{v}_\alpha}{dt} + \mathbf{F}_\alpha^{\text{e}} \right) + \tilde{\boldsymbol{\Omega}}_\alpha \cdot \frac{d(J_\alpha \boldsymbol{\Omega}_\alpha)}{dt}. \tag{5.22}$$

Substituting (5.22) into (5.20), we find the combined fluid–particle momentum equation:

$$\int_{\Omega_0} \rho \left(\frac{d\mathbf{u}}{dt} - \mathbf{f} \right) \cdot \tilde{\mathbf{u}} dV + \int_{\Omega_0} \boldsymbol{\sigma} : \mathbf{e}[\tilde{\mathbf{u}}] dV$$

$$+ \sum_\alpha \tilde{\mathbf{v}}_\alpha \cdot \left(m_\alpha \frac{d\mathbf{v}_\alpha}{dt} - \mathbf{F}_\alpha^{\text{e}} \right) + \sum_\alpha \tilde{\boldsymbol{\Omega}}_\alpha \cdot \frac{d(J_\alpha \boldsymbol{\Omega}_\alpha)}{dt} = 0. \tag{5.23}$$

The stress tensor $\boldsymbol{\sigma}$ in (5.23) can be replaced by its constitutive relation. The weak formulations for the mass conservation (5.1) and the constitutive equation (5.4) can also be similarly obtained by multiplying their corresponding test functions and integrating over the fluid domain.

In summary, the weak formulation for the combined equations of the fluid–particle system is the following:

Find $(\mathbf{U}, p, \boldsymbol{\tau}_{\mathrm{p}}) \in \mathcal{V} \times \mathcal{Q} \times \mathcal{T}$ such that, $\forall \tilde{\mathbf{U}} \in \mathcal{V}_0$, $\forall \tilde{p} \in \mathcal{Q}$, and $\forall \tilde{\boldsymbol{\tau}} \in \mathcal{T}_0$:

$$
\int_{\Omega_0} \rho \left(\frac{d\mathbf{u}}{dt} - \mathbf{f} \right) \cdot \tilde{\mathbf{u}} \, dV - \int_{\Omega_0} p(\nabla \cdot \tilde{\mathbf{u}}) \, dV
$$

$$
+ \int_{\Omega_0} (2\mu_2 \mathbf{e}[\mathbf{u}] + \boldsymbol{\tau}_{\mathrm{p}}) : \mathbf{e}[\tilde{\mathbf{u}}] \, dV + \sum_{\alpha} \left(m_\alpha \frac{d\mathbf{v}_\alpha}{dt} - \mathbf{F}_\alpha^{\mathrm{e}} \right) \cdot \tilde{\mathbf{v}}_\alpha
$$

$$
+ \sum_{\alpha} \frac{d(J_\alpha \boldsymbol{\Omega}_\alpha)}{dt} \cdot \tilde{\boldsymbol{\Omega}}_\alpha = 0, \tag{5.24}
$$

$$
\int_{\Omega_0} (\nabla \cdot \mathbf{u}) \tilde{p} \, dV = 0, \tag{5.25}
$$

$$
\int_{\Omega_0} \left[\lambda \left(\frac{d\boldsymbol{\tau}_{\mathrm{p}}}{dt} - (\nabla \mathbf{u})^T \cdot \boldsymbol{\tau}_{\mathrm{p}} - \boldsymbol{\tau}_{\mathrm{p}} \cdot \nabla \mathbf{u} \right) + \boldsymbol{\tau}_{\mathrm{p}} - 2\mu_1 \mathbf{e}[\mathbf{u}] \right] : \tilde{\boldsymbol{\tau}} \, dV = 0. \tag{5.26}
$$

It may be noted that in the combined momentum equation (5.24) for the fluid–particle system, the hydrodynamic forces and moments acting on the particles do not explicitly appear. This feature arises naturally, since these forces are internal to the system when the fluid and the solid particles are considered together. The advantage of this combined formulation is that the hydrodynamic forces and moments need not be explicitly computed. More importantly, the schemes based on this formulation are not subject to the numerical instability described in Section 5.2.

5.4 Arbitrary Lagrangian–Eulerian (ALE) mesh movement

As we are interested in a large number of solid particles moving freely in the fluid, the domain occupied by the fluid is irregular and changes with time. To handle the movement of the domain, an arbitrary Lagrangian–Eulerian (ALE) technique – in which the time-dependent physical domain of the fluid motion is mapped onto a fixed computational domain – is very useful. In this section we shall describe this technique and discuss some methods for the control of the mesh movement. A general kinematic theory for the ALE technique was originally introduced by Hughes *et al.* (1981).

The mapping of the points $\mathbf{x}$ of the physical domain $\Omega_0(t)$ onto the points $\boldsymbol{\xi}$ of the computational domain is expressed by a relation of the type $\mathbf{x} = \mathbf{x}(\boldsymbol{\xi}, t)$. The time derivative of this relation keeping $\boldsymbol{\xi}$ fixed gives the velocity $\hat{\mathbf{u}}(\mathbf{x}, t)$ of the points of the computational domain as it appears in the physical domain:

$$\frac{\delta}{\delta t}\mathbf{x}(\boldsymbol{\xi}, t) \equiv \left(\frac{\partial \mathbf{x}}{\partial t}\right)_{\boldsymbol{\xi}} = \hat{\mathbf{u}}. \tag{5.27}$$

Since it will be convenient to think of the discretization as carried out on the computational domain, we refer to $\hat{\mathbf{u}}$ as the *mesh velocity*. In general, this mesh velocity is only constrained at the boundaries of the fluid domain: it has to follow the motion of the particles and the motion of the confining flow geometry. In the interior of the domain, however, it is largely arbitrary.

The material time derivative of the velocity at a given physical point $\mathbf{x}$ at the time instant t becomes

$$\frac{d}{dt}\mathbf{u}(\mathbf{x}, t) = \frac{\partial \mathbf{u}}{\partial t} + (\mathbf{u} \cdot \nabla)\,\mathbf{u} = \frac{\delta \mathbf{u}}{\delta t} + [(\mathbf{u} - \hat{\mathbf{u}}) \cdot \nabla]\,\mathbf{u}, \tag{5.28}$$

where

$$\frac{\delta \mathbf{u}}{\delta t} = \left(\frac{\partial}{\partial t}\mathbf{u}\,(\mathbf{x}(\boldsymbol{\xi}, t), t)\right)_{\boldsymbol{\xi}}, \tag{5.29}$$

is the time derivative keeping the coordinate $\boldsymbol{\xi}$ in the computational domain fixed[1].

If the computational domain coincides with the physical domain, $\boldsymbol{\xi} = \mathbf{x}$, and is fixed in time, we have $\hat{\mathbf{u}} = 0$, and the time derivative (5.29) reduces to the local Eulerian time derivative. If the mesh velocity coincides with the velocity of the material particles, $\hat{\mathbf{u}} = \mathbf{u}$, and the time derivative (5.29) reduces to the Lagrangian (or material) time derivative.

If the deformation of the domain is prescribed, or somewhat predictable, the mesh velocity in the interior can simply be expressed as an algebraic function of the motion of the nodes at the boundary, such as the ones used in Huerta and Liu (1988) and Nomura and Hughes (1992). For more complicated motions of the particles, the mesh motion in the interior of the fluid can be assumed to satisfy an elliptic partial differential equation, such as Laplace's equation, to guarantee a smooth variation:

$$\nabla \cdot (k^e \nabla \hat{\mathbf{u}}) = 0 \tag{5.30}$$

where k^e is a function introduced to control the deformation of the domain

---

[1] A simple interpretation of (5.28) is that, during dt, a point of the computational domain has moved by $\hat{\mathbf{u}}dt$. Thus, in order to take $\partial/\partial t$ with $\mathbf{x}$ fixed, one has to "backtrack" by the same amount.

in such a way that the region away from the particles absorbs most of the deformation, while the region next to the particles is relatively stiff and retains its shape better. This mesh movement scheme was used by Hu (1996). It should be noted that the three components of the mesh velocity $\hat{\mathbf{u}}$ are not coupled and can be solved separately. The boundary conditions for the mesh velocity on the particle surface are

$$\hat{\mathbf{u}} = \mathbf{v}_\alpha + \boldsymbol{\Omega}_\alpha \times (\mathbf{x} - \mathbf{X}_\alpha), \quad \mathbf{x} \in \partial\Omega_\alpha, \quad \alpha = 1, 2, \ldots, N. \tag{5.31}$$

It is possible to use different boundary conditions for certain flow problems. For example, with circular or spherical particles, the mesh can be allowed to slip on the particle surface. In this case, the nodes on the particle surface move with the particle translational velocity, but do not need to rotate with the particle.

Similarly, if the particles are accelerating, a mesh acceleration field, $\hat{\mathbf{a}}(\mathbf{x}, t)$, can be defined as

$$\nabla \cdot (k^e \nabla \hat{\mathbf{a}}) = 0 \tag{5.32}$$

with the boundary conditions on the particle surfaces given by

$$\hat{\mathbf{a}} = \dot{\mathbf{v}}_\alpha + \dot{\boldsymbol{\Omega}}_\alpha \times (\mathbf{x} - \mathbf{X}_\alpha) + \boldsymbol{\Omega}_\alpha \times \boldsymbol{\Omega}_\alpha \times (\mathbf{x} - \mathbf{X}_\alpha), \quad \mathbf{x} \in \partial\Omega_\alpha, \quad \alpha = 1, 2, \ldots, N \tag{5.33}$$

where $\dot{\mathbf{v}}_\alpha = d\mathbf{v}_\alpha/dt$ and $\dot{\boldsymbol{\Omega}}_\alpha = d\boldsymbol{\Omega}_\alpha/dt$. This mesh acceleration field is used when a second-order explicit scheme is needed to discretize equation (5.27) for the mesh movement.

A similar mesh movement scheme was described by Johnson and Tezduyar (1995). In their implementation, the domain is modeled as a linear elastic solid, and therefore the equations of linear elasticity are used to calculate the mesh velocity in the interior of the domain on the basis of the given boundary deformation. Thus, in that scheme, the components of the mesh displacement are coupled and have to be solved together. They also used a variable stiffness coefficient to control the mesh deformation in such a way that most of the mesh deformation was absorbed by the larger elements while the small elements were stiffer and better maintained their shape.

The weak formulations for the equations for the mesh velocity (5.30) and the mesh acceleration (5.32) can be written as:

Find $\hat{\mathbf{u}} \in \mathcal{V}_{m1}$ and $\hat{\mathbf{a}} \in \mathcal{V}_{m2}$, such that $\forall \tilde{\hat{\mathbf{u}}} \in \mathcal{V}_{m0}$ and $\forall \tilde{\hat{\mathbf{u}}} \in \mathcal{V}_{m0}$

$$\int_{\Omega_0} \left(k^e \nabla \hat{\mathbf{u}} \cdot \nabla \tilde{\hat{\mathbf{u}}} \right) dV = 0, \qquad \int_{\Omega_0} \left(k^e \nabla \hat{\mathbf{a}} \cdot \nabla \tilde{\hat{\mathbf{a}}} \right) dV = 0, \tag{5.34}$$

where the function spaces are defined as

$$\mathcal{V}_{m1} = \left\{ \hat{\mathbf{u}} \in \mathcal{H}^1(\Omega_0)^3; \ \hat{\mathbf{u}} = \mathbf{v}_\alpha + \boldsymbol{\Omega}_\alpha \times (\mathbf{x} - \mathbf{X}_\alpha), \ \mathbf{x} \in \partial\Omega_\alpha(t) \right\}, \tag{5.35}$$

$$\mathcal{V}_{m2} = \{\hat{\mathbf{a}} \in \mathcal{H}^1(\Omega_0)^3; \ \hat{\mathbf{a}} = \dot{\mathbf{v}}_\alpha + \dot{\boldsymbol{\Omega}}_\alpha \times (\mathbf{x} - \mathbf{X}_\alpha)$$

$$+ \boldsymbol{\Omega}_\alpha \times \boldsymbol{\Omega}_\alpha \times (\mathbf{x} - \mathbf{X}_\alpha), \quad \mathbf{x} \in \partial\Omega_\alpha(t) \}, \tag{5.36}$$

and

$$\mathcal{V}_{m0} = \{\hat{\mathbf{u}} \in \mathcal{H}^1(\Omega_0)^3; \ \hat{\mathbf{u}} = 0, \ \mathbf{x} \in \partial\Omega_\alpha(t)\}. \tag{5.37}$$

5.5 Temporal discretization – finite difference scheme

Due to the special nature of the temporal coordinate, the time derivatives in the system of equations are usually discretized by finite difference methods simpler than those used for the spatial discretization. In this section, we will introduce a finite difference scheme to replace the time derivatives in the combined fluid–particle system of equations (5.24)–(5.26). We shall use a fully implicit discretization in which all the terms in equations (5.24)–(5.26) are evaluated at the time instant $t = t^{n+1}$. First, the time derivative in (5.29) can be discretized as

$$\frac{\delta \mathbf{u}}{\delta t}(\mathbf{x}, t^{n+1}) = a \frac{\mathbf{u}(\mathbf{x}, t^{n+1}) - \mathbf{u}(\bar{\mathbf{x}}, t^n)}{\Delta t} - b \frac{\delta \mathbf{u}}{\delta t}(\bar{\mathbf{x}}, t^n) \tag{5.38}$$

where $\Delta t = t^{n+1} - t^n$ is the time step, and the mesh nodes are moved explicitly according to an integrated version of (5.27):

$$\mathbf{x} = \bar{\mathbf{x}} + \Delta t \, \hat{\mathbf{u}}(\bar{\mathbf{x}}, t^n) + \frac{1}{2}\Delta t^2 \hat{\mathbf{a}}(\bar{\mathbf{x}}, t^n). \tag{5.39}$$

The approximation in (5.38) is first–order accurate in time when $a = 1$, $b = 0$. It can be improved to second-order accuracy by selecting $a = 2, b = 1$, which is a variation of the well-known Crank–Nicholson scheme. With this step, the material time derivative (5.28) can be written as

$$\frac{d\mathbf{u}}{dt}(\mathbf{x}, t^{n+1}) = a\frac{\mathbf{u}(\mathbf{x}, t^{n+1}) - \mathbf{u}(\bar{\mathbf{x}}, t^n)}{\Delta t} - b\frac{\delta \mathbf{u}}{\delta t}(\bar{\mathbf{x}}, t^n)$$

$$+ \left[(\mathbf{u}(\mathbf{x}, t^{n+1}) - \hat{\mathbf{u}}(\mathbf{x}, t^{n+1})) \cdot \nabla\right] \mathbf{u}(\mathbf{x}, t^{n+1}). \tag{5.40}$$

Similarly, the time derivatives of the particle velocities in (5.5) and (5.6) can be discretized as

$$\frac{d\mathbf{v}_\alpha}{dt}(t^{n+1}) = a \frac{\mathbf{v}_\alpha^{n+1} - \mathbf{v}_\alpha^n}{\Delta t} - b \frac{d\mathbf{v}_\alpha}{dt}(t^n), \tag{5.41}$$

and

$$\frac{d(J_\alpha \boldsymbol{\Omega}_\alpha)}{dt}(t^{n+1}) = a \frac{(J_\alpha \boldsymbol{\Omega}_\alpha)^{n+1} - (J_\alpha \boldsymbol{\Omega}_\alpha)^n}{\Delta t} - b\frac{d(J_\alpha \boldsymbol{\Omega}_\alpha)}{dt}(t^n). \tag{5.42}$$

However, it is sufficient to discretize the equations (1.29) for the particle positions and orientations using an explicit finite difference scheme:

$$\mathbf{X}_\alpha^{n+1} = \mathbf{X}_\alpha^n + \Delta t \mathbf{v}_\alpha^n + \frac{1}{2}\Delta t^2 \dot{\mathbf{v}}_\alpha^n, \tag{5.43}$$

$$\mathbf{\Theta}_\alpha^{n+1} = \mathbf{\Theta}_\alpha^n + \Delta t \mathbf{\Omega}_\alpha^n + \frac{1}{2}\Delta t^2 \dot{\mathbf{\Omega}}_\alpha^n. \tag{5.44}$$

As mentioned earlier, in the weak formulations of (5.24), (5.25), and (5.26), the spatial domain and all the functions in the integrals are evaluated at the time instant $t = t^{n+1}$. The time derivatives in (5.24) and (5.26) are replaced by expressions like (5.40) to find

$$\int_{\Omega^{n+1}} \rho \left(\frac{a}{\Delta t}\mathbf{u} + (\mathbf{u} - \hat{\mathbf{u}}) \cdot \nabla \mathbf{u} \right) \cdot \tilde{\mathbf{u}}\, dV - \int_{\Omega^{n+1}} p(\nabla \cdot \tilde{\mathbf{u}})\, dV$$

$$+ \int_{\Omega^{n+1}} (2\mu_2 \mathbf{e}[\mathbf{u}] + \boldsymbol{\tau}_\mathrm{p}) : \mathbf{e}[\tilde{\mathbf{u}}]\, dV + \frac{a}{\Delta t} \sum_\alpha m_\alpha \mathbf{v}_\alpha \cdot \tilde{\mathbf{v}}_\alpha$$

$$+ \frac{a}{\Delta t} \sum_\alpha J_\alpha \mathbf{\Omega}_\alpha \cdot \tilde{\mathbf{\Omega}}_\alpha$$

$$= \int_{\Omega^{n+1}} \rho \left(\frac{a}{\Delta t}\mathbf{u}(\bar{\mathbf{x}}, t^n) + b\frac{\delta \mathbf{u}}{\delta t}(\bar{\mathbf{x}}, t^n) + \mathbf{f} \right) \cdot \tilde{\mathbf{u}}\, dV$$

$$+ \sum_\alpha \left(\frac{a}{\Delta t} m_\alpha \mathbf{v}_\alpha^n + b m_\alpha \frac{d\mathbf{v}_\alpha}{dt}(t^n) + \mathbf{F}_\alpha^\mathrm{e} \right) \cdot \tilde{\mathbf{v}}_\alpha$$

$$+ \sum_\alpha \left(\frac{a}{\Delta t}(J_\alpha \mathbf{\Omega}_\alpha)^n + b\frac{d(J_\alpha \mathbf{\Omega}_\alpha)}{dt}(t^n) \right) \cdot \tilde{\mathbf{\Omega}}_\alpha, \tag{5.45}$$

and

$$\int_{\Omega^{n+1}} \left[\lambda \left(\frac{a}{\Delta t}\boldsymbol{\tau}_\mathrm{p} + (\mathbf{u} - \hat{\mathbf{u}}) \cdot \nabla \boldsymbol{\tau}_\mathrm{p} - (\nabla \mathbf{u})^\mathrm{T} \cdot \boldsymbol{\tau}_\mathrm{p} - \boldsymbol{\tau}_\mathrm{p} \cdot \nabla \mathbf{u} \right) \right.$$

$$\left. + \boldsymbol{\tau}_\mathrm{p} - 2\mu_1 \mathbf{e}[\mathbf{u}] \right] : \tilde{\boldsymbol{\tau}}\, dV$$

$$= \int_{\Omega^{n+1}} \lambda \left(\frac{a}{\Delta t}\boldsymbol{\tau}_\mathrm{p}(\bar{\mathbf{x}}, t^n) + b\frac{\delta \boldsymbol{\tau}_\mathrm{p}}{\delta t}(\bar{\mathbf{x}}, t^n) \right) : \tilde{\boldsymbol{\tau}}\, dV. \tag{5.46}$$

Since the domain of integration and all the functions in the integrals, unless specified otherwise, are evaluated at the current time instant t^{n+1}, the temporal discretizations in (5.24) and (5.26) are fully implicit and unconditionally stable. The functions inside the integrals of the right-hand side of (5.45) and (5.46) are known from the solution computed at the previous time step. Although the domain on which these functions are defined, which is the old domain $\Omega^n = \Omega_0(t^n)$, is not the same as the integration domain,

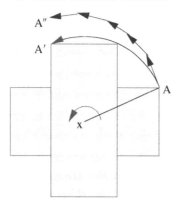

Fig. 5.1. Distortion of the particle shape due to the improper update of the nodes on the particle surface.

which is the new domain $\Omega^{n+1} = \Omega_0(t^{n+1})$, the integration can be conceived as over the fixed computational domain $\boldsymbol{\xi}$. Note in particular that the integration variable in the integrals is $\mathbf{x}$, so that $\bar{\mathbf{x}}$ in the right-hand side of (5.46) should be understood as a function of $\mathbf{x}$. The location of the grid in the new domain, $\mathbf{x}$, and its corresponding one in the old domain, $\bar{\mathbf{x}}$, is the same in the computational domain, with equation (5.39) providing the mapping between points in the two domains.

In using (5.39), one should be careful in the update of the nodes on the surface of a particle, especially with the first-order scheme. If one simply uses the velocity due to the rigid body motion, the body shape will become more and more distorted, as depicted in Fig. 5.1 for the case of a rotating rectangular particle. After the particle rotates by $90°$, the numerically updated position of the corner A will be at A'' rather than the correct position A'. This is a purely numerical artifact due to the linearization of the rotational motion of the rigid particle in Cartesian coordinates. To maintain the shape of the rigid body during the simulation, the nodes on the particle surface should be simply reset to the surface at each time step.

5.6 Spatial discretization – Galerkin finite element scheme

In this section, we shall discuss the approximation of the weak formulations of (5.25), (5.45), and (5.46) by a Galerkin finite element formulation, and the proper choices of the interpolation functions for the fluid velocity, pressure, and stress.

The fluid domain $\Omega^{n+1} = \Omega_0(t^{n+1})$ is first approximated by a finite element triangulation T_h, where h is the typical mesh size. To fit the surface

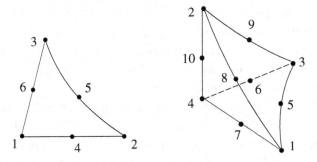

Fig. 5.2. Curved six-node triangle (a) and 10-node tetrahedron (b). The curved line/surface is next to the particle surface.

of the particles, curved P_2 quadratic elements are appropriate. For two-dimensional problems, these elements are triangles with six basis functions that are second-order polynomials defined on the three vertices and three midnodes on each side of the triangle, as shown in Fig. 5.2. For three-dimensional problems, these elements are tetrahedra with 10 basis functions which are second-order polynomials defined on the four vertices and six midnodes on each edge of the tetrahedron. The curved quadratic line segments in these elements approximate the local surface curvature of the particle, as indicated in the figure.

The trial function spaces, $\mathcal{V}$, $\mathcal{Q}$, and $\mathcal{T}$, are approximated by their corresponding finite-dimensional counterparts defined on the triangulation T_h. In this particular implementation, a mixed type of finite element is used, where different interpolation functions are chosen for different unknown variables. The fluid velocity is approximated by P_2 functions, namely piecewise quadratic functions which are taken to be continuous over the domain. Thus, over a finite element, the velocity is locally interpolated with its values on all six nodes in two dimensions or 10 nodes in three dimensions. The pressure is approximated by piecewise linear and continuous P_1 functions. The discrete solution for the components of the stress tensor is also piecewise linear and continuous (P_1). In a finite element, these functions are locally interpolated only with their values on the vertices. This P_1/P_2 element for the pressure and velocity is known to satisfy the so-called Babuska–Brezzi (LBB) stability condition or *inf–sup* condition. A detailed discussion of this condition can be found in Oden and Carey (1994).

One of the advantages of this type of mixed finite element formulation is that it reduces the cost of mesh generation in comparison with equal-order (linear) interpolation functions for both velocity and pressure. For a desired accuracy of the numerical solution, a coarser mesh or fewer elements are needed with the quadratic velocity interpolation (P_2 elements) in comparison with linear interpolating functions (P_1 elements). For simulations of a large number of particles, where the total number of elements is large and the frequency of mesh regeneration (to be discussed in the next section) also high, this cost saving could be considerable.

Another advantage of using P_2 elements for the velocity field is that, when two particles are in relative motion close to each other, in the lubrication limit the velocity profile in the gap that separates them is parabolic. Therefore, P_2 elements will capture the exact solution in this region even with a single layer of elements. With the linear P_1 elements, a single element across the gap between two particles in near-contact not only is inaccurate, but also would cause mesh locking and a failure of the numerical calculation[1]. In such a situation, a few layers of elements are needed in the gap region between two moving particles, and special finite element mesh generators are required for this purpose (Johnson *et al.*, 1999).

On a given finite element mesh and with the finite element interpolation functions described before, the weak formulations (5.25), (5.45), and (5.46) reduce to a nonlinear system of algebraic equations. This nonlinear system can be solved by a Newton–Raphson algorithm. In each step of this iterative procedure, one solves a linear system of the form

$$\begin{pmatrix} \mathbf{A}_1 & \mathbf{A}_2 & \mathbf{G}_1 & \mathbf{B}_1 \\ \mathbf{A}_3 & \mathbf{A} & \mathbf{G} & \mathbf{B}_2 \\ \mathbf{Q}_1 & \mathbf{Q} & \mathbf{D} & \mathbf{0} \\ \mathbf{B}_1^{\mathrm{T}} & \mathbf{B}_2^{\mathrm{T}} & \mathbf{0} & \mathbf{0} \end{pmatrix} \begin{pmatrix} \mathbf{U} \\ \mathbf{u} \\ \boldsymbol{\tau} \\ \mathbf{p} \end{pmatrix} = \begin{pmatrix} \mathbf{f}_U \\ \mathbf{f}_u \\ \mathbf{f}_\tau \\ \mathbf{f}_{\mathrm{p}} \end{pmatrix} \tag{5.47}$$

where all the submatrices $\mathbf{A}$, $\mathbf{B}$, $\mathbf{G}$, $\mathbf{Q}$ are sparse. In general, the system is not symmetric. $\mathbf{U}$ is the vector combining all the translational and angular velocities of the solid particles; $\mathbf{u}$, $\boldsymbol{\tau}$ and p represent, respectively, the vectors collecting all the fluid velocity, stress, and pressure unknowns at the grid points in the fluid. In (5.47) the fluid velocity unknowns on the particle surface are eliminated with the particle velocities using the no-slip relation for rigid body motion. The vectors in the right-hand side of the equation derive from the residual of the corresponding equations in the course of the nonlinear iterations.

---

[1] In a two-dimensional situation, for example, when there is only one triangular element across the gap between two particles (with two vertices on the surface of one particle and the remaining one on that of the other), the velocities in the element will be completely determined by the motion of the particles. Therefore, in general, the continuity equation in the element will not be satisfied, which causes the equation for the determination of the pressure to become singular. This circumstance would lead to a failure of the computation.

The weak formulations for the mesh velocity and acceleration (5.34) can also be approximated on finite-dimensional function spaces based on linear polynomials (P_1 element). The final linear systems of algebraic equations is

$$\mathbf{H}\hat{u} = \mathbf{f}_{m1}, \qquad \mathbf{H}\hat{a} = \mathbf{f}_{m2} \qquad (5.48)$$

where $\mathbf{H}$ is symmetric and positive definite. In (5.48), $\hat{u}$ and $\hat{a}$ represent, respectively, the vectors collecting all the mesh velocity and acceleration unknowns at grid points in the fluid. The vectors in the right-hand sides arise from the application of the boundary conditions.

The algebraic systems (5.47) and (5.48) are coupled since the matrices $\mathbf{A}$ in (5.47) depend on the mesh velocity field $\hat{u}$. Thus, (5.47) and (5.48) need to be solved iteratively at each time step. The linearized system of algebraic equations (5.47) can be solved with an iterative solver using a preconditioned generalized minimal residual scheme (GMRES) or biconjugate gradient stabilized algorithm (BICGSTAB), see Saad (1996). These schemes are suitable for the nonsymmetric matrix in the system (5.47). In simulations with the GMRES scheme, the size of the Krylov subspace should be sufficiently large to ensure good convergence; in our experience, 20 is a suitable choice. In order to make the iterative solver converge, using a proper preconditioner is essential. The preconditioners, such as ILU(0) or ILU(t) (incomplete LU factorization without or with controlled fill-ins), when applied to the global system, are found to be quite robust and efficient. For more efficient implementations, one may take advantage of the structure of the system in (5.47), and design different preconditioners for different parts of the equations within the system. The design of more efficient and reliable preconditioners, especially for parallel computation, is still a topic of active research (Saad, 1996; Sarin and Sameh, 1998; Little and Saad, 1999). The symmetric and positive definite systems of (5.48) can be solved iteratively with the conjugate gradient method. A preconditioner like ILU(0) can be used to improve the convergence.

5.7 Mesh generation

The first task in simulating the motion of a fluid–particle system is to generate a finite element grid based on the initial positions of the particles in the domain. A possible approach is the following. The mesh generator first creates a uniform grid on all the particle surfaces and boundary sections. It then checks the distance between the boundary nodes belonging to different particles or boundary sections. If this distance is smaller than the local grid spacing, the mesh generator performs a local refinement by inserting nodes on the corresponding boundary sections. The purpose of this

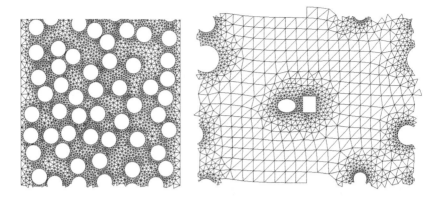

Fig. 5.3. Two-dimensional finite element meshes. Mesh (a) is periodic in one direction, mesh (b) in both directions.

boundary grid refinement is to eventually generate a fine mesh in the regions where it is needed. With the complete boundary grid information, the mesh generator next generates the elements in the interior of the domain using the Delaunay–Voronoi method (see George, 1991). Finally, the midnodes are added on the edges of the mesh to form the P_2/P_1 mixed elements used in the scheme.

In computing solid–liquid flows with a large number of solid particles, it is often necessary to use periodic boundary conditions in one or more directions. At a periodic boundary, the particles leave and enter the simulation domain. The finite element mesh generator should automatically take care of periodic boundaries without introducing artificial cuts through the particles, which may give rise to very unsatisfactory elements. Techniques for the periodic finite element mesh generation are discussed by Patankar and Hu (1996), Johnson and Tezduyar (1999), and Maury (1999).

The mesh generator has a local refinement capability in regions formed by approaching particles or between a particle and the surrounding wall, as mentioned earlier. There is always at least one layer of elements in these regions, and the mesh size is designed to be of the order of the minimum gap size between the approaching particles. The local refinement in the gaps between particles is essential to capture the particle collision process that is to be discussed later in Section 5.12.

Figure 5.3(a) displays a two-dimensional mesh with 50 circular disks in a periodic domain between two channel walls. In the figure straight lines are used to connect three vertices of a triangle. However, the curved P_2 triangles (with six nodes) are actually used in the simulation to fit the curved particle

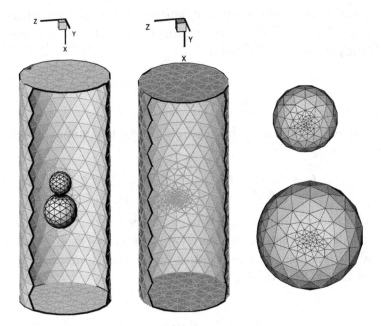

Fig. 5.4. Surface meshes on two spheres and a cylindrical tube. See also http://www.cambridge.org/comp_mult_flow.

surfaces. Figure 5.3(b) shows a mesh with particles with different shapes and sizes in a doubly periodic domain. Figure 5.4 shows the mesh on the surface of two spheres and of a circular tube.

During a typical simulation, the calculation is started by generating a finite element mesh with the automatic mesh generator based on the initial particle positions in the domain. Using this mesh, one can generate and then solve the system of algebraic equations (5.47) and (5.48). At the new time step, the old finite element mesh will be moved using (5.39) according to the mesh velocity and acceleration fields obtained from the previous time step. This updated mesh is checked for the quality of its elements. If unacceptable element distortion is detected, a new mesh is generated with the automatic mesh generator. This new mesh may not have any correspondence with the old mesh. The solution from the old mesh has to be projected onto the new mesh by the methods described in the next section. Once the solution is projected, the computation can proceed normally.

The quality of the mesh can be measured by checking the change in the element volume and/or aspect ratio in comparison with its value in the initial undeformed mesh. The changes of the element volume and aspect ratio are defined as

$$f_1^e = \left| \log \left(\frac{V^e}{V_0^e} \right) \right|, \text{ and } f_2^e = \left| \log \left(\frac{S^e}{S_0^e} \right) \right| \tag{5.49}$$

where V^e and S^e are the volume and aspect ratio of the e-th element, respectively, and V_0^e and S_0^e the corresponding values in the initial undeformed mesh. The aspect ratio is defined as

$$S^e = \frac{(\text{maximum edge length})^3}{V^e}. \tag{5.50}$$

The global quality of the mesh is measured by the maximum element deformation,

$$f_1 = \max_{1 \le e \le N_{\text{el}}} (f_1^e), \quad \text{and} \quad f_2 = \max_{1 \le e \le N_{\text{el}}} (f_2^e) \tag{5.51}$$

where N_{el} is the total number of elements in the mesh. Usually, remeshing is considered when either one of these two parameters exceeds 1.39, which corresponds to volumes or aspect ratios larger than four times or smaller than $1/4$ of the original values.

5.8 Projection scheme

At the beginning of each time step, on the basis of the solution at the previous time step, the finite element mesh is explicitly moved using equation (5.39) and the particle positions are explicitly updated using equations (5.43) and (5.44). If the updated mesh is too distorted, a new mesh is generated as described in the previous section. The solution of the fluid mechanical problem must be sought on the new mesh and, for this purpose, it is necessary to project the flow field defined on the old mesh onto the new one. Projection errors will be introduced in this step and need to be minimized. There are a number of schemes to perform this projection, two of which will be discussed in this section.

Let us assume that the mesh nodes moved from $\bar{\mathbf{x}}$ at t^n to $\mathbf{x}$ at t^{n+1} according to equation (5.39), and let $\mathbf{y}$ denote the position of the nodes of the new mesh. The meshes $\mathbf{y}$ and $\mathbf{x}$ cover the same domain Ω^{n+1} occupied by the fluid at t^{n+1}. Each node $\mathbf{y}$ can be thought of as coming from a point $\bar{\mathbf{y}}$ of the mesh at time t^n. The relation between $\mathbf{y}$ and $\bar{\mathbf{y}}$ is given by a relation similar to (5.39), namely

$$\mathbf{y} = \bar{\mathbf{y}} + \Delta t \hat{\mathbf{u}}(\bar{\mathbf{y}}, t^n) + \frac{1}{2}\Delta t^2 \hat{\mathbf{a}}(\bar{\mathbf{y}}, t^n). \tag{5.52}$$

In order to carry out the fluid dynamical step and update the flow fields at the time level t^{n+1}, it is necessary to find their values in the mesh $\bar{\mathbf{y}}$. Let $\phi(\bar{\mathbf{x}}, t^n)$ denote the generic flow field at time t^n on the mesh $\bar{\mathbf{x}}$, and let

$\phi'(\bar{\mathbf{y}}, t_n)$ be the same field on the mesh $\bar{\mathbf{y}}$ at the same time. The mapping between the meshes at t^n and t^{n+1}, being affine, is one-to-one, and therefore, in principle, there is a unique relation between the points of the two meshes at t^n. The practical difficulty in effecting a direct projection from $\phi(\bar{\mathbf{x}}, t^n)$ to $\phi'(\bar{\mathbf{y}}, t^n)$ is that we do not have the velocity $\hat{\mathbf{u}}$ and the acceleration $\hat{\mathbf{a}}$ on the mesh $\bar{\mathbf{y}}$ required by (5.52). Thus, it is more convenient to find the relation between the nodes $\mathbf{y}$ and $\mathbf{x}$, after which one can backtrack using (5.52) to find the relation between $\bar{\mathbf{y}}$ and $\bar{\mathbf{x}}$. Another way to view this step is to define the projection on the computational domain.

The first projection scheme is a direct local interpolation scheme. To find the flow field information at each node in the new mesh, there are three steps involved. First, one needs to locate the element in the old mesh $\mathbf{x}$ where a given new node $\mathbf{y}$ lies. An example of a search in a two-dimensional problem is depicted in Fig. 5.5. The search is based on evaluating the local area coordinates of $\mathbf{y}$ with respect to an element in the old mesh assumed, for the time being, to be linear:

$$
\begin{aligned}
r &= \frac{(x - x_1)(y_3 - y_1) - (x_3 - x_1)(y - y_1)}{(x_2 - x_1)(y_3 - y_1) - (x_3 - x_1)(y_2 - y_1)}, \\
s &= \frac{(x_2 - x_1)(y - y_1) - (x - x_1)(y_2 - y_1)}{(x_2 - x_1)(y_3 - y_1) - (x_3 - x_1)(y_2 - y_1)},
\end{aligned}
\tag{5.53}
$$

where $\mathbf{y} = (x, y)$ are the coordinates of the given node in the new mesh; $\mathbf{x}_a = (x_a, y_a)$, $(a = 1, 2, 3)$ are the coordinates, in the updated mesh $\mathbf{x}$, of the three vertices of an element (e) encountered during the search. If the local coordinates calculated from (5.53) satisfy

$$
0 \le r \le 1, \ 0 \le s \le 1 \quad \text{and} \quad 0 \le 1 - r - s \le 1,
\tag{5.54}
$$

then the node $\mathbf{y} = (x, y)$ belongs to this element e. Otherwise, the values of the local coordinates and the information on element neighbors are used to shift the search to the next element, using the rules indicated in Fig. 5.6. This search scheme can be easily extended to three dimensions. It is basically a one-dimensional search scheme and is very efficient.

The second step in this interpolation scheme is to calculate the exact local coordinates for the node within the located element. During the search step, the elements are assumed to be linear with straight sides. However, they may be curved. For curved high-order elements, the calculation for the local coordinates involves solving a set of nonlinear equations,

$$
\mathbf{y} = \sum_a \mathbf{x}_a N_a(r, s),
\tag{5.55}
$$

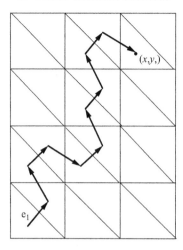

Fig. 5.5. Diagram of a search scheme to find the element where a given point (x, y) lies. The search starts in the element e_1. The line indicates the steps in the search process.

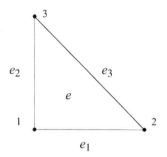

Fig. 5.6. The search proceeds to the next element e_2 if $r < 0$, or e_1 if $s < 0$, or e_3 if $1 - r - s < 0$.

where the summation is over the number of nodes in the element; $\mathbf{x}_a$ and N_a are the coordinates of the nodes and the corresponding interpolation functions in the element, respectively[1]. Once the local coordinates (r, s) for the new node are obtained, the interpolation of a variable at this node can be easily achieved by using the local interpolation functions and the nodal values of the variable on the located element, that is

$$\bar{\phi}'(\mathbf{y}, t^n) = \sum_a \bar{\phi}(\mathbf{x}_a, t^n) N_a(r, s), \qquad (5.56)$$

---

[1] Since the search was based on the assumption of linear edges, in a few cases the point $\mathbf{y}$ might correspond to an element adjacent to the one identified with the previous procedure. In this case, the solution of (5.55) will return values of r or s slightly outside the range (5.54), and this will signal the correct element in which the point in question is located.

where we write $\phi(\bar{\mathbf{x}}(\mathbf{x}), t^n) \equiv \bar{\phi}(\mathbf{x}, t^n)$ and $\phi'(\bar{\mathbf{y}}(\mathbf{y}), t^n) \equiv \bar{\phi}'(\mathbf{y}, t^n)$.

One can estimate the numerical error produced in this type of projection. A simple analysis (see Patankar, 1997) shows that the projection error is $O(h^2)$ for linear elements and $O(h^3)$ for quadratic elements.

The second projection scheme uses a global least-squares method. With the previous notation, the projection is done by minimizing the difference between the function representations on the old mesh and on the new mesh, or

$$\text{Minimize} \int_{\Omega^{n+1}} [\bar{\phi}'(\mathbf{y}, t^n) - \bar{\phi}(\mathbf{x}, t^n)]^2 \, dV \qquad (5.57)$$

where the integration is performed over the new mesh. The weak formulation of (5.57) can be written as:

Given $\bar{\phi}(\mathbf{x}, t^n)$, find $\bar{\phi}'(\mathbf{y}, t^n)$, such that $\forall \tilde{\phi}(\mathbf{y})$,

$$\int_{\Omega^{n+1}} \bar{\phi}'(\mathbf{y}, t^n) \tilde{\phi}(\mathbf{y}) \, dV = \int_{\Omega^{n+1}} \bar{\phi}(\mathbf{x}, t^n) \tilde{\phi}(\mathbf{y}) \, dV \qquad (5.58)$$

where $\bar{\phi}'(\mathbf{y}, t^n)$ and $\tilde{\phi}(\mathbf{y})$ belong to the appropriate function spaces for the flow variables.

If the integration in the right-hand side of (5.58) is performed numerically by Gaussian quadrature, the values of $\bar{\phi}(\mathbf{x}, t^n) = \phi(\bar{\mathbf{x}}, t^n)$ at the quadrature points are needed. These values are calculated using the local interpolation scheme described above. Once the function is approximated by the finite-dimensional finite element space, (5.58) reduces to a set of linear algebraic equations and can be solved by iterative methods such as the conjugate gradient method.

The global least-squares projection scheme generally performs better than the local interpolation scheme. This may be due to the fact that the right-hand side of (5.58) is the same as the terms in the right-hand side of (5.45) and (5.46). If the right-hand side of (5.58) is evaluated exactly, the global least-squares projection scheme would be exact. Thus the projection would not introduce any additional error, which is expressed by saying that it is consistent.

5.9 Explicit–implicit solution procedure

So far we have described all the major steps needed for simulations of fluid–solid systems. In this section, we summarize a solution procedure in which the particle positions and the mesh nodes in the fluid domain are updated explicitly, while the particle velocities and the fluid flow field are determined implicitly.

Scheme 2. Explicit–implicit scheme (ALE particle mover)
   Initialization: $t^0 = 0$ and $n = 0$ (n is the time step index)

   Initialize particle variables $\mathbf{v}_\alpha(0)$, $\mathbf{\Omega}_\alpha(0)$, $\mathbf{X}_\alpha(0)$, $\mathbf{\Theta}_\alpha(0)$ for $\alpha = 1, 2, \ldots, N$.
   Generate initial mesh $\mathbf{x}_0$ based on particle positions and orientations, $\mathbf{X}_\alpha(0)$, $\mathbf{\Theta}_\alpha(0)$.
   Initialize flow field $\mathbf{u}(\mathbf{x}_0, 0)$, $p(\mathbf{x}_0, 0)$, $\boldsymbol{\tau}_{\mathrm{p}}(\mathbf{x}_0, 0)$.

   Do $n = 0, 1, 2, \ldots, M$ (over time steps)

   (i) Select time step Δt^{n+1}: $t^{n+1} = t^n + \Delta t^{n+1}$
   (ii) Using the particle velocity $\mathbf{v}_\alpha^n$, $\mathbf{\Omega}_\alpha^n$, explicitly update particle positions and orientations:

$$\mathbf{X}_\alpha^{n+1} = \mathbf{X}_\alpha^n + \Delta t^{n+1}\mathbf{v}_\alpha^n + \tfrac{1}{2}(\Delta t^{n+1})^2\dot{\mathbf{v}}_\alpha^n$$
$$\mathbf{\Theta}_\alpha^{n+1} = \mathbf{\Theta}_\alpha^n + \Delta t^{n+1}\mathbf{\Omega}_\alpha^n + \tfrac{1}{2}(\Delta t^{n+1})^2\dot{\mathbf{\Omega}}_\alpha^n.$$

   (iii) Update the mesh:

$$\mathbf{y}(t^{n+1}) = \bar{\mathbf{x}}(t^n) + \hat{\mathbf{u}}(\bar{\mathbf{x}}, t^n)\Delta t^{n+1} + \frac{1}{2}(\Delta t^{n+1})^2\hat{\mathbf{a}}(\bar{\mathbf{x}}, t^n).$$

   (iv) Check mesh quality. If the updated mesh $\mathbf{y}(t^{n+1})$ is too distorted:

   Generate a new mesh $\mathbf{x}(t^{n+1})$.
   Project the flow field from $\mathbf{y}(t^{n+1})$ onto $\mathbf{x}(t^{n+1})$.

   (v) Iteratively solve the flow field $\mathbf{u}(\mathbf{x}(t^{n+1}), t^{n+1})$, $p(\mathbf{x}(t^{n+1}), t^{n+1})$, $\boldsymbol{\tau}_{\mathrm{p}}(\mathbf{x}(t^{n+1}), t^{n+1})$, the mesh velocity $\hat{\mathbf{u}}(\mathbf{x}(t^{n+1}), t^{n+1})$, and the particle velocities $\mathbf{v}_\alpha^{n+1}$, $\mathbf{\Omega}_\alpha^{n+1}$.
   (vi) Update $\dot{\mathbf{v}}_\alpha^{n+1}$, $\dot{\mathbf{\Omega}}_\alpha^{n+1}$, $(\delta/\delta t)\mathbf{u}(\mathbf{x}(t^{n+1}), t^{n+1})$, $(\delta/\delta t)\boldsymbol{\tau}_{\mathrm{p}}(\mathbf{x}(t^{n+1}), t^{n+1})$ from equations (5.41), (5.42) and (5.38) respectively.
   (vii) Solve the mesh acceleration $\hat{\mathbf{a}}(\mathbf{x}(t^{n+1}), t^{n+1})$.

   End-Do

   The choice of the time step Δt^{n+1} in the scheme depends on many factors. It can be used to limit the maximum distance each particle is allowed to travel in that time step, to restrict the maximum change in the particle velocity, or to avoid particle–particle and particle–wall collisions. The time step should also be restricted to capture unsteady dynamical behavior of the fluid motion, such as vortex shedding.

This explicit–implicit scheme is second-order accurate in time, and numerically stable. It was first described in Hu *et al.* (1992), and since then it has been used in a number of studies of fluid–particle systems.

5.10 Space-time finite element method

Another method to solve problems with moving boundaries is to use space-time finite element methods (see Hughes and Hulbert, 1988; Tezduyar *et al.*, 1992a,b; and Hansbo 1992). In the space-time approach, along with the spatial coordinates, the temporal coordinate is also discretized using finite element methods. The deformation of the spatial domain with time is reflected simply in the deformation of the mesh in the temporal coordinate.

A space-time finite element scheme for solving fluid–particle systems was developed by Tezduyar *et al.* (1992a, 1992b) Johnson (1995), and Johnson and Tezduyar (1996, 1997). At each time step, their scheme iteratively solves the flow field, the forces and moments acting on the particles, and the particle velocities. In this section, we shall describe the space-time finite element scheme for Newtonian fluids based on the work of Johnson.

In the space-time formulation for the solution of the flow field, the time interval $(0, T)$ is partitioned into subintervals $I^n = (t^n, t^{n+1})$, $n = 0, 1, \ldots, N$. Denote by Ω_0^n and Ω_0^{n+1} the fluid domains at t^n and t^{n+1}, and by P^n the surface described by the boundary $\partial\Omega_0(t)$ as time traverses I^n. In two dimensions, these domains can be visualized as surfaces as shown in Fig. 5.7. A space-time slab Q^n is defined as the domain formed by the three surfaces Ω_0^n, Ω_0^{n+1}, and P^n.

In Johnson's implementation, for each space-time slab, the finite element interpolation functions for both fluid velocity and pressure are taken to be piecewise linear and continuous in space, and piecewise linear but discontinuous in time. The stabilized space-time formulation for the fluid flow can be written as:

For each space-time slab Q^n, given $\mathbf{u}^n_-$, find $\mathbf{u}$ and p in their appropriate function spaces, such that $\forall \tilde{\mathbf{u}}$ and $\forall \tilde{p}$ in their corresponding variational

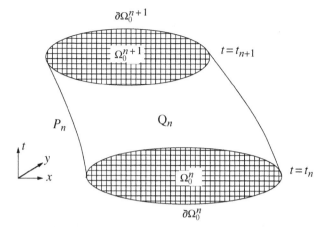

Fig. 5.7. A space-time slab.

spaces,

$$\int_{Q^n} \rho \left(\frac{d\mathbf{u}}{dt} - \mathbf{f} \right) \cdot \tilde{\mathbf{u}} \, dQ + \int_{Q^n} 2\mu \mathbf{e}[\mathbf{u}] : \mathbf{e}[\tilde{\mathbf{u}}] \, dQ - \int_{Q^n} p (\nabla \cdot \tilde{\mathbf{u}}) \, dQ$$

$$+ \int_{Q^n} (\nabla \cdot \mathbf{u}) \tilde{p} \, dQ + \int_{\Omega_0^n} \rho \left(\mathbf{u}_+^n - \mathbf{u}_-^n \right) \cdot \tilde{\mathbf{u}}_+^n \, dV$$

$$+ \sum_e \int_{Q_e^n} \delta \rho (\nabla \cdot \mathbf{u})(\nabla \cdot \tilde{\mathbf{u}}) \, dQ$$

$$+ \sum_e \int_{Q_e^n} \frac{\tau}{\rho} \left[\rho \left(\frac{d\mathbf{u}}{dt} - \mathbf{f} \right) - \nabla p - 2\nabla \cdot (\mu \mathbf{e}[\mathbf{u}]) \right]$$

$$\cdot \left[\rho \left(\frac{\partial \tilde{\mathbf{u}}}{\partial t} + \mathbf{u} \cdot \nabla \tilde{\mathbf{u}} \right) - \nabla \tilde{p} - 2\nabla \cdot (\mu \mathbf{e}[\tilde{\mathbf{u}}]) \right] dQ = 0, \qquad (5.59)$$

where the summation is over the number of finite elements in the mesh,

$$\mathbf{u}_+^n = \lim_{\epsilon \to 0} \mathbf{u}(t^n + \epsilon), \qquad \mathbf{u}_-^n = \lim_{\epsilon \to 0} \mathbf{u}(t^n - \epsilon) \qquad (5.60)$$

and

$$\int_{Q^n} (\ldots) \, dQ = \int_{t^n}^{t^{n+1}} dt \int_{\Omega_0(t)} (\ldots) \, dV. \qquad (5.61)$$

Note that, for each space-time slab, as the elements deform with time, the

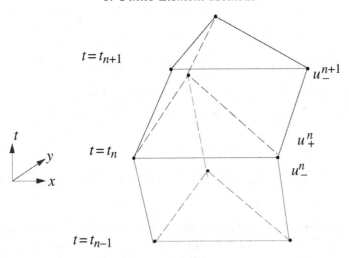

Fig. 5.8. Deformation of an element with time, and the velocity variables at different times.

flow variables are defined both on the mesh at time t^n and on the mesh at t^{n+1}, which are denoted as $\mathbf{u}_+^n$ and $\mathbf{u}_-^{n+1}$, as depicted in Fig. 5.8. Normally, at the same time level, the flow variables that belong to different space-time slabs are not exactly equal, $\mathbf{u}_+^n \neq \mathbf{u}_-^n$. The continuity of the velocity field across the space-time slabs is only enforced in a weak sense, by the fifth term in the formulation (5.59). As a consequence, for each space-time slab, the system of equations to be solved is twice as big in comparison with the semidiscrete finite element method.

In the formulation (5.59), the first three terms constitute the Galerkin form of the fluid momentum equation. The fourth term is the Galerkin form of the continuity equation. The fifth term enforces, in a weak sense, the continuity of the velocity field across the space-time slabs, as mentioned above. The last term is the least-squares approximation to the momentum balance equation. This term provides the necessary stability for convection-dominated flows. This term also provides stability when equal-order interpolation functions are used for velocity and pressure. For piecewise linear velocity interpolation, inside each element, $\nabla \cdot (\mu \mathbf{e}[\mathbf{u}])$ and $\nabla \cdot (\mu \mathbf{e}[\tilde{\mathbf{u}}])$ are identically zero. The stability parameter τ (see Franca *et al.*, 1992) is chosen to be

$$\tau = \left[\left(\frac{2}{\Delta t} \right)^2 + \left(\frac{2|\mathbf{u}|}{h_e} \right)^2 + \left(\frac{4\nu}{h_e^2} \right)^2 \right]^{-1/2}, \qquad (5.62)$$

where ν is the kinematic viscosity and h_e is a suitable measure of the

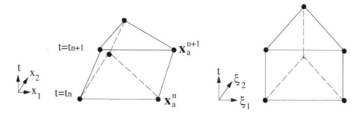

Fig. 5.9. Mapping of a space-time finite element into the parent element.

element length, for example the maximum edge length of the element under consideration. The sixth term in the equation is the least-squares form of the incompressibility constraint. This term enhances the stability of the formulation in high Reynolds number flows. The stability parameter δ (see Franca et $al.$, 1992) is defined as

$$\delta = \frac{h_e}{2} |\mathbf{u}| \min \left(1, \frac{|\mathbf{u}| h_e}{6\nu} \right). \qquad (5.63)$$

With the chosen finite element spaces, the formulation (5.59) reduces to a system of nonlinear equations, which can be solved by a Newton–Raphson scheme. Each step of the linearized system within the Newton–Raphson scheme is solved by iterative schemes such as GMRES. The solver for the fluid flow can be implemented in matrix-free form. Simulations of the fluid–particle system can be efficiently performed on parallel computers, see Johnson (1995).

The advantage of the space-time finite element method is its generality. One can frame the ALE finite element technique discussed before as a special case of the space-time method. To appreciate this fact, let us consider the mapping from a space-time finite element (linear in time) into the parent element (see Fig. 5.9):

$$\mathbf{x} = \mathbf{x}(\boldsymbol{\xi}, t) = \sum_a \left[\mathbf{x}_a^n N_a(\boldsymbol{\xi})(1 - \tau) + \mathbf{x}_a^{n+1} N_a(\boldsymbol{\xi}) \tau \right] \qquad (5.64)$$

where the summation is over all the nodes of the element at one time level, the N_a's are spatial finite element shape functions, and $\tau = (t - t_n)/\Delta t$. Taking the time derivative of the mapping (5.64),

$$\frac{\partial \mathbf{x}}{\partial t} = \sum_a \left[\frac{\mathbf{x}_a^{n+1} - \mathbf{x}_a^n}{\Delta t} N_a(\boldsymbol{\xi}) \right] = \hat{\mathbf{u}} \qquad (5.65)$$

one gets the velocity of the mesh motion in this space-time slab. Similarly, the fluid velocity can be expressed using the same shape functions:

$$\mathbf{u}\left(\mathbf{x}(\boldsymbol{\xi},t),t\right) = \bar{\mathbf{u}}(\boldsymbol{\xi},t) = \sum_a \left[\bar{\mathbf{u}}_a^n N_a(\boldsymbol{\xi})(1-\tau) + \bar{\mathbf{u}}_a^{n+1} N_a(\boldsymbol{\xi})\tau\right] \qquad (5.66)$$

Thus the time derivative of the velocity can be expressed as

$$\left.\frac{\partial \bar{\mathbf{u}}}{\partial t}\right|_\xi = \sum_a \left[\frac{\bar{\mathbf{u}}_a^{n+1} - \bar{\mathbf{u}}_a^n}{\Delta t} N_a(\boldsymbol{\xi})\right]. \qquad (5.67)$$

Since

$$\left.\frac{\partial \bar{\mathbf{u}}}{\partial t}\right|_\xi = \left.\frac{\partial}{\partial t}\mathbf{u}\left(\mathbf{x}(\boldsymbol{\xi},t),t\right)\right|_\xi = \left.\frac{\partial \mathbf{u}}{\partial t}\right|_\mathbf{x} + \left(\frac{\partial \mathbf{x}}{\partial t}\cdot \nabla\right)\mathbf{u}\left(\mathbf{x}(\boldsymbol{\xi},t),t\right),$$

using (5.65) we have,

$$\left.\frac{\partial \mathbf{u}}{\partial t}\right|_\mathbf{x} = \left.\frac{\partial \bar{\mathbf{u}}}{\partial t}\right|_\xi - (\hat{\mathbf{u}}\cdot\nabla)\mathbf{u}. \qquad (5.68)$$

This is identical to the ALE formulation (5.29). As discussed by Hansbo (1992) and Behr and Tezduyar (1994), the first-order ALE formulation amounts to using elements that give a discontinuous piecewise constant approximation in time and a continuous linear approximation in space.

The use of linear interpolation functions for both the fluid velocity and the pressure offers simplicity in the finite element implementation. However, the cost of mesh generation would be high, since many more elements are needed for a simulation of a given accuracy. The linear interpolation function for the fluid velocity field may cause problems in accurately resolving the flow field between particles, as mentioned before.

5.11 Distributed Lagrange multiplier/fictitious domain (DLM/FD) particle mover

In the numerical methods described in the previous sections, the flow field has to be solved in the complex domain occupied by the fluid. In this section, we will discuss a different kind of numerical method to simulate the fluid–particle systems, based on the work of Glowinski *et al.* (1999). We shall describe the method for particles in a Newtonian fluid; an extension to viscoelastic fluids can be found in Singh *et al.* (1999).

The basic idea is to embed the irregular physical domain into a larger, simpler domain. The fluid flow problem is posed on this extended domain, a *fictitious domain*, both outside and inside the particles. This larger domain allows a simple uniform (regular) grid to be used in the discretization of the governing equations, which in turn allows specialized fast solution techniques. The extended domain is also time-independent so the same mesh can be used for the entire simulation, eliminating the need for repeated remeshing and projection. Fictitious-domain methods were first introduced by Hyman (1952), and were discussed by Saul'ev (1963) and Buzbee *et al.* (1971). Glowinski *et al.* (1999) developed a Lagrange-multiplier-based fictitious domain method for the simulation of fluid–solid mixtures. In their distributed Lagrange multiplier/fictitious domain (DLM/FD) approach, the velocity inside a particle is constrained to match the rigid-body motion of the particle. This constraint is enforced by using a distributed Lagrange multiplier, which represents the additional body force needed to maintain the rigid-body motion inside the particle, much like the pressure in incompressible fluid flows maintains the constraint of incompressibility.

Let $\Omega = \Omega_0(t) \cup P(t)$ be the extended computational domain covering both the fluid and all the particles. The extended domain Ω is fixed and does not change with time. The space for the velocity field on this extended domain is defined as

$$\mathcal{W} = \left\{ (\mathbf{u}, \mathbf{v}_\alpha, \mathbf{\Omega}_\alpha) \equiv \mathbf{U} \mid \mathbf{u} \in \mathcal{H}^1(\Omega)^3, \mathbf{v}_\alpha \in R^3, \mathbf{\Omega}_\alpha \in R^3; \mathbf{u} = \mathbf{u}_g, \mathbf{x} \in \partial\Omega_u \right\}. \tag{5.69}$$

This space is similar to the combined velocity space (5.16) used in Section 5.3. However, the rigid-body motion inside the particles is not enforced in this space. The space for the pressure is $\mathcal{Q}(\Omega)$, the same as the one defined in (5.17), however, on the extended domain.

The derivation of the DLM/FD method starts from the combined momentum equation (5.24). It is easy to verify that, when the fluid velocity $\mathbf{u}$ and its variation $\tilde{\mathbf{u}}$ satisfy the rigid body motion relations

$$\mathbf{u} = \mathbf{v}_\alpha + \mathbf{\Omega}_\alpha \times (\mathbf{x} - \mathbf{X}_\alpha), \qquad \tilde{\mathbf{u}} = \tilde{\mathbf{v}}_\alpha + \tilde{\mathbf{\Omega}}_\alpha \times (\mathbf{x} - \mathbf{X}_\alpha) \tag{5.70}$$

then, inside the α-th particle,

$$\int_{\Omega_\alpha} \rho \left(\frac{d\mathbf{u}}{dt} - \mathbf{f} \right) \cdot \tilde{\mathbf{u}} dV = \frac{\rho}{\rho_s} \left(m_\alpha \frac{d\mathbf{v}_\alpha}{dt} - \mathbf{F}_\alpha^e \right) \cdot \tilde{\mathbf{v}}_\alpha + \frac{\rho}{\rho_s} \frac{d(J_\alpha \mathbf{\Omega}_\alpha)}{dt} \cdot \tilde{\mathbf{\Omega}}_\alpha \tag{5.71}$$

and

$$\nabla \cdot \mathbf{u} = 0, \qquad e[\mathbf{u}] = 0. \tag{5.72}$$

In (5.71) the body forces acting on the fluid and on the particle are both due to gravity. For the motion of particles in a Newtonian fluid, by adding (5.71) to (5.24), using the relation (5.72), and introducing a distributed Lagrange multiplier to enforce the rigid-body motion inside the particles, we can write the weak formulation of the combined fluid–particle system in the *extended* domain as (see Glowinski *et al.*, 1999):

Find $\mathbf{U} \in \mathcal{W}$, $p \in \mathcal{Q}(\Omega)$, and $\boldsymbol{\lambda} = \{\boldsymbol{\lambda}_1, \ldots, \boldsymbol{\lambda}_N\} \in \Lambda(t)$, such that $\forall \tilde{\mathbf{U}} \in \mathcal{W}_0$, $\forall \tilde{p} \in \mathcal{Q}(\Omega)$ and $\forall \tilde{\boldsymbol{\lambda}} \in \Lambda(t)$,

$$
\int_\Omega \rho \left(\frac{d\mathbf{u}}{dt} - \mathbf{f} \right) \cdot \tilde{\mathbf{u}} \, dV - \int_\Omega p(\nabla \cdot \tilde{\mathbf{u}}) \, dV + \int_\Omega 2\mu \mathbf{e}[\mathbf{u}] : \mathbf{e}[\tilde{\mathbf{u}}] \, dV
$$

$$
+ \left(1 - \frac{\rho}{\rho_s} \right) \sum_\alpha \left[\left(m_\alpha \frac{d\mathbf{v}_\alpha}{dt} - \mathbf{F}_\alpha^e \right) \cdot \tilde{\mathbf{v}}_\alpha + \frac{d(I_\alpha \boldsymbol{\Omega}_\alpha)}{dt} \cdot \tilde{\boldsymbol{\Omega}}_\alpha \right]
$$

$$
= \sum_{\alpha=1}^N \left\langle \boldsymbol{\lambda}_\alpha, \tilde{\mathbf{u}} - \tilde{\mathbf{v}}_\alpha - \tilde{\boldsymbol{\Omega}}_\alpha \times (\mathbf{x} - \mathbf{X}_\alpha) \right\rangle_{\Omega_\alpha(t)} \tag{5.73}
$$

$$
\int_\Omega (\nabla \cdot \mathbf{u}) \tilde{p} \, dV = 0, \tag{5.74}
$$

$$
\sum_{\alpha=1}^N \left\langle \tilde{\boldsymbol{\lambda}}_\alpha, \mathbf{u} - \mathbf{v}_\alpha - \boldsymbol{\Omega}_\alpha \times (\mathbf{x} - \mathbf{X}_\alpha) \right\rangle_{\Omega_\alpha(t)} = 0. \tag{5.75}
$$

In (5.73) the variational space $\mathcal{W}_0$ is the same as $\mathcal{W}$, except that $\mathbf{u} = 0$ on $\partial \Omega_u$. The factor $(1 - \rho/\rho_s)$ in front of the particle terms comes from the expression (5.71). In both (5.73) and (5.75), the space, $\Lambda(t)$, for the Lagrange multiplier can be selected in $\mathcal{H}^1 (P(t))^3$ and $\langle \cdot, \cdot \rangle$ is the standard inner product on $\Lambda(t)$. Therefore, the differences between (5.73) and (5.24) are that (5.73) is defined on a simpler extended domain and the rigid-body constraint is enforced weakly by a distributed Lagrange multiplier term. If the rigid-body motion (5.70) is enforced in the strong sense, then equation (5.73) reduces to (5.24).

Patankar *et al.* (1999) presented a modification of the DLM/FD method described above. The modified formulation uses a distributed Lagrange multiplier to enforce a zero rate-of-deformation tensor within the particle domain instead of imposing rigid-body motion. An equivalent interpretation is to consider the rigid particles as a special kind of viscoelastic fluid with the density of the solid particles and the viscosity of the surrounding fluid. The elastic stress field inside the viscoelastic fluid has the ability to adjust itself to balance the traction acting on the surface of the particle due to the flow field outside, at the same time enforcing a velocity field such that $\mathbf{e}[\mathbf{u}] = 0$ or $\nabla \cdot \mathbf{e}[\mathbf{u}] = 0$ inside the particle domain. This viscoelastic fluid is effectively a linear elastic solid with an infinite shear modulus. Thus, in this modified formulation, the fluid–solid system is treated as a system consisting of two immiscible fluids similar in some sense to the point of view taken in

Chapter 2. Therefore, the equations of motion for the particles, (5.5) and (5.6), are no longer needed.

For fluid–solid systems, the weak formulation (5.73)–(5.75) is discretized in space by finite element methods. The fluid velocity $\mathbf{u}$ is approximated in a piecewise linear fashion on a regular finite element triangulation T_h, where h is the mesh size. The pressure p is approximated similarly on a regular triangulation T_{2h} twice as coarse. The approximation for the Lagrange multiplier $\boldsymbol{\lambda}$ is also piecewise linear, however, on a different finite element triangulation T_{P_h}. The finite element triangulation T_{P_h} must be coarser than the triangulation T_h to ensure numerical stability. Too many constraint equations from (5.75) make the overall system too stiff. The triangulation T_{P_h} inside the particles is usually irregular and moves with the particles. The evaluation of the inner product terms in (5.73) and (5.75) requires interpolation between the fixed regular mesh where the fluid velocity $\mathbf{u}$ is defined and the irregular mesh where the Lagrange multiplier $\boldsymbol{\lambda}$ is defined. The inner product terms can be evaluated more conveniently by using the collocation method to enforce the rigid body motion inside the particles, which corresponds to choosing the space for the Lagrange multiplier from a set of delta functions that uniformly cover the domain inside the particles (Glowinski *et al.*, 1999).

The direct finite difference discretization for the time derivatives in equations (5.73)–(5.75) will generate a system of nonlinear algebraic equations. Generally, this system will be difficult to solve numerically, since it combines the three principal difficulties: the incompressibility condition; the advection and diffusion terms; and the constraint of rigid-body motion in the particle domain $P(t)$ and the related distributed Lagrange multiplier. More efficient solvers based on operator-splitting algorithms could be used to decouple these difficulties. A simple first-order accurate Marchuk–Yanenko scheme can be expressed as[1]:

The approximate solution $\phi^{n+1} = \phi(t^{n+1})$ at time t^{n+1} of the system

$$\frac{d\phi}{dt} + A_1(\phi) + A_2(\phi) + A_3(\phi) = f, \tag{5.76}$$

where A_1, A_2 and A_3 are operators, can be obtained from the solution $\phi^n = \phi(t^n)$ at time t^n by solving the three successive problems:

$$\frac{\phi^{n+1/3} - \phi^n}{\Delta t} + A_1(\phi^{n+1/3}) = f_1^{n+1}, \tag{5.77}$$

$$\frac{\phi^{n+2/3} - \phi^{n+1/3}}{\Delta t} + A_2(\phi^{n+2/3}) = f_2^{n+1}, \tag{5.78}$$

[1] For a second-order version see Section 2.2 of Chapter 2.

$$\frac{\phi^{n+1} - \phi^{n+2/3}}{\Delta t} + A_3(\phi^{n+1}) = f_3^{n+1}, \tag{5.79}$$

where $f_1^{n+1} + f_2^{n+1} + f_3^{n+1} = f((n+1)\Delta t)$.

By applying the Marchuk–Yanenko scheme to the fluid–particle system (5.73)–(5.75), we obtain a numerical scheme for simulating the motion of solid particles in a Newtonian fluid:

DLM/FD particle mover

(1) Explicit update of particle positions,

$$\mathbf{X}_\alpha^{n+1} = \mathbf{X}_\alpha^n + \frac{\Delta t}{2}\left(\mathbf{v}_\alpha^n + \mathbf{v}_\alpha^{n-1}\right). \tag{5.80}$$

(2) Fractional step 1: find $\mathbf{u}^{n+1/3}$ and $p^{n+1/3}$ such that, $\forall \tilde{\mathbf{u}}$ and $\forall \tilde{p}$,

$$\int_\Omega \rho\left(\frac{\mathbf{u}^{n+1/3} - \mathbf{u}^n}{\Delta t}\right)\cdot\tilde{\mathbf{u}}\,dV - \int_\Omega p^{n+1/3}(\nabla\cdot\tilde{\mathbf{u}})\,dV + \int_\Omega \tilde{p}(\nabla\cdot\tilde{\mathbf{u}}^{n+1/3})\,dV = 0 \tag{5.81}$$

(3) Fractional step 2: find $\mathbf{u}^{n+2/3}$ such that, $\forall\tilde{\mathbf{u}}$,

$$\int_\Omega \rho\left(\frac{\mathbf{u}^{n+2/3} - \mathbf{u}^{n+1/3}}{\Delta t}\right)\cdot\tilde{\mathbf{u}}\,dV$$

$$+ \int_\Omega \rho\left[(\mathbf{u}^{n+1/3}\cdot\nabla)\mathbf{u}^{n+2/3}\right]\cdot\tilde{\mathbf{u}}dV$$

$$+2a\int_\Omega \mu\mathbf{e}[\mathbf{u}^{n+2/3}]:\mathbf{e}[\tilde{\mathbf{u}}]dV = 0 \tag{5.82}$$

(4) Fractional step 3: find $\mathbf{U}^{n+1} = \left(\mathbf{u}^{n+1}, \mathbf{v}_\alpha^{n+1}, \mathbf{\Omega}_\alpha^{n+1}\right) \in \mathcal{W}$, and $\boldsymbol{\lambda}^{n+1} = \left(\boldsymbol{\lambda}_1^{n+1},\ldots,\boldsymbol{\lambda}_N^{n+1}\right) \in \Lambda^{n+1}$ such that, $\forall\tilde{\mathbf{U}} \in \mathcal{W}_0$ and $\forall\tilde{\boldsymbol{\lambda}} \in \Lambda^{n+1}$,

$$\int_\Omega \rho\left(\frac{\mathbf{u}^{n+1} - \mathbf{u}^{n+2/3}}{\Delta t}\right)\cdot\tilde{\mathbf{u}}\,dV + 2(1-a)\int_\Omega \mu\mathbf{e}[\mathbf{u}^{n+1}]:\mathbf{e}[\tilde{\mathbf{u}}]\,dV$$

$$+ \left(1 - \frac{\rho}{\rho_s}\right)\left[\sum_\alpha\left(m_\alpha\frac{\mathbf{v}_\alpha^{n+1} - \mathbf{v}_\alpha^n}{\Delta t} - \mathbf{F}_\alpha^e\right)\cdot\tilde{\mathbf{v}}_\alpha\right.$$

$$\left.+ \sum_\alpha\frac{(J_\alpha\mathbf{\Omega}_\alpha)^{n+1} - (J_\alpha\mathbf{\Omega}_\alpha)^n}{\Delta t}\cdot\tilde{\mathbf{\Omega}}_\alpha\right]$$

$$= \sum_{1\leq\alpha\leq N}\left\langle\boldsymbol{\lambda}_\alpha^{n+1}, \tilde{\mathbf{u}} - \tilde{\mathbf{v}}_\alpha - \tilde{\mathbf{\Omega}}_\alpha\times(\mathbf{x} - \mathbf{X}_\alpha^{n+1})\right\rangle_{\Omega_\alpha^{n+1}} \tag{5.83}$$

$$\sum_{1\leq\alpha\leq N}\left\langle\tilde{\boldsymbol{\lambda}}_\alpha, \mathbf{u}^{n+1} - \mathbf{v}_\alpha^{n+1} - \mathbf{\Omega}_\alpha^{n+1}\times(\mathbf{x} - \mathbf{X}_\alpha^{n+1})\right\rangle_{\Omega_\alpha^{n+1}} = 0, \tag{5.84}$$

This scheme is due to Glowinski *et al.* (1997, 1999). At each time step, the scheme updates the particle positions explicitly, while it solves the particle velocities and the fluid flow field implicitly. The parameter a satisfies $0 \leq a \leq 1$. The solution of the first fractional step (5.81), $\mathbf{u}^{n+1/3}$, is the $L^2(\Omega)$-projection of the velocity field $\mathbf{u}$ on a divergence-free subspace. The pressure $p^{n+1/3}$ acts as a Lagrange multiplier. This step can be efficiently computed using a Uzawa/conjugate-gradient scheme. The second fractional step (5.82) is a classic linearized advection–diffusion problem that can be solved easily by a least-squares/conjugate-gradient algorithm. The third fractional step, equations (5.83) and (5.84), constitutes a diffusion problem with additional particle equations of motion and the constraint of rigid-body motion. A conjugate gradient algorithm was discussed in Glowinski *et al.* (1999).

5.12 Particle collision

It is not possible to simulate the motion of even a moderately dense suspension of particles without a strategy to handle cases in which particles touch. In various numerical methods, frequent near-collisions force large numbers of mesh points into the narrow gap between close particles and the mesh distorts rapidly, requiring an expensive high frequency of remeshing and projection. Different collision models have been developed to prevent near collisions while still conserving mass and momentum. In this section, we will discuss some of these models.

It is easy to show that, within the limits of a Newtonian fluid model, smooth rigid particles cannot touch: the gap between two particles cannot go to zero within a finite time. To have real collisions of smooth rigid particles, it is necessary for the fluid film between the particles to rupture and film rupture requires physics and mathematics beyond the Navier–Stokes equations. Furthermore, in practical situations, the particles are normally not perfectly smooth and rigid.

The first approach to modeling particle collisions is to provide a finer zone between the particles as they approach each other, and to use smaller time steps. This approach attempts to capture the collision process as exactly as numerically possible without introducing any modeling. Local mesh refinement schemes, such as the ones discussed in Section 5.7, are necessary. Numerical experiments based on local mesh refinements show good stability and robustness properties, see Hu *et al.* (1992, 1996). In these studies, the smallest gap size between the colliding particles was allowed to be as small as 10^{-5} times the particle diameter. Nevertheless, this approach has the drawback that there is no control of the computational cost.

The next approach is to use a solid-body collision model with a coefficient of restitution. This approach only models the collision process of the solid particles, while its consequences on the fluid is neglected. The fluid motion during the particle collision is quite complicated as a singularity develops at the time of contact. Johnson and Tezduyar (1996, 1997) have implemented this solid-body collision model. In their model, at each time step, once the total forces on the particles are obtained, the particle velocities and positions as given by (5.5) and (1.29) are explicitly updated by

$$\mathbf{v}_\alpha^{n+1} = \mathbf{v}_\alpha^n + \frac{\Delta t}{2m_\alpha} \left(\mathbf{F}_\alpha^{n+1} + \mathbf{F}_\alpha^n \right), \qquad (5.85)$$

and

$$\mathbf{X}_\alpha^{n+1} = \mathbf{X}_\alpha^n + \frac{\Delta t}{2} \left(\mathbf{v}_\alpha^{n+1} + \mathbf{v}_\alpha^n \right), \qquad (5.86)$$

for $\alpha = 1, 2, \ldots, N$, where $\mathbf{F}_\alpha^n$ and $\mathbf{F}_\alpha^{n+1}$ are the total forces acting on particle α at the time steps t^n and t^{n+1}, respectively. If, with the new positions of the particles, particle α is found to overlap with particle β, a collision occurs between these two particles. The velocities after the collision are modified according to

$$\hat{\mathbf{v}}_\alpha^{n+1} = \mathbf{v}_\alpha^{n+1} + \left[V_{n\alpha}^n - V_{n\alpha}^{n+1} - \frac{(1+e)m_\beta}{m_\alpha + m_\beta} \left(V_{n\alpha}^n + V_{n\beta}^n \right) \right] \mathbf{n}_\alpha, \qquad (5.87)$$

$$\hat{\mathbf{v}}_\beta^{n+1} = \mathbf{v}_\beta^{n+1} + \left[V_{n\beta}^n - V_{n\beta}^{n+1} - \frac{(1+e)m_\alpha}{m_\alpha + m_\beta} \left(V_{n\alpha}^n + V_{n\beta}^n \right) \right] \mathbf{n}_\beta, \qquad (5.88)$$

where e is the coefficient of restitution, $V_{n\alpha} = \mathbf{v}_\alpha \cdot \mathbf{n}_\alpha$, $V_{n\beta} = \mathbf{v}_\beta \cdot \mathbf{n}_\beta$, and $\mathbf{n}_\alpha = -\mathbf{n}_\beta$ is the unit normal vector pointing from the center of particle α to that of particle β. In deriving (5.87) and (5.88), it is assumed that the linear momentum of the two particles is conserved and the tangential forces are zero during the collision process. The velocity correction due to the collision is applied in an iterative fashion. The new particle positions (5.86) are updated with these corrected velocities, and checked again for other possible collisions.

Different collision models have been developed for the coupled solvers for the fluid and solid systems, for example in the ALE particle mover and DLM/FD particle mover. These collision models aim to capture the collision process for both the solid particles and the fluid motion by introducing short-range forces as additional body forces acting on the particles. A *security zone* is defined around the particles such that when the gap between particles is smaller than the security zone a repelling force is activated. This force can be thought of as representing, for example, surface roughness. The repelling force pushes the particles out of the security zone into the region in which fluid forces computed numerically govern. The different strategies differ in the nature of the repelling forces and how they are computed.

The scheme used by Glowinski *et al.* (1999) introduces a short-range repulsive force between particles near contact. This force takes an explicit form:

$$\mathbf{G}^{\mathrm{p}}_{\alpha,\beta} = \begin{cases} \mathbf{0}, & d_{\alpha\beta} > R_\alpha + R_\beta + \delta \\ \varepsilon_{\mathrm{p}} \left(\mathbf{X}_\alpha - \mathbf{X}_\beta\right) \left(R_\alpha + R_\beta + \delta - d_{\alpha\beta}\right)^2, & d_{\alpha\beta} \leq R_\alpha + R_\beta + \delta \end{cases}$$
(5.89)

where $d_{\alpha\beta}$ is the distance between the centers of the α and β particles, R_α is the radius of the α-th particle, δ is the force range, and ε_{p} is a large positive stiffness parameter. A similar repulsive force is introduced to handle the collision between a particle and the wall:

$$\mathbf{G}^{\mathrm{W}}_{\alpha\beta} = \begin{cases} \mathbf{0}, & d_{\alpha\beta} > 2R_\alpha + \delta \\ \varepsilon_{\mathrm{w}} \left(\mathbf{X}_\alpha - \mathbf{X}_{\alpha\beta}\right) \left(2R_\alpha + \delta - d_{\alpha\beta}\right)^2, & d_{\alpha\beta} \leq 2R_\alpha + \delta \end{cases}.$$
(5.90)

In (5.90) an imaginary particle of the same size, located symmetrically on the other side of the wall at $\mathbf{X}_{\alpha\beta}$, is introduced. The distance $d_{\alpha\beta}$ is between the particle α and its image with respect to the wall segment β. Another positive stiffness parameter ε_{w} controls the magnitude of this force.

Thus, the extra body force on the α-th particle, due to the collision with all the other particles and the walls, can be expressed as

$$\mathbf{G}^{(c)}_\alpha = \sum_{\beta=1,\beta\neq\alpha}^{N} \mathbf{G}^{\mathrm{p}}_{\alpha\beta} + \sum_{\beta=1}^{N_{\mathrm{wall}}} \mathbf{G}^{\mathrm{W}}_{\alpha\beta}.$$
(5.91)

This collision force is used in the combined momentum equations for the fluid and the solid to determine the fluid velocity field and the particle velocities. In this collision scheme, the choice of the stiffness parameters, ϵ_{p} and ϵ_{w}, is crucial. If the parameters are too small, the collision force will be too small, and collisions would be not prevented. On the other hand, if the parameters are too large, the repulsive force will be too strong, and the particles will bounce too much during the collision. In general, the optimum value may vary from case to case. In this collision scheme, there is no control on the minimum distance between the colliding particles and the particles may still overlap.

Another strategy due to Maury (1997) uses the lubrication force to separate touching particles and also requires the touching particles to transfer tangential as well as normal momentum.

In the collision scheme used in the ALE particle mover by Hu *et al.* (1996), the expression of the repelling force is not specified. One can imagine that the particles are not smooth, and there is a contact force between them when they approach within a distance of the size of the particle roughness. The magnitude of these contact forces are iteratively determined by requiring that the particles do not overlap and remain at the edge of the security zone. In this collision model, after the new particle position is explicitly updated using

$$\mathbf{X}_\alpha^{n+1} = \mathbf{X}_\alpha^n + \mathbf{v}_\alpha^n \Delta t + \frac{\Delta t^2}{2} \dot{\mathbf{v}}_\alpha^n, \tag{5.92}$$

the collisions between the particles and between the particle and the walls are first detected. If particle α is found to contact particle β, then a contact force $\mathbf{G}_{\alpha\beta}$, satisfying the relation $\mathbf{G}_{\alpha\beta} = -\mathbf{G}_{\beta\alpha}$, is introduced. The new positions of the particles are modified according to

$$\hat{\mathbf{X}}_\alpha^{n+1} = \mathbf{X}_\alpha^{n+1} + \left(\sum_{\beta=1,\beta\neq\alpha}^{N+N_{\text{wall}}} \mathbf{G}_{\alpha\beta} \right) \frac{\Delta t^2}{2m_\alpha}. \tag{5.93}$$

where N_{wall} is the number of boundary segments. The contact forces on particle α include all contributions from the neighboring particles and the boundary segments. For particles not in contact, $\mathbf{G}_{\alpha\beta} = 0$. The values of all the contact forces are iteratively determined such that the new distance between the particles originally in contact equals the size of the security zone:

$$d_{\alpha\beta} = \left| \hat{\mathbf{X}}_\alpha^{n+1} - \hat{\mathbf{X}}_\beta^{n+1} \right| = R_\alpha + R_\beta + \delta. \tag{5.94}$$

The procedure is applied iteratively, since the new particle positions may create new contact points. The final values of the contact forces contribute to a collision force on each particle,

$$\mathbf{G}_\alpha^{(c)} = \sum_{\beta=1,\beta\neq\alpha}^{N+N_{\text{wall}}} \mathbf{G}_{\alpha\beta} \tag{5.95}$$

which is introduced as an additional body force in the calculation of the particle motion.

Maury (1999) further developed the idea presented above into a more solid mathematical framework. The objective of the collision model is to find a modified particle configuration set (positions and orientations)

$$Y = (\mathbf{X}_1, \mathbf{X}_2, \ldots, \mathbf{X}_N, \mathbf{\Theta}_1, \mathbf{\Theta}_2, \ldots, \mathbf{\Theta}_N) \in R^{6N} \tag{5.96}$$

to minimize a functional

$$\Psi(Y) = \sum_{d_{\alpha\beta} < \delta} [d_{\alpha\beta}(Y) - \delta]^2 . \qquad (5.97)$$

The optimum configuration set Y_e can be obtained by performing a steepest descent algorithm on $\Psi(Y)$. This strategy was developed for two-dimensional smooth bodies of arbitrary shape.

It should be noted that the collision strategies that use a security zone to prevent close contact of the particles tend to keep the particles farther apart than they ought to be, resulting in lower particle volume fractions in the fluid–solid mixture. The size of the security zone should be kept as small as possible, which requires a balance between accuracy of the numerical scheme and computational cost.

5.13 Sample applications

The ALE particle mover described in Section 5.9 has been used in a number of studies investigating the behavior of solid particles in various flows of both Newtonian and viscoelastic fluids. Hu *et al.* (1992) first simulated two-dimensional sedimentation of circular and elliptic cylinders confined in a channel. Feng *et al.* (1994a, 1994b) studied the motion and interaction of circular and elliptical particles in sedimenting, Couette, and Poiseuille flows of a Newtonian fluid. Huang *et al.* (1994) examined the turning couples on an elliptic particle settling in a channel. Hu (1995) studied the rotation of a circular cylinder settling close to a solid wall. Feng *et al.* (1995) analyzed the mechanisms for the lifting of flying capsules in pipelines. Later Hu (1996) reported the results of two-dimensional direct numerical simulation of the motion of a large number of circular particles in a Newtonian fluid at particle Reynolds numbers around 100. Feng *et al.* (1996) also studied the sedimentation of circular particles in an Oldroyd-B fluid. Huang *et al.* (1997) examined the motion of particles in Couette and Poiseuille flows of viscoelastic and shear-thinning fluids. Huang *et al.* (1997) also investigated the effects of viscoelasticity and shear thinning on the stable orientation of ellipses falling in a viscoelastic fluid. Patankar (1997) investigated the rheology of suspensions of particles in both Newtonian and viscoelastic fluids. Zhu (1999) studied extensively the migration and interaction of spheres in various three-dimensional flows.

In this section we shall present results of some of the direct numerical simulations using the ALE particle mover, mostly due to Patankar (1997) and Zhu (1999).

5.13.1 Sedimentation of a single sphere in a tube

As a first example to illustrate the performance of the ALE particle mover we consider the drag on a sphere settling along the centerline of an infinitely long cylindrical tube filled with a Newtonian fluid with viscosity μ. The fluid inertia is turned off so that the numerical results can be compared with the exact Stokes-flow solution available for this problem. We define the ratio of the sphere diameter d to the tube diameter D as the blockage ratio and the ratio of the terminal drag F_D exerted on the sphere in the tube to the drag F_∞ in an infinite domain as the wall factor $K = F_D/F_\infty$. At the terminal velocity V, the drag on the sphere equals its effective weight:

$$F_D = \frac{\pi}{6}d^3(\rho_s - \rho)g, \tag{5.98}$$

and the drag F_∞ is given by the Stokes drag law:

$$F_\infty = 3\pi\mu V d. \tag{5.99}$$

The wall factor K is directly calculated from (5.98) and (5.99). Figure 5.10 shows a comparison between the numerical results and the theoretical ones of Haberman and Sayre (1958) for various blockage ratios. The agreement is

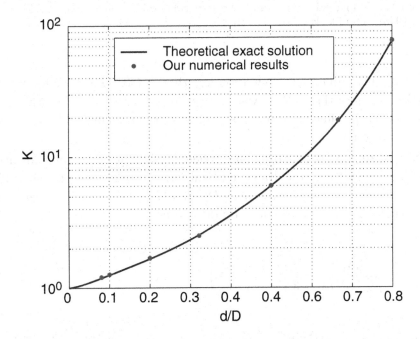

Fig. 5.10. Wall factor $K = F_D/F_\infty$ as a function of the blockage ratio d/D for the settling of a sphere of diameter d in a circular tube of diameter D in Stokes flow conditions. The comparison is between the numerical results of the ALE particle mover and the theoretical ones of Haberman and Sawyer (1958).

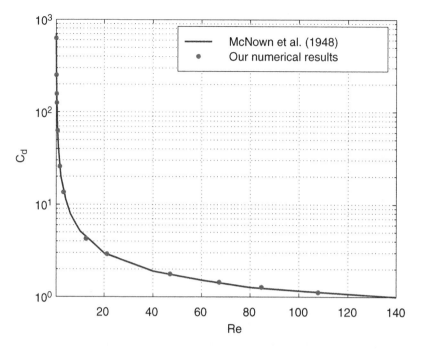

Fig. 5.11. Drag coefficient as a function of Reynolds number. Comparison is between the numerical results and the experimental measurements of McNown *et al.* (1948). The blockage ratio of the flow is $d/D = 0.3125$.

excellent, with a maximum deviation less than 0.4%. These results, as well all the other ones that follow, were proven to be insensitive (less than 1%) to further mesh refinement, an enlargement of the domain, and a reduction of the time step.

The next test checks the drag on a sphere settling at finite Reynolds number. Figure 5.11 shows the drag coefficient C_D as a function of the Reynolds number Re at a blockage ratio of $d/D = 0.3125$. The drag coefficient and the Reynolds number of the flow are defined as

$$C_D = \frac{F_D}{\rho V^2 \pi d^2/8} = \frac{4}{3}\frac{gd}{V^2}\left(\frac{\rho_s}{\rho} - 1\right), \qquad \text{and} \qquad Re = \frac{\rho V d}{\mu} \qquad (5.100)$$

where F_D is given by (5.98). A comparison of the calculated results with the experimental measurements by McNown *et al.* (1948) is shown in Fig. 5.11; the maximum deviation between the two is less than 0.5%.

5.13.2 Interaction of a pair of particles in a Newtonian fluid: Drafting–kissing–tumbling

One very important mechanism that controls the particle microstructure in flows of a Newtonian fluid is the so-called *drafting, kissing, and tumbling* (Hu *et al.*, 1993). There is a wake with low pressure at the back of a sedimenting particle. If a trailing particle is caught in this wake, it experiences a reduced drag and thus falls faster than the leading particle. This is called drafting, after the well-known bicycle racing strategy that is based on the same principle. The increased speed of fall impels the trailing particle into a kissing contact with the leading particle. Kissing particles form a long body that is unstable in a Newtonian fluid when the line of centers is along the stream. The same couples which force a long body to float broadside-on cause the two kissing particles to tumble. Tumbling particles in a Newtonian fluid induce an anisotropy in the configuration of the suspended particles since on average the line of centers between particles must be across the stream.

Figure 5.12 displays a sequence of the numerically simulated process of drafting–kissing–tumbling. Two spheres are dropped in tandem into an infinitely long tube filled with a Newtonian fluid. The fluid properties are selected as $\rho = 1.0$ g cm^{-3} and $\mu = 1.0$ poise, the particle density is

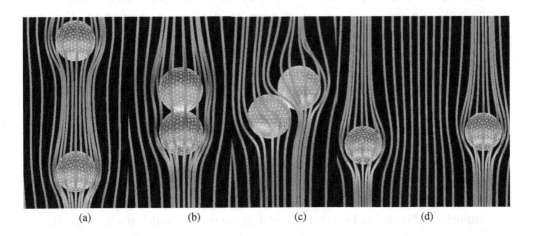

(a)   (b)   (c)   (d)

Fig. 5.12. Interaction between a pair of particles settling in a Newtonian fluid–drafting–kissing–tumbling. (a) streamlines at $t = 1.08$ s (drafting); (b) streamlines at $t = 1.45$ s (kissing); (c) streamlines at $t = 1.67$ s (tumbling); (d) streamlines at $t = 2.58$ s. See also http://www.cambridge.org/comp_mult_flow.

$\rho_s = 2.0\,\mathrm{g\,cm^{-3}}$, the particle diameter is $d = 2$ cm, and the tube diameter is $D = 20$ cm. This case corresponds to a particle Reynolds number of 22.

5.13.3 Interaction of particles in a viscoelastic fluid: Chaining

The interaction of two particles in a viscoelastic fluid is quite different from that in the Newtonian fluid. The particles still undergo drafting and kissing. However, due to the fact that the broadside-on sedimentation of a long body in a viscoelastic fluid is stable under certain conditions, kissing particles form a long body that is stable and will not tumble, see Joseph (1996). The simulation shown here involves two spheres of the same size released side-by-side in a tube filled with an Oldroyd-B fluid. The ratio of the sphere diameter to the tube diameter is $d/D = 0.2$. The spheres are symmetrically placed with respect to the axis of the tube and with a center-to-center separation of 1.5 times of the sphere diameter. The material properties for this fluid are μ = 30 poise, $\lambda = 0.1$ s, $\rho = 0.868$ g cm^{-3}, and the viscosity ratio $\mu_2/\mu = 1/8$. The density ratio of the solid to the fluid is $\rho_s/\rho = 2$, and the diameter of the sphere is $d = 2$ cm.

Figure 5.13 displays snapshots of the positions and orientations of the particles at various time instants. Figure 5.14 plots the particle trajectories, where y is the direction along the initial particle separation. It is observed that, after the particles are released, they attract each other. This attraction is caused by the strong shear flow on the outside surfaces of the particles, which produces a high pressure that pushes them toward each other. Once they are almost in contact, they momentarily form a long body. However broadside-on orientation of a long body is unstable in a viscoelastic fluid, and therefore it tends to turn longside along the direction of the fall. Thus the two spheres turn and eventually settle one behind the other in a chain.

If more particles are involved in the system, they tend to form longer chains. Due to the fact that a longer chain falls faster than a shorter one, a sufficiently long chain falls so fast that an isolated particle cannot catch up with it. Thus, there exists a maximum chain length, or a maximum number of particles in a chain (Patankar and Hu, 1997). The chaining of the particles in a viscoelastic fluid only occurs when the elastic behavior of the fluid dominates. When both fluid inertia and the fluid elasticity are important, the particles tend to form clusters (Hu and Patankar 1999).

5.13.4 Lubrication in pressure-driven particulate flows

Direct numerical simulation is also very useful to study the global behavior of a fluid–particle suspension. One can examine the short-time rheology of the

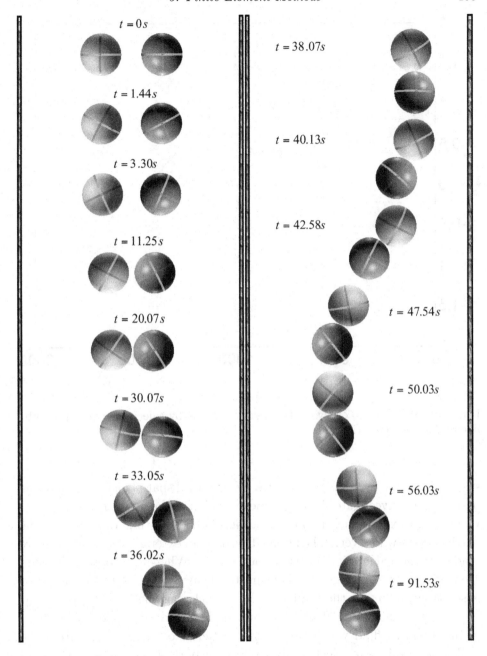

Fig. 5.13. Snapshots of the sedimentation of two spheres in a viscoelastic fluid. See also http://www.cambridge.org/comp_mult_flow.

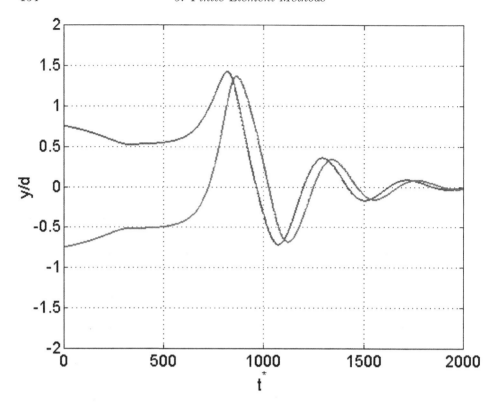

Fig. 5.14. Trajectories of two spheres released side-by-side in a cylindrical tube filled with a viscoelastic fluid.

suspension for a given microstructure (with a fixed spatial distribution of the particles), or investigate the long-time evolution of the microstructure. Here we show an example of the investigation of the long-time evolution of the particulate flow in a vertical channel driven by an externally applied pressure gradient; the calculation is two-dimensional. When the applied pressure gradient either assists gravity causing the heavy particles to fall faster or is so strong to pump the fluid and the particles against gravity, the shear stress at the channel wall is sufficiently high to induce a velocity gradient in the adjacent fluid which causes the particles to migrate away from the wall. The result is in effect a lubricated transport of the particles (Hu and Patankar, 1999). Figure 5.15 shows a typical case of this lubricated flow. It shows snapshots of 90 heavy particles (disks), with a density ratio of $\rho_s/\rho = 1.1$, in a Newtonian fluid being pumped upward against gravity at the initial instant and at a time instant later. For this simulation, the channel width is 12 times the particle diameter, and the volume (area) fraction of the particles

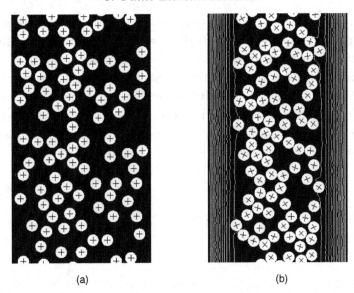

(a)   (b)

Fig. 5.15. Snapshots of the particle positions in a pressure-driven channel flow. White lines represent isolines of velocity in the x-direction. (a) Initial positions of the particles; (b) particle positions at $t = 16.4$ s.

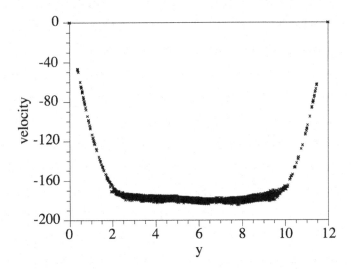

Fig. 5.16. Velocity component along the channel as a function of the coordinate across the channel.

is $\alpha = 26.8\%$. The Reynolds number of the flow is $Re = \rho V d/\mu = 12.26$, where the scale $V = (\rho_s - \rho)g d^2/(4\mu)$ is used for the slip velocity between the solid and fluid is used. The nondimensional pressure gradient is defined as $C_{\mathrm{p}} = |dp/dx|/[\alpha g(\rho_s - \rho)] = 2$.

We observe that the particles migrate towards the center of the channel forming a prominent core. Figure 5.16 shows that the velocity profile of the fluid becomes blunt due to migration of the particles away from the wall and their concentration at the center of the channel. The core nearly moves like a rigid body so that the velocity of the fluid varies almost linearly from the walls to the particle-rich core due to the absence of the particles in that region. This motion is similar to that established when a porous piston that occupies the zone of plug flow rises inside a channel.

6

Lattice Boltzmann models for multiphase flows

6.1 Brief history of the lattice Boltzmann method

In recent years, the lattice Boltzmann method (LBM) (Chen and Doolen, 1998; Succi, 2001) has become a popular numerical scheme for simulating fluid flows and modeling physics in fluids. The lattice Boltzmann method is based on a simplified mesoscopic equation, i.e. the discrete Boltzmann equation. By starting from mesoscopic modeling, instead of doing numerical discretizations of macroscopic continuum equations, the LBM can easily incorporate underlying physics into numerical solutions. By developing a simplified version of the kinetic equation, one avoids solving complicated kinetic equations such as the full Boltzmann equation. The kinetic feature of the LBM provides additional advantages of mesoscopic modeling, such as easy implementation of boundary conditions and fully parallel algorithms. This is arguably the major reason why the LBM has been quite successful in simulating multiphase flows. Furthermore, because of the availability of very fast and massively parallel computers, there is a current trend to use codes that can exploit the intrinsic features of parallelism. The LBM fulfills these requirements in a straightforward manner.

The lattice Boltzmann method was first proposed by McNamara and Zanetti in 1988 in an effort to reduce the statistical noise in lattice gas automaton (LGA) simulations (Rothman and Zaleski, 2004). The difference is that LBM uses real numbers to count particle population, while LGA only allows integers. The result has been phenomenal: the lattice Boltzmann model substantially reduced the statistic noise observed in LGA simulations. This makes the lattice-based model a more practical tool for computational fluid dynamics.

The original LGA model (Frisch, Hasslacher, and Pomeau, 1986) was proposed as a simplified molecular dynamics (MD) tool. Later studies showed that a Boltzmann equation can be used to describe the collective behavior of particles in LGA models (Frisch *et al.*, 1987). The lattice Boltzmann method took one step further by directly using the Boltzmann equation of LGA, the so-called lattice Boltzmann equation, to simulate fluid flows. In the early days, it was not clear how the lattice Boltzmann equation related to the Boltzmann equation of classical gas kinetic theory. Sterling and Chen (1985) first showed that the lattice Boltzmann equation can be regarded as a finite difference discretization of the Boltzmann equation. Later, He and Luo (1997b) proved that the lattice Boltzmann model can be derived rigorously from kinetic theory.

The kinetic nature of the LBM introduces some important features that distinguish it from traditional numerical methods. First, the convection operator of the LBM is in the particle velocity space and is linear. This feature is borrowed from kinetic theory and contrasts with the nonlinear convection terms in other macroscopic fluid equations. The upwind numerical approach can be achieved automatically through particle advection. Simple convection combined with a relaxation process (or collision operator) allows the recovery of the nonlinear macroscopic advection through multiscale expansions up to second-order numerical accuracy in space and time. Second, the incompressible Navier–Stokes (NS) equations can be obtained in the nearly incompressible limit of the LBM. The pressure of the LBM is calculated using an equation of state, avoiding the solution of the Poisson equation for the pressure. To a certain degree, this property is similar to the classic relaxation method. Third, the LBM utilizes a minimal set of discrete velocities in phase space compared with the complete Boltzmann equation. This greatly reduces the computational cost.

6.2 Lattice Boltzmann equations

It has been shown that the lattice Boltzmann equation can be obtained from the continuum Boltzmann equation with discrete velocities by using a small Mach number expansion. The Boltzmann equation for the velocity distribution function, f_a, can be written as follows:

$$\frac{\partial f_a}{\partial t} + \mathbf{e}_a \cdot \nabla f_a = \Omega(\mathbf{x}, t) = \frac{1}{\varepsilon \tau}(f_a^{\text{eq}} - f_a), \qquad (a = 0, 1, \ldots, M). \quad (6.1)$$

Here M is the number of discrete velocities, $f_a^{\text{eq}} = w_a \rho [1 + 3(\mathbf{e}_a \cdot \mathbf{u}) + \frac{9}{2}(\mathbf{e}_a \cdot \mathbf{u})^2 - \frac{3}{2}u^2]$ is the equilibrium distribution function, ρ is the density, $\mathbf{u}$ is the velocity of the fluid, $\mathbf{e}_a$ is the discrete velocity along the a direction, w_a

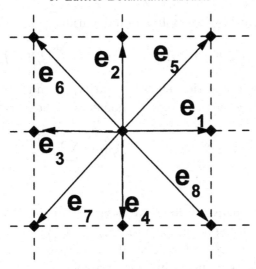

Fig. 6.1. The nine speeds for a two-dimensional square lattice in velocity space.

are suitable weights, τ is the relaxation time, and ε is a small parameter, proportional to the Knudsen number. In the above equation, we have employed the single time relaxation which simplifies the term in the right-hand side.

For a two-dimensional square lattice with nine speeds, we have (Fig. 6.1)

$$\mathbf{e}_a = (\Delta x/\Delta t)(\cos(\pi/2(a-1)), \sin(\pi/2(a-1))) \text{ for } a = 1, 3, 5, 7,$$

$$\mathbf{e}_a = \sqrt{2}(\Delta x/\Delta t)(\cos(\pi/2(a-1) + \pi/4, \sin(\pi/2(a-1)+\pi/4) \text{ for } a = 2, 4, 6, 8;$$

$$\mathbf{e}_0 = 0$$

It was found that it is necessary to take $w_0 = 4/9$, $w_a(a = 1, 3, 5, 7) = 1/9$ and $w_a(a = 2, 4, 6, 8) = 1/36$ in order to recover the macroscopic Navier–Stokes equations.

If in equation (6.1) the time derivative is replaced by a first-order time difference, and the first-order upwind discretization for the convective term $\mathbf{e}_a \cdot \nabla f_a$ is used and a downwind collision term $\Omega(\mathbf{x} - \Delta t \mathbf{e}_a, t)$ for $\Omega(\mathbf{x}, t)$ is used, then we obtain the finite difference equation for f_a:

$$\begin{aligned} f_a(\mathbf{x}, t + \Delta t) &= f_a(\mathbf{x}, t) - \alpha(f_a(\mathbf{x}, t) - f_a(\mathbf{x} - \Delta t \mathbf{e}_a, t)) \\ &\quad - \frac{\beta}{\tau}[f_a(\mathbf{x} - \Delta t \mathbf{e}_a, t) - f_a^{\mathrm{eq}}(\mathbf{x} - \Delta t \mathbf{e}_a, t)], \end{aligned} \quad (6.2)$$

where $\alpha = \Delta t |\mathbf{e}_a|/\Delta x$, $\beta = \Delta x/\varepsilon$, and Δt and Δx are the time step and the grid step, respectively. By choosing $\Delta t = \Delta x = \Delta y = \varepsilon$, equation (6.2) becomes the standard lattice Boltzmann equation:

$$f_a(\mathbf{x} + \Delta t \mathbf{e}_a, t + \Delta t) = f_a(\mathbf{x}, t) + \frac{\Delta t}{\tau}(f_a^{\text{eq}} - f_a). \qquad (6.3)$$

The macroscopic quantities, such as density, ρ, and momentum density, $\rho\mathbf{u}$, are defined as the particle velocity moments of the distribution function, f_a,

$$\rho = \sum_a f_a, \qquad \rho\mathbf{u} = \sum_a f_a \mathbf{e}_a, \qquad (6.4)$$

where $\sum_a \equiv \sum_{a=1}^M$.

If only the physics in the long-wavelength and low-frequency limit are of interest, the lattice spacing $\Delta x = \Delta t |\mathbf{e}_a|$ and the time increment Δt in equation (6.3) can be regarded as small parameters of the same order, ε. Performing a Taylor expansion in time and space, we obtain the following continuum form of the kinetic equation up to second-order in ε for (6.3),

$$\frac{\partial f_a}{\partial t} + \mathbf{e}_a \cdot \nabla f_a + \varepsilon \left(\frac{1}{2} \mathbf{e}_a \mathbf{e}_a : \nabla\nabla f_a + \mathbf{e}_a \cdot \nabla \frac{\partial f_a}{\partial t} + \frac{1}{2} \frac{\partial^2 f_a}{\partial t^2} \right) = \frac{\Omega_a}{\varepsilon}. \quad (6.5)$$

In order to recover the macroscopic hydrodynamic equation, the Chapman–Enskog expansion must be used, which is essentially a formal multiscaling expansion for the small parameter ε. After some tedious algebra, the resulting momentum equation is

$$\rho\left(\frac{\partial \mathbf{u}}{\partial t} + \nabla \cdot (\mathbf{uu}) \right) = -\nabla p + \nu\nabla \cdot \left[\nabla(\rho\mathbf{u}) + \nabla(\rho\mathbf{u})^{\text{T}} \right], \qquad (6.6)$$

which is exactly the same as the Navier–Stokes equation if density variations are small enough.

6.3 Lattice Boltzmann models for multiphase flow

The use of the lattice Boltzmann method to study multiphase flow can be dated back to the early days of the lattice gas automaton. Shortly after the model of Frisch et al. (1986) was published, Rothman and Keller (1988) proposed a color-fluid LGA model for multiphase flow. Several LBM multiphase models have been proposed since Rothman and Keller's work. They include the interparticle potential model (Shan and Chen, 1994), the free-energy model (Swift et al., 1995), the mean-field theory model (He and Luo, 1997) and the index-function model (He et al., 1999; He and Doolen, 2002). This section reviews the theoretical basis for these models. Some applications will be discussed in the next section.

6.3.1 Theory

6.3.1.1 Color-fluid model

The color-fluid model was first proposed by Rothman and Keller (1988) for lattice automata and later extended by Gunstensen *et al.* (1991) to the lattice Boltzmann method. In this model, red and blue "colored" particle or particle distribution functions were introduced to represent different fluids. The lattice Boltzmann equation for each phase is written as:

$$f_a^k(\mathbf{x} + \mathbf{e}_a, t + 1) = f_a^k(\mathbf{x}, t) + \Omega_a^k(\mathbf{x}, t), \qquad (6.7)$$

where k denotes either the red or blue fluid, and

$$\Omega_a^k = (\Omega_a^k)^1 + (\Omega_a^k)^2, \qquad (6.8)$$

is the collision operator. The first term in the collision operator, $(\Omega_a^k)^1$, represents the process of relaxation to the local equilibrium similar to the lattice BGK model (Chen *et al.*, 1991, 1992; Qian, d'Humières, and Lallemand, 1992), while the second collision operator, $(\Omega_a^k)^2$, contributes to the interfacial dynamics and generates a surface tension:

$$(\Omega_a^k)^2 = \frac{A_k}{2} |\mathbf{F}| \left[(\mathbf{e}_a \cdot \mathbf{F})^2 / |\mathbf{F}|^2 - 1/2 \right], \qquad (6.9)$$

where $\mathbf{F}$ is the local color gradient, defined as:

$$\mathbf{F} = \sum_a \mathbf{e}_a [\rho^r(\mathbf{x} + \Delta t \mathbf{e}_a) - \rho^b(\mathbf{x} + \Delta t \mathbf{e}_a)]. \qquad (6.10)$$

The densities of the color fluids are calculated from the distribution functions:

$$\rho^k = \sum_a f_a^k. \qquad (6.11)$$

Note that in a single-phase region of the incompressible fluid model, $\mathbf{F}$ vanishes. Therefore, the second term of the collision operator $(\Omega_a^k)^2$ only contributes to interfaces and mixing regions. The parameter A_k is a free parameter determining the surface tension. The collision operator $(\Omega_a^k)^2$ itself does not cause the phase separation. To maintain interfaces or to separate the different phases, the "colored-fluid" LBM model needs to force the local color momentum, $\mathbf{j} = \sum_a (f^r - f^b)\mathbf{e}_a$, to align with the direction of the local color gradient after collision. In other words, the colored distribution functions at the interfaces need to be redistributed to maximize $-\mathbf{j} \cdot \mathbf{F}$. Intuitively, this step will force colored fluids to move toward fluids with the same colors.

The "colored-fluid" model has two drawbacks. First, the procedure of redistribution of the colored density at each node requires time-consuming calculations of local maxima. Second, the perturbation step with the redistribution of colored distribution functions causes an anisotropic surface tension that induces unphysical vortices near interfaces. Blake *et al.* (1995) modified the model of Gunstensen *et al.* In this model, the recoloring step is replaced by an evolution equation for f_a^k that increases computational efficiency.

6.3.1.2 Models based on interparticle potentials

Using interparticle potentials to model multiphase flows in the LBM was first proposed by Shan and Chen (1993, 1994), although a similar concept had been proposed earlier in the framework of the lattice gas automaton (Chen *et al.*, 1989). Shan and Chen chose to use a nearest-neighbor inter-action model to incorporate the effect of the intermolecular potential. This interaction model can be considered as an approximation to the Lennard-Jones potential. The particle evolution of the Shan–Chen model follows the same governing equation as equation (6.7), with only one collision operator

$$\Omega_a = -\frac{f_a - f_a^{\text{eq}}(\rho, \boldsymbol{U}^{\text{eq}})}{\tau}. \tag{6.12}$$

The equilibrium distribution, however, differs from the single-phase LBM in that the equilibrium velocity is calculated as:

$$\rho \boldsymbol{U}^{\text{eq}} = \sum_a f_a \mathbf{e}_a + \tau \mathbf{F}. \tag{6.13}$$

The additional term, $\mathbf{F}$, represents an interaction force among particles:

$$\mathbf{F}(\mathbf{x}, t) = -\mathcal{G}\psi \sum_a w_a \psi(\mathbf{x} + \mathbf{e}_a, t)\mathbf{e}_a, \tag{6.14}$$

where $\psi(\rho)$ is a function of the density and $\mathcal{G}$ is the strength of the interpar-ticle potential. The weight coefficient w_a is not in the original Shan–Chen model but is required for nonhexagonal lattices. It is worth mentioning that the macroscopic velocity in the Shan–Chen model differs from $\boldsymbol{U}^{\text{eq}}$. The correct velocity is actually given by:

$$\boldsymbol{U}(\mathbf{x}, t) = \sum_a f_a \mathbf{e}_a + \frac{1}{2}\mathbf{F}. \tag{6.15}$$

The momentum equation can only be recovered using this velocity, instead of $\boldsymbol{U}^{\text{eq}}$. This subtle difference is the primary source for the so-called spurious currents observed in some publications (Hou *et al.*, 1997).

The Shan–Chen model has a much better physical meaning than the color-fluid model. The phase separation is governed now by the interparticle potential instead of heuristically maximizing the color-density gradient. In addition, the computation is more efficient and the interface is much smoother than in the color-fluid model.

The main drawback of the Shan–Chen model is the lack of thermodynamics as first pointed out by Swift *et al.* (1995). The thermodynamic inconsistency of the Shan–Chen model can be better explained by examining the pressure tensor. By Taylor-expanding equation (6.14) about **x** and recognizing that

$$\nabla \cdot \boldsymbol{P} = \nabla(\rho RT) - \mathbf{F} \tag{6.16}$$

must be satisfied at equilibrium, one has:

$$\boldsymbol{P} = \left[\rho RT + \frac{\mathcal{G}RT}{2}\psi^2 + \frac{\mathcal{G}(RT)^2}{2}\left(\psi\nabla^2\psi + \frac{1}{2}|\nabla\psi|^2\right)\right]\mathbf{I} - \frac{\mathcal{G}(RT)^2}{2}\nabla\psi\nabla\psi. \tag{6.17}$$

This pressure tensor implies that the Shan–Chen model has the two basic properties of non-ideal gases: an equation of state of the form

$$p_0 = \rho RT + \frac{\mathcal{G}RT}{2}\psi(\rho)^2, \tag{6.18}$$

and the surface tension (from the mechanical definition $\sigma = \int_{-\infty}^{\infty}(p_N - p_T)\,dz$),

$$\sigma^{sc} = \frac{\mathcal{G}RT}{2}\int_{-\infty}^{\infty}|\nabla\psi|^2\,dz. \tag{6.19}$$

Indeed, with proper choices of $\psi(\rho)$ and $\mathcal{G}$, the Shan–Chen model can be used to simulate many phase-separation and interface phenomena (Martys and Chen, 1996; Sankaranarayanan *et al.*, 1999; Shan and Chen, 1993, 1994). However, to be consistent with the equation of state in thermodynamic theory, we must have:

$$\psi = \sqrt{\frac{2(p_0 - \rho RT)}{\mathcal{G}RT}}. \tag{6.20}$$

On the other hand, to be consistent with the thermodynamic definition of the surface tension (Rowlinson and Widom, 1982):

$$\sigma^{\text{theory}} \propto \int_{-\infty}^{\infty}|\nabla\rho|^2\,dz, \tag{6.21}$$

we must have:

$$\psi \propto \rho. \tag{6.22}$$

Since equations (6.20) and (6.22) cannot be satisfied simultaneously, one can claim that the Shan–Chen model is thermodynamically inconsistent.

This thermodynamic inconsistency is due to the assumption that a molecule only interacts with its nearest neighbors. The idea of nearest neighbor interaction originates from the celebrated Ising model but may not be appropriate for describing molecular interactions in dense fluids. To be specific, the nearest-neighbor interaction model only has one characteristic length (the lattice size) and therefore is not sufficient to describe the Lennard-Jones potential where both the short-range repulsion and the long-range attraction are important.

6.3.1.3 Models based on free energy

It was first pointed out by Swift *et al.* (1995) that a successful LBM multiphase model must be consistent with thermodynamics. In other words, the equation of state, pressure tensor, chemical potential, etc. must be derivable from the free energy. For single-component fluids, the free energy has been given in the literature (see, e.g. Cahn and Hilliard, 1958). Obviously, the requirement of being able to derive these quantities from the free energy is a challenging but important constraint for developing consistent LBM multiphase models.

Unfortunately, unlike the thermodynamic theory for the equilibrium state where all variables can be determined from a free energy, there is no nonequilibrium theory relating the evolution of the density distribution to the free energy. In the original free-energy model proposed by Swift *et al.* (1995, 1996), the following constraint was imposed:

$$\sum f_a^{\mathrm{eq}} \mathbf{e}_a \mathbf{e}_a = \boldsymbol{P} + \rho \boldsymbol{UU}, \qquad (6.23)$$

where $\boldsymbol{P}$ is the pressure tensor defined in terms of the free energy, equation (6.25),

$$\boldsymbol{P} = \rho \frac{\delta \Psi}{\delta \rho} - \Psi, \qquad (6.24)$$

where

$$\Psi = \int \left[\psi(\rho, T) + \frac{\kappa}{2} |\nabla \rho|^2 \right] \, d\mathbf{r}, \qquad (6.25)$$

and ψ is the free-energy density. This constraint provides a convenient way to incorporate the free energy and hence has been widely used as a cornerstone in most of the free-energy-based LBM multiphase models (Osborn *et al.*, 1995; Holdych *et al.*, 1998; Palmer and Rector, 2000).

In the original free-energy model by Swift *et al.*, the following formula for f^{eq} is adopted:

$$f_a^{\mathrm{eq}} = A_a + B_a \mathbf{e}_a \cdot \boldsymbol{U} + C_a \boldsymbol{U}^2 + D_a (\mathbf{e}_a \cdot \boldsymbol{U})^2 + \mathbf{e}_a \cdot \mathbf{G}_a \cdot \mathbf{e}_a, \qquad (6.26)$$

where the coefficients A_a, B_a, C_a, D_a, and $\mathbf{G}_a$ are determined by the Chapman–Enskog expansion. More terms may also be included to recover Galilean invariance. Once f_a^{eq} is fixed, it can be substituted into the LBM-BGK model to simulate multiphase flow.

The free-energy model is important in that it relates the lattice Boltzmann method to thermodynamics, a crucial step for extending the LBM multiphase flow model for general applications. It allows one to exploit existing knowledge in the field of thermodynamics for nonuniform fluids.

The primary shortcoming of the original free-energy model is lack of Galilean invariance, which is a direct result of the constraint, equation (6.23). The left-hand side of equation (6.23) is the momentum flux due to molecular motion while the right-hand side is the total pressure tensor. These two properties are in general not equal. To be consistent, the correct constraint should be:

$$\sum f_a^{\mathrm{eq}} \mathbf{e}_a \mathbf{e}_a = \rho R T \mathbf{I} + \rho \boldsymbol{U}\boldsymbol{U}. \qquad (6.27)$$

To understand how the inconsistent definition, equation (6.23), destroys Galilean invariance, let us examine the viscous stress tensor in the Chapman–Enskog expansion (Swift *et al.*, 1996):

$$\boldsymbol{\Pi} = \frac{\nu}{RT} \partial_t \sum_a f_a^{\mathrm{eq}} \mathbf{e}_a \mathbf{e}_a + \frac{\nu}{RT} \nabla \cdot \sum (f_a^{\mathrm{eq}} \mathbf{e}_a \mathbf{e}_a \mathbf{e}_a). \qquad (6.28)$$

The second term on the right-hand side can be further written as:

$$\frac{\nu}{RT} \nabla \cdot \sum f_a^{\mathrm{eq}} (\mathbf{e}_a \mathbf{e}_a \mathbf{e}_a)$$
$$= \nu \{ \nabla(\rho \boldsymbol{U}) + [\nabla(\rho \boldsymbol{U})]^{\mathrm{T}} + \nabla \cdot (\rho \boldsymbol{U}) \mathbf{I} \}$$
$$= \rho \nu \{ \nabla \boldsymbol{U} + [\nabla \boldsymbol{U}]^{\mathrm{T}} \} + \nu \{ \boldsymbol{U} \nabla \rho + [\boldsymbol{U} \nabla \rho]^{\mathrm{T}} + \nabla \cdot (\rho \boldsymbol{U}) \mathbf{I} \}$$
$$= \rho \nu \{ \nabla \boldsymbol{U} + [\nabla \boldsymbol{U}]^{\mathrm{T}} \} - \frac{\nu}{RT} \partial_t \{ \rho R T \mathbf{I} + \rho \boldsymbol{U}\boldsymbol{U}) \} + \mathcal{O}(\boldsymbol{U}^3).$$

Together, we have the following viscous term:

$$\boldsymbol{\Pi} = \rho \nu \{ \nabla \boldsymbol{U} + [\nabla \boldsymbol{U}]^{\mathrm{T}} \} + \frac{\nu}{RT} \partial_t \left\{ \sum f_a^{\mathrm{eq}} \mathbf{e}_a \mathbf{e}_a - \rho R T \mathbf{I} - \rho \boldsymbol{U}\boldsymbol{U}) \right\} + \mathcal{O}(\boldsymbol{U}^3).$$
$$(6.29)$$

The first curly bracket represents the usual viscous stress tensor. The second curly bracket has no physical interpretation and it actually induces the lack of Galilean invariance whenever a density gradient exists. This unphysical term can be cancelled if and only if:

$$\sum f_a^{\mathrm{eq}} \mathbf{e}_a \mathbf{e}_a = \rho R T \mathbf{I} + \rho \boldsymbol{U} \boldsymbol{U}.$$

Any other constraint including equation (6.23) does not achieve an exact cancellation and consequently causes a lack of Galilean invariance.

The inconsistency of the constraint equation (6.23) has been noticed before. Swift et al. (1996) themselves have tried to add density gradient terms to reduce the non-Galilean invariance. This approach was further extended by Holdych et al. (1998). This modification indeed reduces the lack of Galilean invariance to order $\boldsymbol{U}^2$, but it does not eliminate the error to order $\boldsymbol{U}^3$.

There is one scenario in which the model of Swift et al. works consistently: binary fluids with both fluids being ideal gases. In this case, the constraint equation (6.23) is equivalent to equation (6.27) and, consequently, Galilean invariance is guaranteed.

6.3.1.4 Mean-field model

To resolve the difference between various LBM multiphase models, He et al. (1998) proposed an LBM multiphase model based on mean-field theory and Enskog's model for dense fluids. It was demonstrated later that the mean-field model can be derived from the BBGKY theory with appropriate approximations (He and Doolen, 2002). The key to the mean-field model is to use the mean-field theory to describe the long-range attraction among molecules, while using the Enskog theory for dense fluids to account for the short-range repulsion. The mean-field model not only recovers the correct mass, momentum, and energy equations, but it also contains the correct thermodynamics.

Historically, kinetic theory was first developed for studying ideal gas transport (Chapman and Cowling, 1970). To extend its application to phase transitions and multiphase flows, one must incorporate molecular interactions which become increasingly important in most fluids as the density increases. The most rigorous way to incorporate molecular interactions would be to start from the BBGKY equations (Chapman and Cowling, 1970; Reichl, 1998). In the theory of the BBGKY hierarchy, the evolution equation for the single-particle distribution function, $f(\boldsymbol{\xi}_1, \mathbf{r}_1)$, is:

$$\partial_t f + \boldsymbol{\xi}_1 \cdot \nabla_{\mathbf{r}_1} f + \mathbf{F} \cdot \nabla_{\xi_1} f = \int \int \frac{\partial f^{(2)}}{\partial \boldsymbol{\xi}_1} \cdot \nabla_{\mathbf{r}_1} V(r_{12}) \, d\boldsymbol{\xi}_2 d\mathbf{r}_2, \qquad (6.30)$$

where $\mathbf{F}$ is the external force, $\boldsymbol{\xi}_1$ and $\boldsymbol{\xi}_2$ are microscopic velocities, $f^{(2)}(\boldsymbol{\xi}_1, \mathbf{r}_1, \boldsymbol{\xi}_2, \mathbf{r}_2)$ is the two-particle distribution function, and $V(r_{12})$ is the pairwise intermolecular potential. In the BBGKY hierarchy of equations, the time evolution of the n-particle distribution depends on the $(n + 1)$-th particle distribution. Approximations have to be introduced to close this formulation.

Here, we use a simple closure at the level of the two-particle distribution. To do so, let us divide the spatial integral domain in the right-hand side of equation (6.30) into two parts, $\{\mathcal{D}_1 : |\mathbf{r}_2 - \mathbf{r}_1| < d\}$ and $\{\mathcal{D}_2 : |\mathbf{r}_2 - \mathbf{r}_1| \geq d\}$:

$$\int \int \frac{\partial f^{(2)}}{\partial \boldsymbol{\xi}_1} \cdot \nabla_{\mathbf{r}_1} V(r_{12}) \, d\boldsymbol{\xi}_2 d\mathbf{r}_2 = \underbrace{\int_{\mathcal{D}_1} \int \frac{\partial f^{(2)}}{\partial \boldsymbol{\xi}_1} \cdot \nabla_{\mathbf{r}_1} V(r_{12}) \, d\boldsymbol{\xi}_2 d\mathbf{r}_2}_{\mathcal{I}_1}$$

$$+ \underbrace{\int_{\mathcal{D}_2} \int \frac{\partial f^{(2)}}{\partial \boldsymbol{\xi}_1} \cdot \nabla_{\mathbf{r}_1} V(r_{12}) \, d\boldsymbol{\xi}_2 d\mathbf{r}_2}_{\mathcal{I}_2}, \quad (6.31)$$

where d is the effective diameter of molecules. It is known that many intermolecular potentials can be approximated by the Lennard-Jones potential, which possesses a short-range strong repulsive core and a long-range weak attractive tail. In the above partition, the first integral, $\mathcal{I}_1$, describes the short-range molecular interaction dominated by the strong repulsive force, while the second integral, $\mathcal{I}_2$, describes the long-range molecular interaction which causes a weak attractive force.

Since the short-range molecular interaction is dominated by strong repulsion, it is essentially a collision process. The rate of change of the single-particle distribution in this process, $\mathcal{I}_1$, can be well modeled by Enskog's theory for dense fluids (Chapman and Enskog, 1970):

$$\begin{aligned} \mathcal{I}_1 &= \int_{\mathcal{D}_1} \int \frac{\partial f^{(2)}}{\partial \boldsymbol{\xi}_1} \cdot \nabla_{\mathbf{r}_1} V(r_{12}) \, d\boldsymbol{\xi}_2 d\mathbf{r}_2 \\ &= \chi \Omega_0 - b\rho\chi f^{\mathrm{eq}} \\ &\quad \left\{ (\boldsymbol{\xi} - \boldsymbol{U}) \cdot \left[\nabla \ln(\rho^2 \chi T) + \frac{3}{5} \left(C^2 - \frac{5}{2} \right) \nabla \ln T \right] \right. \\ &\quad \left. + \frac{2}{5} \left[2\mathbf{CC} : \nabla \boldsymbol{U} + \left(C^2 - \frac{5}{2} \right) \nabla \cdot \boldsymbol{U} \right] \right\}, \quad (6.32) \end{aligned}$$

where Ω_0 is the ordinary collision term which neglects particle size; $\mathbf{C} = (\boldsymbol{\xi} - \boldsymbol{U})/\sqrt{2RT}$ and $C = |\mathbf{C}|$; ρ, $\boldsymbol{U}$, and T are the macroscopic density, velocity, and temperature, respectively. f^{eq} is the equilibrium distribution function:

$$f^{\mathrm{eq}} = \frac{\rho}{(2\pi RT)^{3/2}} \exp\left[-\frac{(\boldsymbol{\xi} - \boldsymbol{U})^2}{2RT} \right]. \quad (6.33)$$

χ is the density-dependent collision probability,

$$\chi = 1 + \frac{5}{8}b\rho + 0.2869(b\rho)^2 + 0.1103(b\rho)^3 + 0.0386(b\rho)^4 + \cdots, \qquad (6.34)$$

where $b = 2\pi d^3/3m$, with d the diameter and m the molecular mass. Notice that the χ corresponding to the van der Waals equation of state is:

$$\chi = \frac{1}{1 - b\rho}, \qquad (6.35)$$

which only agrees with equation (6.34) to zeroth order.

The rate of change of the single-particle distribution due to long-range molecular interactions, $\mathcal{I}_2$, is neglected in Enskog's original work. It can be very important in real fluids as elucidated in the van der Waals theory of liquids (see, e.g. Rowlinson and Widom, 1982). Modern physics has shown that, for most liquids, the radial distribution function is approximately unity beyond a distance of one molecular diameter (Reichl, 1998). This implies that $f^{(2)}(\boldsymbol{\xi}_1, \mathbf{r}_1, \boldsymbol{\xi}_2, \mathbf{r}_2) \approx f(\boldsymbol{\xi}_1, \mathbf{r}_1)f(\boldsymbol{\xi}_2, \mathbf{r}_2)$ in $\mathcal{D}_2$. This approximation leads to:

$$\begin{aligned}
\mathcal{I}_2 &= \int_{\mathcal{D}_2} \int \frac{\partial f^{(2)}}{\partial \boldsymbol{\xi}_1} \cdot \nabla_{\mathbf{r}_1} V(r_{12}) \, d\boldsymbol{\xi}_2 d\mathbf{r}_2 \\
&= \nabla \left\{ \int_{\mathcal{D}_2} \rho(\mathbf{r}_2) V(r_{12}) \, d\mathbf{r}_2 \right\} \cdot \nabla_{\boldsymbol{\xi}_1} f.
\end{aligned} \qquad (6.36)$$

The term in brackets is exactly the mean-field approximation for the intermolecular potential (Rowlinson and Widom, 1982):

$$V_m = \int_{\mathcal{D}_2} \rho(\mathbf{r}_2) V(r_{12}) \, d\mathbf{r}_2. \qquad (6.37)$$

Its gradient gives the average force acting on a molecule by the surrounding molecules. Assuming the density to be a slowly varying variable, we can expand it in a Taylor series:

$$\rho(\mathbf{r}_2) = \rho(\mathbf{r}_1) + \mathbf{r}_{21} \cdot \nabla \rho + \frac{1}{2}\mathbf{r}_{21}\mathbf{r}_{21} : \nabla\nabla\rho + \cdots, \qquad (6.38)$$

where $\mathbf{r}_{21} = \mathbf{r}_2 - \mathbf{r}_1$. Substituting equation (6.38) into equation (6.37), we have:

$$V_m = -2a\rho - \kappa\nabla^2\rho, \qquad (6.39)$$

where the coefficients a and κ are defined in terms of the intermolecular potential by:

$$a = -\frac{1}{2} \int_{r>d} V(r)\, d\mathbf{r},$$

$$\kappa = -\frac{1}{6} \int_{r>d} r^2 V(r)\, d\mathbf{r}.$$

a and κ are usually assumed to be constant. The integral $\mathcal{I}_2$ subsequently becomes:

$$\mathcal{I}_2 = \nabla V_m \cdot \nabla_{\xi_1} f. \tag{6.40}$$

This form of $\mathcal{I}_2$ suggests that the average long-range intermolecular potential acts on a molecule exactly as an external potential. In other words, the long-range molecular interactions can be modeled as a local point force. It should be mentioned that the above derivations are based on the assumption that the density varies slowly. Although this assumption has been used widely in the literature (Rowlinson and Widom, 1982), it should nevertheless be regarded as an approximation.

Combining Enskog's theory for dense fluids and the mean-field theory for the intermolecular potential, one obtains the following kinetic equation to describe the flow of a dense fluid (He *et al.*, 1998):

$$\partial_t f + \boldsymbol{\xi} \cdot \nabla f + \mathbf{F} \cdot \nabla_{\xi} f = \mathcal{I}_1 + \nabla V_m \cdot \nabla_{\xi} f, \tag{6.41}$$

where the subscripts have been dropped for simplicity. The macroscopic fluid density, ρ, velocity, $\mathbf{U}$, and temperature, T, are calculated as the velocity moments of the distribution function:

$$\rho = \int f\, d\boldsymbol{\xi}, \tag{6.42}$$

$$\rho \mathbf{U} = \int \boldsymbol{\xi} f\, d\boldsymbol{\xi}, \tag{6.43}$$

$$\frac{3\rho R T}{2} = \int \frac{(\boldsymbol{\xi} - \mathbf{U})^2}{2} f\, d\boldsymbol{\xi}. \tag{6.44}$$

It should be pointed out that, unlike $\mathcal{I}_2$, $\mathcal{I}_1$ in general cannot be expressed as the product of a single force and a velocity gradient of the distribution function. As a result, the molecular interaction as a whole cannot be modeled by a single force term either, as noticed by Luo (1998, 2000).

The lattice Boltzmann multiphase model can be further derived by discretizing equation (6.41) in time and space. The governing equation of the mean-field model is written as:

$$f_a(\mathbf{r} + \mathbf{e}_a \Delta t, t + \Delta t) - f_a(\mathbf{r}, \Delta t) = -\chi \frac{f_a - f_a^{\text{eq}}}{\tau_* + \frac{1}{2}} + \frac{\tau_*}{\tau_* + \frac{1}{2}} \Omega_a f^{\text{eq}} \Delta t, \tag{6.45}$$

where, here and in the following, $\tau_* = \tau/\Delta t$ with τ the relaxation parameter;

$$\Omega_a = \frac{(\mathbf{e}_a - \mathbf{U}) \cdot (\mathbf{F} - \nabla V_m)}{RT} - b\rho\chi$$
$$\left\{ (\mathbf{e}_a - \mathbf{U}) \cdot \left[\nabla \ln(\rho^2\chi T) + \frac{3}{5} \left(C_a^2 - \frac{5}{2} \right) \nabla \ln T \right] \right.$$
$$\left. + \frac{2}{5} \left[2\mathbf{C}_a\mathbf{C}_a : \nabla\mathbf{U} + \left(C_a^2 - \frac{5}{2} \right) \nabla \cdot \mathbf{U} \right] \right\}, \tag{6.46}$$

$\mathbf{C}_a = (\mathbf{e}_a - \mathbf{U})/\sqrt{2RT}$ and $C_a = |\mathbf{C}_a|$. The equilibrium distributions have the following form:

$$f_a^{\text{eq}} = w_a \left[1 + \left(\frac{\boldsymbol{\xi}^2}{2RT_0} - \frac{3}{2} \right) \theta + \frac{\boldsymbol{\xi} \cdot \mathbf{U}}{RT_0} + \frac{(\boldsymbol{\xi} \cdot \mathbf{U})^2}{2(RT_0)^2} - \frac{\mathbf{U}^2}{2RT_0} \right]. \tag{6.47}$$

Notice that at least a second-order time integration scheme is necessary for LBM multiphase models (He et $al.$, 1998), otherwise, unphysical properties such as spurious currents arise in simulations.

The macroscopic variables can be calculated using:

$$\rho = \sum_a f_a, \tag{6.48}$$

$$\rho\mathbf{U} = \sum_a f_a\mathbf{e}_a + \frac{\Delta t}{2} [\rho\mathbf{F} - \rho\nabla V_m - \nabla(b\rho^2\chi RT)] \tag{6.49}$$

$$\frac{3\rho RT}{2} = \frac{3\rho RT_0(1+\theta)}{2} = \sum_a f_a \frac{(\mathbf{e}_a - \mathbf{U})^2}{2}. \tag{6.50}$$

The viscosity and thermal conductivity of the above model have the following form:

$$\mu = \tau\rho RT_0\Delta t \left(\frac{1}{\chi} + \frac{2}{5}b\rho \right), \tag{6.51}$$

$$\lambda = \frac{5}{2}\tau\rho R^2 T_0\Delta t \left(\frac{1}{\chi} + \frac{3}{5}b\rho \right). \tag{6.52}$$

For isothermal nearly incompressible flow, the above LBM multiphase model reduces to the model proposed in He and Luo (1997a)

$$f_a(\mathbf{r} + \mathbf{e}_a\Delta t, t + \Delta t) - f_a(\mathbf{r}, \Delta t) = -\chi\frac{f_a - f_a^{\text{eq}}}{\tau_* + \frac{1}{2}} + \frac{\tau_*}{\tau_* + \frac{1}{2}}\frac{(\mathbf{e}_a - \mathbf{U}) \cdot \mathbf{F}}{RT}f^{\text{eq}}\Delta t \tag{6.53}$$

in which the equilibrium distributions have the following form:

$$f_a^{\text{eq}} = w_a \left[1 + \frac{\boldsymbol{\xi} \cdot \mathbf{U}}{RT} + \frac{(\boldsymbol{\xi} \cdot \mathbf{U})^2}{2(RT)^2} - \frac{\mathbf{U}^2}{2RT} \right]. \tag{6.54}$$

The mean-field LBM multiphase model is derived from kinetic theory with the intermolecular potential incorporated intrinsically. From this perspective, it inherits the fundamental feature of the interparticle potential model. At the same time, the mean-field theory guarantees thermodynamic consistency.

The drawback of the mean-field LBM multiphase model is that it cannot simulate multiphase flows with high density ratio. This drawback is likely due to the assumption that the density profile across an interface must be smooth. How to improve LBM multiphase models to simulate high density-ratio flows is still a challenging task. There have been several advances on this front, including using TVD-ENO schemes (Teng *et al.*, 2000) and solving the pressure field separately; see Inamuro *et al.* (2004) and Lee and Lin (2005) for recent research.

6.3.1.5 Phase function model

The LBM multiphase flow model in principle can only be used to study nearly incompressible flow. For many applications where gravity is involved, gravity-induced density variations may be too large and introduce spurious phase separation, especially near the top and bottom of a domain. To resolve this problem, He *et al.* (1999a,b) modified the mean-field model by assuming a constant density for different fluids. The phase evolution is tracked by an imaginary phase function, the so-called "index function". The index function satisfies a nonideal gas equation of state and therefore can maintain original separate phases as flow proceeds. The surface tension is still modeled by intermolecular force via the index function.

The evolution equation for the phase function model is written as:

$$f_a(\mathbf{x} + \mathbf{e}_a \Delta t, t + \Delta t) - f_a(\mathbf{x}, t) = -\frac{f_a(\mathbf{x}, t) - f_a^{\text{eq}}(\mathbf{x}, t)}{\tau_*} \tag{6.55}$$

$$- \frac{(2\tau_* - 1)}{2\tau_*} \frac{(\mathbf{e}_a - \boldsymbol{U}) \cdot \nabla \psi(\phi)}{RT} \Gamma_a(\boldsymbol{U}) \Delta t,$$

$$g_a(\mathbf{x} + \mathbf{e}_a \Delta t, t + \Delta t) - g_a(\mathbf{x}, t) = -\frac{g_a(\mathbf{x}, t) - g_a^{\text{eq}}(\mathbf{x}, t)}{\tau_*} \tag{6.56}$$

$$+ \frac{2\tau_* - 1}{2\tau_*} (\mathbf{e}_a - \boldsymbol{U}) \cdot [\Gamma_a(\boldsymbol{U})(\mathbf{F}_s + \mathbf{g})$$

$$- (\Gamma_a(\boldsymbol{U}) - \Gamma_a(0)) \nabla (p - \rho RT)] \Delta t,$$

where f is the distribution function for phase function ϕ, g is the distribution function for pressure, τ_* is the ratio of relaxation time and time step, $\mathbf{g}$ is the acceleration of gravity, $\Gamma(\boldsymbol{U})$ is a function of the macroscopic velocity $\boldsymbol{U}$:

$$\Gamma(\boldsymbol{U}) = \frac{1}{(2\pi RT)^{D/2}} \exp\left[-\frac{(\boldsymbol{\xi} - \boldsymbol{U})^2}{2RT}\right], \qquad (6.57)$$

with D the dimensionality of the space, and ψ is a function of the phase function:

$$\psi(\phi) = \rho RT(1 + b\rho\chi) - a\rho^2. \qquad (6.58)$$

χ can be determined by an equation of state.

A third-order differencing scheme was used to calculate $\nabla\psi$. The macroscopic variables can be calculated using:

$$\phi = \sum f_a, \qquad (6.59)$$

$$p = \sum g_a - \frac{1}{2}\boldsymbol{U} \cdot \nabla\psi(\rho)\Delta t, \qquad (6.60)$$

$$\rho RT\boldsymbol{U} = \sum e_a g_a + \frac{RT}{2}(\mathbf{F_s} + \mathbf{g})\Delta t. \qquad (6.61)$$

6.3.2 Applications of the lattice Boltzmann multiphase models

The application of the LBM multiphase method goes hand in hand with model development. Successful examples include flow in porous media, Rayleigh–Taylor instability simulation and mixing, boiling dynamics, ternary flow (Chen *et al.*, 2000), and spinodal decomposition (Alexander, Chen, and Grunau, 1993; Osborn *et al.*, 1995).

6.3.2.1 Multiphase flow in porous media

Lattice Boltzmann multiphase fluid flow models have been extensively used to simulate multicomponent flow in porous media to understand the fundamental physics associated with enhanced oil recovery, including permeabilities. The LBM is particularly useful for this problem because of its capability of handling complex geometrical boundaries and varying physical parameters, including viscosity and wettability. In the work by Buckles *et al.* (1994), Martys and Chen (1996), and Ferreol and Rothman (1995), realistic sandstone geometries from oil fields were used. Very complicated flow patterns were observed. The numerical values of the relative permeability as a function of percent saturation of wetting fluid agree qualitatively with experimental data.

Gunstensen and Rothman (1993) studied the linear and nonlinear multicomponent flow regimes corresponding to large and small flow rates, respectively. They found, for the first time, that the traditional Darcy law must be modified in the nonlinear regime because of the capillary effect. Zhang, Zhang, and Chen (2000) studied flow in porous media at the pore scale with lattice Boltzmann simulations on pore geometries reconstructed from computed microtomographic images. Pore scale results were analyzed to give quantities such as permeability, porosity, and specific surface area at various scales and at various locations. With this, some fundamental issues such as scale dependency and medium variability were assessed quantitatively. More specifically, they quantified the existence and size of the well-known concept of representative elementary volume (REV). They suggested that, for heterogeneous media, a better measure may be the so-called "statistical REV," which has weaker requirements than does the deterministic REV. Kang *et al.* (2002a) further incorporated surface chemical reactions to study corrosion in porous media and developed a unified lattice Boltzmann method for flow in multiscale porous media (Kang *et al.*, 2002b).

6.3.2.2 Rayleigh–Taylor instability

The quantitative accuracy of the lattice Boltzmann multiphase model was best demonstrated in studies by He *et al.* (1999a,b) on the Rayleigh–Taylor instability. He *et al.* used a phase-function model and the simulation results agree well with those reported in the existing literature.

Figure 6.2 shows the growth of the Rayleigh–Taylor instability from a small single-mode perturbation in a computational domain of width W. The Atwood number $A = (\rho_h - \rho_l)/(\rho_h + \rho_l)$, where ρ_h and ρ_l are densities of heavy and light fluids, respectively, is 0.5. The Reynolds number $Re = \sqrt{Wg}W/\nu$ is 2048. The simulation was carried out on a 256×1024 grid. The gravity was chosen so that $\sqrt{Wg} = 0.04$. The interface was represented by 19 equally spaced density contours. During the early stages ($t < 1.0$), the growth of the fluid interface remains symmetric up and down. Later, the heavy fluid falls down as a spike and the light fluid rises to form bubbles. Starting from $t = 2.0$, the heavy fluid begins to roll up into two counter-rotating vortices. This phenomenon was first computed by Daly (1967) and studied further by other authors (Youngs, 1984; Tryggvason, 1988; Glimm *et al.*, 1986; Mulder *et al.*, 1992). At a later time ($t = 3.0$), these two vortices become unstable and a pair of secondary vortices appear at the tails of the roll-ups. The roll-ups and vortices in the heavy fluid spike are due to the Kelvin–Helmholtz instability (Daly, 1967). The shapes of the fluid interface in the current study compare well with previous studies (Daly, 1967; Tryggvason, 1988).

In the later stage of the Rayleigh–Taylor instability ($t = 5.0$), the heavy fluid falling down gradually forms one central spike and two side spikes. It is

174 6. Lattice Boltzmann Models

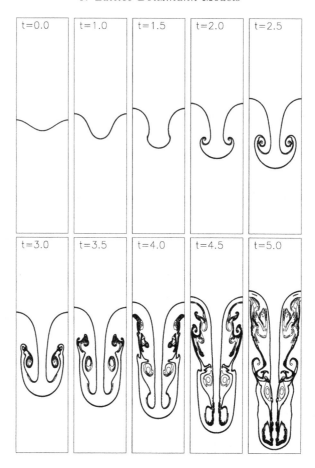

Fig. 6.2. Evolution of the fluid interface from a single-mode perturbation in the Rayleigh–Taylor instability. The Atwood number is 0.5 and the Reynolds number is 2048. A total of 19 density contours are plotted. The time is measured in units of $\sqrt{W/g}$.

interesting to note that only the side spikes of the heavy fluid experience the Kelvin–Helmholtz instability and the interface along these two spikes is stretched and folded into very complicated shapes. The mixing of heavy and light fluid is significant. On the other hand, the interface along the central spike, as well as the fronts of both bubble and spike, remain relatively smooth.

A more interesting study is the simulations of multiple-mode Rayleigh–Taylor instability. The simulation was carried out on a 512×512 grid. The Reynolds number was chosen as 4096 and the Atwood number was fixed at 0.5. Gravity was chosen so that $\sqrt{Wg} = 0.08$. The initial perturbation of the fluid interface was given by:

$$h = \sum_n a_n \cos(k_n x) + b_n \sin(k_n x), \qquad (6.62)$$

where $k_n = 2n\pi/W$ is the wavenumber. The amplitudes, a_n and b_n, were randomly chosen from a Gaussian distribution. A total of 10 modes ($n \in [21, 30]$) was used.

The evolution of this multiple-mode Rayleigh–Taylor instability is shown in Fig. 6.3 at six time instants. The early stage ($t < 1.0$) is characterized by the growth of structures with small wavenumbers. The heavy fluid falls down as slender spikes while the light fluid rises up as small bubbles. The amplitudes of the perturbation have grown much larger than the initial wavelength at $t = 1.0$. The interaction among small structures becomes obvious at $t = 2.0$ and continues throughout the simulation. Three features can be observed during this stage. First, the larger bubbles rise faster than the smaller ones. This is because the bubble at this stage moves proportionally to its size. Second, small bubbles continue to merge into larger ones. At the time of $t = 4.0$, the initial perturbation has totally disappeared and

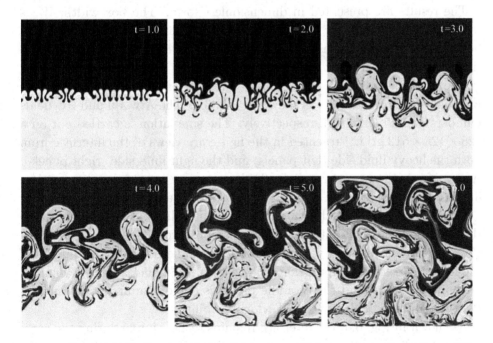

Fig. 6.3. Rayleigh–Taylor instability from a multiple mode perturbation. The Atwood number is 0.5 and the Reynolds number is 4096. The time is measured in units of $\sqrt{W/g}$.

the dominant wavelength becomes $W/2$. Notice that this long wavelength does not exist in the initial perturbation. Third, the interaction among bubbles leads to a turbulent mixing layer. The thickness of this mixing layer increases with time.

The same LBM multiphase flow model was also used to study the three-dimensional Rayleigh–Taylor instability – a task challenging many conventional multiphase schemes. He *et al.* (1999b) carried out a three-dimensional Rayleigh–Taylor instability simulation in a rectangular box with a square horizontal cross-section. The height–width aspect ratio was fixed at 4:1. Gravity pointed downward and surface tension was neglected. For simplicity, the kinematic viscosities were chosen to be the same for both the heavy and light fluids. Periodic boundary conditions were applied at the four sides, while no-slip boundary conditions were applied at the top and bottom walls. The instability develops from the imposed single-mode initial perturbation:

$$\frac{h(x,y)}{W} = 0.05 \left[\cos\left(\frac{2\pi x}{W}\right) + \cos\left(\frac{2\pi y}{W}\right) \right], \tag{6.63}$$

where h is the height of the interface and W is the box width. The origin of the coordinates is at the lower left bottom corner of the box.

The results are presented in dimensionless form. The box width, W, is taken as the length scale and $T = \sqrt{W/g}$ is taken as the time scale. As before, the relevant dimensionless parameters are the Reynolds number and the Atwood number.

The typical evolution of the fluid interface in the three-dimensional Rayleigh–Taylor instability is shown in Fig. 6.4. The Atwood and Reynolds numbers are 0.5 and 1024, respectively. The simulation is carried out on a $128 \times 128 \times 512$ grid. Presented in the figure are views of the interface from both the heavy fluid side (left panels) and the light fluid side (right panels). As expected, the heavy and light fluids penetrate into each other as time increases. The light fluid rises to form a bubble and the heavy fluid falls to generate a spike. Furthermore, there is an additional landmark feature that distinguishes this interface from that observed in a two-dimensional Rayleigh–Taylor instability, namely the four saddle points at the middle of the four sides of the computational box shown in Fig. 6.4(a). The evolution of the interface around these saddle points is one of the unique features of the three-dimensional Rayleigh–Taylor instability.

As shown in Fig. 6.4, the interface remains rather simple during the early stages (Fig. 6.4a) but becomes more complicated as time increases. For this typical case, the Kelvin–Helmholtz instability does not develop until $t = 2.0$ when the first roll-up of the heavy fluid appears in the neighborhood of the

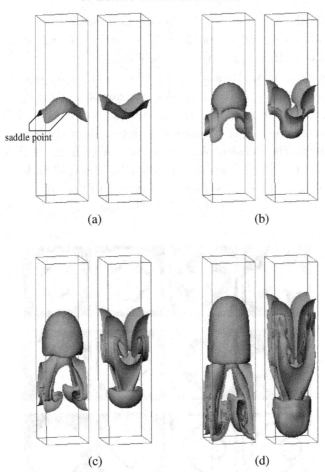

Fig. 6.4. Evolution of the fluid interface from a single-mode perturbation at (a) $t = 1.0$, (b) $t = 2.0$, (c) $t = 3.0$, and (d) $t = 4.0$. The Atwood number is 0.5 and the Reynolds number 1024. Time is measured in units of $\sqrt{W/g}$. The interface is viewed from the heavy fluid side (left panel) and from the light fluid side (right panel). The interfaces in the left panel are shifted $W/2$ in both x- and y-directions for a better view of the bubble.

saddle points (Fig. 6.4b). The roll-up at the edge of the spike starts at a later time (Fig. 6.4c). At $t = 4.0$, these roll-ups have been stretched into two extra layers of heavy fluid folded upwards: one forms a skirt around the spike and the other forms a girdle inside the bubble (Fig. 6.4d). Notice also the small curls inside the skirt. These structures are formed by the Kelvin–Helmholtz instability. From the outside, both the bubble and the spike look like mushrooms. A similar evolution of the interface was observed in other numerical simulations (Tryggvason and Unverdi, 1990; Li, Jin, and

Glimm, 1996) although those calculations were not carried out to this late
stage. This two-layer roll-up phenomenon is another unique feature of the
three-dimensional Rayleigh–Taylor instability.

The complicated structures of the interface in the later stages can be seen
clearly in horizontal cross-sections. As an example, in Fig. 6.4(d) we show
the results at $t = 4.0$ from the same simulation. We found it helpful to
present the interfaces above the saddle points separately from those below.
Figure 6.5 shows the density plots at a number of horizontal planes above
the saddle point. The planes are labelled by the grid index $k \in [1, 512]$
and the plane altitude can be found using $z = k/128$. At the saddle point
level ($k = 213$), the light fluid exhibits the shape of a "rosette" with the

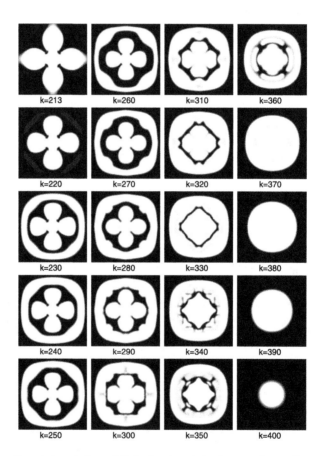

Fig. 6.5. Density plots at $t = 4.0$ in horizontal planes above the saddle points.
Black represents the heavy fluid and white the light fluid. The plots are shifted
$W/2$ in both x- and y-directions for a better view of the bubble. The Atwood
number is 0.5 and the Reynolds number 1024. The plane altitude can be calculated
using $z = k/128$.

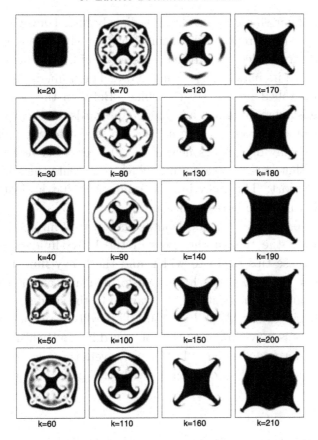

Fig. 6.6. Density plots at $t = 4.0$ in horizontal planes below the saddle points. Black represents the heavy fluid and white the light fluid. The Atwood number is 0.5 and the Reynolds number 1024. The plane altitude can be calculated using $z = k/128$.

petals pointing towards the saddle points. Moving upwards, the "rosette" quickly contracts its petals and becomes embedded in a box of heavy fluid ($k = 220$). At a slightly higher level ($k = 230$), the surrounding heavy fluid is divided into two parts by a ring of light fluid. This ring of light fluid comes from the outer layer of the mushroom of the bubble, while the heavy fluid between the "rosette" and the light fluid ring comes from the roll-up girdle of the heavy fluid around the saddle points. A similar structure persists all the way up to $k = 300$. Above that level, the "rosette" of light fluid in the middle gradually transforms into a "diamond" shape ($k = 320$–360). In the meantime, the girdle of the heavy fluid shrinks and finally disappears at $k = 370$. Above $k = 370$, only a bubble of light fluid is visible.

Figure 6.6 shows density plots for a number of horizontal planes below the saddle points. Across the bottom of the spike, the blob of heavy fluid has a roughly rectangular shape at $k = 20$. At slightly higher levels ($k = 30$–40), this simple geometry is replaced by a configuration with a core of heavy fluid in the middle surrounded by a frame of heavy fluid. The heavy-fluid core forms a cross with its bars oriented diagonally. The frame is a skirt of heavy fluid which has rolled up around the spike. Between the heavy-fluid core and the frame is a gap of light fluid. Moving upwards, the heavy-fluid frame gradually transforms from its original square shape to a roughly circular shape until it finally disappears near $k = 120$. For k between 70 and 110, we can see another layer of heavy fluid which represents the folding back of the heavy-fluid skirt. The wavy interface indicates that an instability may start to develop in these horizontal planes.

Regarding the instability in horizontal planes, it is interesting to notice what happens at the tips of the cross of the heavy-fluid core. Starting from $k = 60$, these tips begin to widen and the heavy fluid begins to roll inward. At $k = 90$, these tips have transformed into an "anchor" shape. The same configuration persists all the way up to the saddle level, although the roll-ups begin to shrink at $k = 170$. This roll-up phenomenon in the horizontal planes is very similar to that observed in the Richtmyer–Meshkov instability (Zhang and Graham, 1998). However, the phenomena are not exactly the same because in our study we do not have an explicit outward force.

6.4 Fluid–wall interactions in the lattice Boltzmann simulations

Usually, the lattice Boltzmann equation evolves on a Cartesian lattice as a consequence of the *coherent discretization* of time t and the single-particle phase space $\Gamma := (\mathbf{r}, \boldsymbol{\xi})$ which couples discretizations of t and Γ together such that the lattice spacing Δx is related to the time-step size Δt by $\Delta x = e\Delta t$, where e is the basic unit of the discrete velocity set $\mathbb{V}_Q := \{\mathbf{e}_a\}$. This makes the LBM method a very simple scheme consisting of two essential steps: collision at each grid point and advection to neighboring grid points. The collision mimics interactions among particles and the advection models the transport of particles according to their momenta. The simplicity and kinetic nature of the LBM are among its appealing features.

The boundary conditions for interactions between flow and solid boundaries are essential to flow physics. Accurate modeling and proper implementation of fluid–solid boundary conditions are crucial for any computational fluid dynamics (CFD) methods. The treatment for fluid–solid interactions in the LBM is a particular area of research interest. For the Boltzmann equation, the diffusive boundary condition leads to the no-slip boundary condition for the hydrodynamic equations (Sone, 2002). The LBM counterpart of the diffusive boundary condition is the bounce-back boundary condition (BBBC), which mimics the particle–boundary interaction for no-slip boundary conditions by reversing the momentum of a particle colliding with a rigid impenetrable wall. The bounce-back condition is easy to implement, and therefore it is most often used in LBM simulations. It can be rigorously shown that the bounce-back condition is indeed second-order accurate in space when the actual boundary position is off the grid point where the bounce-back collision takes place (Ginzbourg and Adler, 1994; Ginzbourg and d'Humiéres, 1996, 2003; He *et al.* 1997c; and Luo, 1997). The BBBC is easy to implement for simple geometries which can be represented by straight lines connecting grid points on a Cartesian mesh. However, this zig-zag approximation is inadequate for complex geometries with arbitrary curvatures. A common strategy to overcome this problem is to use interpolations and there are two approaches to doing so. One is to maintain the regular Cartesian mesh and apply interpolations to track the position of solid–fluid boundaries; thus the bounce-back boundary condition is applied only at the boundary locations which are off the Cartesian grid points (Bouzidi *et al.*, 2002; Yu *et al.*, 2002). This approach is similar to immersed boundary methods (see e.g. Mittal and Iaccarino, 2005). The other is to use a body-fitted mesh and employs interpolations throughout the entire mesh (He *et al.*, 1996, 1997c; He and Doolen, 1997b). When applied to moving boundaries, the former approach is much simpler than the latter one because it does not require remeshing. We shall discuss the former approach in what follows.

Because the bounce-back boundary condition plays a central role in the LBM treatment of fluid–solid interaction, we shall elucidate its basic idea and numerical implementation. The fluid–boundary interaction can be most easily understood within the framework of the generalized lattice Boltzmann method (GLBM) with multiple relaxation times (MRT) (d'Humières, 1992; d'Humières *et al.*, 2001, 2002; Lallemand and Luo, 2000). Thus, we shall first briefly discuss the GLBM in Section 6.4.1. Among its features, the GLBM has superior numerical stability (d'Humières *et al.*, 2002; Lallemand and Luo, 2000; Yu *et al.*, 2005) over the popular lattice BGK (LBGK) equation (Chen *et al.*, 1992; Qian *et al.*, 1992). Moreover, the GLBM can achieve viscosity-independent boundary conditions which is impossible for the LBGK equation, and this is often the source of misunderstanding and confusion. We shall then discuss the bounce-back boundary condition with interpolation to treat curved boundaries (Bouzidi *et al.*, 2002) in Section 6.4.2, followed by a discussion of an extension to moving boundaries

(Lallemand and Luo, 2003) in Section 6.4.3. Finally we show a number of test cases to demonstrate the validity of the LBM treatment of fluid–solid boundaries in Section 6.5, which include flow in porous media in three dimensions (3D) (Pan et al., 2005) and an impulsively started cylinder asymmetrically placed in a two-dimensional channel at rest (Lallemand and Luo, 2003).

6.4.1 The generalized lattice Boltzmann method with multiple relaxation times

On a D-dimensional lattice space $\Delta x \mathbb{Z}_D$ with discrete time $t^n \in \Delta t \mathbb{N}_0 := \Delta t \{0, 1, 2, \ldots\}$, the lattice Boltzmann equation can be generally written as

$$\mathbf{f}(\mathbf{r}_j + \mathbf{e}_a \Delta t, \, t^n + \Delta t) = \mathbf{f}(\mathbf{r}_j, \, t^n) + \mathbf{C}(\mathbf{f}(\mathbf{r}_j, \, t^n)) + \mathbf{F}. \tag{6.64}$$

In the above equation, the bold-face symbols denote column vectors in $\mathbb{R}^Q$:

$$\mathbf{f}(\mathbf{r}_j, \, t^n) := (f_0(\mathbf{r}_j, \, t^n), \, f_1(\mathbf{r}_j, \, t^n), \, \cdots, \, f_N(\mathbf{r}_j, \, t^n))^\mathrm{T},$$
$$\mathbf{f}(\mathbf{r}_j + \mathbf{e}\Delta t, \, t^n + \Delta t) := (f_0(\mathbf{r}_j, \, t^n + \Delta t), \, f_1(\mathbf{r}_j + \mathbf{e}_1\Delta t, \, t^n + \Delta t), \cdots,$$
$$f_N(\mathbf{r}_j + \mathbf{e}_N\Delta t, \, t^n + \Delta t))^\mathrm{T}$$
$$\mathbf{C} := (C_0, \, C_1, \, \cdots, \, C_N)^\mathrm{T}, \qquad \mathbf{F} := (F_0, \, F_1, \, \cdots, \, F_N)^\mathrm{T},$$

where $\{f_a\}$, $\{C_a\}$, and $\{F_a\}$ are the single-particle distribution functions, the collision terms, and the external forcing terms, respectively, and $\mathbb{V}_Q := \{\mathbf{e}_a\} = -\mathbb{V}_Q$ is a symmetric discrete velocity set of Q elements. We always assume $\mathbb{V}_Q$ includes the zero velocity $\mathbf{e}_0 := \mathbf{0}$. The lattice Boltzmann equation is thus a trajectory in a $Q \times N$ discrete phase space with discrete time t^n, where N is the number of spatial grid points under consideration.

We can construct a linear transformation from space $\mathbb{F}$ to some other space that may be more convenient for the description of collision processes. In particular, we shall use the space $\mathbb{M}$ of physically meaningful moments of $\{f_a\}$, such that $\mathbf{m} = \mathsf{M} \cdot \mathbf{f}$ and $\mathbf{f} = \mathsf{M}^{-1} \cdot \mathbf{m}$. The essence of the GLBM is that the collision process is executed in the moment space $\mathbb{M}$, whereas the advection process is done in the velocity space $\mathbb{F}$. In the GLBM, collision is modeled by a simple relaxation process:

$$\mathbf{C} = -\mathsf{M}^{-1} \cdot \mathsf{S} \cdot [\mathbf{m} - \mathbf{m}^{(\mathrm{eq})}], \tag{6.65}$$

where S is a $Q \times Q$ *diagonal* matrix of relaxation rates $\{s_a\}$.

On the basis of the kinetic theory of gases (cf. e.g. Hirschfelder, Curtis, and Bird, 1954; Harris, 1971) and by using the symmetries of the discrete velocity set $\mathbb{V}_Q$, M is constructed as follows. The components of the row vector $\mathbf{m}_b^\mathrm{T}$ of M are polynomials of the x and y components of the velocities

$\{\mathbf{e}_a\}$, $e_{a,x}$ and $e_{a,y}$. The vectors $\mathbf{m}_b^{\mathrm{T}}$, $b = 1, 2, \ldots, Q$, are constructed by using the Gram-Schmidt orthogonalization procedure (d'Humières, 1992; d'Humières *et al.*, 2001, 2002; Lallemand and Luo, 2000). It is interesting to point out that the moments, $m_a = \mathbf{m}_a^{\mathrm{T}} \cdot \mathbf{f}$, restricted to a finite number of velocities, are the low-order moments of the continuous velocity distribution function in kinetic theory, including the mass density ρ (the zeroth-order moment), the flow momentum $\mathbf{j} := \rho \mathbf{u}$ (the first-order moments), energy ρe and stresses σ_{ij} (the second-order moments), heat fluxes $\mathbf{q}$ (the third-order moments) and high-order fluxes (d'Humières, 1992; d'Humières *et al.*, 2001, 2002; Lallemand and Luo, 2000). The representations of a locate state by $\mathbf{f} \in \mathbb{F}$ and $\mathbf{m} \in \mathbb{M}$ are fully equivalent, and $\mathbb{M}$ is the one-to-one mapping between the two representations.

From kinetic theory, the equilibria of the nonconserved moments $\{m_a^{(\mathrm{eq})}\}$ are functions of the conserved moments, i.e. ρ and $\mathbf{j}$ for *athermal* fluids. In addition, the equilibria are (second-order) polynomials of $\mathbf{j}$ in the small Mach number limit. The GLBM reduces to its lattice BGK counterpart if all the relaxation rates are set to a single relaxation time τ, i.e. $s_a = 1/\tau_*$, and with appropriately chosen equilibria (d'Humières *et al.*, 2002; Lallemand and Luo, 2000). It should be stressed that the relaxation rates are not independent in some cases as the isotropy constraints often lead to relationships between some relaxation rates (Lallemand and Luo, 2000; Bouzidi *et al.*, 2001). Obviously, the usual LBGK models cannot satisfy such constraints, and consequently are less isotropic. The additional computational cost due to the linear transformations between the spaces $\mathbb{F}$ and $\mathbb{M}$ is rather insignificant ($< 20\%$), provided some care is taken in programming (d'Humières *et al.*, 2002). It is certainly worth the extra programming effort due to the GLBM compared to the simple LBGK equation because GLBM is far superior in terms of numerical stability, insensitivity to spurious acoustic waves, and isotropy in certain models (d'Humières *et al.*, 2001).

6.4.2 The bounce-back boundary condition

The treatment for a curved boundary is a combination of the bounce-back scheme and interpolations (Bouzidi *et al.*, 2002). For the sake of simplicity, this boundary condition is illustrated in Fig. 6.7 for an idealized situation in two dimensions. Consider a wall located at an arbitrary position $\mathbf{r}_{\mathrm{W}}$ between two grid sites $\mathbf{r}_{\mathrm{A}}$ and $\mathbf{r}_{\mathrm{S}}$, with $\mathbf{r}_{\mathrm{S}}$ situated in the nonfluid region – the shaded area depicted in Fig. 6.7. The parameter q defines the fraction of a grid spacing intersected by the boundary lying in fluid region, i.e. $q := |\mathbf{r}_{\mathrm{A}} - \mathbf{r}_{\mathrm{W}}|/\Delta x$. It is well understood that the bounce-back boundary

conditions place the wall somewhere beyond the last fluid node (Ginzbourg
and d'Humières, 1996, 2003), and for the MRT-GLBM model with appropriate choice of the relaxation rates, it is about one-half grid spacing beyond the
last fluid node, i.e. $q = 1/2$, as shown in Fig. 6.7(a). That is, even though
the bounce-back collision occurs on the node $\mathbf{r}_A$, the actual position of the
wall is located at $\mathbf{r}_W$, which is about one-half grid spacing beyond the last
fluid node $\mathbf{r}_A$. Thus one could intuitively picture the bounce-back boundary
as follows: the particle with velocity $\mathbf{e}_1$, starting from $\mathbf{r}_A$, travels from left to
right, hits the wall at $\mathbf{r}_W$, reverses its momentum, then returns to its starting point $\mathbf{r}_A$. This imaginary particle trajectory is indicated by the thick
bent arrow in Fig. 6.7(a). The total distance traveled by the particle is one
grid spacing Δx during the bounce-back collision. Therefore, one can imagine that the bounce-back collision either takes one time step or no time at

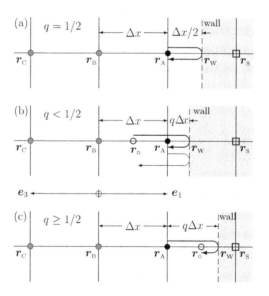

Fig. 6.7. Illustration of the boundary conditions for a rigid wall located arbitrarily
between two grid sites in one dimension. The thin solid lines are the grid lines,
the dashed line is the boundary location situated arbitrarily between two grids.
Shaded disks (•) are the fluid nodes, and the disks (●) are the fluid nodes next to
boundary. Circles (○) are located in the fluid region but not on grid nodes. The
square boxes (□) are within the nonfluid region. The thick arrows represent the
trajectory of a particle interacting with the wall, described in equations (6.66a)
and (6.66b). The distribution functions at the locations indicated by disks are used
to interpolate the distribution function at the location marked by the circles (○).
(a) $q := |\mathbf{r}_A - \mathbf{r}_W|/\Delta x = 1/2$. This is the perfect bounce-back condition – no
interpolations needed. (b) $q < 1/2$. (c) $q \geq 1/2$.

all, corresponding to two implementations of the bounce-back scheme: the so-called "link" and "node" implementations (Ladd, 1994a,b), respectively. The difference between these two implementations is that the bounced-back distribution (e.g. f_3 in Fig. 6.7a) at the boundary nodes in the "link" implementation is one time step behind that in the "node" implementation. This difference vanishes for steady state calculations, although these two implementations of the bounce-back boundary conditions have different stability characteristics, because the "link" implementation destroys the parity symmetry on a boundary node, whereas the "node" implementation preserves it. However, one cannot determine *a priori* the advantage of one implementation over the other, because the features of the two implementations also strongly depend on the precise location of the boundary and the local flow structure. It is also important to stress that the actual boundary location is not affected by the two different implementations of the bounce-back scheme (Ginzbourg and d'Humières, 1996; Bouzidi *et al.*, 2002).

Assuming that the picture for the simple bounce-back scheme is correct, let's first consider the situation depicted in Fig. 6.7(b) in detail for the case of $q < 1/2$. At time t^n the distribution function of the particles with velocity pointing to the wall ($\mathbf{e}_1$ in Fig. 6.7) at the grid point $\mathbf{r}_A$ (a fluid node) would end up at the point $\mathbf{r}_O$ located at a distance $(1 - 2q)\Delta x$ away from the grid point $\mathbf{r}_A$ after the bounce-back collision, as depicted by the thin bent arrow in Fig. 6.7(b). Because $\mathbf{r}_O$ is not a grid point, the value of f_3 at the grid point $\mathbf{r}_A$ needs to be reconstructed. However, noticing that f_1 starting from point $\mathbf{r}_O$ would become f_3 at the grid point $\mathbf{r}_A$ after the bounce-back collision with the wall, we can construct the value of f_1 at the point $\mathbf{r}_O$ by, for instance, a quadratic interpolation involving values of f_1 at the three locations: $f_1(\mathbf{r}_A), f_1(\mathbf{r}_B) = f_1(\mathbf{r}_A - \mathbf{e}_1 \Delta t)$, and $f_1(\mathbf{r}_C) = f_1(\mathbf{r}_A - 2\mathbf{e}_1 \Delta t)$. In a similar manner, for the case of $q \geq 1/2$ depicted in Fig. 6.7(c), we can construct $f_3(\mathbf{r}_A)$ by a quadratic interpolation involving $f_3(\mathbf{r}_O)$ that is equal to $f_1(\mathbf{r}_A)$ before the bounce-back collision, and the values of f_3 at the nodes after collision and advection, i.e. $f_3(\mathbf{r}_B)$, and $f_3(\mathbf{r}_C)$. Therefore the interpolations are applied differently for the two cases:

- For $q < 1/2$, interpolate before propagation and bounce-back collision.
- For $q \geq 1/2$, interpolate after propagation and bounce-back collision.

We do so to avoid the use of extrapolations in the boundary conditions for the sake of numerical stability. This leads to the following interpolation formulas which combine the advection and bounce-back together in one step

(Lallemand and Luo, 2003):

$$f_{\bar{a}}(\mathbf{r}_{\mathrm{A}}, t) = q\,(1 + 2\,q) f_a(\mathbf{r}_{\mathrm{A}} + \mathbf{e}_a \Delta t, t) + (1 - 4\,q^2) f_a(\mathbf{r}_{\mathrm{A}}, t)$$

$$-q\,(1 - 2\,q) f_a(\mathbf{r}_{\mathrm{A}} - \mathbf{e}_a \Delta t, t) + 3 w_a(\mathbf{e}_a \cdot \mathbf{u}_{\mathrm{w}}), \quad q < \frac{1}{2}, \tag{6.66a}$$

$$f_{\bar{a}}(\mathbf{r}_{\mathrm{A}}, t) = \frac{1}{q\,(2\,q + 1)} f_a(\mathbf{r}_{\mathrm{A}} + \mathbf{e}_a \Delta t, t) + \frac{(2\,q - 1)}{q} f_{\bar{a}}(\mathbf{r}_{\mathrm{A}} - \mathbf{e}_a \Delta t, t)$$

$$-\frac{(2\,q - 1)}{(2\,q + 1)} f_{\bar{a}}(\mathbf{r}_{\mathrm{A}} - 2\mathbf{e}_a \Delta t, t) + \frac{3 w_a}{q\,(2\,q + 1)}(\mathbf{e}_a \cdot \mathbf{u}_{\mathrm{w}}), \quad q \geq \frac{1}{2}, \tag{6.66b}$$

where the coefficients $\{w_a\}$ are model dependent (He and Luo, 1997a,b), $f_{\bar{a}}$ denotes the distribution function of the velocity $\mathbf{e}_{\bar{a}} := -\mathbf{e}_a$, and $\mathbf{u}_{\mathrm{w}}$ is the velocity of the wall at the point $\mathbf{r}_{\mathrm{W}}$ in Fig. 6.7.

It should be mentioned that for linear GLBM, the multiple-reflection scheme (Ginzbourg and d'Humières, 2003) can set the boundary location for quadratic flows (such as the Poiseuille flow) exactly one half grid spacing beyond the last fluid node, regardless of the flow direction relative to the underlying lattice. It should be emphasized that equation (6.66a) only involves the position of a boundary relative to the computational mesh of Cartesian grids, and therefore it can be readily applied in 3D (d'Humières et al., 2001, 2002; Yu et al., 2002).

The interpolations only improve the geometric representation of a curved boundary, which is independent of the LBM. It can be analytically shown that the exact location where the no-slip boundary conditions are satisfied depends on the value of the relaxation parameter τ and the specific implementation of the BBBCs for the LBGK models (He et al., 1997c). The τ-dependence of the boundary location is particularly severe when $\tau > 1$ for the LBGK equation (He et al., 1997). This deficiency of the LBGK equation can be easily overcome by using the MRT-LBM with two relaxation times:

$$s_{\mathrm{e}} = \frac{1}{\tau_*} = \frac{2}{1 + 6\nu_*}, \qquad s_{\mathrm{o}} = 8\frac{(2 - s_{\mathrm{e}})}{(8 - s_{\mathrm{e}})}, \tag{6.67}$$

where $\nu_* = \nu/(\mathbf{e}^2 \Delta t)$ is the dimensionless viscosity, and s_{e} and s_{o} are the relaxation rates for the (nonconserved) even-order moments and odd-order moments, respectively. With two-relaxation times (Ginzbourg et al., 2004; Ginzbourg, 2005), the viscosity-dependence of boundary location is completely eliminated for the linear LBM models valid for Stokes flows, i.e. the nonlinear terms in terms of $\mathbf{j}$ in equilibria are neglected. For the full MRT-LBM models with the nonlinear terms and the relaxation rates given by equation (6.67), the effect due to the viscosity dependence of the boundary location is much reduced (Ginzbourg and d'Humières, 2003; Pan *et al.*, 2005).

6.4.3 Treatment of moving boundaries

To extend the above boundary conditions to a moving boundary illustrated in Fig. 6.8, provided that the velocity of the moving wall, $\mathbf{u}_w$, is not too fast compared to the sound speed c_s in the system, one must also solve the following problem. When a grid point moves out of the nonfluid region into the fluid region to become a fluid node (indicated by $\square$ in Fig. 6.8), one must specify a number of unknown distribution functions on this node. We use a second-order extrapolation to compute the unknown distribution functions along the direction of a chosen discrete velocity $\mathbf{e}_a$ which maximizes the quantity $\hat{n}\cdot\mathbf{e}_a$, where $\hat{n}$ is the out-normal vector to the wall at the point (marked by $\diamond$ in Fig. 6.8) through which the node moves into the fluid region. For example, the unknown distribution functions $\{f_a(\mathbf{r})\}$ at node $\mathbf{r}$ as depicted in Fig. 6.8 can be given by

$$f_a(\mathbf{r}) = 3f_a(\mathbf{r}') - 3f_a(\mathbf{r}'') + f_a(\mathbf{r}'' + \Delta t \mathbf{e}_6).$$

Obviously the method to compute values of the unknown distribution functions (on the nodes which move from a nonfluid to a fluid region) is not unique. One could, for instance, compute the equilibria at $\mathbf{r}$ by using the velocity $\mathbf{u}_w$ of the moving boundary and the averaged density in the system ρ_0 or an otherwise obtained locally averaged density, and use the equilibria for the unknown distribution functions. Alternatively, one could

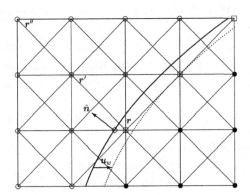

Fig. 6.8. Illustration of a moving boundary with velocity $\mathbf{u}_w$. The circles ($\circ$) and disks ($\bullet$) denote the fluid and nonfluid nodes, respectively. The squares ($\square$) denote the solid nodes to become fluid nodes at one time step Δt ($= 1$). The solid and dotted curves are the wall boundary at time t^n and $t^n + \Delta t$, respectively.

also systematically update the distribution functions in the non-fluid regions by performing collisions as in the fluid regions while velocity is kept at the moving velocity $\mathbf{u}_{\mathrm{w}}$ of the solid object. All these schemes produce similar results.

As depicted in Fig. 6.7, the momentum transfer at the boundary along the direction of $\mathbf{e}_a$ is:

$$\delta \mathbf{j}_a = [f_a(\mathbf{r}_{\mathrm{W}}, t) + f_{\bar{a}}(\mathbf{r}_{\mathrm{A}}, t)] \mathbf{e}_a \Delta t. \tag{6.68}$$

The above formula gives the momentum flux through any boundary normal to $\mathbf{e}_a$ situated between point $\mathbf{r}_{\mathrm{W}}$ and point $\mathbf{r}_{\mathrm{A}}$, as depicted in Fig. 6.7. In the simulations, we use the above formula to evaluate the momentum exchange in the interaction between fluid and solid bodies.

6.5 Some test cases

In this section we shall present several test cases to demonstrate the LBM simulations with the bounce-back boundary conditions and interpolations. We choose the flow through a porous medium in three dimensions (Pan et al., 2005) as an example for a fixed boundary problem, and the flow past an impulsively started cylinder in a two-dimensional channel (Lallemand and Luo, 2003) as an example of a moving boundary problem.

6.5.1 Flow in porous media

We use the 19-velocity model in 3D (D3Q19) for the porous medium simulations. The relaxation rates in the MRT-D3Q19 model are chosen according to equation (6.67). We first consider the case of flow through an idealized porous medium, i.e. a periodic body-centered cubic (BCC) array of spheres of equal radius a in a box of L^3. The permeability for this porous medium can be obtained analytically (Hasimoto, 1959; Sangani and Acrivos, 1982). We measure the fluid permeability κ according to Darcy's law:

$$u_{\mathrm{d}} = -\frac{\kappa}{\rho \nu} \left(\nabla_z p + \rho g \right), \tag{6.69}$$

where the Darcy velocity u_{d} is obtained as the volume-averaged velocity over the system (Pan et al., 2001), ρg is the strength of the forcing, and z is the vertical coordinate parallel to the forcing direction.

We evaluate the viscosity dependence of the computed permeability by using different LBM schemes. Figure 6.9(a) shows the normalized permeability κ/κ^* for the BCC array of spheres with $\chi := 4a/\sqrt{3}L = 0.85$, where κ^* is the analytic value (Hasimoto, 1959; Sangani and Acrivos, 1982),

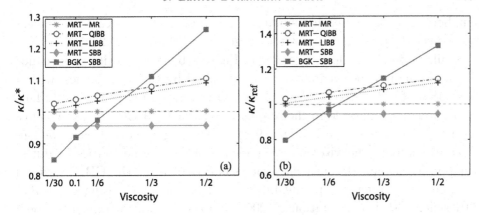

Fig. 6.9. The normalized permeability vs the viscosity ν. (a) κ/κ^* for the BCC array of spheres. The porosity is $\chi = 0.85$ and the resolution is 32^3. (b) $\kappa/\kappa_{\rm ref}$ for a portion of the GB1b sphere-packing porous medium. The resolution is 64^3. MR, QIBB, LIBB, SBB denote multireflection, quadratic interpolated bounce-back, linear interpolated bounce-back, and standard bounce-back, respectively. (Courtesy of Pan.)

at $\tau = 1/s_\nu = 0.6$, 0.8, 1.0, 1.5, and 2.0. The resolution used is 32^3, corresponding to the sphere radius $\zeta \approx 11.8$ and the pore throat $\eta \approx 2.8$ in lattice units. The results clearly show that the values of κ/κ^* obtained by the MRT scheme with BBBCs are much less viscosity-dependent than those obtained by the LBGK scheme.

We next consider the flow through a complex porous medium, which is intended to model an experimental medium GB1b (Hilpert, McBride, and Miller, 2001; Hilpert and Miller, 2001). The GB1b medium was represented by a random sphere packing (Hilpert and Miller, 2001) generated by using the algorithm of Yang *et al.* (1996) with a specified porosity ϕ and the probability density function of the grain size distribution, measured from a real porous medium (Yang *et al.*, 1996; Pan *et al.*, 2001).

The porosity of the simulated GB1b medium is 0.36; the mean grain diameter is 0.1149 mm and the relative standard deviation of the grain size was 4.7%. The entire sphere packing includes about 10 000 spheres in a cube of size 13 mm^3. We only use a small subset of the entire GB1b sphere packing here, which contained about 23 spheres. The sphere-packing algorithm generates the center locations and radii of the spheres, which are used to determine the exact boundary locations with respect to the lattice, i.e. the parameter q needed for the interpolation formulas of equations (6.66). In the simulations of the flow through the sphere-packing GB1b, the nonlinear lattice Boltzmann equation was used. We keep the average Reynolds number $Re < 0.1$ for each simulation, but higher local Reynolds numbers are expected due to the complex pore geometry (Pan *et al.*, 2005).

Similarly to Fig. 6.9(a), we show in Fig. 6.9(b) the viscosity dependence of the normalized permeability $\kappa/\kappa_{\mathrm{ref}}$ obtained by various schemes, with a resolution of 64^3, corresponding to an averaged grain radius $\zeta = 11.2$. This resolution is comparable to the resolution 32^3 for the BCC array of spheres with $\chi = 0.85$, as shown in Fig. 6.9(a), in which $\zeta = 11.8$. Hence, we can directly compare the viscosity dependence of the permeabilities in different porous media. Because the exact solution of κ for the GB1b medium is unknown, a reference permeability κ_{ref} is obtained by using the MRT-MR scheme with a resolution of 200^3 (Ginzbourg and d'Humières, 2003; Pan *et al.*, 2005).

In all cases we are able to show unequivocally that MRT-LBM is far superior over the popular LBGK equation in terms of accuracy and stability for porous media simulations (Pan *et al.*, 2005).

6.5.2 An impulsively started cylinder moving in a 2D channel

We conduct numerical simulations to investigate the accuracy of the bounce-back boundary conditions for a moving boundary. We use a cylinder asymmetrically placed in a channel in two dimensions as the basic configuration. With different boundary conditions and initial conditions, we can simulate different flow situations in the channel (Lallemand and Luo, 2003). The flow simulations can be carried out in two frames of reference. First, the position of the cylinder is fixed on the computational mesh, so that the boundary of the cylinder is also at rest. And, second, the cylinder is moving at a constant velocity with respect to the mesh, so that the boundary of the cylinder is moving. However, the relative motion between the cylinder and the flow in the channel remains the same in the two cases by matching the boundary conditions and forcing in the two frames of reference. By directly comparing the results obtained from these two different settings, one can test not only the accuracy of the moving boundary conditions, but also the Galilean invariance of the LBM (Lallemand and Luo, 2003).

At time $t = 0$, the fluid in the channel is at rest and the cylinder is impulsively started with a constant speed $U_c = 0.04$ in the x-direction. This is equivalent to the following initial conditions on the flow: the fluid is given a uniform speed $U_f = -0.04 = -U_c$, so are the lower and upper walls given a constant speed $U_w = -0.04 = -U_c$, and the cylinder is at rest. Periodic boundary conditions are applied in the x-direction in both cases. The former case is in the frame of reference at rest: the boundary of the

cylinder is moving with respect to the mesh. The latter is in the frame of
reference moving with the cylinder, so that the boundary of the cylinder is
fixed while the fluid is moving with the opposite velocity. In both cases the
channel has dimensions $L \times W = 1001 \times 101$, the radius $r = 12$, $H = 54$,
and the initial position of the cylinder is $(x_0, y_0) = (60.3, 54)$. The viscosity
ν is such that the Reynolds number $Re = 2rU_c/\nu$ equals 200.

Figure 6.10 shows the drag and lift experienced by the cylinder as functions of time. Shown in the figure are three sets of results. First, the
flow was computed with the fixed cylinder. The computations with the
cylinder moving with respect to the mesh were performed by using either
the second-order extrapolations or the equilibrium distribution functions for

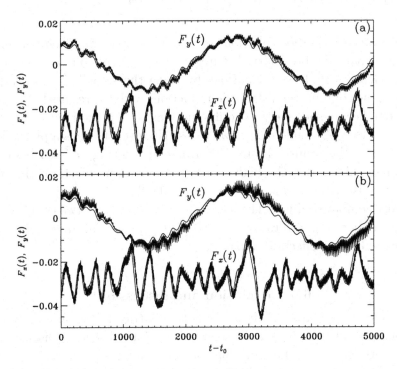

Fig. 6.10. An impulsively started cylinder moving with a constant speed $U_c = 0.04$ in the x-direction in a 2D channel with periodic boundary condition in the x-direction. Reynolds number $Re = 200$. Total force (F_x, F_y) on the cylinder measured as functions of time t after an initial run time $t_0 = 14000$. The smooth lines are the results obtained in the frame of reference moving with the cylinder (the fixed boundary calculation) and the fluctuating lines are the results obtained in the frame of reference at rest (the moving boundary calculations). (a) The second-order extrapolations are used to compute the unknown distribution functions, and (b) the equilibrium distribution functions are used for the unknown distribution functions.

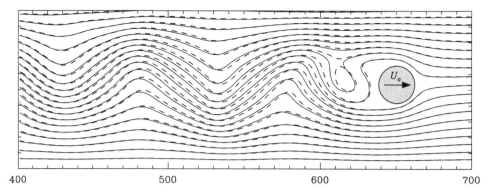

400   500   600   700

Fig. 6.11. The stream functions of the flow at $t = 15\,000$ correspond to the two calculations in Fig. 6.10(a). Solid lines and dashed lines correspond to results with fixed and moving boundary of the cylinder, respectively.

the unknown distribution functions on the nodes emerging from within the cylinder next to the boundary. Figures 6.10(a) and (b) compare the results of the moving boundary calculations by using the second-order extrapolations and the equilibrium distribution functions with the results by using the fixed boundary, respectively. The fluctuations due to the movement of the boundary are comparable in the two cases. However, there is a phase difference between the results obtained by using the moving or fixed boundary. This phase shift is due the higher order non-Galilean effects in the sound speed and viscosities (Lallemand and Luo, 2000).

   We also compare in Fig. 6.11 the stream function for the two computations corresponding to Fig. 6.10(a) at $t = 15\,000$. The phase shift due to the higher order non-Galilean effects is apparent.

6.6 Conclusion and discussion

In this chapter, we have briefly presented the lattice Boltzmann method for multiphase fluid flows and fluid–solid interaction. We have discussed the LBM treatment of the fluid–fuid interface and fluid–solid boundary within the framework of the mean-field theory and the generalized LBM with multiple relaxation times. The implementation of the LBM has been demonstrated with a number of examples. Future research directions in the lattice Boltzmann method for multiphase flows include developing stable numerical methods for multiphase flows with large density variation and simulating suspensions in liquids at high Reynolds numbers.

7

Boundary integral methods for Stokes flows

Boundary integral methods are powerful numerical techniques for solving multiphase hydrodynamic and aerodynamic problems in conditions where the Stokes or potential-flow approximations are applicable. Stokes flows correspond to the low Reynolds number limit, and potential flows to the high Reynolds number regime where fluid vorticity can be neglected. For both Stokes and potential flows, the velocity field in the system satisfies linear governing equations. The total flow can thus be represented as a superposition of flows produced by appropriate point sources and dipoles at the fluid interfaces.

In the boundary integral approach the flow equations are solved directly for the velocity field at the fluid interfaces, rather than in the bulk fluid. Thus, these methods are well suited for describing multiphase systems. Examples of systems for which boundary integral algorithms are especially useful include suspensions of rigid particles or deformable drops under Stokes-flow conditions. Applications of boundary integral methods in fluid dynamics, however, cover a broader range. At one end of this range are investigations of the hydrodynamic mobility of macromolecules; at the other end are calculations of the flow field around an airplane wing in a potential flow approximation. Here we will not address the potential flow case, limiting ourselves to Stokes flow.

7.1 Introduction

In the present chapter we discuss boundary integral methods for multiphase flows in the Stokes-flow regime. We review the governing differential equations, derive their integral form, and show how to use the resulting boundary integral equations to determine the motion of particles and drops. Specific

193

issues that are relevant for the numerical implementation of these equations are also described.

The boundary integral formulation of the Stokes-flow problem relies on the linearity of the governing equations: the nonlinear inertial term in the momentum transport equation (1.8) vanishes in the low Reynolds number limit. The theoretical framework for the boundary integral techniques involves recasting the linear differential flow equations into the form of integrals of the single-layer and double-layer Green's functions over the boundaries of the fluid domain. In specific applications, such boundaries may include container walls and the interfaces of particles or drops suspended in a fluid.

By solving the boundary integral equations, the velocities and stresses on the interfaces are obtained without evaluating these quantities in the bulk. For many systems, the evolution is completely determined by the velocity and stress distribution on the boundaries. Examples include the dynamics of suspensions of rigid particles and flow of emulsions of deformable drops (provided that bulk transport of surface active species, such as surfactants, is unimportant). The rheological viscoelastic response of suspensions and emulsions can also be expressed solely in terms of interfacial quantities.

Suspension and emulsion flows often occur at low particle-scale Reynolds numbers, because of the small length scales characteristic of their microstructure. Hence, boundary integral algorithms for Stokes flows are well suited for dynamical simulations of microscale flow in fluid–fluid and fluid–solid dispersions. According to the boundary integral approach, only fluid boundaries require discretization; thus the number of nodes needed to resolve the flow field to a given accuracy is reduced compared to algorithms which require discretization of the entire three-dimensional fluid volume. The advantages of boundary integral methods are particularly marked in problems where a high numerical resolution is necessary.

Examples of applications of boundary integral methods for Stokes flows include simulations of drop breakup (e.g. Stone, 1994; Cristini *et al.*, 1998, 2003; Patel *et al.*, 2003; Bazhlekov *et al.*, 2004), studies of interactions and coalescence of deformable drops (e.g. Loewenberg and Hinch, 1997; Zinchenko *et al.*, 1997; Bazhlekov *et al.*, 2004; Nemer *et al.*, 2004), and investigations of emulsion rheology (e.g. Kennedy, Pozrikidis, and Skalak, 1994; Loewenberg and Hinch, 1996; Zinchenko and Davis, 2002). Boundary integral methods are also being used to simulate microscale flows in biological systems, for example to calculate hydrodynamic mobilities of biomolecules (Zhao and Pearlstein, 2002), and to relate the mobility of protein molecules to their conformations and hydration levels (Squire and Himmel, 1979; Harding, 1995; Ferrer *et al.*, 2001; Zhou, 2001; Halle and

Davidovic, 2003). Boundary integral algorithms have also been used to simulate the dynamics of red blood cells and capsules (e.g. Diaz *et al.*, 2000, 2001).

7.2 Preliminaries

We focus here on boundary integral methods for incompressible multiphase flows under creeping-flow conditions, where fluid inertia is negligible. In this limit, the flow field is governed by the stationary Stokes equations. Before we can proceed with a description of the methods we thus need to establish some results concerning flows of this type. For a comprehensive discussion of Stokes flows in single-phase and multiphase systems we refer the reader to the monographs by Happel and Brenner (1983) and Kim and Karrila (1991).

7.2.1 Stokes flow limit

The regime where fluid motion is dominated by viscous forces can be identified by comparing the relative magnitudes of different terms in the Navier–Stokes equation (1.8). The comparison can be formulated in terms of the time scales characterizing the fluid motion. One of these time scales is the vorticity-diffusion time

$$\tau_{\rm v} = \frac{\rho L^2}{\mu}, \tag{7.1}$$

were L is the characteristic length scale, such as the size of a particle. Unbalanced forces associated with the acceleration of fluid elements relax by viscous dissipation on this time scale. There are also two time scales imposed by the external forcing: $\tau_{\dot\gamma} = \dot\gamma_0^{-1}$ associated with the characteristic strain rate $\dot\gamma_0$, and τ_0 imposed by time variations of the flow field due, for example, to the variation of an applied force.

The nonlinear inertial term $\mathbf{u} \cdot \nabla \mathbf{u}$ in the momentum equation (1.8) is unimportant under conditions where unbalanced forces relax on a much shorter time scale than the convective time scale. The ratio of the corresponding time scales $\tau_{\rm v}$ and $\tau_{\dot\gamma}$ is the Reynolds number Re, defined in equation (1.37), which must therefore be small for $\mathbf{u} \cdot \nabla \mathbf{u}$ to be unimportant. The time derivative $\partial \mathbf{u}/\partial t$ in the Navier–Stokes equation is negligible if the ratio between the Reynolds and Strouhal numbers (1.37) and (1.36) is small, namely

$$\frac{\tau_{\rm v}}{\tau_0} \ll 1. \tag{7.2}$$

Under these conditions, the motion of an incompressible fluid is governed
by the stationary Stokes equations, which include the incompressibility constraint

$$\boldsymbol{\nabla} \cdot \mathbf{u} = 0, \tag{7.3}$$

and the linearized stress-balance equation

$$\boldsymbol{\nabla} \cdot \boldsymbol{\sigma} = -\mathbf{f}, \tag{7.4}$$

or, upon substituting the constitutive relation (1.5) for a Newtonian fluid,

$$-\boldsymbol{\nabla} p + \mu \nabla^2 \mathbf{u} = -\mathbf{f}. \tag{7.5}$$

As remarked in Section 1.3, if the force $\mathbf{f}$ can be derived from a scalar potential $\mathcal{U}$, the force term can be absorbed into a modified pressure. Thus,
potential force fields do not affect the flow directly, but only through a modification of the pressure. When the potential is harmonic, which is the most
frequent case, upon taking the divergence of the momentum equation (7.4)
and recalling that the velocity is divergenceless, we conclude that the pressure is harmonic as well:

$$\nabla^2 p = 0. \tag{7.6}$$

The conditions $Re \ll 1$ and (7.2), are often encountered in the case of
highly viscous fluids (e.g. heavy oils) or flows at small length scales L.
For aqueous systems, with $L \sim 100 \ \mu\text{m}$, the condition $Re \sim 10^{-2}$ requires
velocities of the order of $100 \ \mu\text{m/s}$. It may be concluded that the primary
application of the boundary integral methods described in this chapter is
in the study of microscale flow in fluid–fluid and fluid–solid dispersions of
small particles.

7.2.2 Stokes-flow problems in multiphase systems

In our discussion of Stokes flows in multiphase dispersions we will focus
on systems composed of a particulate phase suspended in an unbounded
continuous-phase fluid. Boundary integral methods are applicable also to
other geometries, such as networks of channels in porous materials and emulsions with bicontinuous morphology. Most of the techniques presented herein
can be applied to such systems directly or with only minor modifications;
however, a discussion of the issues specific for these systems is beyond the
scope of this chapter.

The essential features of the boundary integral approach are most easily
presented for a single drop (or particle) embedded in an unbounded external

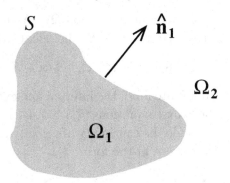

Fig. 7.1. Fluid regions Ω_1 and Ω_2.

flow $\mathbf{u}^\infty$. We focus on this case to avoid unnecessary complexity in our derivation of the boundary integral equations. The corresponding results for many-particle systems can be obtained by combining single-particle results, as discussed in Section 7.4.

Consider an individual particle or drop occupying a finite region Ω_1, suspended in a continuous-phase fluid occupying the complementary infinite domain Ω_2; the boundary between these regions is denoted by S. The unit vector normal to S, pointing out of Ω_i, is denoted by $\hat{\mathbf{n}}_i$. Accordingly,

$$\hat{\mathbf{n}}_2 = -\hat{\mathbf{n}}_1. \tag{7.7}$$

The geometry is illustrated in Figure 7.1.

The flow field $\mathbf{u}$ in the whole space $\Omega_1 \cup \Omega_2$ can conveniently be represented in the form

$$\mathbf{u}(\mathbf{x}) = \chi_1(\mathbf{x})\mathbf{u}_1(\mathbf{x}) + \chi_2(\mathbf{x})[\mathbf{u}_2(\mathbf{x}) + \mathbf{u}^\infty(\mathbf{x})], \tag{7.8}$$

where

$$\chi_i(\mathbf{x}) = \begin{cases} 1, & \mathbf{x} \in \Omega_i, \\ 0, & \mathbf{x} \notin \Omega_i \end{cases} \tag{7.9}$$

is the characteristic function of the region Ω_i. In the above relation, the flow field in the outer region Ω_2 is decomposed as

$$\mathbf{u}^{\text{out}}(\mathbf{x}) = \mathbf{u}_2(\mathbf{x}) + \mathbf{u}^\infty(\mathbf{x}), \tag{7.10}$$

where $\mathbf{u}^\infty$ is the imposed flow which satisfies the Stokes equations and is nonsingular in the whole space (but does not vanish at infinity), and $\mathbf{u}_2$ is the disturbance (or "scattered") flow field which is nonsingular in the region Ω_2, and satisfies the boundary condition

$$\mathbf{u}_2(\mathbf{x}) \to 0, \qquad |\mathbf{x}| \to \infty. \tag{7.11}$$

The flow field $\mathbf{u}_1$ is nonsingular in the region Ω_1. Assuming that the flow field $\mathbf{u}$ is continuous across the interface S (the usual situation), we have

$$\mathbf{u}_1(\mathbf{x}) = \mathbf{u}_2(\mathbf{x}) + \mathbf{u}^\infty(\mathbf{x}), \qquad \mathbf{x} \in S. \qquad (7.12)$$

For a liquid drop in an infinite fluid, the internal and external flow fields $\mathbf{u}_1$ and $\mathbf{u}^{\text{out}}$ both satisfy the Stokes equations (7.3), (7.4). Boundary conditions at the drop interface S include the continuity of velocity (7.12) and the stress balance (1.22) which we rewrite in the form

$$\mathbf{t}_1(\mathbf{x}) + \mathbf{t}^{\text{out}}(\mathbf{x}) = \mathbf{f}_S(\mathbf{x}), \qquad \mathbf{x} \in S, \qquad (7.13)$$

neglecting the mass transfer term $\dot{m}$. Here we have introduced the surface tractions defined by relations

$$\mathbf{t}_i = \hat{\mathbf{n}}_i \cdot \boldsymbol{\sigma}_i, \qquad \mathbf{t}^\infty = \hat{\mathbf{n}}_2 \cdot \boldsymbol{\sigma}^\infty, \qquad \mathbf{t}^{\text{out}} = \mathbf{t}_2 + \mathbf{t}^\infty \qquad (7.14)$$

and we have denoted by $\mathbf{f}_S$ the total surface force. For a typical system this force consists of the surface-tension contribution and of a gravity term due to the use of the modified pressure as in (1.26):

$$\mathbf{f}_S = \mathbf{f}_\gamma + \mathbf{f}_g = \nabla \cdot \left[\left(\hat{\boldsymbol{I}} - \hat{\mathbf{n}}_1\hat{\mathbf{n}}_1\right)\gamma\right] - [(\rho_2 - \rho_1)\mathbf{g} \cdot \mathbf{x}]\,\hat{\mathbf{n}}_1. \qquad (7.15)$$

It will be recalled from Section 1.3 that, when the divergence term in the right-hand side is expanded, the normal component of this equation involves the local curvature of the interface.

We note that the interfacial term $\nabla \cdot \left[\left(\hat{\boldsymbol{I}} - \hat{\mathbf{n}}_1\hat{\mathbf{n}}_1\right)\gamma\right]$, while usually sufficient, is not the most general form of an interfacial stress density encountered in emulsion flows. For example, high molecular-weight surfactants adsorbed on the interface may produce surface stresses associated with the interfacial viscous dissipation. In such systems, the adsorbed surfactant behaves as a two-dimensional "interfacial" viscous fluid. More about the interfacial stresses can be learned from the monographs on interfacial rheology, such as the one by Edwards, Brenner, and Wasan (1991).

A rigid particle can translate and rotate, but it cannot deform. Thus, the internal flow $\mathbf{u}_1$ in the region occupied by the particle is equal to the rigid-body velocity field,

$$\mathbf{u}_1(\mathbf{x}) = \mathbf{u}^{\text{rb}}(\mathbf{x}) \equiv \mathbf{U}^{\text{rb}} + \boldsymbol{\Omega}^{\text{rb}} \times (\mathbf{x} - \mathbf{x}_0). \qquad (7.16)$$

Here $\mathbf{U}^{\text{rb}}$ is the translational velocity of the particle, $\boldsymbol{\Omega}^{\text{rb}}$ is its rotational velocity, and $\mathbf{x}_0$ is an arbitrarily chosen reference point (center of rotation). We note for future reference that the rigid-body velocity field satisfies the Stokes equations with $p = $ constant. The velocity field (7.16) is related to the outside flow (7.10) by the continuity condition (7.12). In place of the local stress balance (7.13), we now have a situation in which the hydrodynamic force and torque acting on the particle equal the negative of the external force **F** and torque **L**. Since the unit normal $\hat{n}_2$ is *into* the particle, with the definition (7.14) of the tractions, we have that the integrals over the particle surface:

$$\mathbf{F} = \int_S \mathbf{t}^{\text{out}}(\mathbf{x})\,\mathrm{d}S = \int_S \mathbf{t}_2(\mathbf{x})\,\mathrm{d}S, \qquad (7.17)$$

$$\mathbf{L} = \int_S (\mathbf{x} - \mathbf{x}_0) \times \mathbf{t}^{\text{out}}(\mathbf{x})\,\mathrm{d}S = \int_S (\mathbf{x} - \mathbf{x}_0) \times \mathbf{t}_2(\mathbf{x})\,\mathrm{d}S \qquad (7.18)$$

are the opposite of the hydrodynamic force and torque on the particle, and therefore represent the external force and torque applied to the particle. In equations (7.17) and (7.18) the traction $\mathbf{t}^{\text{out}}$ has been replaced with $\mathbf{t}_2$, because the traction $\mathbf{t}^\infty$, associated with the nonsingular imposed flow field $\mathbf{u}^\infty$, does not contribute to the force and torque as can readily be shown by an application of the divergence theorem.

In order to determine the particle motion, its translational and angular velocities $\mathbf{U}^{\text{rb}}$ and $\mathbf{\Omega}^{\text{rb}}$ need to be evaluated for a given incident flow $\mathbf{u}^\infty$ and external force **F** and torque **L**. Alternatively, for a given external flow, one may be interested in the force and torque necessary to produce particle velocities $\mathbf{U}^{\text{rb}}$ and $\mathbf{\Omega}^{\text{rb}}$. The first approach is called the *mobility formulation* of the problem, and the second the *resistance formulation*. Due to the linearity of the Stokes equations, the velocities $\mathbf{U}^{\text{rb}}$ and $\mathbf{\Omega}^{\text{rb}}$ and the force and torque **F** and **L** are linearly related. In the absence of external flow, for example, the mobility and resistance problems can be expressed in the matrix form

$$\begin{bmatrix} \mathbf{U}^{\text{rb}} \\ \mathbf{\Omega}^{\text{rb}} \end{bmatrix} = \begin{bmatrix} \mu^{tt} & \mu^{tr} \\ \mu^{rt} & \mu^{rr} \end{bmatrix} \cdot \begin{bmatrix} \mathbf{F} \\ \mathbf{L} \end{bmatrix}, \qquad \begin{bmatrix} \mathbf{F} \\ \mathbf{L} \end{bmatrix} = \begin{bmatrix} \zeta^{tt} & \zeta^{tr} \\ \zeta^{rt} & \zeta^{rr} \end{bmatrix} \cdot \begin{bmatrix} \mathbf{U}^{\text{rb}} \\ \mathbf{\Omega}^{\text{rb}} \end{bmatrix},$$
$$(7.19)$$

where ζ^{pq} and μ^{pq}, with $p, q = t, r$, refer to the translational and rotational components of the resistance and mobility tensors. The 6×6 resistance and mobility matrices are inverse to each other, and, therefore, the mobility and resistance formulations are equivalent. In the presence of external flow, a similar formulation involves additional terms that are linear in $\mathbf{u}^\infty$.

In a typical situation, one would know the force and possibly the couple applied to the particle – for example, its weight in a sedimentation problem – and be interested in the resulting motion. Equations (7.19) show that, to solve this problem, one needs to know the components of mobility tensors

μ^{pq}. Conversely, when the problem consists in the determination of the force and torque necessary to provide a given motion, the solution requires a knowledge of the resistance tensors ζ^{pq}.

7.2.3 Lorentz integral identities

Lorentz integral identities provide the fundamental building blocks from which the boundary integral equations for specific systems (such as liquid drops or solid particles immersed in a fluid) are derived. The Lorentz identities relate the fluid flow $u_i(\mathbf{x})$ $(i = 1, 2)$ in the bounded and unbounded bulk regions Ω_1 and Ω_2 to the distribution of velocity and stress on the boundary S. When the field point $\mathbf{x}$ is on the surface S, the Lorentz identities yield integral relations between the interfacial velocity and the inside or outside stress. The boundary integral equations for a specific system are obtained by combining these relations with the boundary conditions discussed in Section 7.2.2.

In order to derive *Lorentz's reciprocal theorem*, we consider two solutions $(\mathbf{u}, \boldsymbol{\sigma})$, $(\tilde{\mathbf{u}}, \tilde{\boldsymbol{\sigma}})$, of the Stokes equations:

$$\nabla \cdot \boldsymbol{\sigma} = -\mathbf{f}, \qquad \nabla \cdot \tilde{\boldsymbol{\sigma}} = -\tilde{\mathbf{f}}, \tag{7.20}$$

in which $\mathbf{f}$ and $\tilde{\mathbf{f}}$ are given force fields. Upon multiplying the first equation by $\tilde{\mathbf{u}}$, the second one by $\mathbf{u}$ and subtracting, it is easy to show that

$$\nabla \cdot (\tilde{\mathbf{u}} \cdot \boldsymbol{\sigma} - \mathbf{u} \cdot \tilde{\boldsymbol{\sigma}}) = \mathbf{u} \cdot \tilde{\mathbf{f}} - \tilde{\mathbf{u}} \cdot \mathbf{f}. \tag{7.21}$$

The proof of this result relies on the symmetry of the stress tensor and incompressibility, as a consequence of which

$$\begin{aligned}
\nabla \cdot (\tilde{\mathbf{u}} \cdot \boldsymbol{\sigma}) &= (\nabla \tilde{\mathbf{u}}) : \boldsymbol{\sigma} + \tilde{\mathbf{u}} \cdot (\nabla \cdot \boldsymbol{\sigma}) & (7.22) \\
&= 2\mu \tilde{e} : e + \tilde{\mathbf{u}} \cdot (\nabla \cdot \boldsymbol{\sigma}), & (7.23)
\end{aligned}$$

with a similar expression for the other term. The contribution $2\mu\tilde{e}:e$ of the rate of deformation tensors (1.6) is the same for both terms and it cancels on subtraction.

Upon integrating (7.21) over a domain Ω and using the divergence theorem, we have

$$\int_\Omega (\mathbf{u} \cdot \tilde{\mathbf{f}} - \tilde{\mathbf{u}} \cdot \mathbf{f}) \, \mathrm{d}^3 x = \int_S \hat{\mathbf{n}} \cdot (\tilde{\mathbf{u}} \cdot \boldsymbol{\sigma} - \mathbf{u} \cdot \tilde{\boldsymbol{\sigma}}) \, \mathrm{d}S, \tag{7.24}$$

which is Lorentz's result; the integration surface S is the boundary of Ω as illustrated in Fig. 7.1. Relation (7.24) is valid for a finite or infinite domain $\Omega = \Omega_1$ or Ω_2. To avoid complications associated with the boundary term at infinity, when $\Omega = \Omega_2$ we always assume that the velocity field $\mathbf{u}(\mathbf{x})$ vanishes sufficiently fast for $|\mathbf{x}| \to \infty$.

7.2.4 Green's functions

Let us now take the field $\mathbf{u}$ to be one of the fields of interest to us, namely $\mathbf{u}_i$, so that $\Omega = \Omega_i$. We also assume that the force $\mathbf{f}$ vanishes or has been absorbed into the modified pressure. For $\tilde{\mathbf{f}}$, we take a force $\tilde{\mathbf{F}}$ concentrated at a point $\mathbf{x}'$, in which case the tilde field satisfies the relation

$$\boldsymbol{\nabla} \cdot \tilde{\boldsymbol{\sigma}} = -\tilde{\mathbf{F}} \, \delta(\mathbf{x} - \mathbf{x}') \,. \tag{7.25}$$

Substituting the above equation into (7.24) we then find

$$\tilde{\mathbf{F}} \cdot \mathbf{u}_i(\mathbf{x}')\chi_i(\mathbf{x}') = \int_S \hat{\mathbf{n}}_i \cdot (\tilde{\mathbf{u}} \cdot \boldsymbol{\sigma} - \mathbf{u} \cdot \tilde{\boldsymbol{\sigma}}) \, \mathrm{d}S, \tag{7.26}$$

where the characteristic function (7.9) for the region Ω_i is given by

$$\chi_i(\mathbf{x}') = \int_{\Omega_i} \delta(\mathbf{x} - \mathbf{x}') \, \mathrm{d}^3 x. \tag{7.27}$$

The relation (7.26) will now be turned into an integral equation for the field $\mathbf{u}_i$.

The first step is to calculate the auxiliary fields denoted by the tilde. It is evident from equation (7.25) and the linearity of Stokes equations that the flow field $\tilde{\mathbf{u}}$ associated with the stress tensor $\tilde{\boldsymbol{\sigma}}$ must have the form

$$\tilde{\mathbf{u}}(\mathbf{x}) = \mu^{-1}\boldsymbol{G}(\mathbf{x}, \mathbf{x}') \cdot \tilde{\mathbf{F}}, \tag{7.28}$$

where the second-rank tensor $\boldsymbol{G}$ is the *single-layer Green's function* for Stokes flow. In order for $\tilde{\mathbf{u}}$ to be divergenceless, it is necessary that

$$\boldsymbol{\nabla} \cdot \boldsymbol{G}(\mathbf{x}, \mathbf{x}') = 0 \,. \tag{7.29}$$

Again by linearity, the pressure and stress fields corresponding to (7.28) can be written in the form

$$\tilde{p}(\mathbf{x}) = \mathbf{Q}(\mathbf{x}, \mathbf{x}') \cdot \tilde{\mathbf{F}}, \tag{7.30}$$

$$\tilde{\boldsymbol{\sigma}}(\mathbf{x}) = \boldsymbol{T}(\mathbf{x}, \mathbf{x}') \cdot \tilde{\mathbf{F}}, \tag{7.31}$$

where $\mathbf{Q}$ is a vector, and $\boldsymbol{T}$ is a third-rank tensor given by the relation

$$\boldsymbol{T}(\mathbf{x}, \mathbf{x}') = \boldsymbol{\nabla} \boldsymbol{G}(\mathbf{x}, \mathbf{x}') + [\boldsymbol{\nabla} \boldsymbol{G}(\mathbf{x}, \mathbf{x}')]^\dagger - \hat{\boldsymbol{I}}\mathbf{Q}(\mathbf{x}, \mathbf{x}'), \tag{7.32}$$

where the dagger indicates the transpose of the first two indices. Accordingly, in component form, we have

$$T_{ijk}(\mathbf{x}, \mathbf{x}') = \partial_i G_{jk}(\mathbf{x}, \mathbf{x}') + \partial_j G_{ik}(\mathbf{x}, \mathbf{x}') - \delta_{ij}Q_k(\mathbf{x}, \mathbf{x}') \tag{7.33}$$

where $\partial_i = \partial/\partial x_i$. The tensor $\boldsymbol{T}$ is the *double-layer Green's function* for Stokes flow. The reason for this denomination, together with an interpretation of the physical meaning of the Green's functions $\boldsymbol{G}$ and $\boldsymbol{T}$, will be seen shortly.

As a consequence of the symmetry of the Laplace operator in the Stokes equation (7.5), it can be shown that the Green's function $\boldsymbol{G}$ has the symmetry property (see, e.g. Kim and Karrila, 1991)

$$\boldsymbol{G}(\mathbf{x}, \mathbf{x}') = \boldsymbol{G}^{\dagger}(\mathbf{x}', \mathbf{x}) \qquad \text{or} \qquad G_{ij}(\mathbf{x}, \mathbf{x}') = G_{ji}(\mathbf{x}', \mathbf{x}), \tag{7.34}$$

where the dagger denotes the transpose. Moreover, the definitions (7.32) or (7.33) of $\boldsymbol{T}$ imply that the double-layer Green's function satisfies the symmetry relation

$$\boldsymbol{T}(\mathbf{x}, \mathbf{x}') = \boldsymbol{T}^{\dagger}(\mathbf{x}, \mathbf{x}'), \qquad \text{or} \qquad T_{ijk}(\mathbf{x}, \mathbf{x}') = T_{jik}(\mathbf{x}, \mathbf{x}'). \tag{7.35}$$

Due to the incompressibility condition (7.29), we have the fluid-volume conservation relation

$$\int_S \hat{\mathbf{n}}_1(\mathbf{x}) \cdot \boldsymbol{G}(\mathbf{x}, \mathbf{x}') \, \mathrm{d}S = 0, \tag{7.36}$$

while the double-layer Green's function satisfies the relation

$$\int_S \hat{\mathbf{n}}_1(\mathbf{x}) \cdot \boldsymbol{T}(\mathbf{x}, \mathbf{x}') \, \mathrm{d}S = -\chi_1(\mathbf{x}')\hat{\boldsymbol{I}}. \tag{7.37}$$

This relation is a consequence of

$$\boldsymbol{\nabla} \cdot \boldsymbol{T}(\mathbf{x}, \mathbf{x}') = -\hat{\boldsymbol{I}}\delta(\mathbf{x} - \mathbf{x}'), \tag{7.38}$$

which follows directly from equations (7.25) and (7.31). Note that, in order to derive (7.37), it is expedient to integrate (7.38), over the region Ω_1 rather than Ω_2 to avoid considering the boundary term at infinity. It should also be noted that the integral on the right-side of relation (7.27) is undefined for $\mathbf{x}' \in S$. Thus, evaluation of the double-layer integral (7.37) at points lying on the surface S is not straightforward. We will return to this problem in Section 7.3.1.

The specific form of the Green's functions depends on the domain in which the inhomogeneous Stokes problem (7.25) is solved and on the related boundary conditions. The simplest case, which is of special importance to us, is the free-space solution that decays at infinity. This solution is known as the *Oseen tensor* and is given by (see, e.g. Kim and Karrila, 1991)

$$\boldsymbol{G}(\mathbf{r}) = \frac{1}{8\pi r}(\hat{\boldsymbol{I}} + \hat{\mathbf{r}}\hat{\mathbf{r}}), \tag{7.39}$$

where $\hat{\mathbf{r}} = \mathbf{r}/r$, $\mathbf{r} = \mathbf{x} - \mathbf{x}'$, and $r = |\mathbf{r}|$. The corresponding expressions for $\mathbf{Q}$ and T are

$$T(\mathbf{r}) = -\frac{3\hat{\mathbf{r}}\hat{\mathbf{r}}\hat{\mathbf{r}}}{4\pi r^2}, \qquad \mathbf{Q}(\mathbf{r}) = \frac{\hat{\mathbf{r}}}{4\pi r^2}. \tag{7.40}$$

It may be noted that equation (7.38) is a linear inhomogeneous equation, of which T, as given by equation (7.40), is a particular solution. Any other solution of (7.38) (with different boundary conditions) can be obtained by adding a homogeneous solution to the T given by (7.40). As a consequence, the expressions for G and $\mathbf{Q}$ will also be modified.

Upon substituting (7.28) and (7.31) into the integral identity (7.26) and canceling the vector $\tilde{\mathbf{F}}$ (which appears in all the terms and is arbitrary) we find[1]

$$\mathbf{u}_i(\mathbf{x}')\chi_i(\mathbf{x}') = \int_S \hat{\mathbf{n}}_i(\mathbf{x}) \cdot \left[\boldsymbol{\sigma}_i(\mathbf{x}) \cdot \mu^{-1}\mathbf{G}(\mathbf{x}, \mathbf{x}') - \mathbf{u}_i(\mathbf{x}) \cdot T(\mathbf{x}, \mathbf{x}')\right] \, \mathrm{d}S.$$
$$\tag{7.41}$$

If $\boldsymbol{\sigma}_i$ and $\mathbf{u}_i$ are known on the integration surface S, this relation permits one to find the velocity field $\mathbf{u}_i$ at any other point in the i-th domain. Alternatively, by considering the limit $\mathbf{x}' \to \mathbf{x}_S \in S$, we will find an integral relation between $\boldsymbol{\sigma}_i$ and $\mathbf{u}_i$ on S which will enable us to calculate one quantity if the other one is known. This is the basic idea of the *boundary integral method*, so called as it relies on an integral equation on the boundary of the domain of interest. However, as anticipated before, taking the limit $\mathbf{x}' \to \mathbf{x}_S \in S$ is not straightforward. The difficulties are associated with the presence of the discontinuous characteristic function $\chi_i(\mathbf{x}')$ on the left side of equation (7.41) – the integral on the right side must thus also have a discontinuity as $\mathbf{x}'$ traverses the interface S.

Before we turn to these developments, it is useful to make some comments. The first term on the right side of the identity (7.41) describes the flow produced by a layer of point forces distributed on the surface S with the density $\mathbf{t}_i(\mathbf{x}) = \hat{\mathbf{n}}_i(\mathbf{x}) \cdot \boldsymbol{\sigma}_i(\mathbf{x})$. Hence, this term is called a single-layer integral, which justifies the denomination single-layer Green's function adopted for the tensor G. The interpretation of the second term is more involved. This term corresponds to a flow field produced by a tensorial source distribution $\mathbf{m}(\mathbf{x}) = \hat{\mathbf{n}}_i(\mathbf{x})\mathbf{u}_i(\mathbf{x})$. It can be shown that $\mathbf{m}$ involves a distribution of force dipoles (Kim and Karrila, 1991). Since a surface dipole distribution can be represented as a double layer of point forces, the second term is called a double-layer integral, and the Green's function T is called the double-layer Green's function. We thus see that equation (7.41) is

---

[1] This equation will mostly be used with $\mu = \mu_i$, except on one occasion, equation (7.47). For this reason we do not append a subscript to μ.

analogous to the well-known Green's identity of the theory of the electro-static potential, according to which an arbitrary potential distribution can be represented by the superposition of suitable charge and dipole densities on the surface of the domain of interest.

Boundary integral equations with the infinite-space Green's functions (7.39) and (7.40) are used in simulations of finite groups of drops or parti-cles suspended in an unbounded fluid. In our presentation of the boundary integral techniques we will focus on the infinite-space case. However, our analysis relies only on general properties of Green's functions, such as equa-tions (7.29)–(7.32); thus, the resulting boundary integral equations apply to other geometries as well, provided the appropriate Green's functions are used.

7.3 The boundary integral equations

Our goal is to apply the Lorentz integral identity to obtain boundary integral equations for liquid drops or rigid particles in a form that requires evaluation of the flow field only on the particle interface, but not in the bulk fluids. To this end the Lorentz integral identity (7.41) must be evaluated for the field point $\mathbf{x}'$ on the surface S. At first sight it would seem that the singularity proportional to $|\mathbf{x}-\mathbf{x}'|^{-1}$ of $\boldsymbol{G}$ might cause a problem when $\mathbf{x}' = \mathbf{x}$. However, this is actually not so as this singularity is canceled by the surface element $\mathrm{d}S = \rho \, \mathrm{d}\rho \, \mathrm{d}\theta$, where ρ and θ are local polar coordinates (see Section 7.5.3 below). Therefore the single-layer integral exists in an ordinary sense.

The situation is different for what concerns the second term with $\boldsymbol{T}$. If we refer to the free-space expression (7.40) for it, it is evident that the singu-larity is stronger and will not be cancelled. As remarked before, any other double-layer Green's function appropriate for different boundary conditions will also contain the same singularity. This singularity produces the discon-tinuity of the double-layer integral and the characteristic function $\chi_i(\mathbf{x}')$, and it is thus necessary to take the limit $\mathbf{x}' \to \mathbf{x}_\mathrm{S} \in S$ with some care.

7.3.1 Singularity subtraction

The customary way to deal with the discontinuity problem (see, e.g. Pozrikidis, 1992) is to effect a careful calculation of the limit $\mathbf{x}' \to \mathbf{x}_\mathrm{S} \in S$, where $\mathbf{x}'$ is in either one of the bulk regions Ω_1 or Ω_2. A more direct, although equivalent, approach consists in removing the singularity of the double-layer integrand in (7.41) *before* taking the limit to avoid dealing with discontinuous results. For this purpose we note that, by (7.37), we have the identity

$$\int_S \hat{\mathbf{n}}_i(\mathbf{x}) \cdot [\mathbf{u}_i(\mathbf{x}) \cdot \mathbf{T}(\mathbf{x}, \mathbf{x}')] \, dS = \int_S \hat{\mathbf{n}}_i(\mathbf{x}) \cdot \{ [\mathbf{u}_i(\mathbf{x}) - \mathbf{u}_i(\mathbf{x}')] \cdot \mathbf{T}(\mathbf{x}, \mathbf{x}') \} \, dS$$
$$\mp \chi_1(\mathbf{x}') \mathbf{u}(\mathbf{x}'), \qquad (7.42)$$

where the symmetry of the Green's function (7.35) has been used. The upper sign corresponds to $i = 1$ and the lower one to $i = 2$. The change of sign results from the orientation change of the normal vector. Upon re-expressing the second term of the integral in (7.41) in the form given in the right-hand side of equation (7.42) and noting that $\chi_1 \mp \chi_i = \delta_{i2}$ (where δ_{i2} is the Kronecker delta), we find

$$\mathbf{u}_i(\mathbf{x}')\delta_{i2} = \int_S \hat{\mathbf{n}}_i(\mathbf{x}) \cdot \{ \boldsymbol{\sigma}_i(\mathbf{x}) \cdot \mu^{-1} \mathbf{G}(\mathbf{x}, \mathbf{x}') - [\mathbf{u}_i(\mathbf{x}) - \mathbf{u}_i(\mathbf{x}')] \cdot \mathbf{T}(\mathbf{x}, \mathbf{x}') \} \, dS.$$
$$(7.43)$$

In this result, the asymmetry between the regions Ω_1 and Ω_2, reflected by the presence of the Kronecker delta δ_{i2}, results from the slow decay of the field $\mathbf{T}(\mathbf{x}, \mathbf{x}')$ for $|\mathbf{x} - \mathbf{x}'| \to \infty$, which precludes using the region Ω_2 instead of Ω_1 in equation (7.37) without including the boundary term at infinity.

Two goals have been achieved by the above transformation. In the first place, the order of the singularity of the double-layer integrand has been decreased to $O(r^{-1})$, because $\mathbf{u}(\mathbf{x}') \to \mathbf{u}(\mathbf{x})$ for $\mathbf{x} \to \mathbf{x}'$, which makes the term integrable. Second, the resulting integral is a continuous function of the position $\mathbf{x}'$. The singularity-subtraction procedure not only facilitates the theoretical analysis, but also the numerical integration. Therefore, it should always be carried out before the boundary integral equations are discretized.

7.3.2 First-kind boundary integral formulation

A simple application of equation (7.43) is to the case of a rigid particle, the velocity at the surface of which is given by (7.16). In the presence of the imposed flow $\mathbf{u}^\infty$, the velocity continuity condition (7.12) yields

$$\mathbf{u}_2(\mathbf{x}) = \mathbf{U}^{\mathrm{rb}} + \boldsymbol{\Omega}^{\mathrm{rb}} \times (\mathbf{x} - \mathbf{x}_0) - \mathbf{u}^\infty(\mathbf{x}), \qquad \mathbf{x} \in S. \qquad (7.44)$$

By using this relation in (7.43) with $i = 2$, $\mathbf{x}' \in S$, we then have

$$\mathbf{U}^{\mathrm{rb}} + \boldsymbol{\Omega}^{\mathrm{rb}} \times (\mathbf{x}' - \mathbf{x}_0)$$
$$= \mathbf{u}^\infty(\mathbf{x}') + \mu_2^{-1} \int_S \mathbf{t}_2(\mathbf{x}) \cdot \mathbf{G}(\mathbf{x}, \mathbf{x}') \, dS$$
$$- \int_S \hat{\mathbf{n}}_2 \cdot \{ \left[(\boldsymbol{\Omega}^{\mathrm{rb}} \times (\mathbf{x}' - \mathbf{x}) - \mathbf{u}^\infty(\mathbf{x}) + \mathbf{u}^\infty(\mathbf{x}') \right] \cdot \mathbf{T}(\mathbf{x}, \mathbf{x}') \} \, dS,$$
$$(7.45)$$

where μ_2 is the viscosity of the fluid surrounding the particle.

This equation can be used to find the components of the resistance and mobility tensors, introduced in equation (7.19), by taking $\mathbf{U}^{\mathrm{rb}}$ and $\mathbf{\Omega}^{\mathrm{rb}}$ equal, in turn, to each one of the three unit vectors. Substitution of the corresponding six results for $\mathbf{t}_2$ into the expressions (7.17) and (7.18) generates the corresponding components of the forces and torques $\mathbf{F}$ and $\mathbf{L}$. From these results the components of the resistance tensors can be directly evaluated; the mobility matrix is then obtained by inversion of the 6×6 resistance matrix.

Since the unknown field $\mathbf{t}_2(\mathbf{x})$ appears only inside the integral operator, equation (7.45) is a Fredholm integral equation of the first kind (see, e.g. Arfken and Weber, 2000). Such equations are in fact mathematically ill-posed and, accordingly, often difficult to solve because of numerical instabilities associated with short-wavelength oscillations of the solution. The singularity of the integral kernel $\mathbf{G}(\mathbf{x}, \mathbf{x}')$ in equation (7.45) mitigates the problem because the integral has a large contribution from the region $\mathbf{x} \approx \mathbf{x}'$. The matrix obtained by a discretization of the integral operator with such a kernel is thus diagonally dominated. However, these numerical difficulties suggest that we try to reformulate the problem in terms of an equation of the second kind, a task to which we now turn.

7.3.3 Second-kind boundary integral formulation

It is well known from potential theory that a representation of the field in terms of both surface charges and dipoles is actually redundant, as either charges only, or dipoles only (possibly augmented by a single point charge), can be equivalently used to represent the same potential. A similar result holds for the present case of Stokes flow, and it will permit us to develop the reformulation of the problem that we are seeking.

We consider a liquid drop of viscosity μ_1 immersed in the continuous-phase fluid of viscosity μ_2. The corresponding results for a solid particle will be obtained by taking the high-viscosity limit. To derive the boundary integral equation for the flow field on the drop interface S, we start by writing the Lorentz integral identity (7.43) in region 1 for the fields $\mathbf{u}_1$ and $\mathbf{u}^\infty$, which are both nonsingular inside S. For $\mathbf{u}_1$ we find

$$0 = \int_S \hat{\mathbf{n}}_1(\mathbf{x}) \cdot \left\{ \boldsymbol{\sigma}_1(\mathbf{x}) \cdot \mu_1^{-1} \mathbf{G}(\mathbf{x}, \mathbf{x}') - [\mathbf{u}_1(\mathbf{x}) - \mathbf{u}_1(\mathbf{x}')] \cdot \mathbf{T}(\mathbf{x}, \mathbf{x}') \right\} \, \mathrm{d}S.$$

$$(7.46)$$

In writing a similar relation for $\mathbf{u}^\infty$ we must remember that this field, while regular inside S, satisfies the Stokes equation with viscosity μ_2; therefore,

$$0 = \int_S \hat{\mathbf{n}}_1(\mathbf{x}) \cdot \left\{ \boldsymbol{\sigma}^\infty(\mathbf{x}) \cdot \mu_2^{-1} \mathbf{G}(\mathbf{x}, \mathbf{x}') - [\mathbf{u}^\infty(\mathbf{x}) - \mathbf{u}^\infty(\mathbf{x}')] \cdot \mathbf{T}(\mathbf{x}, \mathbf{x}') \right\} \, \mathrm{d}S.$$

$$(7.47)$$

This application of the Lorentz identity (7.43) with $i = 1$ but $\mu = \mu_2$ is not contradictory: the imposed flow $\mathbf{u}^\infty$ can be viewed as the velocity field produced by external forces in the uniform continuous-phase fluid ($\mu = \mu_2$) when the drop is absent. Since these forces are applied outside the region occupied by the drop, the flow field $\mathbf{u}^\infty$ is nonsingular in the domain Ω_1. In the presence of the drop, only the portion $\mathbf{u}^\infty(\mathbf{x})\chi_2(\mathbf{x})$ has a direct physical meaning, according to decomposition (7.8); this restriction, however, does not affect the validity of the double-layer integral (7.47). The disturbance field $\mathbf{u}_2$ is nonsingular in the outer domain. Accordingly, relation (7.43) must be used with $i = 2$, which yields

$$\mathbf{u}_2(\mathbf{x}') = \int_S \hat{\mathbf{n}}_2(\mathbf{x}) \cdot \left\{ \boldsymbol{\sigma}_2(\mathbf{x}) \cdot \mu_2^{-1} \mathbf{G}(\mathbf{x}, \mathbf{x}') - [\mathbf{u}_2(\mathbf{x}) - \mathbf{u}_2(\mathbf{x}')] \cdot \mathbf{T}(\mathbf{x}, \mathbf{x}') \right\} \, \mathrm{d}S.$$

$$(7.48)$$

To eliminate the contribution of the single-layer Green's function, (7.47) is subtracted from equation (7.48), and the result is combined with equation (7.46) multiplied by the viscosity ratio $\lambda = \mu_1/\mu_2$. The unknown viscous tractions (7.14) and flow components $\mathbf{u}_2$ and $\mathbf{u}^\infty$ are eliminated using the boundary conditions (7.12) and (7.13) and relation (7.7). As a result, a boundary integral equation of the form

$$\mathbf{u}_S(\mathbf{x}') + (\lambda - 1) \int_S \hat{\mathbf{n}}_1(\mathbf{x}) \cdot \left\{ [\mathbf{u}_S(\mathbf{x}) - \mathbf{u}_S(\mathbf{x}')] \cdot \mathbf{T}(\mathbf{x}, \mathbf{x}') \right\} \, \mathrm{d}S$$

$$= \mathbf{u}^\infty(\mathbf{x}') + \int_S \mathbf{f}_S(\mathbf{x}) \cdot \mu_2^{-1} \mathbf{G}(\mathbf{x}, \mathbf{x}') \, \mathrm{d}S \qquad (7.49)$$

is obtained for the interfacial velocity

$$\mathbf{u}_S(\mathbf{x}) = \mathbf{u}_1(\mathbf{x}), \qquad \mathbf{x} \in S. \qquad (7.50)$$

For a given drop configuration and external flow, the quantities on the right-hand side of the boundary integral equation (7.49) are known. In particular, the surface force density $\mathbf{f}_s$ is given by (7.15). The evolution of the system can thus be determined by solving the equation numerically for the interfacial velocity $\mathbf{u}_S$. Once this quantity is known, the position of the interface can be updated by integrating the kinematic relation

$$\frac{\mathrm{d}\mathbf{x}_S}{\mathrm{d}t} = \mathbf{u}_S(\mathbf{x}_S), \qquad (7.51)$$

in which $\mathbf{x}_S$ is a generic interface point. If the interface has a purely geometric nature, only the normal component of this relation is actually needed. In the presence of an adsorbed surfactant, on the other hand, the evolution of the interfacial surfactant distribution depends both on the normal and tangential velocity components.

In the boundary integral equation (7.49), the unknown quantity $\mathbf{u}_S$ appears not only at the integration point $\mathbf{x}$ but also at the field point $\mathbf{x}'$. This contributes to the diagonal dominance of the matrix resulting from the discretization of the equation. For drops with moderate viscosities, the associated numerical problem is well conditioned, and accurate solutions can be obtained at a relatively low numerical cost using numerical methods discussed in Section 7.5.

The problem, however, is ill conditioned for highly viscous drops and for drops with very low viscosities, $\lambda \to \infty$ and $\lambda \to 0$. The roots of this difficulty are apparent on physical grounds. Suppose that the viscosity of the drop is much greater than that of the surrounding fluid. In this situation, the drop will respond readily to an external flow by translating and rotating, while its deformation will occur on a much slower time scale. Conversely, if the outer fluid is much more viscous than the inner one, a drop (or, more appropriately for this case, a bubble) will readily expand or contract since viscous stresses vanish identically in a purely radial motion. Deformations, however, generate stresses in the surrounding fluid and therefore will occur much more slowly. In both cases we expect to find a great disparity between the velocities associated with the deformation modes and those associated with the translation/rotation/expansion modes. This large difference may be expected to affect the accuracy and facility with which the integral equation can be solved. The remedy is to rescale the "slow" modes in such a way that they will not be "swamped" by the "fast" modes. This is the purpose of the next section in which we remove this near-singular behavior.

7.3.4 A better form of the boundary integral equation

The arguments given at the end of the previous section indicate that numerical difficulties may arise for a highly viscous drop, because such a drop readily translates and rotates, but its motion differs little from that of a rigid body. However, it is precisely this difference which governs the deformation of the drop. A similar (although more complex) situation occurs for an expanding bubble. It is therefore expedient to reformulate the problem in such a way that the unknown effectively becomes the *difference* between the actual velocity field and the corresponding rigid-body or bubble modes, rather than the actual velocity field itself. This reformulation can be efficiently described by adopting the language of operators in a scalar product space.

The double-layer integral in equation (7.49) may be considered as defining an integral operator acting on the velocity field $\mathbf{u}$. Recognizing this fact, it is convenient to write

$$[\hat{\mathsf{T}}\mathbf{u}](\mathbf{x}') = \int_S \hat{\mathbf{n}}_1(\mathbf{x}) \cdot \{[\mathbf{u}(\mathbf{x}) - \mathbf{u}(\mathbf{x}')] \cdot \boldsymbol{T}(\mathbf{x}, \mathbf{x}')\} \, \mathrm{d}S, \qquad \mathbf{x}' \in S. \quad (7.52)$$

The operator $\hat{\mathsf{T}}$ acts in the linear space of vector fields $\mathbf{u}$ specified on the surface S. The natural scalar product in this space is

$$(\mathbf{u}, \mathbf{v}) = \int_S \mathbf{u}(\mathbf{x}) \cdot \mathbf{v}(\mathbf{x}) \, \mathrm{d}S. \quad (7.53)$$

Consider now the action of the operator $\hat{\mathsf{T}}$ on the rigid-body surface velocity fields (7.16). It is shown in Appendix A, that

$$\hat{\mathsf{T}}\mathbf{u}^{\mathrm{rb}}(\mathbf{x}) = 0, \quad (7.54)$$

which implies that $\mathbf{u}^{\mathrm{rb}}(\mathbf{x})$ is an eigenvector of $\hat{\mathsf{T}}$ corresponding to the eigenvalue 0. This result reflects the fact that the rigid-body velocity field does not produce any energy dissipation. Since there are six linearly independent rigid-body velocity fields – three translational and three rotational motions in orthogonal directions – we thus see that the null eigenspace of $\hat{\mathsf{T}}$ has dimension (at least) six. It can also be demonstrated (Appendix A) that, for any $\mathbf{u}$, we have

$$(\hat{\mathsf{T}}\mathbf{u}, \hat{\mathbf{n}}_1) = (\mathbf{u}, \hat{\mathbf{n}}_1), \quad (7.55)$$

which implies that the field $\hat{\mathbf{n}}_1(\mathbf{x})$ is an eigenvector of the operator $\hat{\mathsf{T}}^\dagger$ (the adjoint of $\hat{\mathsf{T}}$), corresponding to the eigenvalue 1. Note that the expansion eigenmode $\mathbf{v}(\mathbf{x}) = \hat{\mathbf{n}}_1(\mathbf{x})$ has a nonzero flux through the surface S because

$$\int_S \hat{\mathbf{n}}_1(\mathbf{x}) \cdot \hat{\mathbf{n}}_1(\mathbf{x}) \, \mathrm{d}S = S. \quad (7.56)$$

On the basis of (7.54) and (7.55) we thus conclude that the double-layer operator $\hat{\mathsf{T}}$ has the rigid-body and expansion eigenvalues

$$\alpha_0 = 0, \qquad \alpha_1 = 1. \quad (7.57)$$

A detailed analysis of the problem (Kim and Karrila, 1991) reveals that all other eigenvalues of the operator $\hat{\mathsf{T}}$ are in the range

$$\alpha_0 < \alpha < \alpha_1. \quad (7.58)$$

We can now proceed with our plan to write the surface velocity $\mathbf{u}_S$ as the sum of a rigid-body component, an expansion component, plus a suitably scaled remainder. For this purpose, for a given surface velocity field $\mathbf{u}_S$, we need to identify these components explicitly. It can be shown that the translational and rotational velocities, evaluated with respect to the center of mass of the surface S,

$$\mathbf{x}_0 = S^{-1} \int_S \mathbf{x} \, \mathrm{d}S , \tag{7.59}$$

are given by the expressions (Zinchenko *et al.*, 1997)

$$\mathbf{U}^{\mathrm{rb}} = S^{-1} \int_S \mathbf{u}_S(\mathbf{x}) \, \mathrm{d}S, \qquad \mathbf{\Omega}^{\mathrm{rb}} = \boldsymbol{D}^{-1} \cdot \int_S \bar{\mathbf{x}} \times \mathbf{u}_S(\mathbf{x}) \, dS, \tag{7.60}$$

where

$$\boldsymbol{D} = \int_S (\bar{\mathbf{x}} \cdot \bar{\mathbf{x}})(\hat{\boldsymbol{I}} - \hat{\bar{\mathbf{x}}}\hat{\bar{\mathbf{x}}}) \, dS, \tag{7.61}$$

and $\bar{\mathbf{x}} = \mathbf{x} - \mathbf{x}_0$ and $\hat{\bar{\mathbf{x}}} = \bar{\mathbf{x}}/|\bar{\mathbf{x}}|$. Accordingly, the rigid-body component of the velocity field $\mathbf{u}_S$, which we denote by $\hat{\mathsf{P}}^{\mathrm{rb}}\mathbf{u}_S$, is

$$\hat{\mathsf{P}}^{\mathrm{rb}}\mathbf{u}_S = \mathbf{U}^{\mathrm{rb}} + \mathbf{\Omega}^{\mathrm{rb}} \times (\mathbf{x} - \mathbf{x}_0) , \tag{7.62}$$

with $\mathbf{U}^{\mathrm{rb}}$ and $\mathbf{\Omega}^{\mathrm{rb}}$ given by (7.60). The expansion component of the flow field $\mathbf{u}_S$ is denoted by $\hat{\mathsf{P}}^1\mathbf{u}_S$. It is given by the relation

$$\hat{\mathsf{P}}^1\mathbf{u}_S = \hat{\mathbf{n}}_1(\mathbf{x})S^{-1} \int_S \hat{\mathbf{n}}_1(\mathbf{x}') \cdot \mathbf{u}_S(\mathbf{x}') \, \mathrm{d}S'. \tag{7.63}$$

In the above expressions, $\hat{\mathsf{P}}^{\mathrm{rb}}$ is the projection operator on the rigid-body subspace, and $\hat{\mathsf{P}}^1$ is the projection operator on the expansion eigenmode.

We are now ready to carry out the rescaling mentioned before, starting with the rigid-body component. We decompose the surface velocity $\mathbf{u}_S$ into a rigid-body part and a remainder:

$$\mathbf{u}_S = \hat{\mathsf{P}}^{\mathrm{rb}}\mathbf{u}_S + (1 - \hat{\mathsf{P}}^{\mathrm{rb}}) \, \mathbf{u}_S . \tag{7.64}$$

Now, the non-rigid-body part is rescaled by introducing a rescaled surface velocity $\bar{\mathbf{u}}_S$ according to the relation

$$(1 - \hat{\mathsf{P}}^{\mathrm{rb}}) \, \mathbf{u}_S = 2(\lambda + 1)^{-1}(1 - \hat{\mathsf{P}}^{\mathrm{rb}})\bar{\mathbf{u}}_S. \tag{7.65}$$

The viscosity ratio λ in the prefactor of the above expression assures that $(1 - \hat{\mathsf{P}}^{\mathrm{rb}})\bar{\mathbf{u}}_S$ remains finite in the limit[1] $\lambda \to \infty$. The 1 is added so that $\bar{\mathbf{u}}_S$ is nonsingular in the limit $\lambda \to 0$, and the factor 2 is introduced for later

---

[1] Note that equations (7.49) and (7.54) imply that $\lambda(1 - \hat{\mathsf{P}}^{\mathrm{rb}})\mathbf{u}_S \to \text{const.}$ for $\lambda \to \infty$.

convenience. Equation (7.65) only defines the part of $\bar{\mathbf{u}}_S$ that is orthogonal to the rigid-body mode; it is natural to leave the rigid-body motion component unscaled, $\hat{\mathsf{P}}^{\mathrm{rb}}\mathbf{u}_S = \hat{\mathsf{P}}^{\mathrm{rb}}\bar{\mathbf{u}}_S$. Accordingly, the decomposition of $\mathbf{u}_S$ that we introduce is

$$\mathbf{u}_S = \hat{\mathsf{P}}^{\mathrm{rb}}\bar{\mathbf{u}}_S + 2(\lambda+1)^{-1}(1 - \hat{\mathsf{P}}^{\mathrm{rb}})\bar{\mathbf{u}}_S. \tag{7.66}$$

The inverse of this relation is readily found by inverting equation (7.65):

$$\bar{\mathbf{u}}_S = \hat{\mathsf{P}}^{\mathrm{rb}}\mathbf{u}_S + \frac{1}{2}(\lambda+1)(1 - \hat{\mathsf{P}}^{\mathrm{rb}})\mathbf{u}_S. \tag{7.67}$$

Upon substitution of equation (7.66) into the the boundary integral equation (7.49), we obtain its *purged* form

$$(1 - \varkappa)\bar{\mathbf{u}}_S(\mathbf{x}') + \varkappa\hat{\mathsf{P}}^{\mathrm{rb}}\bar{\mathbf{u}}_S(\mathbf{x}')$$
$$+2\varkappa \int_S \hat{\mathbf{n}}_1(\mathbf{x}) \cdot \left\{ [\bar{\mathbf{u}}_S(\mathbf{x}) - \bar{\mathbf{u}}_S(\mathbf{x}')] \cdot \boldsymbol{T}(\mathbf{x},\mathbf{x}') \right\} \, \mathrm{d}S$$
$$= \mathbf{u}^\infty(\mathbf{x}') + \int_S \mathbf{f}_S(\mathbf{x}) \cdot \mu_2^{-1}\boldsymbol{G}(\mathbf{x},\mathbf{x}') \, \mathrm{d}S, \tag{7.68}$$

where

$$\varkappa = \frac{\lambda - 1}{\lambda + 1}. \tag{7.69}$$

An analysis of the purged boundary integral equation (7.68) indicates that it is well-behaved for all values of viscosity ratio $\lambda > 0$ (including $\lambda \to \infty$). This can be seen by noting that

$$-1 \leq \varkappa \leq 1, \tag{7.70}$$

which implies that all terms in the equation remain finite in the high-viscosity limit, unlike the double-layer term in the unscaled equation (7.49). Moreover, due to the presence of the rigid-body projection $\hat{\mathsf{P}}^{\mathrm{rb}}\bar{\mathbf{u}}_S$, the rigid-body mode (7.16) is associated with the eigenvalue $\bar{\beta} = 1$ of the combined operator corresponding to all three terms on the left side of (7.68). Relations (7.57) and (7.58) thus indicate that the purged equation (7.68) is singular only for $\lambda = 0$ (i.e. $\varkappa = -1$), because of the property (7.55) of the expansion eigenmode. The next step is therefore to remove this remaining singularity by purging the expansion eigenmode.

We focus in this chapter on drops of fixed volume and, therefore, the total fluid flux through the surface S is assumed to vanish. Using the projection-operator notation (7.63), this volume conservation condition can be expressed in the form

$$\hat{\mathsf{P}}^1\bar{\mathbf{u}}_S = 0. \tag{7.71}$$

By subtracting equation (7.71) multiplied by $\varkappa$ from equation (7.68), the fully purged boundary integral equation is obtained:

$$(1 - \varkappa)\bar{\mathbf{u}}_S(\mathbf{x}') + \varkappa(\hat{\mathsf{P}}^{rb} - \hat{\mathsf{P}}^1)\bar{\mathbf{u}}_S(\mathbf{x}')$$

$$+2\varkappa \int_S \hat{\mathbf{n}}_1(\mathbf{x}) \cdot \left\{ [\bar{\mathbf{u}}_S(\mathbf{x}) - \bar{\mathbf{u}}_S(\mathbf{x}')] \cdot \boldsymbol{T}(\mathbf{x},\mathbf{x}') \right\} \, \mathrm{d}S$$

$$= \mathbf{u}^\infty(\mathbf{x}') + \int_S \mathbf{f}_S(\mathbf{x}) \cdot \mu_2^{-1} \boldsymbol{G}(\mathbf{x},\mathbf{x}') \, \mathrm{d}S. \tag{7.72}$$

Using relation (7.55), definition (7.52) of the double-layer operator, and expression (7.53) for the inner product, one can verify that the expansion-mode eigenvalue of the linear operator that combines all terms in the left side of the boundary integral equation has been shifted from $\bar{\beta} = 1 + \varkappa$ in equation (7.68) to $\bar{\beta} = 1$ in equation (7.72). Thus, after the purge, the expansion eigenvalue is nonzero also for $\lambda = 0$ ($\varkappa = -1$).

Recalling that the spectrum of the boundary integral operator is in the range (7.58), we find that the eigenvalues of the combined operator in equation (7.72) are in the range

$$1 - \varkappa < \bar{\beta} < 1 + \varkappa, \tag{7.73}$$

with the sharp inequalities that ensure that the equation is nonsingular for all viscosity ratios, including the limiting values $\lambda = 0$ and $\lambda = \infty$. Thus, after the full purging procedure, which is sometimes referred to as *Wielandt's deflation*, the boundary integral equation (7.72) is well-conditioned in the whole viscosity range.

7.3.5 Rigid-particle limit

The developments of the previous section enable us to present an alternative formulation of the boundary integral equation for a rigid particle. The purged form (7.68) – or, alternatively, (7.72) – of the boundary integral equation is well behaved in the infinite-viscosity limit $\lambda \to \infty$. The purged boundary integral equation therefore can be used to describe the motion of a rigid particle.

By taking the limit $\lambda \to \infty$ of the purged equation (7.72), we obtain a rigid-particle boundary integral equation of the form

$$(\hat{\mathsf{P}}^{rb} - \hat{\mathsf{P}}^1)\bar{\mathbf{u}}_S(\mathbf{x}') + 2 \int_S \hat{\mathbf{n}}_1(\mathbf{x}) \cdot \left\{ [\bar{\mathbf{u}}_S(\mathbf{x}) - \bar{\mathbf{u}}_S(\mathbf{x}')] \cdot \boldsymbol{T}(\mathbf{x},\mathbf{x}') \right\} \, \mathrm{d}S$$

$$= \mathbf{u}^\infty(\mathbf{x}') + \int_S \mathbf{f}_S(\mathbf{x}) \cdot \mu_2^{-1} \boldsymbol{G}(\mathbf{x},\mathbf{x}') \, \mathrm{d}S. \tag{7.74}$$

The (unscaled) rigid-body velocity field $\mathbf{u}_S = \mathbf{u}^{rb}$, corresponding to the rotation and translation of the particle, is related to the auxiliary field $\bar{\mathbf{u}}_S$ by the expression

$$\mathbf{u}_S = \hat{\mathsf{P}}^{rb}\bar{\mathbf{u}}_S, \tag{7.75}$$

which is obtained by taking the limit $\lambda \to \infty$ of the transformation (7.66). The translational and angular particle velocities $\mathbf{U}^{rb}$ and $\mathbf{\Omega}^{rb}$ can thus be obtained by solving the boundary integral equation (7.74) for the auxiliary field $\bar{\mathbf{u}}_S$, and then using relation (7.75) and expressions (7.59)–(7.61) for the projection operator $\hat{\mathsf{P}}^{rb}$. The surface traction in the boundary integral equation (7.74) should be chosen consistently with the constraints

$$\mathbf{F} = \int_S \mathbf{f}_S(\mathbf{x})\,\mathrm{d}S, \qquad \mathbf{L} = \int_S (\mathbf{x}-\mathbf{x}_0)\times\mathbf{f}_S(\mathbf{x})\,\mathrm{d}S, \tag{7.76}$$

where $\mathbf{F}$ and $\mathbf{L}$ are the total force and torque acting on the particle. Since the particle can only translate and rotate but it cannot deform, all other details of the surface force densities $\mathbf{f}_S$ are irrelevant for the particle motion. A change of traction at constant $\mathbf{F}$ and $\mathbf{L}$ results in a variation of the stresses in the rigid particle; the solution $\bar{\mathbf{u}}_S$ is accordingly modified, without, however, changing the resulting rigid-body projection (7.75). This statement can be formally demonstrated using the Lorentz relation (7.43); this is done in Appendix B.

The freedom of choice of the surface force densities $\mathbf{f}_S$ can be used to avoid numerical evaluation of the single-layer integral on the right-hand side of the boundary integral equation (7.74). To this end we select the tractions in such a way that the associated single-layer velocity field in the region Ω_2 is identical to the flow produced by a point force and torque acting at a position $\mathbf{x}_0 \in \Omega_1$ inside the particle. Accordingly, we choose $\mathbf{f}_S$ to satisfy the relation

$$\int_S \mathbf{f}_S(\mathbf{x})\cdot\mathbf{G}(\mathbf{x},\mathbf{x}')\,\mathrm{d}S = \mathbf{G}(\mathbf{x}',\mathbf{x}_0)\cdot\mathbf{F} + \mathbf{R}(\mathbf{x}',\mathbf{x}_0)\cdot\mathbf{L}, \qquad \mathbf{x}\in\Omega_2, \tag{7.77}$$

where

$$\mathbf{R}(\mathbf{x}',\mathbf{x}) = \frac{1}{2}[\mathbf{\nabla}\times\mathbf{G}(\mathbf{x},\mathbf{x}')]^\dagger, \qquad \text{or} \qquad R_{ij}(\mathbf{x}',\mathbf{x}) = \frac{1}{2}\epsilon_{jkl}\partial_k G_{li}(\mathbf{x},\mathbf{x}'). \tag{7.78}$$

Inserting (7.77) into equation (7.74) yields

$$(\hat{\mathsf{P}}^{rb} - \hat{\mathsf{P}}^1)\bar{\mathbf{u}}_S(\mathbf{x}') + 2\int_S \hat{\mathbf{n}}_1(\mathbf{x})\cdot\left\{[\bar{\mathbf{u}}_S(\mathbf{x})-\bar{\mathbf{u}}_S(\mathbf{x}')]\cdot\mathbf{T}(\mathbf{x},\mathbf{x}')\right\}\,\mathrm{d}S$$

$$= \mathbf{u}^\infty(\mathbf{x}') + \mu_2^{-1}\mathbf{G}(\mathbf{x}',\mathbf{x}_0)\cdot\mathbf{F} + \mu_2^{-1}\mathbf{R}(\mathbf{x}',\mathbf{x}_0)\cdot\mathbf{L}. \tag{7.79}$$

The simplified boundary integral equation (7.79), supplemented with the expression (7.75), forms a convenient basis for the construction of numerical simulation algorithms, using techniques described in Section 7.5.

The force and torque in equation (7.79) are given, and the rigid-body velocity field is the unknown. Thus, this is the mobility formulation of the problem, in contrast with the earlier resistance formulation (7.45). The boundary integral formulation (7.79) is more involved than the earlier one given in (7.45), but we have gained a significant advantage when it comes to the numerical solution of the problem. Namely, the unknown function $\bar{u}_S$ appears both at the integration point $\mathbf{x}$ and at the field point $\mathbf{x}'$. As already noted in our discussion of equation (7.49), this feature usually ensures numerical stability of the problem. Technically, equation (7.49), and its purged versions (7.72) and (7.79), are Fredholm equations of the second kind. In contrast, the boundary integral equation in the resistance formulation (7.45) is a Fredholm equation of the first kind, with the unknown function present only in the integral term. As mentioned in Section 7.3.2, short-wave numerical instabilities complicate the accurate solution of first-kind equations.

Equation (7.79) has been obtained by taking the high-viscosity limit of the fully purged boundary integral equation (7.72). Thus, not only is the rigid body mode purged, but also the expansion mode, as indicated by the presence of the projection operators $\hat{P}^{rb}$ and $\hat{P}^1$. In general, for rigid particles only the rigid-body purging is necessary. We retained, however, also the purge of the expansion mode, because the shift of the corresponding eigenvalue to the center of the spectrum of the boundary integral operator is useful in constructing a simple solution algorithm based on a Picard iteration scheme described in Section 7.5.4.

7.4 Multiparticle systems

So far, boundary integral equations were developed for an isolated drop or rigid particle. In the present section, we generalize these results for multiparticle systems. In Sections 7.4.1–7.4.3 we consider the motion of finite groups of liquid drops or rigid particles immersed in an infinite continuous-phase fluid. Examples of problems where such a generalization is important include breakup and coalescence of drops in a dilute emulsion where only a small number of drops interact simultaneously. At higher concentrations a further generalization is necessary, which involves implementation of periodic boundary conditions to mimic the behavior of a macroscopic multiparticle system. This generalization is outlined in Section 7.4.4.

In our discussion, we follow steps that are similar to those applied in the derivation of the boundary integral equations for a single particle. First, the boundary integral equations for a system of liquid drops are obtained by combining Lorentz integral identities specified for the flow fields in the drop-phase and continuous-phase fluids. Next, Wielandt's deflation technique is used to derive equations that are well behaved for low and high drop viscosities. Finally, boundary integral equations for rigid particles are obtained by taking the high-viscosity limit of the liquid-drop results. An outline of these steps is given in the following subsection.

7.4.1 System of viscous drops

We consider a system of M viscous drops immersed in a fluid of viscosity μ_1. For simplicity it is assumed that all drops have an equal viscosity $\mu_2 = \lambda\mu_1$. The regions occupied by the drops are bounded and nonoverlapping. The drop interfaces are denoted by S_α, where $\alpha = 1, \ldots, M$.

To generalize the derivation of the single-drop boundary integral equation (7.49) for this system, we notice that the flow $\mathbf{u}_\alpha^\infty$ incident on drop α is given by the superposition of the external flow $\mathbf{u}^\infty$ and the disturbance flow fields $\mathbf{u}_{2\beta}$ produced by the remaining drops, i.e.

$$\mathbf{u}_\alpha^\infty(\mathbf{x}) = \sum_{\substack{\beta=1 \\ \beta \neq \alpha}}^{M} \mathbf{u}_{2\beta}(\mathbf{x}) + \mathbf{u}^\infty(\mathbf{x}), \qquad \mathbf{x} \in S_\alpha. \tag{7.80}$$

Using this observation, and following the derivation of the single-drop equation (7.49), an integral relation is obtained for the flow field $\mathbf{u}_{S_\alpha}$ on the surface of drop α,

$$\mathbf{u}_{S_\alpha}(\mathbf{x}') + (\lambda - 1) \int_{S_\alpha} \hat{\mathbf{n}}_1(\mathbf{x}) \cdot \left\{ [\mathbf{u}_{S_\alpha}(\mathbf{x}) - \mathbf{u}_{S_\alpha}(\mathbf{x}')] \cdot \boldsymbol{T}(\mathbf{x}, \mathbf{x}') \right\} \mathrm{d}S$$

$$= \mathbf{u}_\alpha^\infty(\mathbf{x}') + \int_{S_\alpha} \mathbf{f}_{S_\alpha}(\mathbf{x}) \cdot \mu_2^{-1} \boldsymbol{G}(\mathbf{x}, \mathbf{x}') \, \mathrm{d}S, \quad \mathbf{x}' \in S_\alpha, \tag{7.81}$$

where $\mathbf{f}_{S_\alpha}$ denotes the surface-force density (7.15) for drop α.

The corresponding relation for the disturbance-flow components $\mathbf{u}_{2,\beta}$ in equation (7.80) is derived in a similar manner, by combining equations (7.46)–(7.48). The resulting expression can be written in the form

$$\mathbf{u}_{2,\beta}(\mathbf{x}') = -(\lambda - 1) \int_{S_\beta} \hat{\mathbf{n}}_1(\mathbf{x}) \cdot \left\{ [\mathbf{u}_{S_\beta}(\mathbf{x}) - \mathbf{u}_{S_\beta}(\mathbf{x}''_\beta)] \cdot \boldsymbol{T}(\mathbf{x}, \mathbf{x}') \right\} \mathrm{d}S$$

$$+ \int_{S_\beta} \mathbf{f}_{S_\beta}(\mathbf{x}) \cdot \mu_2^{-1} \boldsymbol{G}(\mathbf{x}, \mathbf{x}') \, \mathrm{d}S, \qquad \mathbf{x}' \in S_\alpha, \tag{7.82}$$

where the identity

$$\int_{S_\beta} \hat{\mathbf{n}}_1(\mathbf{x}) \cdot \left[\mathbf{A} \cdot \boldsymbol{T}(\mathbf{x}, \mathbf{x}')\right] \, \mathrm{d}S = 0, \qquad \mathbf{x}' \in S_\alpha, \quad \beta \neq \alpha, \tag{7.83}$$

has been used with $\mathbf{A} = \mathbf{u}_{S_\alpha}(\mathbf{x}') - \mathbf{u}_{S_\beta}(\mathbf{x}''_\beta)$ to shift the subtracted velocity from the original point $\mathbf{x}' \in S_\alpha$, as in equation (7.81), to a new position $\mathbf{x}''_\beta \in S_\beta$. Relation (7.83) follows from equation (7.37) and the observation that the field point $\mathbf{x}' \in S_\alpha$ is outside the region bounded by the surface S_β, because the drops do not overlap. The proper choice of the point $\mathbf{x}''_\beta$ is discussed below.

By inserting relations (7.80) and (7.82) into (7.81), we obtain the following set of coupled integral equations for the flow field at the surfaces of particles $\alpha = 1, \ldots, M$:

$$\mathbf{u}_{S_\alpha}(\mathbf{x}') + (\lambda - 1) \int_{S_\alpha} \hat{\mathbf{n}}_1(\mathbf{x}) \cdot \left\{ [\mathbf{u}_{S_\alpha}(\mathbf{x}) - \mathbf{u}_{S_\alpha}(\mathbf{x}')] \cdot \boldsymbol{T}(\mathbf{x}, \mathbf{x}') \right\} \, \mathrm{d}S$$

$$+ (\lambda - 1) \sum_{\substack{\beta=1 \\ \beta \neq \alpha}}^{M} \int_{S_\beta} \hat{\mathbf{n}}_1(\mathbf{x}) \cdot \left\{ [\mathbf{u}_{S_\beta}(\mathbf{x}) - \mathbf{u}_{S_\beta}(\mathbf{x}''_\beta)] \cdot \boldsymbol{T}(\mathbf{x}, \mathbf{x}') \right\} \, \mathrm{d}S$$

$$= \mathbf{u}^\infty(\mathbf{x}') + \sum_{\beta=1}^{M} \int_{S_\beta} \mathbf{f}_{S_\beta}(\mathbf{x}) \cdot \mu_2^{-1} \boldsymbol{G}(\mathbf{x}, \mathbf{x}') \, \mathrm{d}S, \quad \mathbf{x}' \in S_\alpha. \tag{7.84}$$

In the above equation, the single-layer and double-layer integrals with $\beta \neq \alpha$ represent the effect of the surrounding particles on the motion of particle α.

The proper choice of the subtraction points $\mathbf{x}''_\beta$ in a double-layer integral with $\beta \neq \alpha$ will now be discussed. In this case the singularity of the Green's function lies outside the integration surface S_β so that the integral, strictly speaking, is not singular. However, when the distance ϵ between the field point $\mathbf{x}' \in S_\alpha$ and the integration surface $S_\beta \neq S_\alpha$ is small, the integrand involves a rapidly varying factor

$$\hat{\mathbf{n}}_1 \cdot \boldsymbol{T}(\mathbf{x}, \mathbf{x}') = -\frac{3\hat{\mathbf{n}}_1(\mathbf{x}) \cdot \hat{\mathbf{r}} \hat{\mathbf{r}} \hat{\mathbf{r}}}{4\pi r^2}, \tag{7.85}$$

where $\mathbf{r} = \mathbf{x} - \mathbf{x}'$, and $\hat{\mathbf{r}} = \mathbf{r}/r$. This factor has a large $O(\epsilon^{-2})$ amplitude in a small portion s_ϵ of the surface S_β closest to the point $\mathbf{x}'$; outside of this region it decays rapidly, because the vectors $\hat{\mathbf{n}}_1(\mathbf{x})$ and $\hat{\mathbf{r}}$ become nearly perpendicular. The integration over the local region s_ϵ yields a sizable contribution unless the subtracted value $\mathbf{u}_{S_\beta}(\mathbf{x}''_\beta)$ cancels the term $\mathbf{u}_{S_\beta}(\mathbf{x})$ in the square bracket in equation (7.84). Since a sizable local contribution causes numerical difficulties (it requires additional discretization points in order to be accurately resolved), the subtraction point $\mathbf{x}''_\beta$ should be chosen to ensure such a cancellation with sufficient accuracy.

The most natural choice for the subtraction point $\mathbf{x}''_\beta \in S_\beta$ is the position for which the distance $|\mathbf{x}' - \mathbf{x}''_\beta|$ is minimal. A modified approach has also been proposed, where the numerical error is further reduced by using variational criteria to find the optimal position of $\mathbf{x}''_\beta$ (Zinchenko and Davis, 2002). We note that the choice $\mathbf{x}''_\beta = \mathbf{x}'$ is inadequate, because the velocity field may vary rapidly in the gap between the drop or particle surfaces.

7.4.2 Wielandt's deflation

As in the case of a single drop, the system of linear integral equations (7.84) is singular in the limits $\lambda \to \infty$ and $\lambda \to 0$. Difficulties associated with this behavior can be removed using Wielandt's deflation technique introduced in Section 7.3.4 for a single drop. Accordingly, the transformed variable (7.67) is introduced for the velocity field on the surface of each drop, and the relations (7.54) and (7.71) are used. As a result, a set of purged boundary integral equations is obtained:

$$
\begin{aligned}
(1 - \varkappa)\bar{\mathbf{u}}_{S_\alpha}(\mathbf{x}') &+ \varkappa(\hat{\mathsf{P}}^{\mathrm{rb}} - \hat{\mathsf{P}}^1)\bar{\mathbf{u}}_{S_\alpha}(\mathbf{x}') \\
&+ 2\varkappa \int_{S_\alpha} \hat{\mathbf{n}}_1(\mathbf{x}) \cdot \big\{[\bar{\mathbf{u}}_{S_\alpha}(\mathbf{x}) - \bar{\mathbf{u}}_{S_\alpha}(\mathbf{x}')] \cdot \boldsymbol{T}(\mathbf{x}, \mathbf{x}')\big\} \, \mathrm{d}S \\
&+ 2\varkappa \sum_{\substack{\beta=1 \\ \beta \neq \alpha}}^{M} \int_{S_\beta} \hat{\mathbf{n}}_1(\mathbf{x}) \cdot \big\{[\bar{\mathbf{u}}_{S_\beta}(\mathbf{x}) - \bar{\mathbf{u}}_{S_\beta}(\mathbf{x}''_\beta)] \cdot \boldsymbol{T}(\mathbf{x}, \mathbf{x}')\big\} \, \mathrm{d}S \\
&= \mathbf{u}^\infty(\mathbf{x}') + \sum_{\beta=1}^{M} \int_{S_\beta} \mathbf{f}_{S_\beta}(\mathbf{x}) \cdot \mu_2^{-1} \boldsymbol{G}(\mathbf{x}, \mathbf{x}') \, \mathrm{d}S, \quad \mathbf{x}' \in S_\alpha, \quad (7.86)
\end{aligned}
$$

where the relation between the unscaled and scaled velocity fields is given by the equation

$$
\mathbf{u}_{S_\alpha} = \hat{\mathsf{P}}^{\mathrm{rb}}\bar{\mathbf{u}}_{S_\alpha} + 2(\lambda + 1)^{-1}(1 - \hat{\mathsf{P}}^{\mathrm{rb}})\bar{\mathbf{u}}_{S_\alpha}, \tag{7.87}
$$

and $\varkappa$ is defined by equation (7.69). The definitions of the projection operators $\hat{\mathsf{P}}^{\mathrm{rb}}$ and $\hat{\mathsf{P}}^1$ for drop α are given by equations (7.59) to (7.63) with S replaced by S_α.

7.4.3 System of rigid particles

The boundary integral equations describing a system of rigid particles are obtained from the purged liquid-drop equations (7.86) by taking the limit

$\lambda \to \infty$ ($\varkappa \to 1$). In addition, the single-layer integrals are replaced with the corresponding point-force and point-torque contributions, as described in Section 7.3.5. As a result of this procedure, we get the following set of boundary integral equations:

$$(\hat{\mathsf{P}}^{\mathrm{rb}} - \hat{\mathsf{P}}^1)\bar{\mathbf{u}}_{\mathrm{S}_\alpha}(\mathbf{x}') + 2\int_{S_\alpha}\hat{\mathbf{n}}_1(\mathbf{x})\cdot\left\{[\bar{\mathbf{u}}_{\mathrm{S}_\alpha}(\mathbf{x}) - \bar{\mathbf{u}}_{\mathrm{S}_\alpha}(\mathbf{x}')]\cdot\boldsymbol{T}(\mathbf{x},\mathbf{x}')\right\}\,\mathrm{d}S$$

$$+\,2\sum_{\substack{\beta=1\\\beta\neq\alpha}}^{M}\int_{S_\beta}\hat{\mathbf{n}}_1(\mathbf{x})\cdot\left\{[\bar{\mathbf{u}}_{\mathrm{S}_\beta}(\mathbf{x}) - \bar{\mathbf{u}}_{\mathrm{S}_\beta}(\mathbf{x}''_\beta)]\cdot\boldsymbol{T}(\mathbf{x},\mathbf{x}')\right\}\,\mathrm{d}S$$

$$= \mathbf{u}^\infty(\mathbf{x}') + \sum_{\beta=1}^{M}\mu_2^{-1}\left[\boldsymbol{G}(\mathbf{x}',\mathbf{x}_\beta)\cdot\mathbf{F}_\beta + \boldsymbol{R}(\mathbf{x}',\mathbf{x}_\beta)\cdot\mathbf{L}_\beta\right], \quad \mathbf{x}' \in S_\alpha,$$

$$(7.88)$$

where $\mathbf{F}_\beta$ and $\mathbf{L}_\beta$ denote the force and torque acting on the particle β, and $\mathbf{x}_\beta$ is an arbitrary reference point in the region occupied by the particle. As for a single particle, the rigid-body velocity of particle α ($\alpha = 1, \ldots, M$) is obtained from the auxiliary field $\bar{\mathbf{u}}_{\mathrm{S}_\beta}$ using the high-viscosity limit

$$\mathbf{u}_{\mathrm{S}_\alpha} = \hat{\mathsf{P}}^{\mathrm{rb}}\bar{\mathbf{u}}_{\mathrm{S}_\alpha} \tag{7.89}$$

of the relation (7.87), by analogy with equation (7.75) for a single particle.

7.4.4 Periodic boundary conditions

In many applications (for example, the investigation of emulsion and suspension flows), boundary integral equations describing the motion of finite groups of particles in unbounded fluid are insufficient. Today, on a fast workstation, one can simulate systems of, at most, a few hundred drops, and this can be achieved only after implementing complex acceleration algorithms (outlined in Section 7.5.6). A particle cluster of this size is by far inadequate for representing the macroscopic behavior of an emulsion.

A much more efficient method to approximate macroscopic multiparticle media is to use periodic boundary conditions in which the basic cell that includes a relatively small number of particles is periodically replicated in all directions an infinite number of times, thus mimicking a macroscopic system. In simulations with periodic boundary conditions, the boundary integral equations are solved only for the basic cell. The periodic images of the basic cell are included by using periodic Green's functions in the kernels of the boundary integrals.

The periodic Green's functions $\boldsymbol{G}^p$, $\boldsymbol{T}^p$, and $\boldsymbol{Q}^p$ are defined by the inhomogeneous Stokes equations of the form

$$\boldsymbol{\nabla} \cdot \boldsymbol{G}^p(\mathbf{r}) = 0, \tag{7.90}$$

$$\boldsymbol{\nabla} \cdot \boldsymbol{T}^p(\mathbf{r}) = -\hat{\boldsymbol{I}} \left[\sum_{\mathbf{n}} \delta(\mathbf{r} - \mathbf{x_n}) - V_c^{-1} \right], \tag{7.91}$$

where

$$\boldsymbol{T}^p(\mathbf{r}) = \boldsymbol{\nabla} \boldsymbol{G}^p(\mathbf{r}) + [\boldsymbol{\nabla} \boldsymbol{G}^p(\mathbf{r})]^\dagger - \hat{\boldsymbol{I}} \boldsymbol{Q}^p(\mathbf{r}), \tag{7.92}$$

and $\mathbf{r} = \mathbf{x} - \mathbf{x}'$. The point forces in equation (7.91) are positioned at the lattice points

$$\mathbf{x_n} = \sum_{i=1}^{3} n_i \hat{\mathbf{g}}_i, \tag{7.93}$$

where $\hat{\mathbf{g}}_i$ are the basis vectors describing the periodic directions, and $\mathbf{n} = (n_1, n_2, n_3)$ is a set of three integers. The source term in the right-hand side of equation (7.91) includes not only a periodic array of point forces, but also a compensating background force uniformly distributed over the volume $V_c = \hat{\mathbf{g}}_1 \cdot \hat{\mathbf{g}}_2 \times \hat{\mathbf{g}}_3$ of the basic cell. Without this background force, the periodic solution of equations (7.90)–(7.92) would not exist: in the Stokes-flow regime there are no inertial effects, thus the net force acting on a basic cell must vanish.

To assure uniqueness of the solution of the problem, the system of equations (7.90)–(7.92) is supplemented by a condition that replaces the boundary condition at infinity. Usually one imposes the requirement that the spatial average of the velocity field produced by a periodic array of point forces vanishes, $\int_{V_c} \boldsymbol{G}^p \, \mathrm{d}^3 x = 0$.

The periodic Green's functions $\boldsymbol{G}^p$, $\boldsymbol{T}^p$, and $\boldsymbol{Q}^p$ cannot be evaluated by a direct summation of the individual point-force contributions of the form (7.39) and (7.40) without specifying the summation procedure as the series is only conditionally convergent due to the slow r^{-1} decay of the velocity field produced by a point force. Moreover, the direct summation is very inefficient numerically.

An efficient method for evaluating the periodic solutions of equations (7.90)–(7.92) is the Ewald summation technique (Hasimoto, 1959). The underlying physical idea is well captured by Frenkel and Smit (2002) in their description of the technique for Coulombic interactions. In the application to Stokes flow, this idea can be summarized as follows. Instead of adding the point-force contributions directly, each point is first surrounded by a diffuse Gaussian-like distribution of a screening force. This force assures that the flow field produced at each lattice site decays exponentially at large distances. The lattice sum of the screened contributions thus converges rapidly. On the other hand, the contributions of the screening clouds (combined with the neutralizing background) can be efficiently evaluated in Fourier space, because their Fourier transforms decay exponentially for large wavevectors. Rapidly converging series expressions for the periodic Green's functions are obtained by adding the real- and Fourier-space results.

The Ewald sum for the hydrodynamic Green's functions $\boldsymbol{G}^p$, $\boldsymbol{T}^p$, and $\boldsymbol{Q}^p$ can be conveniently expressed in terms of two periodic scalar functions $S_1(\mathbf{r})$ and $S_2(\mathbf{r})$ in the form

$$\boldsymbol{Q}^p(\mathbf{Z}) = -\boldsymbol{\nabla}S_1(\mathbf{r}), \tag{7.94}$$

$$\boldsymbol{G}^p(\mathbf{Z}) = \mu^{-1}[S_1(\mathbf{r})\hat{\boldsymbol{I}} - \boldsymbol{\nabla}\boldsymbol{\nabla}S_2(\mathbf{r})], \tag{7.95}$$

$$\boldsymbol{T}^p(\mathbf{n}) = 3[\boldsymbol{\nabla}S_1(\mathbf{r})\hat{\boldsymbol{I}}]^{\text{sym}} - 2\boldsymbol{\nabla}\boldsymbol{\nabla}\boldsymbol{\nabla}S_2(\mathbf{r}), \tag{7.96}$$

where $[\boldsymbol{w}]^{\text{sym}}$ denotes the fully symmetric part of the third-order tensor $\boldsymbol{w}$ and (see, e.g. Cichocki and Felderhof, 1989)

$$S_1(\mathbf{r}) = \tfrac{1}{4}\pi^{-1}\sum_{\mathbf{n}} r_{\mathbf{n}}^{-1}\text{erfc}(\sigma^{-1}r_{\mathbf{n}}) - \tfrac{1}{4}V_c^{-1}\sigma^2$$

$$+V_c^{-1}\sum_{\mathbf{n}\neq 0} k_{\mathbf{n}}^{-2}\exp(-\tfrac{1}{4}\sigma^2k_{\mathbf{n}}^2 - i\mathbf{k_n}\cdot\mathbf{r}), \tag{7.97}$$

and

$$S_2(\mathbf{r}) = \tfrac{1}{8}\pi^{-1}\sum_{\mathbf{n}}\left[r_{\mathbf{n}}\text{erfc}(\sigma^{-1}r_{\mathbf{n}}) - \pi^{-1/2}\sigma\exp(-\sigma^{-2}r_{\mathbf{n}}^2)\right] + \tfrac{1}{32}V_c^{-1}\sigma^4$$

$$-V_c^{-1}\sum_{\mathbf{n}\neq 0} k_{\mathbf{n}}^{-4}(1 + \tfrac{1}{4}\sigma^2k_{\mathbf{n}}^2)\exp(-\tfrac{1}{4}\sigma^2k_{\mathbf{n}}^2 - i\mathbf{k_n}\cdot\mathbf{r}). \tag{7.98}$$

The first sum in relations (7.97) and (7.98) is over the periodic lattice points (7.93) in the real space, with $r_{\mathbf{n}} = |\mathbf{r} - \mathbf{x_n}|$. The second sum is over the reciprocal lattice points

$$\mathbf{k_n} = \sum_{i=1}^{3} n_i\hat{\mathbf{g}}^i \tag{7.99}$$

in Fourier space, where $\hat{\mathbf{g}}^i$ stands for the reciprocal lattice vectors defined by the relation

$$\hat{\mathbf{g}}^i \cdot \hat{\mathbf{g}}_j = 2\pi\delta_{ij}. \tag{7.100}$$

The zero-wavenumber term $\mathbf{n} = 0$ is omitted from the Fourier sum due to the presence of the neutralizing background force. The parameter $\sigma > 0$ in the above equations characterizes the width of the Gaussian distribution of the screening force. The quantity σ^{-1} is usually called the splitting parameter, and it controls the relative convergence rates for the sums over the real-space lattice and reciprocal lattice. It should be chosen to minimize the total computation time for the problem to which the Ewald sum is applied.

In dynamic simulations of suspension flows the periodic lattice undergoes affine deformation corresponding to the applied linear flow. Thus, periodic Green's functions with nonorthogonal basis vectors $\hat{\mathbf{g}}_i$ are essential in such calculations. In stationary flows the affine deformation produces unbounded lattice distortion, which may lead to ill-conditioned numerical problems. In a shear flow, this difficulty can easily be solved by periodically redefining a deformed unit cell into the cell with the opposite strain, when the angle between the unit vectors in the velocity–gradient plane reaches $\pi/4$. A similar technique can also be used for certain other linear flows (Kraynik and Reinelt, 1992).

7.5 Numerical integration methods

After deriving the boundary integral equations for liquid drops and rigid particles, we focus on discretization techniques for solving these equations numerically. In the present section we discuss basic algorithms for transforming the boundary integral equations into a finite system of linear algebraic equations for the interfacial velocity at the mesh nodes. We also present iterative solution methods that are used to solve these equations efficiently, and outline time-stepping techniques for updating the position of the evolving interface.

7.5.1 Surface discretization

In boundary integral algorithms, the single-layer and double-layer integrals are evaluated using the boundary element approach. Accordingly, the particle or drop interface S is divided into K polygonal (usually triangular) boundary elements

$$S = (\Delta S_1, \ldots, \Delta S_K). \tag{7.101}$$

The elements are specified by a set of N nodes

$$\mathbf{X} = (\mathbf{X}_1, \ldots, \mathbf{X}_N) \tag{7.102}$$

through which the interface is assumed to pass. In a typical representation, the vertices of the surface elements (7.101) coincide with the node positions (7.102). Thus the discretization (7.101) is fully specified by the list of nodes corresponding to each vertex of every element.

The mesh nodes (7.101) not only describe the position of the discretized interface, but they are also used to represent various quantities defined on the interface S, such as the interfacial velocity $\mathbf{u}_S$, surface-force density $\mathbf{f}_S$, and possibly surfactant concentration Γ for surfactant-covered drops. Accordingly, for each quantity $A = \mathbf{u}_S, \mathbf{f}_S, \Gamma, \ldots$, a set of values

$$A_i = A(\mathbf{X}_i), \qquad i = 1, \ldots, N \tag{7.103}$$

is assigned. At each element ΔS_i, a local representation of the surface is introduced (e.g. in terms of low-order polynomials), and the surface fields (7.103) are similarly interpolated.

In boundary integral applications triangular surface elements are often used because this geometry is most appropriate for forming unstructured, irregular meshes on complex evolving surfaces. To obtain the initial surface triangulation of a closed surface, such as a drop, one may start from a regular icosahedron. The sides of the icosahedron are triangular, and they can be divided into smaller triangles to achieve the required resolution as described, e.g. in Loewenberg and Hinch (1996). The triangular elements of the icosahedron should be projected onto the initial surface.

A simple discretization which involves a uniform structured mesh with fixed topology is sufficient only for simple, static geometries, such as a single rigid particle with modest curvature variation. For more complex physical systems (e.g. drops deforming in external flow, or suspensions of many particles) dynamic remeshing is necessary to maintain the required surface resolution. Examples of structural features that need to be resolved to obtain accurate numerical results include high-curvature regions that form spontaneously at interfaces of deformable drops, and near-contact regions in systems that involve relative motion of nearly-touching particles. Two examples of boundary integral simulations of such systems are shown in Figs. 7.2 and 7.3.

Several adaptive algorithms have been developed for discretization of static (Bulpitt and Efford, 1996; Wilson and Hancock, 2000) and evolving surfaces (Unverdi and Tryggvason, 1992; Mavriplis, 1997; Kwak and Pozrikidis, 1998; Cristini et al., 2001; Bazhlekov et al., 2004b). For detailed descriptions of such methods we refer the reader to the original literature.

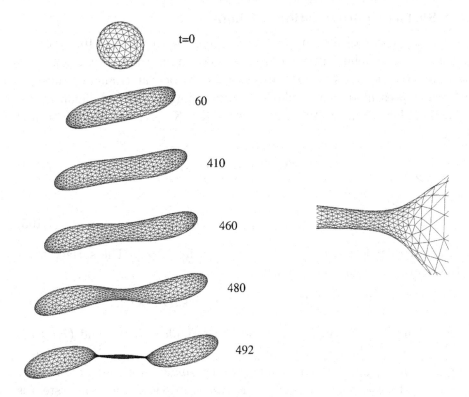

Fig. 7.2. Breakup of an initially spherical drop in shear flow. Elapsed time is indicated in drop relaxation-time units $\tau_d = \mu a/\gamma$. Inset shows neck region for $t = 492$ (Cristini *et al.* 2001).

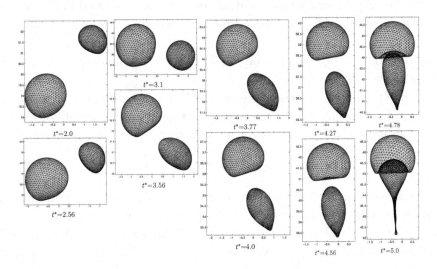

Fig. 7.3. Buoyancy-driven interaction of two drops with radii ratio 0.7 for viscosity $\lambda = 0$ and Bond number $Bo = 7$. Reprinted from Bazhlekov (2003).

7.5.2 Surface approximation methods

The reconstruction of the surface S from the node positions (7.102) requires appropriate interpolation procedures. At the simplest approximation level, the boundary elements (7.101) are assumed to be flat triangles; thus the surface is represented by a polyhedron with straight edges and planar facets. A flat boundary element ΔS_i with vertices $\mathbf{X}_{i_1}, \mathbf{X}_{i_2}, \mathbf{X}_{i_3}$ can be represented as the set of points

$$\mathbf{x} = \alpha_1 \mathbf{X}_{i_1} + \alpha_2 \mathbf{X}_{i_2} + \alpha_3 \mathbf{X}_{i_3}, \qquad (7.104)$$

where

$$\alpha_1 + \alpha_2 + \alpha_3 = 1, \qquad \alpha_1, \alpha_2, \alpha_3 \geq 0, \qquad (7.105)$$

are the barycentric coordinates (see, e.g. Farin, 1988). The surface fields (7.103) can be similarly interpolated,

$$A(\mathbf{x}) = \alpha_1 A(\mathbf{X}_{i_1}) + \alpha_2 A(\mathbf{X}_{i_2}) + \alpha_3 A(\mathbf{X}_{i_3}). \qquad (7.106)$$

The accuracy of the linear interpolation formulas (7.104) and (7.106) is $O(l^2)$, where l is the local mesh edge length. Since $l \sim N^{-1/2}$, the linear interpolation is sufficient for evaluating the single-layer and double-layer boundary integrals with an $O(N^{-1})$ accuracy, provided that the system is characterized by a local length scale $L \gg l$.

Improved accuracy of the surface representation can be achieved by using quadratic or higher-order interpolation formulas as described in Chapter 5. Moreover, such formulas are needed for the evaluation of the normal vector and the local mean curvature. We recall that the mean curvature appears in the boundary integral equation (7.86) for deformable drops through the capillary force distribution (7.15).

A simple technique (Zinchenko *et al.*, 1997) that has frequently been used relies on constructing a local parabolic surface fit in the neighborhood of each mesh node $i = 1, \dots N$. The parabolic surface

$$z' = a_{x'} x' + a_{y'} y' + b_{x'x'} x'^2 + 2b_{x'y'} x'y' + b_{y'y'} y'^2 \qquad (7.107)$$

is constructed in a local coordinate system (x', y', z'), with the mesh point $\mathbf{X}_i$ at the origin. The parameters of the parabola are obtained by a mean-square fit to the nodes adjacent to node i. The direction of the axis z' is iteratively adjusted until it coincides with the normal vector at the node position, $\hat{\mathbf{n}}_1(\mathbf{X}_i)$, which requires that

$$a_{x'} = a_{y'} = 0 \qquad (7.108)$$

within the assumed accuracy. In the new coordinate system, the normal vector and the mean curvature are

$$\hat{\mathbf{n}}_1(\mathbf{X}_i) = \hat{\mathbf{e}}_{z'}, \qquad \kappa(\mathbf{X}_i) = -(b_{x'x'} + b_{y'y'}). \qquad (7.109)$$

The procedure usually converges after several iterations. Since the parabolic surface (7.107) involves five adjustable parameters, each mesh node ought to have at least five nearest neighbors.

In the neighborhood of the mesh point i, the accuracy of the parabolic fit (7.107) is of the order $N^{-3/2}$; the corresponding accuracies of the normal vector and curvature (7.109) are N^{-1} and $N^{-1/2}$, respectively. For a sufficiently symmetric mesh, the accuracy of the curvature at the mesh point $\mathbf{X}_i$ is improved by the factor $N^{-1/2}$, because cubic terms in the power expansion of the surface do not contribute to the discretization error. Several tests (Zinchenko *et al.*, 1997, 1999) indicate that a typical error of the curvature (7.109) varies approximately linearly with N^{-1} in the range of moderate mesh resolutions. However, at higher resolution, the $O(N^{-1/2})$ convergence is observed.

In principle, the accuracy of the curvature calculation can be improved by using a higher order polynomial fit instead of the parabolic approximation (7.107). In practice, the optimal order results from a compromise between the accuracy requirements and the numerical stability of the method. For readers that are interested in constructing boundary integral algorithms of higher order accuracy, we note that advanced surface interpolation and approximation methods were developed for applications in computer-aided design and computer graphics. A number of textbooks (e.g. Farin, 1988; Beach, 1991; Piegl and Tiller, 1997) and review articles (Bohm *et al.*, 1984; Montagnat *et al.*, 2001; Shen and Yoon, 2004) describe these techniques.

7.5.3 Evaluation of the single-layer and double-layer integrals

The accurate and efficient numerical evaluation of the boundary integrals is a key component of boundary integral methods. In the usual approach, the surface integration is performed over the set of triangular boundary elements (7.101). Accordingly, an integral

$$I(\mathbf{x}') = \int_S g(\mathbf{x}', \mathbf{x}) \, dS \qquad (7.110)$$

is represented as a sum

$$I(\mathbf{x}') = \sum_{i=1}^{K} I_i(\mathbf{x}'), \qquad (7.111)$$

where

$$I_i(\mathbf{x}') = \int_{\Delta S_i} g(\mathbf{x}', \mathbf{x}) \, dS. \tag{7.112}$$

Here $g(\mathbf{x}', \mathbf{x})$ generically stands for the integrand of the single-layer or double-layer integral. A similar decomposition also applies to the integrals that are involved in the evaluation of the rigid-body and expansion projections (7.59) to (7.63).

The integrals over the individual boundary elements (7.112) are evaluated using a properly chosen numerical quadrature formula. Accordingly, the integral is approximated by a linear combination of the values of the integrand $g(\mathbf{x}', \mathbf{x})$ at a discrete set of quadrature points $\mathbf{x} = \mathbf{x}_j$,

$$I_i(\mathbf{x}') = \Delta S_i \sum_{j=1}^{n} w_j g(\mathbf{x}', \mathbf{x}_j), \tag{7.113}$$

where n is the order of the quadrature formula, and ΔS_i is the area of the element i. The position of the points $\mathbf{x}_j$ and the corresponding weights w_j depend on the specific quadrature formula adopted.

If the function $g(\mathbf{x}', \mathbf{x})$ is sufficiently smooth on a given surface element, the trapezoidal rule

$$\mathbf{x}_j = \mathbf{X}_{i_j}, \qquad w_j = \tfrac{1}{3}, \qquad j = 1, 2, 3, \tag{7.114}$$

can be applied to obtain results with $O(N^{-1})$ accuracy. Since the quadrature points coincide with the grid nodes in this case, interpolation of surface-defined quantities (such as the velocity field and the surface-tension forces) is unnecessary. Examples of higher order integration methods include surface Gauss quadrature; the weights and quadrature-point positions for this method are listed, e.g., in Pozrikidis (1997). In order to obtain integration results that have higher than $O(N^{-1})$ accuracy, not only should higher-order quadratures be used, but also appropriate surface representations and interpolation techniques.

Neither the trapezoidal rule nor a standard surface Gauss quadrature are sufficient to evaluate the boundary integrals in the domains where the integrands vary rapidly due to the singular nature of the Green's functions (7.39) and (7.40). We recall that the order of the singularity of the double-layer integral in equations (7.86) has been reduced using the singularity-subtractions technique. As described in Section 7.3.1, the resulting regularized integral is nonsingular when the field point $\mathbf{x}'$ is on the integration surface S; however, the integrand undergoes a rapid $O(\epsilon^{-1})$ variation when $\mathbf{x}'$ is at a small but finite distance ϵ from the surface S. The local region where the rapid variation occurs yields a significant contribution to the double-layer integral. Thus the integration in this region has to be resolved accurately.

The integrand of the single-layer integral in equations (7.86) also involves singular and near-singular behavior, according to equation (7.39). As shown in Appendix C, the singularity can be removed using a subtraction technique, but only when the surface force distribution $\mathbf{f}_S$ is normal to the drop interface S. The regularization cannot be performed in the presence of tangential Marangoni stresses produced by surfactant-concentration gradients or temperature variation on the drop interface.

The singular and near-singular boundary integrals can conveniently be evaluated using a polar integration method (Galea, 2004). To illustrate this technique, we consider a singular integrand

$$g(\mathbf{x}', \mathbf{x}) = \frac{\bar{g}(\mathbf{x}', \mathbf{x})}{|\mathbf{x}' - \mathbf{x}|}, \tag{7.115}$$

where $\bar{g}(\mathbf{x}', \mathbf{x})$ is a nonsingular, slowly varying function. The function of the single-layer integral in equation (7.86) has the above form.

In order to integrate the singular function (7.115) over a planar surface element ΔS_i, the integral (7.112) is transformed into a local polar coordinate system (ρ, θ) in the plane P of the triangle ΔS_i. The origin of the new coordinates coincides with the orthogonal projection $\bar{\mathbf{x}}'$ of the point $\mathbf{x}'$ onto P or with $\mathbf{x}'$ itself for $\mathbf{x}' \in P$. In the polar coordinate system, the integral (7.112) with the integrand given in the from (7.115) becomes

$$I_i(\mathbf{x}') = \int_{\Delta S_i} \frac{\tilde{g}(\mathbf{x}'; \rho, \theta)\rho}{(\rho^2 + d^2)^{1/2}} \, d\rho \, d\theta, \tag{7.116}$$

where $d = |\mathbf{x}' - \bar{\mathbf{x}}'|$ is the distance of the point $\mathbf{x}'$ from the integration plane, and $\tilde{g}(\mathbf{x}'; \rho, \theta) \equiv \bar{g}(\mathbf{x}', \mathbf{x})$.

For $\mathbf{x}'$ in the integration plane P, the distance d in equation (7.116) vanishes, and the singularity of the integrand (7.115) is canceled by the Jacobian ρ of the polar surface element. Since $\tilde{g}(\mathbf{x}'; \rho, \theta)$ is a slowly varying function of its arguments, the integral (7.116 can be accurately evaluated by using a product of two Gauss quadratures in the radial and angular directions. The domain of the angular integration should be divided into sectors defined by the vertices of ΔS_i to avoid using the Gauss integration rule for a function with discontinuous derivatives.

If the distance d of point $\mathbf{x}'$ from the integration plane is nonzero, the integrand in (7.116) undergoes a $O(1)$ variation in the region $\rho \sim d$. In order to resolve this variation for configurations with $d \lesssim l$ (where l is the mesh-edge length), the domain of the radial integration should be divided into several subdomains the sizes of which depend on the distance from the origin.

The singularity of the subtracted double-layer integrand in the boundary integral equations (7.86) has a somewhat more complicated form than the first-order singularity (7.115). However, the polar integration method described above can be applied also in this case. After the change of variables, the integration is significantly simplified, because the rapid variation of the integrand occurs only in the radial direction and has a smaller amplitude.

The polar integration method can be used not only for flat triangles, but also for curved surface elements. The integration in this case is performed by projecting the curved element onto the plane defined by its three vertices. Such an integration technique, introduced in Galea (2004), is especially useful for surfaces of known shape. In particular, it can be applied in boundary integral simulations of rigid-particle systems.

An alternative method for evaluating the single-layer and double-layer boundary integrals has been proposed by Bazhlekov *et al.* (2004b). In their approach, the mean-value theorem is used on each boundary element to factor out the interfacial velocity $\mathbf{u}_S$ and surface force density $\mathbf{f}_S$ (assumed to be normal to the interface) from the integrals. The remaining integrands depend only on the normal vector $\hat{\mathbf{n}}_1$ and the Green's functions $\boldsymbol{G}$ and $\boldsymbol{T}$. The simplified surface integrals are transformed into contour integrals over the perimeters of the surface elements ΔS_i, using appropriate generalizations of the Stokes integration formula (Bazhlekov *et al.*, 2004b; Bazhlekov, 2003). The contour integrals are then evaluated by a one-dimensional quadrature.

The error-controlling approximation in the above integration procedure is the use of the mean-value theorem on the surface elements. If the field point $\mathbf{x}'$ is sufficiently far from the integration surface S, the kernel functions vary slowly on the scale set by the mesh edge length l, and the results have $O(N^{-1})$ accuracy. However, if the distance of the point $\mathbf{x}'$ from the integration surface is comparable to l, the accuracy is reduced to the order $O(N^{-1/2})$.

Contour integration methods can also be used to evaluate the mean curvature averaged over the surface domain bounded by the integration contour (Loewenberg and Hinch, 1996). Advantages and disadvantages of this method are discussed by Bazhlekov (2003).

7.5.4 Iterative solution methods

The surface representation techniques, interpolation methods, and numerical integration algorithms outlined above can be used to evaluate the integral terms in the boundary integral equation (7.72) – or, for a multiple drop problem, (7.86) – on a set of N collocation points. In boundary integral applications, the sets of collocation points and mesh nodes are usually chosen to be identical. The discretized boundary integral equations form a linear algebraic equation for the array $\mathbf{U} = (\mathbf{U}_1, \ldots, \mathbf{U}_N)$ containing the values of the interfacial velocity field at the nodes[1]

$$(1 - \varkappa)\mathbf{U} + 2\varkappa\mathsf{A} \cdot \mathbf{U} = \mathbf{b}. \tag{7.117}$$

Here the matrix A represents the combination of the projection operators and the double-layer operator in the left-hand side of equation (7.72), and the array $\mathbf{b}$ represents the sum of the external flow and the single-layer term on the right-hand side of this equation.

In dynamical simulations of multiphase flows, the discretized boundary integral equation (7.117) is usually solved by one of several existing iterative solution methods, some of which are described below. The numerical cost of the standard iterative methods scales as $O(N^2)$ with the number of mesh nodes N, which should be compared with the much larger $O(N^3)$ cost of direct matrix inversion.

The simplest iterative procedure is the Picard method of successive substitutions. In order to ensure convergence of the Picard iteration for all values of the viscosity parameter $\varkappa$, equation (7.117) is first rewritten in the form

$$\mathbf{U} + \varkappa\mathsf{A}' \cdot \mathbf{U} = \mathbf{b}, \tag{7.118}$$

where

$$\mathsf{A}' = 2\mathsf{A} - \mathsf{I}, \tag{7.119}$$

and I is the identity matrix. The discussion of the spectrum of the double-layer operator given before in Section 7.3.4 shows that the eigenvalues α of the matrix A are in the range $0 < \alpha < 1$, as indicated by equations (7.57) and (7.58). It follows that the spectrum of the modified matrix (7.119) is in the range

$$-1 < \alpha' < 1. \tag{7.120}$$

Accordingly, the Picard iteration

$$\mathbf{U}^{(k)} = -\varkappa\mathsf{A}' \cdot \mathbf{U}^{(k-1)} + \mathbf{b}, \qquad k = 1, 2, \ldots \tag{7.121}$$

---

[1] Strictly speaking, due to the eigenvalue purge, the array $\mathbf{U}$ represents the auxiliary field $\bar{\mathbf{u}}_S$, from which the node velocities are reconstructed using projection (7.87).

converges for all physically admissible values of the viscosity parameter $-1 \leq \varkappa \leq 1$.

The initial value of the velocity array $\mathbf{U}^{(0)}$ is arbitrary. In dynamical simulations, it should be set equal to the solution obtained at the preceding time step, in order to accelerate the convergence of the method.

For moderate values of the viscosity ratio (i.e. for $|\varkappa| \ll 1$) several iterations (7.121) usually suffice to obtain well-converged results. The convergence may be slower for highly viscous drops ($\varkappa \approx 1$) and for drops with very low viscosity ($\varkappa \approx -1$), especially in near-contact configurations. To improve the convergence of the velocity evaluation for such systems, Zinchenko *et al.* (1997) have proposed to use a combination of the Picard and biconjugate gradient (Fletcher, 1976) iterations. Another efficient iterative solution technique is the GMRES algorithm (Saad and Schultz, 1986) which has been used in boundary integral calculations of foam permeability (Galea, 2004).

7.5.5 Surface evolution

In boundary integral simulations of the motion of rigid particles, the translational and angular particle velocities are obtained by solving the discretized equation (7.88) and projecting the solution onto the rigid-body modes. The resulting rigid-body velocities are used to update the particle positions at discrete time steps. The time stepping can be performed by any of the standard algorithms, such as the Runge–Kutta method (see, e.g. Press *et al.*, 1992). In simulations of systems of liquid drops, the node positions $\mathbf{X}$ are updated individually, using the node velocities $\mathbf{U}$ obtained from the solution of the discretized boundary integral equation. For surfactant-covered drops, not only do the drop positions, but also the relevant information about the interfacial surfactant concentration need to be updated at each time step. Techniques for incorporating surfactant effects into boundary integral simulations have been described in several papers (e.g. Yon and Pozrikidis, 1998; Bazhlekov *et al.*, 2004a; Vlahovska *et al.*, 2005).

For deformable drops, the size of the time step Δt is limited by the numerical instability associated with the capillary relaxation mechanism. The instability occurs when Δt exceeds the relaxation time τ_{m} for drop deformation on the mesh length scale l; thus, the condition

$$\Delta t \lesssim \tau_{\mathrm{m}} = \gamma^{-1} \mu l, \tag{7.122}$$

where γ denotes the interfacial tension and $\mu = \frac{1}{2}(\mu_1 + \mu_2)$, is required. In contrast, drop evolution typically occurs on a much longer drop-relaxation time scale

$$\tau_{\mathrm{d}} = \gamma^{-1}\mu a \gg \tau_{\mathrm{m}}, \tag{7.123}$$

where a is the size of the drop. An additional time-stepping stability condition can be formulated in terms of the surfactant relaxation time in the presence of surfactants.

It is evident from the above expressions that the restriction (7.122) limits numerical efficiency of dynamical boundary integral simulations of liquid drops, especially at high mesh resolution. The restriction on the size of the time step may become quite severe if local structures (such as pointed ends or a neck) develop on the drop interface – to accurately resolve such local features, a fine mesh with small edge length l is necessary.

In three-dimensional systems, the time-step restriction (7.122) cannot be circumvented by using a standard stiff-integration algorithm (see, e.g. Hall and Watt, 1976) that requires evaluation of the Jacobian matrix

$$\mathbf{J}_{ij} = \partial \mathbf{U}_i / \partial \mathbf{X}_j, \tag{7.124}$$

because of the high numerical cost. In contrast, in axisymmetric simulations such as those reported in Nemer *et al.* (2004), the number of mesh points required to accurately resolve the drop interfaces is much smaller, and in long-time simulations the gain resulting from applying a stiff integration procedure may exceed the cost of the Jacobian evaluation.

An alternative approach to the numerical stiffness problem has been proposed by Bazhlekov *et al.* (2004). Their multiple-step time integration scheme is based on the observation that the numerical instability results from the local response of the system to node displacements. Accordingly, only local single-layer contributions from the nodes surrounding a given node are updated at each time step, which requires $O(N)$ operations. A similar time-saving technique is used for the double-layer contribution. The full $O(N^2)$ velocity calculation is performed only after a certain number of local steps.

The tests presented by Bazhlekov *et al.* (2004b) show that the multiple time-step algorithm yields a substantial acceleration of the calculations for a given number of nodes but at the cost of a decreased accuracy, which reduces the benefits of the method. The development of an efficient stiff integration algorithm that overcomes the limitation (7.122) is thus still an open problem. An alternative to a multiple-step time integration scheme is the development of a stiff integration algorithm with the Jacobian matrix (7.124) approximated by a sparse matrix in which only the elements corresponding to neighboring nodes i and j are nonzero.

A general review of advanced stiff integration methods which may be applicable in boundary integral algorithms is given in the monograph by Hairer and Wanner (1996). There also is an entirely different class of acceleration techniques that are focused on the rapid evaluation of the boundary integrals. We outline these techniques in the following section.

7.5.6 Acceleration methods

Since the matrix A in the discretized boundary integral equation (7.117) is dense, the standard, unaccelerated procedures for solving boundary integral problems are numerically expensive. Even after applying the iterative solution methods of Section 7.5.4, the calculation time scales as N^2 with the number of mesh points N. Iterative unaccelerated methods are adequate for obtaining accurate results for a single drop or a pair of drops (cf. Figs. 7.2 and 7.3), but they are insufficient for simulations of many-drop systems.

In this section we briefly review several acceleration techniques methods that decrease the order at which the numerical cost of a simulation grows with the system size. These methods were originally developed for systems interacting with long-range electrostatic and gravitational forces, but they are applicable to Stokes-flow problems as well. A good introductory description can be found in the book by Frenkel and Smit (2002). The potential gain from implementing acceleration methods in boundary integral algorithms is large. This gain, moreover, grows with increasing system size, because for smaller systems a large prefactor in the cost estimate offsets the benefit associated with the lower order of the method.

The simplest way to decrease the numerical cost of a boundary integral calculation for a periodic system is to use the structure of the Ewald sums (7.97) and (7.98) to optimize the evaluation of the single-layer and double-layer integrals. Accordingly, the short-range real part of the Ewald sums is truncated at a finite distance and applied only locally. The Fourier part is factorized into components that depend only on the coordinates of individual mesh points; thus the corresponding contribution to the boundary integrals can be evaluated at an $O(N)$ cost per Fourier mode. By choosing the value of the splitting parameter so that the costs of evaluating the real-space and Fourier-space contributions are comparable, the total cost of the whole boundary integral calculation can be reduced from $O(N^2)$ to $O(N^{3/2})$ (Frenkel and Smit, 2002).

The numerical efficiency of the boundary integral methods can be improved further by using fast multipole methods, which have an $O(N)$ numerical cost.

These methods involve clustering mesh points into large groups. The field produced by distant groups is then efficiently evaluated using a multipolar expansion of the Stokes-flow field; the standard direct point-to-point calculation is performed only locally. Using an algorithm based on this idea, Zinchenko and Davis (2000, 2002) have achieved an acceleration by a factor of 100 for a system of 200 drops (compared to a standard $O(N^2)$ calculation). A further acceleration can presumably be achieved by a fast multipole algorithm that utilizes a hierarchy of larger and larger groups of mesh nodes (Greengard and Rokhlin, 1987; Cheng *et al.*, 1999). This method, however, has not been applied to Stokes-flow problems so far.

The development of fast multipole methods for Stokes-flow applications requires constructing appropriate multipolar representations of Stokes flow. Such a representation can be defined in a similar way as the multipolar expansion of the electrostatic potential (see, e.g. Jackson, 1999). A convenient multipolar representation of Stokes flow has been introduced by Cichocki *et al.* (1988). The representation includes basis sets of singular and non-singular multipolar solutions of Stokes equations, as well as explicit transformations that relate the basis fields centered at different points (Felderhof and Jones, 1989). For periodic systems, a set of Ewald summation expressions for multipolar Stokes flow fields has been derived by Cichocki and Felderhof (1989).

A separate category of acceleration algorithms relies on the particle–mesh approach. Here, an auxiliary regular spatial mesh is used for evaluating the long-range contributions to the single-layer and double-layer integrals. The force distribution (or the double-layer density) on each surface element ΔS is surrounded by an appropriate distribution of the screening forces on the nearby auxiliary mesh points. The flow field produced by the combination of the original and the screening force distributions decays rapidly with distance, and thus is evaluated only locally at a reduced $O(N)$ numerical cost. The long-range contributions produced by the screening force are evaluated in Fourier space using the $O(N \ln N)$ fast Fourier transform method.

Different variants of the particle–mesh algorithms for potential problems are described in the book by Frenkel and Smit (2002) and in the review article by Deserno and Holm (1998). In application to the Stokesian-dynamics methods for rigid spheres, a particle–mesh approach was used by Sierou and Brady (2001). The particle–mesh techniques have also been employed by Higdon (unpublished) to accelerate boundary integral simulations of two-dimensional emulsion flows. However, the full potential of particle–mesh methods in Stokes-flow applications has yet to be explored.

Appendix

We collect here some proofs that were omitted in the main text.

A. To show that the rigid-body velocity field (7.16) satisfies the eigenvalue problem (7.54), we note that the strain-rate tensor associated with this field vanishes identically. Accordingly, the corresponding stress tensor has only the constant hydrostatic contribution $\boldsymbol{\sigma}^{\mathrm{rb}} = -p_0\hat{\boldsymbol{I}}$. Inserting the fields $\mathbf{u}^{\mathrm{rb}}$ and $\boldsymbol{\sigma}^{\mathrm{rb}}$ into the Lorentz integral identity (7.43) specified for the bounded region Ω_1 yields

$$\int_S \hat{\mathbf{n}}_1(\mathbf{x}) \cdot \left\{ [\mathbf{u}^{\mathrm{rb}}(\mathbf{x}) - \mathbf{u}^{\mathrm{rb}}(\mathbf{x}')] \cdot \mathbf{T}(\mathbf{x}, \mathbf{x}') \right\} \mathrm{d}S$$

$$= -\mu^{-1} p_0 \int_S \hat{\mathbf{n}}_1(\mathbf{x}) \cdot \mathbf{G}(\mathbf{x}, \mathbf{x}') \, \mathrm{d}S$$

$$= 0, \tag{A.1}$$

where the second equality results from the fluid-volume conservation condition (7.36). Recalling definition (7.52) of the double-layer integral operator $\hat{\mathsf{T}}$, relation (7.54) is thus obtained.

To demonstrate that the expansion mode $\mathbf{v} = \hat{\mathbf{n}}_1$ satisfies the adjoint eigenvalue problem (7.55), the Lorentz integral relation (7.43) with $i = 2$ is multiplied by $\hat{\mathbf{n}}_1(\mathbf{x}')$ and integrated with respect to the variable $\mathbf{x}'$ over the surface S. The single-layer term vanishes by the symmetry property (7.34) of the function $\mathbf{G}(\mathbf{x}, \mathbf{x}')$ and incompressibility condition (7.36). Using relation (7.7), we thus get

$$\int_S \mathrm{d}S' \int_S \mathrm{d}S \, \hat{\mathbf{n}}_1(\mathbf{x}) \cdot \left\{ [\mathbf{u}(\mathbf{x}) - \mathbf{u}(\mathbf{x}')] \cdot \mathbf{T}(\mathbf{x}, \mathbf{x}') \right\} \cdot \hat{\mathbf{n}}_1(\mathbf{x}') = \int_S \mathbf{u}(\mathbf{x}) \cdot \hat{\mathbf{n}}_1(\mathbf{x}) \, \mathrm{d}S. \tag{A.2}$$

Recalling the inner product notation (7.53) and the definition (7.52) of the double-layer operator, we find that the above relation is equivalent to (7.55).

In our derivation of (7.55) we have used the Lorentz integral identity (7.43) with $i = 2$ to ensure that the resulting relation is valid for all surface fields $\mathbf{u}$. A velocity distribution $\mathbf{u}$ on the surface S corresponds in this case to a bulk Stokes flow that is nonsingular in the region Ω_2 and vanishes at infinity. Such a flow field exists for an arbitrary boundary condition on the surface S. In contrast, Stokes flows that are nonsingular in the bounded region Ω_1 require the consistency condition $(\hat{\mathbf{n}}_1, \mathbf{u}) = 0$, by the incompressibility of the flow.

B. We now show that the rigid particle motion determined by the boundary integral equation (7.74) depends on the total force and torque (7.76), but not on the details of the surface force distribution. According to the existence theorem for Stokes flows (Kim and Karrila, 1991), for a given surface force distribution $\Delta \mathbf{f}_S$ that satisfies the constraints

$$\int_S \Delta \mathbf{f}_S(\mathbf{x})\, \mathrm{d}S = 0, \qquad \int_S (\mathbf{x} - \mathbf{x}_0) \times \Delta \mathbf{f}_S(\mathbf{x})\, \mathrm{d}S = 0 \qquad (\text{A.3})$$

there exists a unique velocity field $\Delta \mathbf{u}$ that is nonsingular in the region Ω_1, has no projection on the rigid-body mode,

$$\hat{P}^{\mathrm{rb}} \Delta \mathbf{u} = 0, \qquad (\text{A.4})$$

and satisfies the boundary condition

$$\hat{\mathbf{n}}_1 \cdot \Delta \boldsymbol{\sigma} = \frac{1}{2} \Delta \mathbf{f}_S \qquad (\text{A.5})$$

on the surface S. Here $\Delta \boldsymbol{\sigma}$ is the stress corresponding to the velocity $\Delta \mathbf{u}$, and the factor $1/2$ is introduced for convenience. The stress $\Delta \boldsymbol{\sigma}$ and the flow field $\Delta \mathbf{u}$ are related by the Lorentz integral identity (7.43) specified for the region Ω_1. Using the boundary condition (A.5) in the Lorentz relation with $\mu = \mu_2$ yields

$$2 \int_S \hat{\mathbf{n}}_1(\mathbf{x}) \cdot \left\{ [\Delta \mathbf{u}(\mathbf{x}) - \Delta \mathbf{u}(\mathbf{x}')] \cdot \mathbf{T}(\mathbf{x}, \mathbf{x}') \right\} \mathrm{d}S = \int_S \Delta \mathbf{f}_S(\mathbf{x}) \cdot \mu_2^{-1} \mathbf{G}(\mathbf{x}, \mathbf{x}')\, \mathrm{d}S.$$
$$(\text{A.6})$$

By adding the above expression to the boundary integral equation (7.74) and using conditions (A.4) and $\hat{P}^1 \Delta \mathbf{u} = 0$, we find that the change $\mathbf{f}_S \to \mathbf{f}_S + \Delta \mathbf{f}_S$ of the surface traction field produces a corresponding change $\bar{\mathbf{u}}_S \to \bar{\mathbf{u}}_S + \Delta \mathbf{u}$ of the solution of the boundary integral equation. This change, however, does not affect the particle motion, due to relation (7.75) and the constraint (A.4).

C. For the single-layer integral, the singularity-subtraction regularization procedure can be performed only in the special case of the force $\mathbf{f}_S$ normal to the interface S,

$$\mathbf{f}_S(\mathbf{x}) = f_S(\mathbf{x}) \hat{\mathbf{n}}_1(\mathbf{x}). \qquad (\text{A.7})$$

Assuming (A.7), and using the identity (7.36) we find

$$\int_S \mathbf{f}_S(\mathbf{x}) \cdot \mathbf{G}(\mathbf{x}, \mathbf{x}')\, \mathrm{d}S = \int_S [f_S(\mathbf{x}) - f_0] \hat{\mathbf{n}}_1(\mathbf{x}) \cdot \mathbf{G}(\mathbf{x}, \mathbf{x}')\, \mathrm{d}S, \qquad (\text{A.8})$$

where f_0 is an arbitrary constant. For the field point $\mathbf{x}'$ on the integration surface S, the choice

$$f_0 = f_S(\mathbf{x}'), \qquad \mathbf{x}' \in S, \qquad (\text{A.9})$$

ensures the cancellation of the first-order singularity of the Green's function
(7.39). For $\mathbf{x}' \notin S$, it is convenient to choose

$$f_0 = f_S(\mathbf{x}''), \qquad \mathbf{x}' \notin S, \tag{A.10}$$

where $\mathbf{x}'' \in S$ is the position for which the distance $|\mathbf{x}' - \mathbf{x}''|$ is minimized.
Due to this choice the integrand on the right side of equation (A.8) remains
$O(1)$ when the point $\mathbf{x}'$ approaches the surface S (cf. the discussion of
equation (7.85) for the double-layer integral case).

8

Averaged equations for multiphase flow

The previous chapters have been devoted to methods capable of delivering "numerically exact" solutions of the Navier–Stokes equations as applied to various multiphase flow problems. In spite of their efficiency, these methods still require a substantial amount of computation even for relatively simple cases. It is therefore evident that the simulation of more complex flows approaching those encountered in most natural situations or technological contexts (sediment transport, fluidized beds, electric power generation, and many others) cannot be pursued by those means but must be based on a different approach. Furthermore, even if we did have detailed knowledge, e.g., of the motion of all the particles and of the interstitial fluid, most often, for practical purposes, we would be interested in quantities obtained by applying some sort of averaging to this immense amount of information. This observation suggests that it might be advantageous to attempt to formulate equations governing the time evolution of these averages directly. In this approach, rather than aiming at a detailed solution of the Navier–Stokes equations, we would be satisfied with a reduced description based on simplified mathematical models. While one may try to base such models on intuition, a more reliable way is perhaps to start from the exact equations and carry out a process of averaging which would filter out the inessential details retaining the basic physical processes which determine the behavior of the system.

8.1 Introduction

The issue of averaging in multiphase flow is a long-standing one with a history which stretches nearly as far back as for single-phase turbulence. There is, however, an unfortunate difference with that case: while in single-phase turbulence relatively simple ideas, such as the mixing length hypothesis, already give useful results, simple approaches to multiphase averaging have had a much more limited success. As a matter of fact, broadly speaking, it is fair to say that averaged-equations models for multiphase flow are far less developed, and far less faithful to physical reality, than those of single-phase turbulence. There are still considerable difficulties plaguing the modeling of many terms and the equations often appear to be somewhat deficient even at the purely mathematical level.

If the detailed form of the averaged equations presents a number of uncertainties, their general structure for several broad classes of problems is fairly well established and it is on this general structure that the numerical methods for their solution hinge. Hence, one may expect with some confidence that the numerical methods to be described in the following chapters will be broadly applicable to many specific averaged-equations models, be they already in existence, or still to be developed. Before describing these numerical methods, however, it is appropriate to present a straightforward derivation of a general set of averaged equations to better appreciate the origin of the various terms and, thereby, the physical origin of the numerical challenges that they present. We will only outline a simple "general-purpose" derivation and refer the reader to appropriate sources for different approaches, additional details, and more sophisticated developments (see e.g. Drew 1983).

Most readers will be familiar with various forms of averaging, such as time, volume, and ensemble averaging. Each one of these approaches has been applied to multiphase flow and the literature contains various claims as to the superiority of one versus the others. As our aim is simply to show the origin of the basic structure of the averaged equations, and of the problems that arise in trying to derive a closed mathematical model, in the next section we use volume averaging, which is simple and perhaps more natural for this application, although not free of difficulties – conceptual as well as practical – if taken too literally.

The most serious problem with volume averaging is that the notion of separation of scales which it presupposes is seldom satisfied. As expressed by Nigmatulin (1979), separation of scales requires the possibility of defining an "elementary macrovolume ... the characteristic linear dimensions of which are many times greater than the nonuniformities ... , but at the same time much less than the characteristic macrodimension of a problem." It is evident that, when this condition is not satisfied, volume-averaged quantities can certainly be defined, but they would be far from smooth. To obviate this shortcoming, volume averaging has been combined with other forms of averaging, such as time, or double volume averaging. A conceptual device, which really amounts to a combination of volume and ensemble averaging, appeals to the notion of *representative elementary volume* which, as the denomination implies, avoids the "extreme" configurations which may be encountered in an actual snapshot of a real multiphase flow (e.g. two adjacent averaging volumes consisting nearly entirely of different phases) by postulating the existence of some sort of "average" averaging volume. While it may be doubtful whether representative elementary volumes have a physical counterpart in reality, they offer a convenient conceptual model useful for a visualization of the averaging process, and it is in this spirit that we will have recourse to them.

8.2 Volume averaging

In order to derive the volume-averaged equations, we attach to each point $\mathbf{x}$ in space an averaging volume $\mathcal{V}(\mathbf{x})$ bounded by a surface $\mathcal{S}(\mathbf{x})$. We work at the analytical level here and, therefore, our aim is to derive a set of averaged differential equations rather than to discretize space. Thus, we associate an averaging volume with each point $\mathbf{x}$ and the averaging volumes centered at two neighboring points will in general overlap. For simplicity, and again in view of the limited aims that we have in this chapter, we use the same averaging volume for all points even though, for example, the averaging volume to be used in a boundary layer would probably differ from that suitable for the bulk flow.

As generalizations are immediate, it is sufficient to consider the case of two phases, which we distinguish by indices 1 and 2. The averaging volume $\mathcal{V}$ can then be divided into portions $\mathcal{V}_1$ and $\mathcal{V}_2$ occupied by each phase, and likewise for its surface, $\mathcal{S}(\mathbf{x}) = \mathcal{S}_1(\mathbf{x}, t) + \mathcal{S}_2(\mathbf{x}, t)$. The boundary of $\mathcal{V}_J$ ($J = 1$ or 2) will consist of $\mathcal{S}_J(\mathbf{x}, t)$ and of interface(s) $\mathcal{S}_i(\mathbf{x}, t)$ contained inside $\mathcal{V}$. Thus $\mathcal{S}_J + \mathcal{S}_i$ is a closed surface for both $J = 1$ and 2 (Fig. 8.1).

Let F_J be any extensive physical quantity pertaining to the matter which constitutes phase J such as mass, momentum, energy, etc. We express the infinitesimal amount of F_J in terms of a density f_J per unit mass as

$$dF_J = \rho_J f_J \, dV_J. \tag{8.1}$$

A general statement for the balance of the total amount of F_J in the averaging volume $\mathcal{V}$ can then be written as

$$\frac{\partial}{\partial t} \int_{\mathcal{V}_J} \rho_J f_J \, dV_J$$

$$= \int_{\mathcal{S}_J} (-\rho_J f_J \mathbf{u}_J + \boldsymbol{\phi}_J) \cdot \mathbf{n}_J \, dS_J$$

$$+ \int_{\mathcal{S}_i} [-\rho_J f_J (\mathbf{u}_J - \mathbf{w}) + \boldsymbol{\phi}_J] \cdot \mathbf{n}_J \, dS_i + \int_{\mathcal{V}_J} \rho_J \theta_J \, dV_J. \tag{8.2}$$

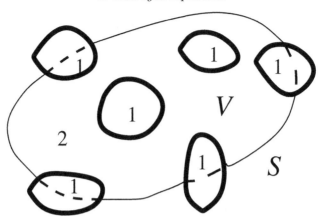

Fig. 8.1. The averaging volume $\mathcal{V} = \mathcal{V}_1 + \mathcal{V}_2$ is constituted by a volume $\mathcal{V}_1$ of phase 1 and a volume $\mathcal{V}_2$ of phase 2. Similarly, the boundary $\mathcal{S}$ of $\mathcal{V}$ consists of a part $\mathcal{S}_1$ (dashed) in contact with phase 1 and a part $\mathcal{S}_2$ in contact with phase 2. The collection of interfaces separating the two phases inside $\mathcal{V}$ is denoted by $\mathcal{S}_i$. Note that $\mathcal{S}_1 + \mathcal{S}_i$ and $\mathcal{S}_2 + \mathcal{S}_i$ are closed surfaces, while none of $\mathcal{S}_1$, $\mathcal{S}_2$ or $\mathcal{S}_i$ is in general closed.

The first term in the right-hand side is the amount of F_J transported by convective or diffusive processes through the part $\mathcal{S}_J$ of the (outer) surface of the averaging volume occupied by the phase J. The second term is the analogous contribution through the interface $\mathcal{S}_i$ contained inside the averaging volume. The unit normal $\mathbf{n}_J$ is directed out of phase J and, therefore, $\mathbf{n}_1 = -\mathbf{n}_2$ on $\mathcal{S}_i$. In writing these surface terms we have distinguished between a nonconvective transport, or surface source, $\boldsymbol{\phi}_J$ (e.g. the stress when F is momentum), and a convective transport. The convective terms in the two integrals differ because the surface of the averaging volume is assumed to be stationary, while the interface in general moves with a local velocity $\mathbf{w}$. The last term is the contribution of the source θ_J of F_J inside $\mathcal{V}$.

We now define the volume average of any quantity q_J by

$$\langle q_J \rangle (\mathbf{x}, t) = \frac{1}{\mathcal{V}_J} \int_{\mathcal{V}_J} q_J(\mathbf{x} + \boldsymbol{\xi}, t) \, d^3\xi \qquad (8.3)$$

where $\boldsymbol{\xi}$ is the integration variable referred to the center $\mathbf{x}$ of the averaging volume. The first and the last terms of (8.2) can then be written directly as volume averages. In order to write the integral over $\mathcal{S}_J$ in the form of a volume average, we use the simple result:

$$\int_{\mathcal{S}_J} \mathbf{A}_J \cdot \mathbf{n}_J \, dS_J = \nabla \cdot \int_{\mathcal{V}_J} \mathbf{A}_J \, d^3\xi, \qquad (8.4)$$

where $\mathbf{A}_J$ is any vector or tensor quantity pertaining to the phase J. A straightforward proof of this relation is given in the Appendix together with a graphical interpretation. Suffice it to say here that this is an exact result of a purely geometrical nature. Finally, we introduce the *volume fraction* of phase J by

$$\alpha_J = \frac{\mathcal{V}_J}{\mathcal{V}}. \tag{8.5}$$

It is evident from this definition that, in the present case where we consider only two phases,

$$\alpha_1 + \alpha_2 = 1. \tag{8.6}$$

With these definitions and result, we may therefore write (8.2) as

$$\frac{\partial}{\partial t}\left(\alpha_J\langle\rho_J f_J\rangle\right) + \boldsymbol{\nabla}\cdot\left(\alpha_J\langle\rho_J f_J \mathbf{u}_J\rangle\right)$$

$$= \boldsymbol{\nabla}\cdot\left(\alpha_J\langle\boldsymbol{\phi}_J\rangle\right) + \alpha_J\langle\rho_J\theta_J\rangle + \frac{1}{\mathcal{V}}\int_{S_i}\left[-\rho_J f_J\left(\mathbf{u}_J - \mathbf{w}\right) + \boldsymbol{\phi}_J\right]\cdot\mathbf{n}_J dS_i. \tag{8.7}$$

By suitably specializing f_J, Φ_J, and θ_J when F is, in turn, mass, momentum, and energy according to Table 8.1, we can now obtain the averaged equations for these quantities.

Other forms of volume averaging exist in which, in place of (8.3), the average quantity is defined by means of a mollifier, or weighting, function g:

$$\alpha_J\langle q_J\rangle(\mathbf{x},t) = \int_{\mathcal{V}_{\infty J}} q_J(\boldsymbol{\xi},t)\, g\left(|\mathbf{x} - \boldsymbol{\xi}|\right) d^3\xi, \tag{8.8}$$

Table 8.1. Specification of the quantities necessary to adapt the general balance equation to the specific cases of mass, momentum, kinetic energy, and total energy; $\boldsymbol{\sigma}$ is the total stress tensor, e the internal energy per unit mass, Φ the dissipation function, and Q the rate of heat generation per unit mass.

Conserved quantity	f	ϕ	θ
Mass	1	0	0
Momentum	ρ	$\boldsymbol{\sigma}$	$\mathbf{g}$
Total energy	$e + \frac{1}{2}\mathbf{u}\cdot\mathbf{u} - \mathbf{x}\cdot\mathbf{g}$	$\mathbf{u}\cdot\boldsymbol{\sigma} - \mathbf{q}$	Q

in which the integral is extended over the entire space occupied by the
J-phase. The function g has compact support, is nonnegative, and satisfies
the normalization condition

$$\int g\left(|\mathbf{x} - \boldsymbol{\xi}|\right) d^3\xi = 1, \tag{8.9}$$

where the integral is over all space. In this context, the volume fraction of
the J-phase at $\mathbf{x}$ is defined by

$$\alpha_J(\mathbf{x}, t) = \int_{\mathcal{V}_{\infty J}} g\left(|\mathbf{x} - \boldsymbol{\xi}|\right) d^3\xi. \tag{8.10}$$

The definition (8.3) used above corresponds to a "top-hat" weighting function
having a sharp cutoff at the edge of the averaging volume. The definition
(8.8) has certain advantages of smoothness, although it leads to more complicated algebra.

8.2.1 Conservation of mass

When F is mass, the physical process responsible for nonconvective transport would be diffusion and those responsible for the volume source might
be, e.g. chemical or nuclear reactions. If these processes are neglected for
simplicity, since, as is evident from (8.1), $f_J = 1$ in the case of mass, the
previous general average equation (8.7) becomes

$$\frac{\partial}{\partial t}\left(\alpha_J\langle\rho_J\rangle\right) + \boldsymbol{\nabla} \cdot \left(\alpha_J\langle\rho_J\mathbf{u}_J\rangle\right) = -\frac{1}{\mathcal{V}}\int_{S_i} \rho_J\left(\mathbf{u}_J - \mathbf{w}\right) \cdot \mathbf{n}_J\, dS_i. \tag{8.11}$$

The convective mass flux across the interface is given by (1.19) and is

$$\dot{m}_J = \rho_J\left(\mathbf{u}_J - \mathbf{w}\right) \cdot \mathbf{n}_J \tag{8.12}$$

and would be nonzero, for example, if the J-phase were to undergo a phase
change, e.g. from liquid to vapor. In this case, the corresponding term in
the equation for the other phase, K say, would have a contribution

$$\dot{m}_K = \rho_K\left(\mathbf{u}_K - \mathbf{w}\right) \cdot \mathbf{n}_K = -\dot{m}_J \tag{8.13}$$

with the last equality justified by the absence of sources or sinks of mass at
the interface and the opposite direction of the two normals. As defined here,
the mass fluxes are positive when the fluid velocity in the direction of the

normal is faster than the interface velocity, i.e. when the phase loses mass. The notation

$$\Gamma_J = -\frac{1}{\mathcal{V}} \int_{S_i} \dot{m}_J \, dS_i, \qquad (8.14)$$

is frequently encountered. Note that, by (8.13), $\Gamma_1 + \Gamma_2 = 0$. Here we encounter a first intimation of the difficulties associated with the formulation of a closed averaged-equations model as a precise calculation of these quantities in practice is difficult: phase change processes depend on details of the temperature distribution and flow field in the immediate neighborhood of the interface, and these quantities may be quite different from their volume-averaged values. Furthermore, it is also necessary to have a good estimate of the amount of interface contained in the averaging volume, and this information is not contained in the volume fraction: the same volume fraction can be the result of a few large bubbles, for example, or many small ones, and S_i would be quite different in the two cases. This fact has stimulated intense work on the modeling of the amount of interface per unit volume (see, e.g. Wu *et al.*, 1999; Hibiki and Ishii, 2000; Ishii *et al.*, 2003, 2004; Ishii and Kim, 2004).

At least formally, the appearance of the average of the product $\langle \rho_J \mathbf{u}_J \rangle$ can be avoided by introducing the *Favre average* of $\mathbf{u}_J$, defined by

$$\tilde{\mathbf{u}}_J = \frac{\langle \rho_J \mathbf{u}_J \rangle}{\langle \rho_J \rangle}, \qquad (8.15)$$

in terms of which

$$\frac{\partial}{\partial t} (\alpha_J \langle \rho_J \rangle) + \nabla \cdot (\alpha_J \langle \rho_J \rangle \tilde{\mathbf{u}}_J) = \Gamma_J. \qquad (8.16)$$

In general, the Favre average permits us to write the equation in a somewhat cleaner form, but it is seldom particularly useful in multiphase flow applications because the Mach number is small in most cases and situations in which the density of a phase might undergo large variations from place to place in the averaging volume would be rather unusual. Thus, in the majority of circumstances, $\langle \rho_J f_J \rangle \simeq \langle \rho_J \rangle \langle f_J \rangle$ and $\tilde{f}_J \simeq \langle f_J \rangle$.

By adding the mass conservation equation written for each phase, we find the conservation equation for the total mass

$$\frac{\partial \rho_m}{\partial t} + \nabla \cdot (\rho_m \mathbf{u}_m) = 0 \qquad (8.17)$$

where we have introduced the (total) mixture density

$$\rho_m = \alpha_1 \langle \rho_1 \rangle + \alpha_2 \langle \rho_2 \rangle \qquad (8.18)$$

and the center-of-mass velocity

$$\rho_m \mathbf{u}_m = \alpha_1 \langle \rho_1 \mathbf{u}_1 \rangle + \alpha_2 \langle \rho_2 \mathbf{u}_2 \rangle. \tag{8.19}$$

If a phase can be considered incompressible, we have $\langle \rho_J \rangle = \rho_J$, $\langle \rho_J \mathbf{u}_J \rangle = \rho_J \langle \mathbf{u}_J \rangle$ and, in the absence of phase change, (8.11) becomes

$$\frac{\partial \alpha_J}{\partial t} + \nabla \cdot (\alpha_J \langle \mathbf{u}_J \rangle) = 0. \tag{8.20}$$

This equation shows that, even though the phase is microscopically incompressible, its average velocity field is not divergenceless. Physically, this corresponds to the variation of the amount of the phase inside the averaging volume by a change of its volume fraction and, therefore, it is not a compressibility effect in the usual sense. The quantity

$$\mathbf{j}_J = \alpha_J \langle \mathbf{u}_J \rangle \tag{8.21}$$

has the physical meaning of volumetric flux of the phase J and is usually referred to as the *superficial velocity*.

When both phases are incompressible, by adding the mass conservation equations (8.20) for each one of them and recalling (8.6), we find

$$\nabla \cdot (\alpha_1 \langle \mathbf{u}_1 \rangle + \alpha_2 \langle \mathbf{u}_2 \rangle) = 0. \tag{8.22}$$

The quantity in parentheses has the physical meaning of the total *volumetric flux* of the mixture:

$$\mathbf{u}_V = \alpha_1 \langle \mathbf{u}_1 \rangle + \alpha_2 \langle \mathbf{u}_2 \rangle. \tag{8.23}$$

It is obvious that, since both phases are incompressible, this quantity must have zero divergence.

8.2.2 Conservation of momentum

According to Table 8.1, in order to specialize the general equation (8.7) to the case of momentum, we take $f_J = \mathbf{u}_J$ and $\theta_J = \mathbf{g}$. The surface source $\boldsymbol{\phi}_J$ is the stress. At a point in the fluid, the stress is written in the usual way $\boldsymbol{\sigma}_J = -p_J \mathbf{I} + \boldsymbol{\tau}_J$ as the sum of a pressure and a viscous part $\boldsymbol{\tau}_J$. In addition to the fluid stress, at the interface of a solid such as a suspended particle, one may have to consider a collisional component $\boldsymbol{\sigma}_c$. While the fluid stress acts continuously over the entire surface $\mathcal{S}_J$, in general the latter only acts intermittently at isolated points (or small areas) of $\mathcal{S}_J$. With these identifications, we write

$$\frac{\partial}{\partial t}\left(\alpha_J \langle \rho_J \mathbf{u}_J\rangle\right) + \nabla\cdot\left(\alpha_J \langle \rho_J \mathbf{u}_J \mathbf{u}_J\rangle\right)$$
$$= \nabla\cdot\left(\alpha_J\langle\boldsymbol{\sigma}_J\rangle\right) + \nabla\cdot\left(\alpha_J\langle\boldsymbol{\sigma}_c\rangle\right) + \alpha_J\langle\rho_J\rangle\mathbf{g}$$
$$+\frac{1}{\mathcal{V}}\int_{S_i}\boldsymbol{\sigma}_J\cdot\mathbf{n}_J\,dS_i - \frac{1}{\mathcal{V}}\int_{S_i}\rho_J\mathbf{u}_J\left(\mathbf{u}_J-\mathbf{w}\right)\cdot\mathbf{n}_J\,dS_i. \quad (8.24)$$

The last term accounts for the momentum that a phase gains or loses as some of its matter is lost or gained. We will disregard this term until Chapter 11.

It appears natural at first sight to identify the interfacial integral of $\boldsymbol{\sigma}_J$ with the force that the other phase exerts on the phase J. Actually this identification must be done with some care. Indeed, consider a system at rest without forces. Then $\langle\boldsymbol{\sigma}_J\rangle$ would be a constant so that $\nabla\cdot(\alpha_J\boldsymbol{\sigma}_J) = \langle\boldsymbol{\sigma}_J\rangle\cdot\nabla\alpha_J$ which, if correct, would imply that this term represents a source of momentum solely due to the spatial arrangement of the phases. This is evidently a physically incorrect result[1]. The correct procedure is to combine the interfacial integral with the first term in the right-hand side, writing

$$\nabla\cdot\left(\alpha_J\langle\boldsymbol{\sigma}_J\rangle\right) + \frac{1}{\mathcal{V}}\int_{S_i}\boldsymbol{\sigma}_J\cdot\mathbf{n}_J\,dS_i = \alpha_J\nabla\cdot\langle\boldsymbol{\sigma}_J\rangle + \mathbf{F}_J \quad (8.25)$$

which defines $\mathbf{F}_J$. This is the quantity which should be properly identified with the fluid dynamic force acting on the phase J, and which is then given by

$$\mathbf{F}_J = \frac{1}{\mathcal{V}}\int_{S_i}\boldsymbol{\sigma}_J\cdot\mathbf{n}_J\,dS_i + \langle\boldsymbol{\sigma}_J\rangle\cdot\nabla\alpha_J = \frac{1}{\mathcal{V}}\int_{S_i}\left(\boldsymbol{\sigma}_J-\langle\boldsymbol{\sigma}_J\rangle\right)\cdot\mathbf{n}_J\,dS_i. \quad (8.26)$$

The last step is a consequence of the exact relation

$$\nabla\alpha_J = \frac{1}{\mathcal{V}}\int_{S_J}\mathbf{n}_J\,dS_J = -\frac{1}{\mathcal{V}}\int_{S_i}\mathbf{n}_J\,dS_i, \quad (8.27)$$

in which the first equality follows from (8.4) upon taking $\mathbf{A}_J$ to be the identity two-tensor, and the second one from the fact that $S_J + S_i$ forms a closed surface.

A simple argument supporting the identification of $\mathbf{F}_J$ with the interphase force is the following. Consider a system in which both phases are incompressible, and suppose that this system is subjected to an increase of the static pressure. Since the phases are incompressible, this cannot have an effect on the interphase force. However, if S_i is an open surface (i.e. if the surface of the averaging volume cuts across the other phase as in Fig. 8.1), the interfacial integral in (8.25) will in general be affected by the

---

[1] This argument seems to have been put forward first by Harlow and Amsden (1975).

pressure change. If this integral were identified with the interphase force, this conclusion would be unphysical. But, from (8.27), this is precisely the case in which $\nabla \alpha_J \neq 0$ and, as is evident from the last equality of (8.26), the combination of the two terms will indeed be unaffected by the pressure change. On the other hand, if $\mathcal{S}_i$ is a closed surface (or the collection of closed surfaces), the interfacial integral in (8.26) will be unchanged by the addition of a constant to the pressure but, in this case, $\nabla \alpha_J$ also vanishes by (8.27). Another and perhaps more compelling justification for (8.26) is available for disperse flow and will be given in the next section. The general question of what constitutes a proper expression of the interphase force is another issue which has been much debated in the literature.

On the basis of the relation (8.27), one sometimes encounters a surface average stress $\overline{\boldsymbol{\sigma}}_J$ defined by

$$\overline{\boldsymbol{\sigma}}_J \cdot \nabla \alpha_J = -\frac{1}{V} \int_{\mathcal{S}_i} \boldsymbol{\sigma}_J \cdot \mathbf{n}_J \, dS_i \tag{8.28}$$

so that the momentum equation takes the form

$$\frac{\partial}{\partial t} \left(\alpha_J \langle \rho_J \mathbf{u}_J \rangle \right) + \nabla \cdot \left(\alpha_J \langle \rho_J \mathbf{u}_J \mathbf{u}_J \rangle \right)$$
$$= \alpha_J \nabla \cdot \langle \boldsymbol{\sigma}_J \rangle + \nabla \cdot \left(\alpha_J \langle \boldsymbol{\sigma}_c \rangle \right) + \left(\langle \boldsymbol{\sigma}_J \rangle - \overline{\boldsymbol{\sigma}}_J \right) \cdot \nabla \alpha_J + \alpha_J \langle \rho_J \rangle \mathbf{g}. \tag{8.29}$$

Actually, it is not obvious that equation (8.28) is a consistent definition as it provides only three relations among the six scalar components of $\overline{\boldsymbol{\sigma}}_J$. If an analogous relation is used to define a surface average pressure $\overline{p}_J$, rather than the whole tensor $\overline{\boldsymbol{\sigma}}_J$, as is more often done, one has the opposite problem in that the single scalar $\overline{p}_J$ must now satisfy three in general independent scalar equations[1]. For this reason, for the general considerations of this chapter, we prefer to avoid the form (8.29) writing instead

$$\frac{\partial}{\partial t} \left(\alpha_J \langle \rho_J \mathbf{u}_J \rangle \right) + \nabla \cdot \left(\alpha_J \langle \rho_J \mathbf{u}_J \mathbf{u}_J \rangle \right) = \alpha_J \nabla \cdot \langle \boldsymbol{\sigma}_J \rangle + \nabla \cdot \left(\alpha_J \langle \boldsymbol{\sigma}_c \rangle \right) + \alpha_J \langle \rho_J \rangle \mathbf{g} + \mathbf{F}_J. \tag{8.30}$$

When ρ_J is approximately constant, we may write identically

$$\langle \mathbf{u}_J \mathbf{u}_J \rangle = \langle \mathbf{u}_J \rangle \langle \mathbf{u}_J \rangle + \langle \left(\mathbf{u}_J - \langle \mathbf{u}_J \rangle \right) \left(\mathbf{u}_J - \langle \mathbf{u}_J \rangle \right) \rangle. \tag{8.31}$$

The last term, sometimes referred to as the *streaming stress*, is analogous to the Reynolds stress of single-phase fluid mechanics. The difficulty in attempting to relate this stress to the average fields is familiar from single-phase turbulence and, if anything, it is increased by the presence of another

---

[1] Obviously these difficulties disappear in one space dimension, where (8.28) would legitimately define the single relevant component $\overline{\sigma}_{xx}$ or interfacial pressure $\overline{p}_J$.

phase. In addition, we recognize that other closure problems appear in trying to express the interphase forces as functions of the primary average fields which, in most cases, will include the volume fractions, the average velocities, and the average pressure. The closure of the averaged equations is one of the major difficulties facing the modeling of multiphase flows.

8.2.3 Conservation of energy

From the last line in Table 8.1, with the aid of the relation

$$\rho e \mathbf{u} - \mathbf{u} \cdot \boldsymbol{\sigma} = (\rho e + p)\, \mathbf{u} - \mathbf{u} \cdot \boldsymbol{\tau} = \rho h \mathbf{u} - \mathbf{u} \cdot \boldsymbol{\tau}, \qquad (8.32)$$

where $h = e + p/\rho$ is the enthalpy per unit mass, one can readily write down a formal statement of the total energy balance according to equation (8.7), namely

$$\frac{\partial}{\partial t}\left(\alpha_J \left\langle \rho_J \left(e_J + \frac{1}{2}u_J^2 - \mathbf{x} \cdot \mathbf{g}\right)\right\rangle\right) + \boldsymbol{\nabla} \cdot \left[\alpha_J \left\langle \rho_J \left(h_J + \frac{1}{2}u_J^2 - \mathbf{x} \cdot \mathbf{g}\right) \mathbf{u}_J\right\rangle\right]$$
$$= \boldsymbol{\nabla} \cdot (\alpha_J \langle \mathbf{u}_J \cdot \boldsymbol{\tau}_J - \mathbf{q}_J\rangle) + \alpha_J \langle \rho_J Q_J\rangle$$
$$+ \frac{1}{V} \int_{S_i} \left[-\rho_J \left(h_J + \frac{1}{2}u_J^2 - \mathbf{x} \cdot \mathbf{g}\right)(\mathbf{u}_J - \mathbf{w}) + (\mathbf{u}_J \cdot \boldsymbol{\tau}_J - \mathbf{q}_J)\right] \cdot \mathbf{n}_J \, dS_i.$$
$$(8.33)$$

An alternative form may be written down by casting the potential energy as a source term:

$$\frac{\partial}{\partial t}\left(\alpha_J \left\langle \rho_J \left(e_J + \frac{1}{2}u_J^2\right)\right\rangle\right) + \boldsymbol{\nabla} \cdot \left[\alpha_J \left\langle \rho_J \left(h_J + \frac{1}{2}u_J^2\right) \mathbf{u}_J\right\rangle\right]$$
$$= \boldsymbol{\nabla} \cdot (\alpha_J \langle \mathbf{u}_J \cdot \boldsymbol{\tau}_J - \mathbf{q}_J\rangle) + \alpha_J \langle \rho_J (Q_J + \mathbf{g} \cdot \mathbf{u})\rangle$$
$$+ \frac{1}{V} \int_{S_i} \left[-\rho_J \left(h_J + \frac{1}{2}u_J^2\right)(\mathbf{u}_J - \mathbf{w}) + \mathbf{u}_J \cdot \boldsymbol{\tau}_J - \mathbf{q}_J\right] \cdot \mathbf{n}_J \, dS_i. \quad (8.34)$$

In applications where internal energy differences are relatively small, while the kinetic and potential energies undergo relatively large changes, such as in a pipeline, it is important to retain all three contributions to the total energy and either one of these equivalent forms may be used as a starting point for a closed model of the energy equation (see, e.g. Bendiksen *et al.*, 1991). In many other applications, however, mechanical energy effects are small and it is preferable to use a balance equation for the internal energy (see Table 8.2). In standard single-phase fluid mechanics, the derivation of this equation involves subtracting from the total energy equation the mechanical energy balance equation obtained by forming the dot product of the momentum equation with the velocity (see, e.g. Landau and Lifshitz, 1987). In the

Table 8.2. Specification of the quantities necessary to adapt the general
balance equation to the various forms of the energy balance; h is the
enthalpy per unit mass, and Φ the dissipation function.

Conserved quantity	f	ϕ	θ
Kinetic energy	$\frac{1}{2}\mathbf{u} \cdot \mathbf{u}$	$\mathbf{u} \cdot \boldsymbol{\sigma}$	$p\boldsymbol{\nabla} \cdot \mathbf{u} - \Phi + \rho \mathbf{g} \cdot \mathbf{u}$
Mechanical energy	$\frac{1}{2}\mathbf{u} \cdot \mathbf{u} - \mathbf{x} \cdot \mathbf{g}$	$\mathbf{u} \cdot \boldsymbol{\sigma}$	$p\boldsymbol{\nabla} \cdot \mathbf{u} - \Phi$
Internal energy	e	$-\mathbf{q}$	$\Phi - p\boldsymbol{\nabla} \cdot \mathbf{u}$
Total energy	$e + \frac{1}{2}\mathbf{u} \cdot \mathbf{u} - \mathbf{x} \cdot \mathbf{g}$	$\mathbf{u} \cdot \boldsymbol{\sigma} - \mathbf{q}$	Q

present two-phase situation, this procedure leads to a somewhat complicated
calculation and it is more efficient to start directly from one of the differential
forms of the internal energy equation written in terms either of the internal
energy itself:

$$\frac{\partial}{\partial t}(\rho e) + \boldsymbol{\nabla} \cdot (\rho e \mathbf{u}) = -\boldsymbol{\nabla} \cdot \mathbf{q} - p\boldsymbol{\nabla} \cdot \mathbf{u} + \Phi + \rho Q \qquad (8.35)$$

which, by (8.32), may also be written as

$$\frac{\partial}{\partial t}(\rho e) + \boldsymbol{\nabla} \cdot (\rho h \mathbf{u}) = -\boldsymbol{\nabla} \cdot \mathbf{q} + \mathbf{u} \cdot \boldsymbol{\nabla} p + \Phi + \rho Q, \qquad (8.36)$$

or in terms of the enthalpy:

$$\frac{\partial}{\partial t}(\rho h) + \boldsymbol{\nabla} \cdot (\rho h \mathbf{u}) = -\boldsymbol{\nabla} \cdot \mathbf{q} + \frac{dp}{dt} + \Phi + \rho Q, \qquad (8.37)$$

in which $dp/dt = \partial p/\partial t + \mathbf{u} \cdot \boldsymbol{\nabla} p$. The corresponding averaged forms are
readily obtained; for example

$$\frac{\partial}{\partial t}\left(\alpha_J \langle \rho_J e_J \rangle\right) + \boldsymbol{\nabla} \cdot [\alpha_J \langle \rho_J h_J \mathbf{u}_J \rangle]$$
$$= -\boldsymbol{\nabla} \cdot (\alpha_J \langle \mathbf{q}_J \rangle) + \alpha_J \langle \mathbf{u}_J \cdot \boldsymbol{\nabla} p_J + \Phi_J + \rho_J Q_J \rangle$$
$$- \frac{1}{V}\int_{S_i} [\rho_J h_J (\mathbf{u}_J - \mathbf{w}) + \mathbf{q}_J] \cdot \mathbf{n}_J \, dS_i \qquad (8.38)$$

and

$$\frac{\partial}{\partial t}\left(\alpha_J \langle \rho_J h_J \rangle\right) + \boldsymbol{\nabla} \cdot [\alpha_J \langle \rho_J h_J \mathbf{u}_J \rangle]$$
$$= -\boldsymbol{\nabla} \cdot (\alpha_J \langle \mathbf{q}_J \rangle) + \alpha_J \left\langle \frac{dp_J}{dt} + \Phi_J + \rho_J Q_J \right\rangle$$
$$- \frac{1}{V}\int_{S_i} [\rho_J h_J (\mathbf{u}_J - \mathbf{w}) + \mathbf{q}_J] \cdot \mathbf{n}_J \, dS_i. \qquad (8.39)$$

Numerically, one could calculate the internal energy as the difference between the total and mechanical energies. This procedure is however dangerous in both very low-speed and supersonic flows where, due especially to modelling deficiencies, it may induce temperature fluctuations.

The closure of the energy equation is the least developed aspect of all the averaged balance equations. A significant difficulty arises from the non-linearities contained in the exact formulations just presented. In addition to the velocity terms, difficulties arise due to the dissipation function and to terms such as $\langle \mathbf{u}_J \cdot \nabla p_J \rangle$ or $\langle dp_J/dt \rangle$. Furthermore, interfacial heat fluxes often develop on thin boundary layers which render problematic the closure of a contribution such as $\mathbf{q}_J \cdot \mathbf{n}_J$ in terms of bulk quantities. Since phase change depends on such heat fluxes, the modeling of evaporation and condensation suffers as well. Thus, by necessity, the models used in practice contain many crude assumptions and a great deal of empiricism, as will be seen later in some examples. A detailed discussion of these issues would be lengthy and out of place in this text. The formal results presented here are sufficient to gain an understanding of the origin of the terms in the specific forms of the energy equations encountered later.

8.2.4 Simplified models

The difficulty of closing the average equations and the complexity of the full averaged formulation make simplified models attractive for some applications. In order to sketch a derivation of the two most common ones we add the general averaged equation (8.7) for phase 1 to that for phase 2 and use the fact that, due to the opposite directions of the normals, conservation of mass at the interface requires that

$$\dot{m} \equiv [\rho_1 (\mathbf{u}_1 - \mathbf{w})] \cdot \mathbf{n}_1 = - [\rho_2 (\mathbf{u}_2 - \mathbf{w})] \cdot \mathbf{n}_2. \tag{8.40}$$

In this way we find

$$\frac{\partial}{\partial t} (\alpha_1 \langle \rho_1 f_1 \rangle + \alpha_2 \langle \rho_2 f_2 \rangle) + \nabla \cdot (\alpha_1 \langle \rho_1 f_1 \mathbf{u}_1 \rangle + \alpha_2 \langle \rho_2 f_2 \mathbf{u}_2 \rangle)$$

$$= \nabla \cdot (\alpha_1 \langle \boldsymbol{\phi}_1 \rangle + \alpha_2 \langle \boldsymbol{\phi}_2 \rangle) + \alpha_1 \langle \rho_1 \theta_1 \rangle + \alpha_2 \langle \rho_2 \theta_2 \rangle$$

$$+ \frac{1}{\mathcal{V}} \int_{S_i} [(f_1 - f_2) \dot{m} - (\boldsymbol{\phi}_1 - \boldsymbol{\phi}_2) \cdot \mathbf{n}_2] \, dS_i. \tag{8.41}$$

In terms of the mean density ρ_m and center-of-mass velocity $\mathbf{u}_m$ defined in (8.18) and (8.19), as applied to mass conservation, this equation reduces to equation (8.17). For momentum, omitting surface-tension effects which would cause a discontinuity in the stress across the interface[1].

$$\frac{\partial}{\partial t}(\rho_m \mathbf{u}_m) + \nabla \cdot (\rho_m \mathbf{u}_m \mathbf{u}_m + \mathbf{J})$$

$$= \nabla \cdot (\alpha_1 \langle \boldsymbol{\sigma}_1 \rangle + \alpha_2 \langle \boldsymbol{\sigma}_2 \rangle) + \rho_m \, \mathbf{g} + \frac{1}{\mathcal{V}} \int_{S_i} (\mathbf{u}_1 - \mathbf{u}_2) \, \dot{m} \, dS_i \qquad (8.42)$$

where $\mathbf{J}$ is the (generalized) "drift flux" defined by

$$\rho_m \, \mathbf{J} = \alpha_1^2 \left(\langle \rho_1 \rangle \langle \rho_1 \mathbf{u}_1 \mathbf{u}_1 \rangle - \langle \rho_1 \mathbf{u}_1 \rangle \langle \rho_1 \mathbf{u}_1 \rangle \right)$$

$$+ \alpha_2^2 \left(\langle \rho_2 \rangle \langle \rho_2 \mathbf{u}_2 \mathbf{u}_2 \rangle - \langle \rho_2 \mathbf{u}_2 \rangle \langle \rho_2 \mathbf{u}_2 \rangle \right)$$

$$+ \alpha_1 \alpha_2 \left(\langle \rho_1 \rangle \langle \rho_2 \mathbf{u}_2 \mathbf{u}_2 \rangle + \langle \rho_2 \rangle \langle \rho_1 \mathbf{u}_1 \mathbf{u}_1 \rangle - 2 \langle \rho_1 \mathbf{u}_1 \rangle \langle \rho_2 \mathbf{u}_2 \rangle \right). \qquad (8.43)$$

In the *homogeneous flow model* (see, e.g. Wallis, 1969; Ishii, 1975; Toumi, 1996) the assumption is made that the two phases move with the same velocity and correlation terms are usually ignored so that $\langle \rho_J \mathbf{u}_J \mathbf{u}_J \rangle \simeq \langle \rho_J \rangle \langle \mathbf{u}_J \rangle \langle \mathbf{u}_J \rangle$, etc. In this case $\mathbf{J} = 0$ and the momentum equation reduces to

$$\frac{\partial}{\partial t}(\rho_m \mathbf{u}_m) + \nabla \cdot (\rho_m \mathbf{u}_m \mathbf{u}_m) = \nabla \cdot \bar{\boldsymbol{\sigma}} + \rho_m \, \mathbf{g} \qquad (8.44)$$

where $\bar{\boldsymbol{\sigma}}$ is an approximation to the combined stress of the phases, e.g. having a Newtonian form with an effective viscosity $\mu_{\text{eff}} = \alpha_1 \mu_1 + \alpha_2 \mu_2$ or similar. The volume fractions enter the model algebraically, either through an effective equation of state (e.g. when the volume of dispersed bubbles is related to the local pressure), or through the assumption of local thermodynamic equilibrium as in a vapor–liquid flow. The homogeneous flow model therefore consists of the combined continuity and momentum equations (8.17) and (8.42), a rule to calculate the volume fractions and, possibly, a combined energy equation.

In the *drift flux model* correlation terms are also usually ignored so that $\mathbf{J}$ simplifies to

$$\rho_m \, \mathbf{J} = \alpha_1 \alpha_2 \left(\langle \mathbf{u}_2 \rangle - \langle \mathbf{u}_1 \rangle \right) \left(\langle \mathbf{u}_2 \rangle - \langle \mathbf{u}_1 \rangle \right). \qquad (8.45)$$

A model is then adopted for the "slip velocity" $\langle \mathbf{u}_2 \rangle - \langle \mathbf{u}_1 \rangle$, for example assuming that particles are always at their terminal velocity with respect to the fluid, or using various correlations (see, e.g. Ishii, 1975; Ishii *et al.*, 1976; Ishii and Zuber, 1979; Hibiki and Ishii, 2002, 2003). This model consists therefore of two continuity equations and a single combined momentum

---

[1] Note however that the total force exerted by surface tension on a closed surface vanishes (Prosperetti and Jones 1984; Hesla *et al.*, 1993). Of course, whenever S_i is not closed, a (generally small) error is introduced with this step.

equation. The model may be completed with a combined energy equation or two energy equations, one for each phase.

Frequently these models are used in their one-dimensional formulation which introduces additional terms in the right-hand side of the momentum and energy equations needed to describe, e.g. drag and heat transfer at the walls of the conduit. These effects are usually modelled in terms of algebraic correlations.

Aside from the reduced computational cost, a great advantage of these simplified formulations lies in the fact that many of the phase interaction terms cancel upon combining the equations for the individual phases, or reduce to more manageable forms. Thus, the closure problem is simplified. Furthermore, the troublesome issue of lack of hyperbolicity, to be discussed in Section 8.4, disappears as the use of a single momentum equation makes the characteristics of the system real and puts the equations in conservation form.

8.3 Disperse flow

Multiphase flows in which particles, drops, or bubbles are suspended in a fluid phase are termed *disperse flows*. The next chapter is devoted to numerical methods for the simplest ones of these flows, fluid–particle systems. The simplification arises because, in many practical cases such as fluidized beds, sediment transport and others, one may neglect (or parameterize) the complex phenomena of coalescence, agglomeration, and fragmentation and focus principally on the momentum interaction between the phases. A further simplification occurs in the case of gas–particle systems due to the great difference between the density of the phases in the majority of situations.

The next chapter describes a Lagrangian–Eulerian method, in which the fluid phase is described in a conventional averaged Eulerian framework, while single (or groups of) particles are followed in a Lagrangian fashion. As explained in detail later, this approach is suitable when both the particle size and their volume fraction is small. Chapters 10 and 11 describe, instead, numerical methods for an Eulerian–Eulerian averaged description, in which both phases are treated as continua, which is appropriate for larger particles and dense systems. These two situations share some common features which can be described within the framework set up in the previous section. Furthermore, the framework itself can be better understood by considering this specific example.

The total hydrodynamic force exerted by the fluid on a particle (or drop or bubble) is given by

$$\hat{\mathbf{f}} = \oint_s \boldsymbol{\sigma} \cdot \mathbf{n} \, ds, \tag{8.46}$$

where **n** is the outward unit normal on the surface s of the particle. Here and in the rest of this section, quantities carrying no subscript refer to the fluid phase.

The force on a body immersed in a steady uniform fluid stream is of course one of the classical problems of fluid mechanics, but here there is the added complexity that the flow surrounding the particle in general possesses a nontrivial spatial and temporal structure both at the microlevel – due, e.g., to the influence of other particles or neighboring boundaries – and at the macrolevel, due e.g., to pressure gradients induced by gravity or flow, or large-scale nonuniformities of the macroscopic flow. For modelling purposes, it is desirable to separate processes occurring at these two very different levels. As a guide to how this might be done, it is useful to start by considering the simple case of a single particle in a fluid subject to a gravitational field. The stress $\boldsymbol{\sigma}$ in (8.46) represents the entire stress in the fluid and, therefore, it includes the pressure field due to gravity. We can separate out this effect by writing $\boldsymbol{\sigma} = -\rho \mathbf{g} \cdot \mathbf{x} + \boldsymbol{\sigma}^r$ to find

$$\hat{\mathbf{f}} = -\rho v \mathbf{g} + \oint_s \boldsymbol{\sigma}^r \cdot \mathbf{n} \, ds, \tag{8.47}$$

where v is the particle volume. At this point, in trying to find an expression for the fluid force on the particle, we can disregard the effect of gravity and focus on the stress due to the local flow, which we have denoted by $\boldsymbol{\sigma}^r$ to emphasize its "rapid" spatial variation – on the particle scale as opposed to the macroscale of the whole system.

A natural extension of this procedure to the general situation is to set $\boldsymbol{\sigma} = \boldsymbol{\sigma}^s + \boldsymbol{\sigma}^r$, where $\boldsymbol{\sigma}^s$ is the "ambient" stress characterized, when the macroscopic length scale is sufficiently large, by a slow spatial variation when judged from the particle scale. The basic assumption here is that the rapidly varying stress $\boldsymbol{\sigma}^r$ would produce approximately the same effect on the particle independently of the structure of the macroflow. In other words, the local interaction between the particle and its surroundings as characterized, for example, by the local mean velocity and particle concentration, would be approximately independent of the large-scale structure of the flow, such as the presence of a macroscopic pressure gradient. If then in (8.46) we expand $\boldsymbol{\sigma}^s$ in a Taylor series around the particle centroid **y**:

$$\boldsymbol{\sigma}^s(\mathbf{y} + \mathbf{r}) \simeq \boldsymbol{\sigma}^s(\mathbf{y}) + (\mathbf{r} \cdot \boldsymbol{\nabla}) \, \boldsymbol{\sigma}^s, \tag{8.48}$$

upon substitution in (8.46), we find[1]

$$\hat{\mathbf{f}} \simeq v \nabla \cdot \boldsymbol{\sigma}^s + \mathbf{f}, \tag{8.49}$$

where the "local" force $\mathbf{f}$ is given by

$$\mathbf{f} = \oint_s \boldsymbol{\sigma}^r \cdot \mathbf{n}\, ds. \tag{8.50}$$

The decomposition (8.49) presents a clear similarity with (8.25) and helps understand that definition of the interphase force. By analogy with (8.47), we may refer to the first term in (8.49) as the "generalized buoyancy" force.

Before continuing with an analysis of the local force $\mathbf{f}$, let us return to the interfacial stress acting on the continuous phase according to Eq. (8.26):

$$\mathbf{F} = -\frac{1}{V} \int_{S_i} [\boldsymbol{\sigma} - \langle \boldsymbol{\sigma} \rangle(\mathbf{x}_c)] \cdot \mathbf{n} dS_i . \tag{8.51}$$

The minus sign accounts for the fact that, in the present discussion, the normal is out of the disperse phase and therefore into the continuous phase, and we have written $\langle \boldsymbol{\sigma} \rangle(\mathbf{x}_c)$ to emphasize the fact that this term is evaluated at the center of the averaging volume as is evident from the steps leading to (8.26). In an averaging context it is reasonable to identify the slowly-varying component of the stress $\boldsymbol{\sigma}^s$ with the volume-average stress $\langle \boldsymbol{\sigma} \rangle$ and therefore we may write

$$\mathbf{F} = -\frac{1}{V} \int_{S_i} [\boldsymbol{\sigma}^r + \langle \boldsymbol{\sigma} \rangle - \langle \boldsymbol{\sigma} \rangle(\mathbf{x}_c)] \cdot \mathbf{n} dS_i . \tag{8.52}$$

The last two terms do not precisely cancel as the first one is evaluated at the center of each one of the particles contributing to the total interface S_i, while the second one is evaluated at the center of the averaging volume. Ignoring this very minor difference, we find

$$\mathbf{F} \simeq -\frac{1}{V} \sum_\alpha \mathbf{f}^\alpha = -n\overline{\mathbf{f}}, \tag{8.53}$$

in which the sum is over all the particles in the averaging volume, n is the particle number density and $\overline{\mathbf{f}}$ the mean force per particle. Here we have neglected the fact that the surface of the averaging volume may cut through some particles; a justification will be given presently.

---

[1] Some authors and many commercial CFD codes adopt a procedure which is essentially equivalent to expanding in this way the pressure, rather than the total stress. Although this procedure seems to lack justification, particularly in the light of equation (8.64) below and the related discussion, the attendant error is probably minor in most cases.

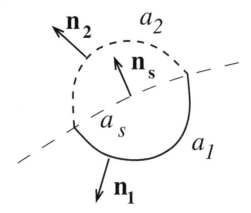

Fig. 8.2. A particle cut across by the surface of the averaging volume.

In considering the stress contribution on the particle phase, it is more convenient to go back to the relevant terms in the balance equation (8.2), namely

$$\boldsymbol{\Sigma}_S = \frac{1}{\mathcal{V}} \left(\int_{S_S} \boldsymbol{\sigma}_S \cdot \mathbf{n}\, dS + \int_{S_i} \boldsymbol{\sigma}_S \cdot \mathbf{n}\, dS_i \right), \tag{8.54}$$

where $\boldsymbol{\sigma}_S$ is the stress in the solid phase and the unit normal is out of the particles on the interfaces S_i and out of the averaging volume on the surface S_S of the latter. At the surface of a particle, the normal component of the stress is continuous and, therefore, $\boldsymbol{\sigma}_S \cdot \mathbf{n} = \boldsymbol{\sigma} \cdot \mathbf{n}$. The contribution of a particle entirely inside the averaging volume is, therefore, the total hydrodynamic force on the particle[1].

A particle which is cut through by the surface of the averaging volume (Fig. 8.2) gives to $\boldsymbol{\Sigma}_S$ a contribution from the stress at its surface through the second integral, and one from its internal stress through the first one:

$$\int_{a_1} \boldsymbol{\sigma}_S \cdot \mathbf{n}_1\, da_1 + \int_{a_s} \boldsymbol{\sigma}_S \cdot \mathbf{n}_S\, da_S = \int_{a_1} \boldsymbol{\sigma} \cdot \mathbf{n}_1\, da_1 + \int_{a_s} \boldsymbol{\sigma}_S \cdot \mathbf{n}_S\, da_S. \tag{8.55}$$

---

[1] The same statement is also applicable to a bubble or drop when the surface tension coefficient is constant, in spite of the fact that in this case there is a difference between the two normal stresses induced by the capillary pressure. Indeed, it can be shown that the total capillary force on a closed surface vanishes irrespective of the shape; see footnote 1 on p. 250.

This situation lends itself to detailed and subtle analyses (see, e.g. Nigmatulin, 1979; Prosperetti and Jones, 1984; Hwang and Shen, 1989; Drew and Lahey, 1993; Prosperetti and Zhang, 1996; Jackson, 2000) which would be out of place here. We will assume that, in this case, the contribution of the particle to (8.54) is simply $(v_1/v)\hat{\mathbf{f}}$, where v_1 is the portion of the particle volume inside the averaging volume. This assumption can be justified by considering the momentum equation for the two parts, of mass m_1 and m_2, into which the particle is cut:

$$m_1 (\mathbf{a}_1 - \mathbf{g}) = \int_{s_1} \boldsymbol{\sigma} \cdot \mathbf{n}_1 \, ds_1 + \int_{s_s} \boldsymbol{\sigma}_s \cdot \mathbf{n}_s \, ds_s \qquad (8.56)$$

and

$$m_2 (\mathbf{a}_1 - \mathbf{g}) = \int_{s_2} \boldsymbol{\sigma} \cdot \mathbf{n}_2 \, ds_2 - \int_{s_s} \boldsymbol{\sigma}_s \cdot \mathbf{n}_s \, ds_s. \qquad (8.57)$$

Upon adding we find

$$(m_1 + m_2) (\mathbf{a} - \mathbf{g}) = \int_{s_1 + s_2} \boldsymbol{\sigma} \cdot \mathbf{n} \, da = \hat{\mathbf{f}} \qquad (8.58)$$

where $(m_1 + m_2) \mathbf{a} = m_1 \mathbf{a}_1 + m_2 \mathbf{a}_2$. If particle rotation is neglected, $\mathbf{a}_1 \simeq \mathbf{a}_2 \simeq \mathbf{a}$ and, therefore,

$$m_1 (\mathbf{a}_1 - \mathbf{g}) \simeq m_1 (\mathbf{a} - \mathbf{g}) \simeq \frac{m_1}{m_1 + m_2} \hat{\mathbf{f}} = \frac{v_1}{v} \hat{\mathbf{f}}. \qquad (8.59)$$

With these developments, (8.54) becomes

$$\boldsymbol{\Sigma}_S = \frac{1}{V} \sum \hat{\mathbf{f}} = \frac{N}{V} \overline{\hat{\mathbf{f}}}, \qquad (8.60)$$

where it is understood that $\hat{\mathbf{f}}$ is weighted by the fraction of volume inside the averaging volume for the incomplete particles, N is the number of particles in the averaging volume, and $\overline{\hat{\mathbf{f}}}$ is the average force per particle. Thus, upon introducing the number density and the decomposition (8.49) of the force, we find

$$\boldsymbol{\Sigma}_S = n\overline{\hat{\mathbf{f}}} = n \left(v \boldsymbol{\nabla} \cdot \langle \boldsymbol{\sigma} \rangle + \overline{\mathbf{f}} \right), \qquad (8.61)$$

or, in terms of the disperse phase volume fraction,

$$\boldsymbol{\Sigma}_S = \alpha_D \boldsymbol{\nabla} \cdot \langle \boldsymbol{\sigma} \rangle + n\overline{\mathbf{f}}. \qquad (8.62)$$

Depending on circumstances, this expression for the average stress on the particle phase may need to be augmented by the collisional contribution $\boldsymbol{\nabla} \cdot (\alpha_D \boldsymbol{\sigma}_c)$ as noted before.

It is now necessary to model the local force on the particles. In classical fluid mechanics, the hydrodynamic force on an object immersed in a steady flow is usually decomposed into a drag $\mathbf{f}_d$ and a lift $\mathbf{f}_l$ component parallel and perpendicular, respectively, to the undisturbed streamlines. In the presence of unsteadiness, inviscid-flow theory shows that there is an additional effect, usually referred to as added, or virtual, mass $\mathbf{f}_a$, due to the inertia of the fluid which comes into play in the presence of a relative acceleration between the particle and the fluid. Low Reynolds number theory points to the existence of a history, or memory, force component $\mathbf{f}_h$, which is due to the convection and diffusion of vorticity in the object's wake. In a general flow situation, these effects act simultaneously and we write

$$\mathbf{f} = \mathbf{f}_d + \mathbf{f}_a + \mathbf{f}_h + \mathbf{f}_l. \qquad (8.63)$$

Since the various terms in this relation are ultimately all due to the fluid stress on the particle surface, and since the Navier–Stokes equations are nonlinear, it is not obvious that the various contributions simply combine additively, although this assumption is invariably made in the literature. Additivity can, however, be proven in the case of very low or very high Reynolds numbers. In the former limit, for a rigid sphere of mass m_p, radius a, volume v, and velocity $\mathbf{v}$ immersed in a nonuniform and time-dependent ambient flow, Maxey and Riley (1983) and Gatignol (1983) derived an equation of motion of the form[1]:

$$m_p \frac{d\mathbf{v}}{dt} = m_p \mathbf{g} + \rho v \left(\frac{D\mathbf{u}}{Dt} - \mathbf{g} \right) + \mathbf{f}_d + \mathbf{f}_a + \mathbf{f}_h, \qquad (8.64)$$

in which

$$\mathbf{f}_d = 6\pi a \mu \left(\mathbf{u} - \mathbf{v} + \frac{a^2}{6} \nabla^2 \mathbf{u} \right) \qquad (8.65)$$

is the Stokes drag on a sphere as corrected by Faxén for the effect of nonuniformities of the flow field,

$$\mathbf{f}_a = \frac{1}{2} \rho v \left[\frac{D}{Dt} \left(\mathbf{u} + \frac{1}{10} a^2 \nabla^2 \mathbf{u} \right) - \frac{d\mathbf{v}}{dt} \right] \qquad (8.66)$$

is the added mass force, and

$$\mathbf{f}_h = 6a^2 \rho \sqrt{\pi \nu} \int_0^t \frac{dt'}{(t-t')^{1/2}} \left[\frac{D}{Dt'} \left(\mathbf{u} + \frac{1}{6} a^2 \nabla^2 \mathbf{u} \right) - \frac{d\mathbf{v}}{dt'} \right] \qquad (8.67)$$

---

[1] The development of this equation builds on the earlier contributions by Boussinesq (1885), Basset (1888), Tchen (1947), Corrsin and Lumley (1956), among others. More recent analyses and refinements of their results have been given by Lovalenti and Brady (1993a, b, c), Asmolov (2002), and others; a review is given by Michaelides (1997); see also the book by Crowe *et al.* (1998).

is Basset's memory force. The terms proportional to $\nabla^2 \mathbf{u}$ in (8.66) and (8.67) are corrections to the classical expressions (see, e.g. Landau and Lifshitz, 1987) approximately accounting for the finite extent of the particle. These results are justified for small values of the particle Reynolds number

$$Re_p = \frac{2a|\mathbf{u} - \mathbf{v}|}{\nu}, \tag{8.68}$$

in which limit the lift force vanishes identically and, therefore, is not present in the particle equation of motion (8.64); furthermore, $D/Dt = \partial/\partial t + \mathbf{u} \cdot \nabla$, while d/dt denotes the time derivative following the particle[1].

In the derivation, it is assumed that the particle diameter is much smaller than the smallest relevant length scale L of the ambient flow. In other words, the conditions hypothesized are such that the fluid velocity remains nearly uniform over a region of several particle diameters. In this sense, we may refer to the field $\mathbf{u}$ as the ambient, or background, flow and think of it as the fluid velocity evaluated at the position occupied by the particle center in the hypothetical situation in which the particle were absent.

The result for the hydrodynamic force in (8.64) supports the breakup (8.49) of this quantity into a virtual buoyancy and a local force. Indeed, since the fluid velocity $\mathbf{u}$ satisfies the Navier–Stokes equations, the second term in the right-hand side of (8.64) precisely equals $v\nabla \cdot \boldsymbol{\sigma}$ and corresponds therefore to the virtual buoyancy term in (8.49). Furthermore, the Faxén-type corrections appearing in (8.65), (8.66), and (8.67) are of order $(a/L)^2$, and are therefore small when the assumptions leading to (8.64) are justified. This fact supports the earlier statement that the local force would be expected to be approximately independent of the large-scale structure of the flow.

The lowest order inertial correction to the Stokes drag law in a uniform flow was found by Oseen (1927), who showed that

$$\mathbf{f}_d = 6\pi\mu a\,(\mathbf{u} - \mathbf{v})\left(1 + \frac{3}{16}Re_p\right), \tag{8.69}$$

where Re_p is the Reynolds number of the relative motion defined in (8.68). Several more terms have been calculated by the method of matched asymptotic expansion (Proudman and Pearson, 1957), but convergence is slow and the resulting expression of limited utility. Much of what is known in the more common regime of intermediate Reynolds numbers is based on experimental and numerical studies. In practice, the drag force is expressed

---

[1] In the low Reynolds number limit the two derivatives D/Dt and d/dt are indistinguishable. In writing these equations we have anticipated the high Reynolds number results which show the proper fluid acceleration to be given by $D\mathbf{u}/Dt$.

in terms of a drag coefficient C_d defined by

$$\mathbf{f}_d = \frac{1}{2}\rho A C_d |\mathbf{u} - \mathbf{v}| (\mathbf{u} - \mathbf{v}), \tag{8.70}$$

where A is the particle frontal area, equal to πa^2 for a sphere of radius a. In the multiphase flow literature, a widely used expression for C_d is that due to Schiller and Naumann (see, e.g. Clift *et al.*, 1978):

$$C_d = \frac{24}{Re_p} \left(1 + 0.15 Re_p^{0.687}\right) \tag{8.71}$$

which accurately describes the drag on a sphere up to a Reynolds number of about 1000. The first term $24/Re_p$ reproduces the Stokes formula.

Lovalenti and Brady (1993a) have shown that the form of the history force given in equation (8.67) is only valid for times shorter than $\nu/|\mathbf{u}-\mathbf{v}|^2$ so that vorticity does not have time to diffuse out to distances of the order of the Oseen scale a/Re_p. In general the expression is more complex (Mei, 1994; Lawrence and Mei, 1995; Lovalenti and Brady, 1993b, 1993c, 1995). The simple expression in (8.67) is also strongly modified when the particle shape is not spherical (Lawrence and Weinbaum, 1988). It will be appreciated that, numerically, the history force presents some difficulties due to the need to store the relative acceleration from the initial up to the current time. Furthermore, the particle acceleration at the current time appears both in the left-hand side of the particle momentum equation and in the integral as well. A simple way to deal with the latter problem is to use a quadrature formula over the last time step which includes the current time as one of its nodes. In this way, the current acceleration can be moved to the left-hand side and combined with the inertia term. Another approach for the computation of this term has been developed by Michaelides (1992).

The opposite case of very high Reynolds number (inviscid flow) also lends itself to analysis. When the variation of the free-stream flow over the particle radius is much smaller than the particle–fluid relative velocity, Auton *et al.* (1988; see also Auton, 1987) showed explicitly that the contributions of the lift, added mass, and effective buoyancy forces combine additively. They give an expression for the fluid force on a sphere which implies an equation of motion of the form

$$m_p \frac{d\mathbf{v}}{dt} = m_p \mathbf{g} + \rho v \left(\frac{D\mathbf{u}}{Dt} - \mathbf{g}\right) + \frac{1}{2}\rho v \left(\frac{D\mathbf{u}}{Dt} - \frac{d\mathbf{v}}{dt}\right)$$

$$+ \frac{1}{2} v \rho \left(\nabla \times \mathbf{u}\right) \times \left(\mathbf{v} - \mathbf{u}\right). \tag{8.72}$$

Note that here again the fluid velocity **u** is to be understood as the ambient velocity and to be evaluated at the particle center. Since this result has been derived for an inviscid flow, no drag term appears. The third term in the right-hand side is the added mass force, with the same form as in (8.66) except for the finite-size correction. The applicability of this expression to intermediate Reynolds numbers and more complicated flows has been established by numerical means by Magnaudet *et al.* (1995) and Bagchi and Balachandar (2003b).

The last term in (8.72) is the high Reynolds number expression for the lift force. In the opposite limit of low Reynolds number, for particles with fore–aft symmetry, it vanishes due to the time-reversal invariance of the Stokes equations (Bretherton, 1962). Saffman (1965, 1968) calculated it at the next order in Re_p for a sphere in a unidirectional simple shear flow with the result

$$f_l \simeq 6.46\mu(v - u_x)a^2 \sqrt{\frac{1}{\nu} \left| \frac{\partial u_x}{\partial z} \right|}. \qquad (8.73)$$

Here the fluid and the particle move steadily in the x-direction with $u_x = u_x(z)$ and the force is in the direction of increasing u_x. Some limitations of Saffman's early work (absence of walls and of streamline curvature, small shear Reynolds number $a^2|\partial u_x/\partial z|/\nu$, and others) were removed by later work (e.g. McLaughlin, 1991, 1993; Asmolov and McLaughlin, 1999; Asmolov, 1999; Magnaudet, 2003).

The proper form of the lift force at intermediate Reynolds numbers is much less established than that for drag or added mass. In the literature, models in which the Saffman form is adopted way beyond any reasonable domain of validity are not uncommon. The available experimental evidence (Naciri, 1992; Ganapathy and Katz, 1995) suggests that both (8.73) and the form implicit in (8.72) give a poor quantitative representation of this force. Furthermore, it is now established that the same amount of local vorticity in a parallel or in a curved flow results in different lift forces (Bagchi and Balachandar, 2002; Magnaudet, 2003). Recent studies of this force are the papers by Catlin (2003) and Bluemink *et al* (2008).

In the case of many interacting particles, the previous expressions need to be corrected. A relation for $\mathbf{F}_d$, the drag force per unit volume acting on the particle phase, is built on the previous single-particle form by writing[1]

$$\mathbf{F}_d = n\alpha_F^{-K}\mathbf{f}_d, \qquad (8.74)$$

---

[1] The accuracy of such a representation has been questioned in the literature and it has been suggested that a drift flux correction should be added to the velocity difference in equation (8.70); for example, see Balzer *et al.* (1995) and Simonin (1996). This drift flux correction can affect the drag force appreciably in dilute particle-laden turbulent flows, but its influence on more concentrated suspensions is not yet entirely clear.

where the factor α_F^{-K}, in which α_F is the fluid volume fraction, is an approximation to the so-called hindrance function accounting for the fact that, in a particle assembly, the mean drag per particle is greater than for an isolated particle. In this context, $\mathbf{u}$ and $\mathbf{v}$ in (8.70) are interpreted as the average fluid and particle velocities. On empirical grounds, Wen and Yu (1966) use the Schiller–Naumann formula (8.71) to obtain $\mathbf{f}_\mathrm{d}$ from (8.70) with $K = 1.65$, but they calculate the single-particle Reynolds number in terms of the difference between the mean particle velocity and the volumetric flux $\mathbf{u}_V$ defined in (8.23). Thus, since $\mathbf{u}_V - \mathbf{v} = \alpha_F (\mathbf{u} - \mathbf{v})$, in place of (8.68), they use

$$Re_p = \frac{2a\alpha_F |\mathbf{u} - \mathbf{v}|}{\nu}. \tag{8.75}$$

Another parameterization of the drag force is based on the observation of Richardson and Zaki (1954) that their measurements of the average sedimentation velocity $\mathbf{v}_\mathrm{sedim}$ could be correlated by

$$\mathbf{v}_\mathrm{sedim} = \alpha_F^N \mathbf{v}_t \tag{8.76}$$

where $\mathbf{v}_t$ is the terminal velocity of an isolated particle and the exponent N depends on the single-particle Reynolds number at terminal velocity $Re_t = 2av_t/\nu$:

$$N = \begin{cases} 4.65 & \text{for } Re_t < 0.2 \\ 4.35 Re_t^{-0.03} & \text{for } 0.2 < Re_t < 1 \\ 4.45 Re_t^{-0.1} & \text{for } 1 < Re_t < 500 \\ 2.39 & \text{for } 500 < Re_t \end{cases} \tag{8.77}$$

This procedure is equivalent to taking $K = 2N$ in (8.74), with a drag coefficient implicitly defined by

$$\frac{1}{2}\pi a^2 \rho_F C_\mathrm{d} v_t^2 = v(\rho_p - \rho_F) g, \tag{8.78}$$

which expresses the balance of drag, gravity, and buoyancy forces for an isolated particle. It may also be observed that, in a fluid–particle suspension, the velocity of an individual particle will, in general, be different from the mean particle velocity. The difference between these two quantities is a measure of the intensity of the fluctuating motion of the particles. In general, one would expect the drag coefficient to depend on these velocity fluctuations, but only *ad hoc* corrections for this effect exist (for example, see Zhang and Reese, 2003), which have not been validated either experimentally or computationally[1].

---

[1] For a summary of the literature on drag coefficient correlations, see Li and Kuipers (2003) and, for some recent computational results of flow through a stationary bed, see Hill *et al.* (2001a, b) and Hill and Koch (2002).

For many particles, the added mass force is usually written as

$$\mathbf{F}_a = \frac{1}{2}C_a \alpha_D \alpha_F \rho_F \left[\frac{\partial \mathbf{u}}{\partial t} + \mathbf{u} \cdot \nabla \mathbf{u} - \left(\frac{\partial \mathbf{v}}{\partial t} + \mathbf{v} \cdot \nabla \mathbf{v} \right) \right], \qquad (8.79)$$

where C_a is the added mass coefficient which reduces to the single-sphere potential flow result 1 for $\alpha_D \to 0$. Multiplication of the single-particle result by $\alpha_D \simeq nv$ is equivalent to summing over the particles present in the unit volume. The additional factor α_F can be understood if the fluid velocity in the single-particle expression is replaced by the volume flux $\mathbf{u}_V$, and the derivatives of α_F that arise with this procedure (and which, very likely, have only a small effect) are dropped. The dependence of C_a on volume fraction has been well studied for potential flow both in the dilute (van Wijngaarden, 1976) and dense cases (Biesheuvel and Spoelstra, 1989; Sangani *et al.*, 1991; Zhang and Prosperetti, 1994) and, to some extent, also for finite Reynolds numbers (ten Cate and Sundaresan, 2005). In summary, one may state that the approximation $C_a \simeq 1$ is adequate[1].

The lift force is treated in a similar way by writing

$$\mathbf{F}_l = C_l \rho \alpha_D \alpha_F \left(\nabla \times \mathbf{u} \right) \times (\mathbf{u} - \mathbf{v}), \qquad (8.80)$$

where C_l is the lift coefficient. When $\alpha_D \simeq nv$ is small, $\alpha_F \simeq 1$, and this form reduces to $n\mathbf{f}_l$, with $\mathbf{f}_l$ as in (8.72), provided $C_l = \frac{1}{2}$. Presumably, for larger particle concentrations, C_l will depend on the volume fraction and possibly Re_p, but this dependence has not been studied in detail.

Sangani *et al.* (1991) have calculated numerically the coefficient of the history force in an oscillatory flow around many particles finding approximately a factor α_F^{-2} multiplying the last term in (8.64). In view of the numerical problems posed by the evaluation of this contribution to the force, this term is frequently neglected and there are estimates of its magnitude which often justify this approximation (Thomas, 1992; Vojir and Michaelides, 1994). On the other hand, recent simulations of oscillatory fluid flow through stationary assemblies of particles (ten Cate and Sundaresan, 2005) suggest that at oscillation frequencies typical of fluctuations seen in unstable liquid fluidized beds, the history force can be as large as the added mass force. Thus, the rationale for retaining the added mass force and neglecting the history force is not necessarily obvious.

---

[1] There is an often quoted result of Zuber (1964) according to which the added mass coefficient can be approximated as $(1 + 2\alpha_D)/(1 - \alpha_D)$. The added mass coefficient to which this result refers is different from the one appearing in (8.79) and it can actually be shown that taking $C_a = 1$ is equivalent to using Zuber's formula (Zhang and Prosperetti, 1994).

As explained in the next chapter, for typical gas–particle flows the lift, added mass, and history forces are often dwarfed by the gravitational and drag forces and, therefore, can very frequently be ignored. Indeed, due to the typical factor of the order of 1000 by which the densities of the two phases differ, the relative velocity tends to be large and the relative acceleration much smaller than the acceleration of gravity. As for the lift force, the curl operation introduces the inverse of the length L characteristic of the macroscopic (average) flow, which makes $|\mathbf{f}_l| \simeq O(a/L)\, |\mathbf{f}_d|$.

In addition to the force, a full closure of the volume-averaged equations for disperse flows requires expressions for the average stresses. There are substantial differences in the form of the closure relations adopted in the point-particle, Lagrangian–Eulerian and in the Eulerian–Eulerian models and, for this reason, they will be described in connection with these topics in the chapters that follow.

8.4 Well-posedness of Eulerian–Eulerian models

We now turn to a topic different from the ones treated so far in this chapter and somewhat controversial, namely the mathematical structure of the Eulerian–Eulerian models and, in particular, their *well-posedness*. The notion of well-posedness was introduced by Hadamard (see, e.g. Zauderer, 1989; Renardy and Rogers, 1993; Garabedian, 1998) in his analysis of the nature of mathematical modeling. According to Hadamard's classification, well-posedness requires the existence and uniqueness of the solution of the model and also what he termed *continuous dependence on the data*. This requirement is motivated by the fact that any model is bound to contain approximations and errors, be they in the initial or boundary conditions, or in the equations themselves. In order for the model predictions to be meaningful, it is necessary to ensure that small changes in the data lead to correspondingly small changes in the solution. For systems of partial differential equations, continuous dependence on the data is intimately related to *hyperbolicity* of the equations (to be defined presently), as the solution of the initial value problem for nonhyperbolic systems fails to depend continuously on the initial conditions. It is very unfortunate that, as it happens, the vast majority of averaged equations models in current use are not hyperbolic.

8.4.1 A simple example

To illustrate these issues, let us start by considering two simple examples of systems of linear, constant-coefficient partial differential equations in two variables:

$$\frac{\partial U}{\partial t} - \frac{\partial V}{\partial x} = -2bU, \qquad \frac{\partial V}{\partial t} \mp \frac{\partial U}{\partial x} = -2cV, \qquad (8.81)$$

or, more compactly,

$$\mathsf{A} \cdot \frac{\partial \mathbf{U}}{\partial t} + \mathsf{B} \cdot \frac{\partial \mathbf{U}}{\partial x} = \mathsf{D} \cdot \mathbf{U} \qquad (8.82)$$

where $\mathbf{U}^{\mathrm{T}} = (U, V)$, the dot is the usual row-by-column product of matrix theory and

$$\mathsf{A} = \begin{vmatrix} 1 & 0 \\ 0 & 1 \end{vmatrix}, \qquad \mathsf{B} = \begin{vmatrix} 0 & -1 \\ \mp 1 & 0 \end{vmatrix}, \qquad \mathsf{D} = -2 \begin{vmatrix} b & 0 \\ 0 & c \end{vmatrix}. \qquad (8.83)$$

Here b and c are nonnegative constants. Upon eliminating, for example, V between the two equations (8.81), we find

$$\frac{\partial^2 U}{\partial t^2} \mp \frac{\partial^2 U}{\partial x^2} + 2(b+c)\frac{\partial U}{\partial t} + 2bU = 0, \qquad (8.84)$$

with an analogous equation for V. For $b = c = 0$, with the upper sign, this is the wave equation, and the Laplace equation with the lower sign. (In this latter case, in most applications, the variable t would be a spatial variable.) For simplicity, we refer to these two cases as the wave and Laplace equations even when b and c are nonzero.

Let $\mathbf{U}(x, t)$ be a solution of the initial value problem corresponding to a certain initial condition $\mathbf{U}(x, 0)$. In order to study the dependence of this solution on the initial data, we consider another solution $\mathbf{U}'(x, t)$ corresponding to slightly modified initial data $\mathbf{U}'(x, 0)$. Loosely speaking, the solution $\mathbf{U}$ will depend continuously on the data if the perturbation $\mathbf{u} = \mathbf{U}' - \mathbf{U}$ remains small provided $\mathbf{u}(x, 0) = \mathbf{U}'(x, 0) - \mathbf{U}(x, 0)$ is small. Since the problem is linear, we can expand $\mathbf{u}(x, 0)$ in a Fourier series or integral, separately evolve each mode in time, and then recombine them to reconstruct $\mathbf{u}(x, t)$. Thus, without loss of generality, we may consider a perturbation of the initial data of the special form

$$\mathbf{u}(x, 0) = \mathbf{u}_0 \exp(ikx), \qquad (8.85)$$

where $\mathbf{u}_0^{\mathrm{T}} = (u_0, v_0)$ is a constant vector and k the (real) wavenumber. The modified solution due to more general initial data can be found by superposition of these elementary solutions. Note that assigning $\mathbf{u}(x, 0)$ is equivalent to assigning Cauchy data to equation (8.84) as, from the second of (8.81), $[\partial u/\partial t]_{t=0} = (ikv_0 - 2bu_0) \exp ikx$. The linearity of the problem implies that $\mathbf{u}$ satisfies the same equation (8.82) as $\mathbf{U}$ and, since the coefficients are

constant, we look for solutions in the form of linear combinations of terms
of the type

$$\mathbf{w}_0 \exp\left[ik(x - \lambda t)\right] \tag{8.86}$$

with $\mathbf{w}_0$ constant vectors determined by the initial conditions and the parameter λ adjusted so that the differential equation is satisfied.

Upon substitution into (8.83) and division by k, we find a homogeneous
linear system in the components of $\mathbf{w}_0$ the solvability condition for which is

$$\det\left(-\lambda\mathsf{A} + \mathsf{B} - \frac{1}{ik}\mathsf{D}\right) = 0 \tag{8.87}$$

or, explicitly,

$$\lambda^2 + 2i\frac{b+c}{k}\lambda - \frac{4bc}{k^2} \mp 1 = 0. \tag{8.88}$$

This equation determines two possible values λ_1 and λ_2; for simplicity, we
assume that b and c are such that $\lambda_1 \neq \lambda_2$. $\mathbf{u}(x,t)$ will then be the sum of
two terms of the form (8.86) with the amplitudes $\mathbf{w}_{01}$ and $\mathbf{w}_{02}$ given by linear
combinations of u_0 and v_0. The explicit forms are easily derived by imposing
the initial condition (8.85) but they are not needed here, the only important
point being that both $\mathbf{w}_{01}$ and $\mathbf{w}_{02}$ will be "small" provided $\mathbf{u}_0$ is.

It is evident from (8.86) that the perturbation will grow indefinitely if
$\mathrm{Im}\lambda > 0$ for $k > 0$ or $\mathrm{Im}\lambda < 0$ for $k < 0$. By using general properties of
the roots of algebraic equations, it can be shown that boundedness of the
solution requires

$$b + c \geq 0, \qquad 4bc \pm k^2 \geq 0. \tag{8.89}$$

The first condition is satisfied as we have assumed that both b and c are
nonnegative. The second condition will be satisfied for any k for the wave
equation (upper sign), but only for sufficiently small k for the Laplace equation. These conclusions can be verified directly on the explicit expression
for the roots of (8.87):

$$\lambda_1 = -i\frac{b+c}{k} + \sqrt{-\left(\frac{b-c}{k}\right)^2 \pm 1}, \qquad \lambda_2 = -i\frac{b+c}{k} - \sqrt{-\left(\frac{b-c}{k}\right)^2 \pm 1}. \tag{8.90}$$

As $k \to \infty$, these solutions reduce to $\lambda_1 = \sqrt{\pm 1}$, $\lambda_2 = -\sqrt{\pm 1}$ which are
the solutions of

$$\det\left(-\lambda\mathsf{A} + \mathsf{B}\right) = 0. \tag{8.91}$$

For the upper sign (wave equation), $\lambda_{1,2} = \pm 1$ and the perturbation remains bounded for all t. However, for the lower sign (Laplace equation),

$\lambda_{1,2} = \pm i$ so that $\exp[ik(x - \lambda t)] = \exp[k(ix \pm t)]$ and, for a fixed t, one of the exponentials becomes larger and larger as $|k|$ is increased. It is clear that increasing $|k|$ does not change the "magnitude" of the initial perturbation, which is only determined by $|\mathbf{u}_0|$, but has a dramatic effect on its subsequent evolution. Thus, if (8.82) were a mathematical model of a physical system, we would conclude that two slightly different initial data (i.e. a small $\mathbf{u}_0$) could result in predictions of the system behavior differing by an arbitrarily large amount depending on the initial wavenumber content of the perturbation. The model would then fail the test of continuous dependence on the data. It should be stressed that the key point here is not the exponential growth of the solution with t, but rather that, *for a given* t, small initial perturbations of the data can produce arbitrarily large effects if their Fourier expansion contains sufficiently large wavenumbers[1].

It is clear from (8.89) that instability for the Laplace case is only encountered for short wavelengths. Indeed, for $k \to 0$, the roots (8.90) are

$$\lambda_1 \simeq -\frac{2ib}{k}, \qquad \lambda_2 \simeq -\frac{2ic}{k} \qquad (8.92)$$

for both the wave and the Laplace equations. Thus, provided b and c are both positive, long wavelengths will be stable even for the Laplace equation[2]. If the Fourier spectrum of $\mathbf{u}(x, 0)$ only contained components with a sufficiently long wavelength, there would be no unbounded growth of $\mathbf{u}(x, t)$.

Equation (8.91) is the *characteristic equation* for the system (8.82) and the values of λ that it determines are its *characteristic directions* or *characteristics*. As shown by (8.86), the lines $dx/dt = \lambda$ give the directions in the (x, t) plane along which high-k initial data propagate. A general system for which all the roots of (8.91) are real and distinct, as for the wave equation considered here, is called *strictly hyperbolic*[3]. Since the characteristic equation is real, complex roots occur in complex conjugate pairs which implies that, in the nonhyperbolic case, one of the exponential terms (8.86) will always grow unboundedly for $k \to \infty$.

The previous example illustrates the explosive growth of the solution of a linear nonhyperbolic system for short waves, i.e. $k \to \infty$, and the consequent

---

[1] One may contrast this situation with that encountered in the theory of chaotic systems. It is well known that such systems exhibit the striking feature that solutions corresponding to two neighboring initial conditions evolve away from each other exponentially fast. However, for a given fixed t, the difference can be made arbitrarily small by choosing the initial data close enough. Here the situation is different: for a given t and $\mathbf{u}_0$, $\mathbf{u}(x, t)$ can still be made arbitrarily large by increasing $|k|$.

[2] Conversely, if one or both of b and c were negative, the solution of the wave equation will be unstable.

[3] The case in which the characteristic equation has repeated roots can be considerably more complicated; see, e.g. LeVeque (2002) for an introduction.

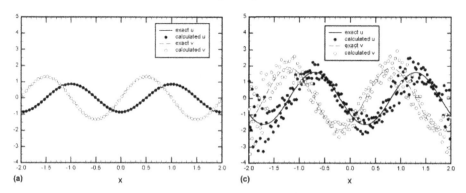

Fig. 8.3.   Numerical solution of the linear nonhyperbolic system (8.82) with B
given by (8.93) at $t = 0.5$ (left) and 0.8 compared with the exact solution (solid
lines); • u component, ○ v component (from Hwang, 2003).

failure to depend continuously on the initial data. It also illustrates that
the behavior of moderate or long waves may be different: depending on the
values of the parameters in the equations, the solution of a nonhyperbolic
system could be stable (and that of a hyperbolic one unstable). In other
words, *in general*, lack of hyperbolicity causes *only short wavelengths* to be
unstable.

An example of the numerical consequences of lack of hyperbolicity in the
linear case is shown in Fig. 8.3, in which the two panels depict at $t = 0.5$
and $t = 0.8$ the numerical solution of a system of the form (8.82) with the
same A, D $= 0$, and

$$\mathsf{B} = \begin{vmatrix} \lambda_r & \lambda_i \\ -\lambda_i & \lambda_r \end{vmatrix}, \tag{8.93}$$

with $\lambda_r = 1$, $\lambda_i = 0.5$ (Hwang, 2003). As implied by the notation, the
characteristic roots for this system are given by $\lambda_r \pm i\lambda_i$. With the initial
condition $u(x,0) = 0$, $v(x,0) = \cos kx$ and $k = \pi$, the exact solution (solid
lines) is $u = \sinh(k\lambda_i t) \sin[(k(x - \lambda_r t)]$, $v = \cosh(k\lambda_i t) \cos[(k(x - \lambda_r t)]$. It
is seen here that, while the numerical solution agrees well with the analytic
one at the earlier time, noticeable errors later develop. Their origin lies in
the fact that, although the initial data only contains a long wave, numerical errors excite short-wave components which then amplify the faster the
shorter their wavelength. At $t = 0.8$, these short-wave components have had
sufficient time to grow so much as to severely contaminate the solution.

Since the shortest wave which can be represented on a mesh spacing Δx
is of the order of $2\Delta x$, one could avoid this contamination by using a Δx
so large that only stable modes can be resolved. Of course, this procedure

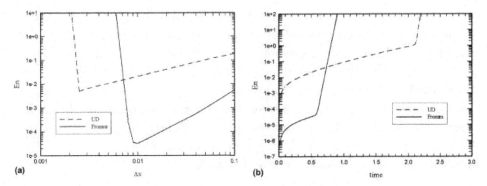

Fig. 8.4. *Left*: effect of the grid spacing on the solution error at $t = 1$ on the linear nonhyperbolic system (8.82) with B given by (8.93). *Right*: time evolution of the error $\Delta x = 0.01$; the solid line is for integration with the Fromm method, the broken line for the upwind method (from Hwang, 2003).

has the possible shortcoming of producing an inaccurate numerical solution due to discretization errors. The point is demonstrated in the left panel of Fig. 8.4, which shows the effect of the grid spacing on a measure of the solution error at $t = 1$ (the two lines correspond to two different solution methods: solid line, Fromm method; broken line, upwind). As Δx is decreased, at first the solution error decreases due to a decrease of the discretization error but, as soon as unstable modes become possible, it rapidly increases. In the presence of this phenomenon, it is evidently impossible to prove grid independence of the numerical solution, one of the cornerstones of good computational practice.

Since, for a given Δx, the corresponding largest k-mode needs some time to grow, it is possible that the solution be accurate for a certain amount of time even in the presence of unstable modes. This feature is demonstrated in the right panel of Fig. 8.4 which shows the time evolution of the error for $\Delta x = 0.01$.

8.4.2 Quasilinear systems

Let us now turn to a general quasilinear system of first-order partial differential equations of the form

$$\mathsf{A} \cdot \frac{\partial \mathbf{U}}{\partial t} + \mathsf{B} \cdot \frac{\partial \mathbf{U}}{\partial x} = \mathbf{C}, \tag{8.94}$$

where $\mathbf{U} = [U_1(x,t), U_2(x,t), \ldots, U_N(x,t)]^{\mathrm{T}}$ is a vector containing the N unknowns, $\mathsf{A} = \mathsf{A}(\mathbf{U}, x, t)$ and $\mathsf{B} = \mathsf{B}(\mathbf{U}, x, t)$ are square matrices the elements of which will depend on the U's and, in general, also on x, t; $\mathbf{C} = \mathbf{C}(\mathbf{U}, x, t)$ is another vector the elements of which have a similar dependence[1]. The vast majority of Eulerian averaged-equations models have this form when higher order spatial derivatives (e.g. due to viscous or surface tension effects) are disregarded.

Proceeding as before, we consider two different solutions of the initial value problem, $\mathbf{U}$ and $\mathbf{U}+\mathbf{u}$, corresponding to slightly different initial data. As long as $\mathbf{u}$ is suitably small, dropping nonlinear terms and the explicit indication of the dependence on x and t, we have

$$\left[\mathsf{A}\left(\mathbf{U}+\mathbf{u}\right)\cdot\frac{\partial}{\partial t}\left(\mathbf{U}+\mathbf{u}\right)\right]_i = \sum_{j=1}^{N} A_{ij}\left(\mathbf{U}+\mathbf{u}\right)\frac{\partial}{\partial t}(U_j+u_j)$$

$$\simeq \sum_{j=1}^{N}\left[A_{ij}(\mathbf{U}) + \sum_{k=1}^{N}\frac{\partial A_{ij}}{\partial U_k}u_k\right]\left(\frac{\partial U_j}{\partial t}+\frac{\partial u_j}{\partial t}\right)$$

$$\simeq \sum_{j=1}^{N}A_{ij}(\mathbf{U})\frac{\partial U_j}{\partial t} + \sum_{j=1}^{N}A_{ij}(\mathbf{U})\frac{\partial u_j}{\partial t} + \sum_{k=1}^{N}\left(\sum_{j=1}^{N}\frac{\partial A_{ij}}{\partial U_k}\frac{\partial U_j}{\partial t}\right)u_k$$

$$= \left[\mathsf{A}\left(\mathbf{U}\right)\cdot\frac{\partial \mathbf{U}}{\partial t}\right]_i + \left[\mathsf{A}\left(\mathbf{U}\right)\cdot\frac{\partial \mathbf{u}}{\partial t}\right]_i + [\mathsf{D}_A(\mathbf{U})\cdot\mathbf{u}]_i \qquad (8.95)$$

where the elements of the matrix D_A are the quantities in parentheses in the last term of the next-to-last line. The same calculation repeated for the x-derivative gives a similar result and, for the right-hand side, we similarly have

$$C_i(\mathbf{U}+\mathbf{u}) \simeq C_i(\mathbf{U}) + \sum_{k=1}^{N}\frac{\partial C_i}{\partial U_k}u_k \simeq C_i(\mathbf{U}) + [\mathsf{D}_C\cdot\mathbf{u}]_i. \qquad (8.96)$$

Upon collecting all the terms, we see that those which do not contain $\mathbf{u}$ reproduce the original equation (8.94) and therefore add up to to 0 given that, by hypothesis, $\mathbf{U}$ is a solution. What is then left is an equation of the same form as (8.82):

$$\mathsf{A}\cdot\frac{\partial \mathbf{u}}{\partial t} + \mathsf{B}\cdot\frac{\partial \mathbf{u}}{\partial x} = \mathsf{D}\cdot\mathbf{u} \qquad (8.97)$$

in which $\mathsf{D} = \mathsf{D}_C - \mathsf{D}_A - \mathsf{D}_B$. This then is the equation which governs the space-time evolution of the perturbation $\mathbf{u}$ to the original solution $\mathbf{U}$,

---

[1] A systems of this type differs from the most general nonlinear system in that A, B, and $\mathbf{C}$ only depend (possibly nonlinearly) on $\mathbf{U}$ but not on the derivatives; the dependence on the derivatives is linear.

at least as long as **u** remains so small that the quadratic and higher terms dropped in the derivation remain negligible.

The study of the perturbation equation (8.97) is now more complicated because the elements of the coefficient matrices are no longer constant in general, but depend on (x, t) through the space-time dependence of the unperturbed solution **U** and, possibly, also directly. Thus, no solutions of the simple exponential form (8.86) will be available in general. However, intuitively speaking, one can still consider approximate solutions of this form provided the wavelength $2\pi/k$ and period $2\pi/(k\lambda)$ are much smaller than the smallest length and time scales of the unperturbed solution **U**. In particular, as long as no shock waves develop, this condition is generally verified for infinitesimally short wavelengths, which will therefore still evolve according to (8.86) with a frequency $k\lambda$ determined from the characteristic equation (8.91). Thus, again, we find a lack of continuous dependence on the initial data and explosive growth of infinitesimally short waves when the system is not hyperbolic.

A case of particular interest is that of a constant **U** which, for the applications of present interest, would correspond to a steady uniform flow. In this case, if we assume that A, B, D do not depend explicitly on (x, t), as is often the case, the perturbation equation (8.97) has constant coefficients and the exponential solutions (8.86) are exact. The corresponding parameters λ are again given by (8.87) and the characteristic directions by (8.91). On the basis of the previous examples, one would then be led to expect that, even if the system fails to be hyperbolic, sufficiently long waves will be stable – and all the more so in the presence of the artificial dissipation introduced by discretization.

This belief is widespread in the literature and it is often argued that the commonly encountered lack of hyperbolicity of averaged-equations Eulerian–Eulerian multiphase flow models should not be a matter of concern because, by its very nature, a volume-averaged model will be inaccurate at spatial scales shorter than the size of the averaging volume and, therefore, any pathology of the equations at short scales is both expected and inconsequential.

Unfortunately, this argument loses much of its force in view of the rather unexpected result that, for a very large class of models, *the stability condition is independent of wavelength*. As a consequence, given that absence of hyperbolicity causes short waves to be unstable, all wavelengths will be unstable (Jones and Prosperetti, 1985; Prosperetti and Satrape, 1990). We refer to this result as the **stability–hyperbolicity theorem**.

The class of models for which this result has been proven consists of one-dimensional systems in which the two phases are individually incompressible.

In this hypothesis, the continuity equation (8.20) for the phase J ($J = 1, 2$) is

$$\frac{\partial \alpha_J}{\partial t} + \frac{\partial}{\partial x}(\alpha_J u_J) = 0, \tag{8.98}$$

where, here and in the following, we omit the explicit indication of averages. The momentum equation has the general form

$$\rho_J \left[\frac{\partial}{\partial t}(\alpha_J u_J) + \frac{\partial}{\partial x}(\alpha_J u_J^2) \right] + \alpha_J \frac{\partial p}{\partial x} = \sum_{K=1}^{2} \left(h_{JK} \frac{\partial u_K}{\partial t} + k_{JK} \frac{\partial u_K}{\partial x} \right)$$
$$+ m_J \frac{\partial \alpha_J}{\partial t} + n_J \frac{\partial \alpha_J}{\partial x} + \rho_J \alpha_J A_J, \tag{8.99}$$

where, in a typical model, the A_J's might describe, for example, drag and body forces, the h_{JK}'s and k_{JK}'s effects such as added mass (see, e.g. equation 8.79) or Reynolds stresses, and the m_J's and n_J's the effect, e.g. of differences between the volume average and interface pressures (see equation 8.28). The only thing that is assumed about these quantities is that they depend on volume fractions, velocities, and densities, but not on their derivatives. The presence of one and the same pressure field p in the momentum equations is not very restrictive as, in the usual two-pressure models, a closure relation of the type

$$p_1 - p_2 = F(\alpha_{1,2}, u_{1,2}, \rho_{1,2}), \tag{8.100}$$

is available, and the form (8.99) is then readily recovered by setting

$$p = \frac{1}{2}(p_1 + p_2), \tag{8.101}$$

so that

$$p_{1,2} = p \pm \frac{1}{2}(p_1 - p_2) = p \pm \frac{1}{2}F, \tag{8.102}$$

which reduces the momentum equations to the form (8.99).

The linear stability analysis of the steady uniform flow described by (8.98) and (8.99) may be carried out in the standard way described earlier in the derivation of (8.97); details can be found in Jones and Prosperetti (1985) and Prosperetti and Satrape (1990). The stability condition is found to have the form

$$\frac{F}{A} \geq \left(K - \frac{D}{B} \right)^2, \tag{8.103}$$

where A, B, D, K, F are algebraic functions of the various parameters appearing in the momentum equations but not of k. The solution of the characteristic equation (8.91) is found in the form

$$\lambda = \frac{1}{2}K \pm \left(\frac{F}{A}\right)^{1/2},$$ (8.104)

where A and F are the same as in (8.103). It is evident that, if the characteristics are complex, $F/A < 0$ and (8.103) cannot be satisfied so that the steady uniform solution will be unstable. Since the stability condition is independent of k, not only short wavelengths, but all wavelengths will be unstable if the model fails to be hyperbolic. On the other hand, it is also evident that the characteristics may be real, but (8.103) may fail to be verified which would render the solution unstable. This would seem the proper way in which the many instabilities present in multiphase flow should be reflected in averaged equations models.

The momentum equation (8.99) does not include terms containing nonlinear combinations of derivatives nor derivatives of order higher than one. Terms of the former type would have no effect on the stability of small disturbances to a uniform flow. Higher order derivatives can be incorporated (Jones and Prosperetti, 1987), with the expected result that they make the stability properties dependent on the wavelength of the perturbation (cf. equation 8.87) and may stabilize short waves. However since, as $k \to 0$, the effect of higher derivatives decreases, one recovers the previous result in the sense that, if the model truncated to only first-order derivatives is not hyperbolic, long waves will be unstable. There is therefore a "short-circuit" between long and short waves that destabilizes the former, whenever the latter are unstable due to lack of hyperbolicity.

The conclusion is that a two-phase model which, possibly truncated to contain only first-order derivatives and algebraic terms, takes the form (8.98), (8.99), cannot describe a stable steady uniform flow unless it is hyperbolic.

The analysis given in the cited references gives explicit formulas from which the characteristics and stability properties of the model (8.98), (8.99) can be calculated in terms of the parameters entering the momentum equations. We do not reproduce them here as they are somewhat lengthy, but it is worth showing explicitly the simplest case, in which the right-hand sides of the momentum equations are reduced to the algebraic drag functions $\rho_{1,2}\alpha_{1,2}A_{1,2}$. One finds that the model is nonhyperbolic and hence unstable whatever the form of the functions $A_{1,2}$ if $u_1 \neq u_2$ as, in this case, equation (8.103) reduces to

$$(u_1 - u_2)^2 \leq 0$$ (8.105)

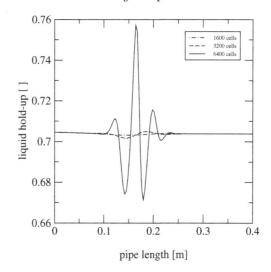

Fig. 8.5. The liquid volume fraction along a tube at the same instant of time for different levels of grid refinement for conditions such that the model is not hyperbolic; – · – · - 1600 cells; – – – 3200 cells; ——— 6400 cells (from Issa and Kempf, 2003).

while (8.104) is

$$\lambda = \left(\frac{\alpha_1}{\rho_1} + \frac{\alpha_2}{\rho_2}\right)^{-1} \left[\frac{\alpha_1}{\rho_1}u_1 + \frac{\alpha_2}{\rho_2}u_2 \pm i \left(\frac{\alpha_1\alpha_2}{\rho_1\rho_2}\right)^{1/2}(u_1 - u_2)\right]. \qquad (8.106)$$

This remark points to the crucial importance of the differential terms in the right-hand side of (8.99), i.e. in the closure of the momentum equations.

It is interesting to illustrate with a specific example given by Issa and Kempf (2003) the numerical consequences of the situation that we have just discussed. The example consists of a one-dimensional gas–liquid stratified flow in a circular tube governed by a system of equations of the form (8.98), (8.99); the reader is referred to the original reference for details. The problem is solved by finite differences with donor-cell differencing. Figure 8.5 shows, at the same instant of time, the liquid volume fraction α_L along the tube for different levels of grid refinement for conditions such that the model is not hyperbolic. As the number of cells is increased, a bigger and bigger fluctuation of α_L is found. Furthermore, as shown in Fig. 8.6, the growth rate of α_L increases without bound as the grid is refined since, as noted before, a finer grid permits the appearance of shorter wavelengths. For other values of the flow velocities, the model becomes hyperbolic and, in certain conditions, can be unstable. Figure 8.7 shows the behavior of

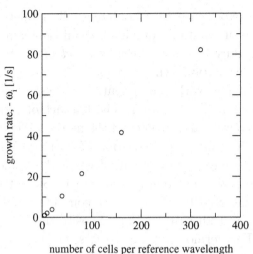

Fig. 8.6. Growth rate of the liquid volume fraction fluctuation as a function of grid refinement (increasing to the right) for conditions such that the model is not hyperbolic (from Issa and Kempf, 2003).

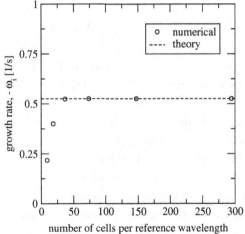

Fig. 8.7. Growth rate of the liquid volume fraction fluctuation as a function of grid refinement (increasing to the right) for conditions such that the model is hyperbolic but unstable; the dashed line is the exact analytical result (from Issa and Kempf, 2003). Unlike the previous figure, as the grid is refined, the numerical result stabilizes around the correct value.

the growth rate of an unstable solution as the grid is refined. With a sufficiently fine discretization, the numerical result faithfully reproduces the analytic prediction (dashed line) and it does not change as the grid is refined further.

It is a common experience with nonhyperbolic models that the numerical solution may well be stable provided a sufficient amount of numerical dissipation is introduced (see, e.g. Pokharna *et al.*, 1997). It is also quite possible to generate solutions which appear to be physically acceptable and in reasonable agreement with experiment. For these reasons, a widespread point of view among many practitioners is that lack of hyperbolicity is more of a nuisance than a critical deficiency of the models. This belief is strengthened by the fact that, often, the addition of terms which make the model hyperbolic seems to have only a small effect on the numerical solution, at least provided the grid is not refined very much. Since the level of grid refinement is often dictated by the available computational resources, rather than by convergence studies, it would perhaps be unfair to be overly critical of this attitude. Furthermore, in many cases, the nature of the solution is very strongly influenced by nondifferential terms and the growth rate of the instability is so slow that it can be very effectively counteracted even by a small amount of the ever-present numerical dissipation. Nevertheless, the examples shown before prove that the generation of a high-quality numerical solution of a nonhyperbolic model and proof of its grid independence might prove difficult.

The stability–hyperbolicity result just discussed only refers to the linear stability of the models. The nonlinear problem is much more difficult and, although many results are available for specific quasilinear systems such as gas dynamics, shallow water flow, and others, not much has been done for the case of multiphase flow, which exhibits peculiar difficulties (see, e.g. LeVeque, 2002). For example, it is well known that the solution of quasilinear systems of the type (8.94) can develop shock waves. The standard way to deal with this phenomenon relies on the possibility of writing the governing equations in conservation form, which is not possible for the multiphase flow equations due, if nothing else, to the volume fractions multiplying the pressure gradient in the momentum equation (8.99).

There are indications arising from very simplified nonlinear nonhyperbolic models of two-phase flow which suggest that the exact analytic solution develops a piecewise smooth form, i.e. consisting of a succession of smooth parts separated by discontinuities. While such a solution may perhaps have a physical interpretation (e.g. a crude model of slug flow), the broader implications of this result are unclear.

An additional issue is the difficulty of developing higher order numerical schemes for nonhyperbolic models; some recent progress in this direction is described in the paper by Hwang (2003) already cited.

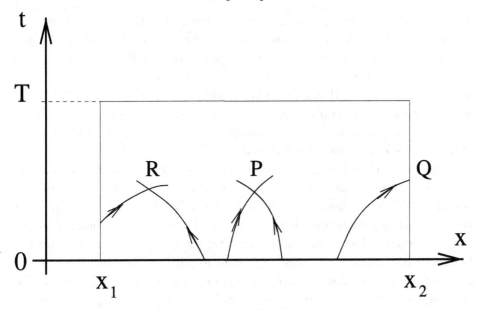

Fig. 8.8. In a hyperbolic problem, information propagates along the characteristic directions.

8.5 Boundary conditions

The natural setting for many multiphase flow calculations is that of an initial/boundary value problem: the situation of the system is prescribed at the initial time $t = 0$, and one is interested in its evolution, possibly to steady state, and usually under the action of specified conditions at the boundaries, such as a varying imposed flow rate, or others.

Suppose that the mathematical model consists of a quasilinear system of the general form (8.94) with N equations, and that the solution is sought over a time interval $0 \leq t \leq T$ in a domain $x_1 < x < x_2$ (Fig. 8.8). Most numerical calculations are carried out over a finite physical domain, and therefore we may assume that both x_1 and x_2 are finite[1]. In thinking about boundary conditions it is useful to refer to the concept of characteristic lines defined by $dx/dt = \lambda$, with the λ's the roots of (8.91).

If the system is strictly hyperbolic with N distinct characteristic directions, as we assume[2], information propagates along the characteristics. With

---

[1] The exception being the use of a transformation to map an infinite or semi-infinite physical domain onto a finite computational one. This case is unusual in multiphase flow and the considerations that follow can be extended to apply to it as well.

[2] As already remarked earlier, in some cases the argument is applicable also when there are repeated roots of the characteristic equation. For simplicity we do not consider this case here; the reader is referred, e.g. to LeVeque (2002).

N unknowns, the determination of the solution at any point of the domain
of interest requires N pieces of information which are carried by the N characteristics meeting at that point. For a point such as P in Fig. 8.8, all the
characteristics issue from $t = 0$, and therefore those N initial conditions
must be assigned on this line. This is also the number of characteristics
directed into the domain of interest through each point of the line $t = 0^3$.
When as many pieces of data are assigned on the boundary as the number
of unknowns, one speaks of *Cauchy data*.

The characteristic ending at a point such as Q located on the boundary x_2
carries one piece of information, thus obviating the need for one boundary
condition. Thus, if there are $s \leq N$ characteristics reaching each point of
the boundary x_2, evidently there are $N - s$ pieces of information missing at
that boundary, which must be supplied by prescribing an equal number of
boundary conditions. The same argument applies to the other boundary at
x_1. Some points of the (x, t) region of interest may lie at the intersection
of some characteristics issuing from $t = 0$ and others issuing from one, or
both, boundaries (e.g. the point R in Fig. 8.8). In this case, the information
carried by the characteristic issuing from the boundary must also be supplied
through a boundary condition.

In conclusion, if we regard the lines $t = 0$ and $t = T$ as also part of the
boundary of the domain of interest (which is of course the viewpoint one
would take from a purely mathematical perspective), we may say that in
the hyperbolic case, at each boundary point, as many pieces of information
must be supplied as there are characteristics directed *into* the domain of
interest issuing from that point.

In the nonhyperbolic case, at least some of the characteristics meeting
at the generic point of the domain of interest have a complex slope and
therefore issue from points off the real x-axis (Fig. 8.9). Data assigned
on the real axis must therefore be continued into the complex plane, which
introduces pathologies such as $\exp(ikx)$ becoming $\exp k(ix \pm y)$ and blowing
up as $|k| \to \infty$. Indeed, it is well known that, for a nonhyperbolic system,
the Cauchy problem is ill-posed in general. In order to have a well-posed
problem it is necessary to prescribe data over the entire boundary of the
domain i.e. in this case, also at the "future time" T. Very loosely speaking,
it may be said that the role of this "future" piece of information is to limit
the "explosion" of the extension of the data off the real axis.

---

[3] Since we are propagating the solution forward in time, $dt > 0$ and no characteristic from any
point of the domain of interest can be directed back toward the line $t = 0$. By the same
argument, there are no characteristics issuing from $t = T$ and running backward into the
domain of interest so that no information needs to be prescribed at the final time $t = T$, as is
only too natural.

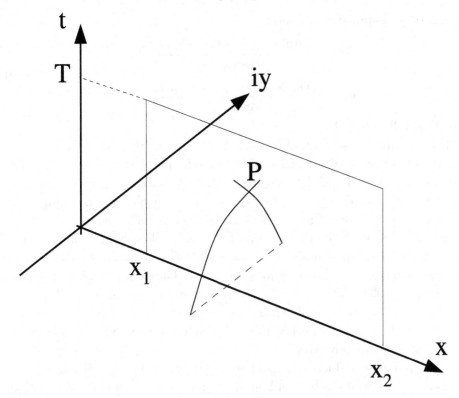

Fig. 8.9. In a nonhyperbolic problem, at least some of the characteristics through a generic point of the domain of interest issue from points off the real axis. The data assigned on the real axis must therefore be continued into the complex plane.

An extremely simple example is the Laplace equation, which may be written in system form as (8.81) with $b = c = 0$:

$$\frac{\partial U}{\partial t} - \frac{\partial V}{\partial x} = 0, \qquad \frac{\partial V}{\partial t} + \frac{\partial U}{\partial x} = 0. \tag{8.107}$$

With the Cauchy data $U(x,0) = A \exp{(ikx)}$, $V(x,0) = B \exp{(ikx)}$, the solution is

$$U = (A \cosh kt + iB \sinh kt)\, e^{ikx}, \qquad V = (-iA \sinh kt + B \cosh kt)\, e^{ikx}, \tag{8.108}$$

and exhibits the expected noncontinuous dependence on the data as $|k| \to \infty$. With the Dirichlet data $U(x,0) = A \exp{(ikx)}$, $V(x,T) = C \exp{(ikx)}$, on the other hand, the solution is

$$\left.\begin{aligned} U &= \frac{A\cosh k(T-t) + iC\sinh kt}{\cosh kT}\, e^{ikx} \\[2mm] V &= \frac{iA\sinh k(T-t) + C\cosh kt}{\cosh kT}\, e^{ikx} \end{aligned}\right\} \tag{8.109}$$

which is well behaved.

These simple considerations show that, in the nonhyperbolic case, problems arise not because a solution of the Cauchy problem does not exist, but because it may well exhibit features – such as noncontinuous dependence on the data – which make it difficult to ascribe a physical meaning to it. Indeed, the well-known Cauchy–Kowalewskaya theorem asserts the existence of a *local* solution of the Cauchy problem (i.e. in the present context, valid for sufficiently small t) under the hypothesis of analyticity of the data and of the coefficients of the equations (see, e.g. Garabedian, 1998; Renardy and Rogers, 1993). Thus, the fact that one can generate numerically a solution of the initial value problem in the nonhyperbolic case in itself does not give any indication of whether the mathematical formulation makes sense mathematically or physically.

The interconnectedness of hyperbolicity and boundary conditions complicates the analysis of the type of boundary (and sometimes initial) conditions suitable for a given multiphase flow model. Rather than attempting a general treatment here, we refer the reader to several specific examples discussed in the following chapters.

Let us return now to the hyperbolic case and consider as an example a simple one-dimensional drift-flux model consisting of equations for the conservation of mass for a liquid (index L) and a gas (index G) phase (Fjelde and Karlsen, 2002):

$$\frac{\partial}{\partial t}(\alpha_J \rho_J) + \frac{\partial}{\partial x}(\alpha_J \rho_J u_J) = 0, \qquad J = L, G, \tag{8.110}$$

and a total momentum balance equation

$$\frac{\partial}{\partial t}(\alpha_L \rho_L u_L + \alpha_G \rho_G u_G) + \frac{\partial}{\partial x}\left(\alpha_L \rho_L u_L^2 + \alpha_G \rho_G u_G^2 + p\right) = -\rho_m g - H u_V, \tag{8.111}$$

in which H is a constant interphase drag parameter, ρ_m is the mixture density defined in (8.18), and u_V is the volumetric flux defined in (8.23). We assume the liquid to be incompressible, a linear relation between the gas pressure and density

$$p = c_G^2 \rho_G, \tag{8.112}$$

in which c_G is a constant, and a linear relation for the gas "slip" velocity:

$$u_G = Ku_V + S \tag{8.113}$$

in which K and S are other constants. In particular, S has the physical meaning of the gas rise velocity in a stagnant fluid, e.g. in the form of long Taylor bubbles. With the assumption $\rho_G \ll \rho_L$, the system can be written in the quasilinear form (8.94) with

$$\mathbf{A} = \begin{vmatrix} \rho_G & \alpha_G\rho_G/p & 0 \\ -\rho_L & 0 & 0 \\ 0 & 0 & \alpha_L\rho_L \end{vmatrix}, \qquad \mathbf{C} = \begin{vmatrix} 0 \\ 0 \\ -\rho_m g - Hu_V \end{vmatrix} \tag{8.114}$$

$$\mathbf{B} = \begin{vmatrix} (u_G - K\alpha_G u_L)\alpha_G\rho_G\rho_L C^2/p & \alpha_G\rho_G u_G/p & K\alpha_G^2\alpha_L\rho_G\rho_L C^2/p \\ -\rho_L u_L & 0 & \alpha_L\rho_L \\ 0 & 1 & \alpha_L\rho_L u_L \end{vmatrix}, \tag{8.115}$$

$$\mathbf{U} = \begin{vmatrix} \alpha_G \\ p \\ u_L \end{vmatrix} \tag{8.116}$$

where

$$C = \sqrt{\frac{p}{\alpha_G(1 - K\alpha_G)\rho_L}} \tag{8.117}$$

is the speed of the linear pressure waves supported by the system. The characteristic roots are readily found from (8.91) and are

$$\lambda = u_G, \qquad \lambda_\pm = u_L \pm C. \tag{8.118}$$

The system can be put in characteristic form in the standard way (see, e.g. Whitham, 1974) to find

$$\left(\frac{dp}{dt}\right)_G + \rho_L C^2 \left(\frac{d\alpha_G}{dt}\right)_G = 0 \tag{8.119}$$

$$\left(\frac{dp}{dt}\right)_\pm \pm \rho_L(u_G - u_L)C \left(\frac{d\alpha_G}{dt}\right)_\pm - \alpha_L\rho_L(u_G - u_L \mp C)\left(\frac{du_L}{dt}\right)_\pm$$

$$= (\rho_m g + Hu_V)(u_G - u_L \mp C), \tag{8.120}$$

where $(d/dt)_G = \partial/\partial t + u_G(\partial/\partial x)$, $(d/dt)_\pm = \partial/\partial t + (u_L \pm C)(\partial/\partial x)$ are the directional derivatives along the characteristic lines. Relations of this type are sometimes referred to as *compatibility relations*.

Consider the inlet boundary $x = 0$. As long as $0 < u_L$, the characteristics u_G and $u_L + C$ are directed into the flow domain $x > 0$. It is therefore necessary to prescribe two boundary conditions, for example the superficial velocities of the two phases which, by (8.113) and $\alpha_G + \alpha_L = 1$, amounts to a prescription of u_L and α_G. As long as $0 < u_L < C$, the third characteristic $u_L - C$ is directed out of the flow domain and, therefore, no additional condition, e.g. for the pressure, can be prescribed. Rather, the pressure at the boundary should be calculated from (8.120) taken with the lower signs. A numerical implementation would require a one-sided approximation for the spatial derivatives. Conversely, for $0 < u_L < C$, at the outlet boundary there are two outgoing characteristics and one incoming one, $u_L - C$. Now only one piece of data can be imposed, e.g. the outlet pressure. The boundary values of α_G and u_L should be determined from (8.119) and (8.120) taken with the upper signs.

The situation would be different if $u_L > C$, because now three boundary conditions should be prescribed at the inlet and none at the outlet; a situation of this type is described by Edwards *et al.* (2000).

Appendix – Proof of equation (8.4)

A more explicit statement of the relation (8.4) is

$$\int_{S_J(\mathbf{x})} \mathbf{A}_J(\mathbf{x} + \boldsymbol{\xi}) \cdot \mathbf{n}_J \, dS_{\xi J} = \nabla_x \cdot \int_{\mathcal{V}_J(\mathbf{x})} \mathbf{A}_J(\mathbf{x} + \boldsymbol{\xi}) \, d^3 \xi_J, \qquad (\text{A.1})$$

in which $\mathbf{x}$ is the position of the center of the averaging volume $\mathcal{V}$. As in Section 8.2, here S_J and $\mathcal{V}_J$ are, respectively, the portion of the surface of $\mathcal{V}$ and of $\mathcal{V}$ itself occupied by the phase J.

The volume integral in (8.4) can be extended to the entire $\mathcal{V}$ by introducing the characteristic function χ_J of the phase J:

$$\int_{\mathcal{V}_J(\mathbf{x})} \mathbf{A}_J(\mathbf{x} + \boldsymbol{\xi}) \, d^3 \xi = \int_{\mathcal{V}(\mathbf{x})} \mathbf{A}_J(\mathbf{x} + \boldsymbol{\xi}) \, \chi_J(\mathbf{x} + \boldsymbol{\xi}) \, d^3 \xi, \qquad (\text{A.2})$$

in which $\chi_J(\mathbf{y}) = 1$ if $\mathbf{y} \in \mathcal{V}_J$ while $\chi_J(\mathbf{y}) = 0$ otherwise. If we assume that $\mathcal{V}$ is independent of $\mathbf{x}$, in view of the dependence of the integrand on the sum $\mathbf{x} + \boldsymbol{\xi}$, we have

$$\nabla_x \cdot \int_{\mathcal{V}_J(\mathbf{x})} \mathbf{A}_J(\mathbf{x} + \boldsymbol{\xi}) \, d^3 \xi = \int_{\mathcal{V}(\mathbf{x})} \nabla_x \cdot [\mathbf{A}_J \, \chi_J] \, d^3 \xi = \int_{\mathcal{V}(\mathbf{x})} \nabla_\xi \cdot [\mathbf{A}_J \, \chi_J] \, d^3 \xi,$$
$$(\text{A.3})$$

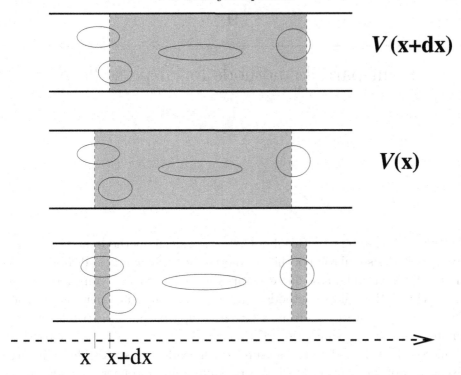

Fig. 8.10. A graphical interpretation of the result (8.4): the difference between the integrals over $\mathcal{V}(x + dx)$ and $\mathcal{V}(x)$ equals dx times the integral over the surface of the averaging volume.

from which, applying the divergence theorem and recognizing that $\chi_J dS = dS_J$, the result immediately follows. More detailed proofs can be found in Slattery (1967), Whitaker (1969), and Gray and Lee (1977).

This result has an immediate graphical interpretation which is given in Fig. 8.10. In one space dimension the theorem (A.1) is

$$\frac{1}{dx}\left[\int_{\mathcal{V}_J(x+dx)} \mathbf{A}_J\, d^3\xi - \int_{\mathcal{V}_J(x)} \mathbf{A}_J\, d^3\xi\right] = \int_{\mathcal{S}_J(x)} \mathbf{A}_J \cdot \mathbf{n}_J\, dS_{\xi J} \qquad (A.4)$$

The figure renders it evident that the difference between the integrals over $\mathcal{V}(x + dx)$ and $\mathcal{V}(x)$ equals dx times the integral over the surface of the averaging volume.

9

Point-particle methods for disperse flows

In the first chapters of this book we have seen methods suitable for a first-principles simulation of the interaction between a fluid and solid objects immersed in it. The associated computational burden is considerable and it is evident that those methods cannot handle large numbers of particles. In this chapter we develop an alternative approach which, while approximate, permits the simulation of thousands, or even millions, of particles immersed in a flow. The key feature which renders this possible is that the exchanges of momentum (and also possibly mass and energy) between the particle and the surrounding fluid are *modeled*, rather than directly resolved. This implies an approximate representation that is based on incorporating assumptions into the development of the mathematical model.

One of the most common approaches used today to model many particle-laden flows is based on the "point-particle approximation," i.e. the treatment of individual particles as mathematical point sources of mass, momentum, and energy. This approximation requires an examination of the assumptions and limitations inherent to this approach, aspects that are given consideration in this chapter. Point-particle methods have relatively wide application and have proven a useful tool for modeling many complex systems, especially those comprised of a very large ensemble of particles. Details of the numerical aspects inherent to point-particle treatments are highlighted.

We start by putting point-particle methods into the context established earlier in this text and, in particular, in the previous chapter. Our treatment will be restricted to particulate systems without mass or energy exchange at the particle surface. This allows us to focus on momentum exchange. Point methods in this regime then require a description of particle dynamics via solution of an equation of motion that models the forces acting on a particle. Without substantial loss of generality, the present chapter will focus on spherical particles.

There has been a substantial body of research devoted to the analysis and development of an equation that can be used to describe the motion of a sphere suspended within an unsteady and nonuniform ambient flow. A brief summary of the development of the equation of motion was given in the previous chapter, where references to more detailed treatments were provided for the interested reader. As discussed, the equation of motion is crucially important for accurately describing the particle interactions with a surrounding carrier flow. Theoretical developments, however, are limited to relatively narrow parameter ranges (e.g. small particle Reynolds numbers). Empirical input via the use, for example, of correlations for various forces is unavoidable within point-particle approximations.

Examples that follow are drawn from gas–solid flows in which the equation of motion for a rigid sphere in a nonuniform and time-dependent flow field can be simplified compared to the more general case. This presents some operational advantages within codes and also permits the investigation of theories for describing the transport characteristics of particulates in gas–solid systems. Applications to canonical flows are used to illustrate the usefulness of the method and also to identify limitations and areas in which future improvements are needed.

9.1 Point-force approximation

In order to develop a model suitable to describe the carrier fluid in the point-particle approximation, we start from the averaged equations developed in the previous chapter and consider the limit in which the volume of each particle becomes negligibly small, so that the disperse-phase volume fraction vanishes while the fluid volume fraction becomes unity. Since the particles are assumed to be smaller than any structure of the flow, the averaging volume can also be taken very small and, as a consequence, the volume-averaged quantities may actually be interpreted as point values. Accordingly, in what follows, we drop the explicit indication of averages.

If, as we assume, the fluid is incompressible, and there is no mass exchange at the particle surfaces, the average continuity equation (8.11) simply becomes

$$\boldsymbol{\nabla} \cdot \mathbf{u} = 0. \tag{9.1}$$

Similarly, the momentum equation (8.30) becomes

$$\rho \left(\frac{\partial \mathbf{u}}{\partial t} + \boldsymbol{\nabla} \cdot (\mathbf{uu}) \right) = -\boldsymbol{\nabla} p + \mu \nabla^2 \mathbf{u} - \mathbf{F}, \tag{9.2}$$

in which the body force has been absorbed into a modified pressure in the usual way (see Section 1.3 in Chapter 1). The last term in this equation represents the force per unit volume exerted by the particles on the fluid, which we approximate by a superposition of Dirac's delta functions centered at the location $\mathbf{x}_p^n$ of each particle:

$$\mathbf{F} = \sum_{n=1}^{N_p} \mathbf{f}^n(\mathbf{x}_p^n)\,\delta(\mathbf{x} - \mathbf{x}_p^n). \tag{9.3}$$

Here the summation is taken over all the N_p particles. Thus, in nonconservation form, the fluid momentum equation to be solved becomes

$$\frac{\partial \mathbf{u}}{\partial t} + \mathbf{u} \cdot \nabla \cdot \mathbf{u} = -\frac{1}{\rho}\nabla p + \nu\nabla^2\mathbf{u} - \frac{1}{\rho}\sum_{n=1}^{N_p} \mathbf{f}^n(\mathbf{x}_p^n)\,\delta(\mathbf{x} - \mathbf{x}_p^n). \tag{9.4}$$

The position $\mathbf{x}_p^n$ of the n-th particle is found from the kinematic relationship

$$\frac{d\mathbf{x}_p^n}{dt} = \mathbf{v}_p^n, \tag{9.5}$$

where $\mathbf{v}_p^n$ is the particle velocity, while Newton's equation for the particle is written as

$$\upsilon\rho_p\frac{d\mathbf{v}_p^n}{dt} = \upsilon\rho_p\mathbf{g} + \mathbf{f}^n(\mathbf{x}_p^n). \tag{9.6}$$

Subject to a prescription of the force $\mathbf{f}^n$ (and of course the appropriate initial and boundary conditions), the system (9.1)–(9.6) comprises the set of equations that we will solve in order to model a fluid–particle flow for situations in which we use a point-particle treatment of the dispersed phase. The actual form of the expression used for $\mathbf{f}^n$ has a strong effect on the results. In part we have discussed the issue in the previous chapter and more will be said about it in the next section. Thus, we envision a particle tracking method in which (9.6) is solved for each particle n from an assembly of N_p particles. The local fluid velocity and acceleration necessary to specify the forces are to be determined by solving everywhere in the domain the continuity (9.1) and momentum (9.4) equations governing the fluid motion. The effect of the particulate phase on the fluid flow is only through the action of the forces acting at the particle centers.

   The previous formulation appears relatively simple and the point-particle approximation offers advantages for many systems of interest for which methods offering higher fidelity are not practical. In this context, modeling disperse flows comprised of very large numbers of particles, accounting for particle–particle interactions and particle–wall interactions, and extensions to include the effects of mass and heat transfer are relatively straightforward tasks compared to other methods. Modeling droplet coalescence, for example, is a problem of substantial interest for which point-particle computations are playing a useful role (e.g. see Post and Abraham, 2002, Villedieu and Simonin, 2004, and references therein). While useful in a wide range of areas, the operational simplicity of point methods is obtained at the cost of modeling certain aspects of particle–fluid and/or particle–particle exchanges. The presumption that the particle may be modeled as a mathematical point leads to restricted ranges over which point methods may be applied. In addition, even for situations where this presumption is justified, care is needed in the interpretation of the results.

In the next section an overview of the equation describing the motion of a rigid sphere is presented. The goal is to provide context for interpreting the usefulness and range of expressions used for the force $\mathbf{f}^n$. Following is a discussion of the point treatment and the implications for solution of (9.1) and (9.4). In practice, the numerical techniques used to impose the coupling force in (9.4) must also be considered in our estimate of the accuracy of the solutions.

9.2 Particle equation of motion

In Section 8.3 of the previous chapter we wrote an equation of motion for a particle immersed in a fluid in the form

$$m_p \frac{d\mathbf{v}}{dt} = m_p \mathbf{g} + \rho v \left(\frac{D\mathbf{u}}{Dt} - \mathbf{g} \right) + \mathbf{f}_{\mathrm{d}} + \mathbf{f}_{\mathrm{l}} + \mathbf{f}_{\mathrm{a}} + \mathbf{f}_{\mathrm{h}} + \mathbf{f}_{\mathrm{additional}} \quad (9.7)$$

in which $\mathbf{f}_{\mathrm{d}}$, $\mathbf{f}_{\mathrm{l}}$, $\mathbf{f}_{\mathrm{a}}$, and $\mathbf{f}_{\mathrm{h}}$ are, respectively, the drag, lift, added mass, and history forces. We have included an additional term $\mathbf{f}_{\mathrm{additional}}$ to account for other forces that might be important for a given application, such as electrostatic interactions or wall collisions. Such a superposition is typical: when point-particle methods are applied to flows in which other forces are thought to be important, these forces are often assumed to act independently with respect to the forces already considered and added to (9.7) or similar equations. The reader is referred to Loth (2000) for an additional discussion of the equation of motion and the parameterization of some other forces.

In the previous chapter we showed explicit expressions for the various forces appearing in (9.7) valid for very low or very high particle Reynolds numbers. The restricted range of these exact results supplies part of the motivation for more *ad hoc* treatments in which the forces are represented using models that incorporate some empirical input.

We have already seen how the drag force is usually expressed in terms of an empirical drag coefficient C_d:

$$\mathbf{f}_d = -\frac{3}{4}\rho v \frac{C_d}{d}|\mathbf{v} - \mathbf{u}|\,(\mathbf{v} - \mathbf{u})\,. \tag{9.8}$$

According to its derivation, this relation is applicable to a particle immersed in a uniform stream with velocity $\mathbf{u}$ far from the particle. In using it, as we will, for the case of a particle in a flow with a complex spatial structure, we must assume therefore that the particle diameter is so much smaller than the smallest relevant scale of the flow that it makes sense to think of $\mathbf{u}$ as a "background" or "far-field" velocity, i.e. the fluid velocity undisturbed by the presence of the particle. In particular, since the flow disturbance induced by other particles has the same length scale as the particle under consideration, it is necessary to be able to ignore the mutual interactions of the particles. This circumstance restricts the applicability of the model to dilute particle assemblies, i.e. situations in which the particle number density is relatively small.

For the added mass we have cited the result of Auton *et al.* (1988), valid for very large particle Reynolds numbers:

$$\mathbf{f}_a = \frac{1}{2}\rho v \left(\frac{D\mathbf{u}}{Dt} - \frac{d\mathbf{v}}{dt}\right), \tag{9.9}$$

in which $D\mathbf{u}/Dt = \partial\mathbf{u}/\partial t + \mathbf{u}\cdot\nabla\mathbf{u}$. For the history force we have seen Basset's result, which we write here omitting the finite-size correction proportional to $\nabla^2\mathbf{u}$:

$$\mathbf{f}_h = \frac{3}{2}d^2\rho\sqrt{\pi\nu} \int_{t_0}^{t} \frac{dt'}{(t - t')^{1/2}} \left(\frac{D\mathbf{u}}{Dt'} - \frac{d\mathbf{v}}{dt'}\right). \tag{9.10}$$

While, as we have pointed out in the previous chapter, there is a considerable uncertainty as to the proper form of these forces for general situations, it is a fortunate circumstance that considerable simplifications can be introduced in the important case of gas–solid flows. In typical particle-laden gas flows, the density of the particle material, ρ_p, is large compared to the fluid density ρ, with typical values of ρ_p/ρ of the order of 1000. In this case, the fluid density multiplying the added mass and the history force contributions, under most circumstances, renders them negligible compared with the drag force. For a rough estimate, we may consider a relative velocity of the order of the terminal velocity $v_t = \sqrt{(4/3)(\rho_p/\rho)dg/C_d}$, obtained by balancing gravity and drag (see equation (1.54) in Chapter 1), and compare it with the added mass force (9.9) in which the relative acceleration a_r is scaled by the acceleration of gravity:

$$\frac{|\mathbf{f}_a|}{|\mathbf{f}_d|} \simeq \frac{\frac{1}{2}\rho v g}{\frac{3}{4}\rho v (C_d/d) v_t^2} \frac{a_r}{g} = \frac{1}{2}\frac{\rho}{\rho_p}\frac{a_r}{g}. \tag{9.11}$$

Situations in which the relative acceleration would be so large as to compensate for the density ratio may be expected to be very rare indeed. As for the history force, we may estimate its magnitude relative to the added mass force by assuming comparable accelerations, with a relaxation time of the order of τ. In this way we find

$$\frac{|\mathbf{f}_h|}{|\mathbf{f}_a|} \simeq 18\sqrt{\frac{\nu\tau}{d^2}}. \tag{9.12}$$

This ratio is therefore of the order of the ratio of the diffusion length to the particle diameter, which may be expected to be of order one as, due to the physical mechanism which causes this force, $\tau \sim d^2/\nu$. Thus, we may conclude that the history force is also negligible with respect to the drag force. The same conclusion can be drawn for the lift force, both because it is also proportional to the fluid density, and because the length scale for the ambient flow vorticity may be expected to be much greater than the particle size. Finally, the buoyancy force is also evidently negligible.

On the basis of these arguments, the particle equation of motion (9.7) reduces to

$$\rho_p \frac{d\mathbf{v}^n}{dt} = \rho_p \mathbf{g} + \mathbf{f}_d^n, \tag{9.13}$$

with $\mathbf{f}_d^n$ given by (9.8). It is convenient to rewrite this equation in terms of the particle response time, which physically represents the time scale over which the particle relative velocity relaxes to zero under the action of the drag force. From the balance $\rho_p v_p/\tau_p^n \sim f_d$ we find

$$\tau_p^n = \frac{4}{3}\frac{\rho_p}{\rho}\frac{d}{C_d^n}\frac{1}{v_r^n}, \tag{9.14}$$

where v_r is the relative velocity (see equation (1.54) of Chapter 1). If the Schiller and Naumann (1935) relation (8.71) is used to express C_d, we have

$$\tau_p^n = \frac{\rho_p}{\rho}\frac{d^2}{18\nu}\frac{1}{1 + 0.15 Re_p^{n0.687}}. \tag{9.15}$$

The equation of motion (9.13) then takes the form

$$\rho_p \frac{d\mathbf{v}^n}{dt} = \rho_p \mathbf{g} - \rho_p \frac{\mathbf{v}^n - \tilde{\mathbf{u}}^n}{\tau_p^n}. \tag{9.16}$$

Here we write the fluid velocity as $\tilde{\mathbf{u}}$ to emphasize the fact that the fluid velocity to be used in the expression of the drag force is the background fluid velocity as discussed above. This same background velocity must be used to calculate the particle Reynolds number Re_p. In the limit of vanishing particle Reynolds numbers, the particle response time given by (9.15) reduces to the so-called Stokes response time, $\tau_{ps} = (\rho_p d^2)/(18\nu\rho)$. Thus, an equivalent form for the equation of motion as considered in the current context of gas–solid flows can be written as

$$\rho_p \frac{d\mathbf{v}^n}{dt} = \rho_p \mathbf{g} - \rho_p \frac{\mathbf{v}^n - \tilde{\mathbf{u}}^n}{\tau_{ps}} f_\mathrm{d}, \qquad f_\mathrm{d} = 1 + 0.15 Re_p^{n^{0.687}}. \tag{9.17}$$

For the current development restricted to gas–solid flows in which the force acting on the particle has been reduced to the drag contribution as written above, the flow is assumed quasisteady in that the expression for the force is taken identical to that measured on a sphere in a steady, uniform free-stream. Such an approximation requires that the adjustment of the flow in the vicinity of the particle occur rapidly compared to variations in the free-stream flow, an approximation that is increasingly accurate when the particle diameter d is small compared to the relevant length scales in the carrier fluid flow. In turbulent flows, this requires that the particle diameter be small compared to the Kolmogorov length scale, $\eta = (\nu^3/\varepsilon)^{1/4}$, where ε represents the dissipation rate of the fluid turbulence.

Finally, whilst the equations of motion for a solid sphere in a time-dependent, spatially uniform flow, such as the one derived by Maxey and Riley (1983) given in equation (8.64) of the previous chapter, are formally restricted in their range of applicability, exploration of properties of their solutions in limiting cases remains of general interest. As shown later, such solutions allow the development of theories for describing important properties of particulate transport, e.g. the kinetic energy of a cloud of particles. Numerical simulations of particle interactions with turbulent fluid flows can be benchmarked against these analytical relations, in turn shedding light on some of the interesting interactions possible using point-particle methods.

9.3 Numerical treatment of the particulate phase

9.3.1 Interpolation

Our focus now shifts to the numerical treatment of the equations used to determine the particle velocity and displacement.

Regardless of the precise form of the equation of motion, in particle-tracking applications we require the value of fluid properties at the location of the particle. Inspection of the equation of motion (9.16) shows that a knowledge of the flow properties at the location of the particle is required to determine the particle velocity. In general, it is only by chance that the particle position will coincide with a mesh point at which the fluid solution is available as part of the computation of the underlying flow field. Therefore, methods for the interpolation of the flow fields at the particle position are required.

In applications, these methods are based on the use of local approximations to obtain estimations away from the mesh points on which the underlying solution is computed. To summarize such approaches, in the following we assume that the fluid flow is computed on a network of mesh points in the x-, y-, and z-directions using N_x, N_y, and N_z points, respectively. We will denote the coordinates of the grid points along the axes as x_1, y_m, and z_n (and so $l = 1, \ldots, N_x$, etc.). Given this notation, a useful way to express the velocity of the fluid at the particle position in a three-dimensional field is

$$\tilde{u}_i(x_p, y_p, z_p, t) = \sum_{x_1} \sum_{y_m} \sum_{z_n} a_i(x_1, y_m, z_n, t) f_1(x_p) g_m(y_p) h_n(z_p), \quad (9.18)$$

where the index i indicates the Cartesian component and we denote the coordinates of the position of the particle as (x_p, y_p, z_p). The form of the interpolation (9.18) expresses the velocity as a weighted summation over the grid of basis functions f_1, g_m, and h_n and coefficients a_i. The particular form of these quantities depends on the method. The ranges of the summations in (9.18) are typically taken over a subset of the grid – an important consideration consistent with the local nature of practical interpolation methods since summation over the entire network of points would require on the order of $3N_x N_y N_z$ operations for the evaluation of the fluid velocity for a single particle.

A common choice for velocity interpolation is the use of polynomials of various order that are applied locally. The simplest approach is to assume that the fluid properties exhibit a linear variation between the grid points, in which case the general relation (9.18) can be written as,

$$u_i(x_p, y_p, z_p, t) = \sum_{x_1} \sum_{y_m} \sum_{z_n} u_i(x_1, y_m, z_n, t) L_{l,1}(x_p) L_{m,1}(y_p) L_{n,1}(z_p)$$

$$(9.19)$$

where the basis function in the x-direction for linear interpolation is

$$L_{l,1}(x_p) = \begin{cases} (x_p - x_{l-1})/\Delta x & \text{for } x_{l-1} \leq x_p \leq x_l \\ (x_{l+1} - x_p)/\Delta x & \text{for } x_l \leq x_p \leq x_{l+1} \end{cases} \quad (9.20)$$

and is zero for all other values of x_1. In (9.20) the spacing between the grid points in the x-direction that bound the x-coordinate of the particle, x_p, is denoted $\Delta x = x_{l+1} - x_l$. Similar formulas apply in the y- and z-directions. Linear interpolation represents an approximation of the function (e.g. the fluid velocity) between the grid points using first-order polynomials and when evaluated at the grid points the right-hand side of (9.19) yields the fluid velocity.

The approach above can readily be generalized to Lagrange interpolation of higher order. As an example in one dimension, with this method a function $f(x)$ is approximated by a polynomial of order n as

$$f(x) \approx p_n(x) = \sum_{j=0}^{n} f(x_j) L_{j,n}(x) , \qquad (9.21)$$

where the $L_{j,n}(x)$'s are Lagrangian polynomials of order n given by

$$L_{j,n}(x) = \Pi_{k=0,k\neq j}^{n} \frac{x - x_k}{x_j - x_k} . \qquad (9.22)$$

Thus, (9.21) shows that the basic idea is to define the polynomial of order n that will be used to approximate $f(x)$ as a sum of $n + 1$ Lagrangian polynomials (or Lagrangian basis functions), each of order n. According to their definition (9.22), the Lagrangian interpolants take on a value of 1 at the grid point of interest and are zero at all other grid points. This in turn allows the coefficients in the general expression (9.18) to be simply the function value at that node point. The interpolating polynomial (9.21) is unique for a given function $f(x)$ and a Taylor series analysis can be used to show that the error, i.e. $f(x) - p_n(x)$, decreases as $\mathcal{O}(\Delta^{n+1})$ where Δ is the spacing between the grid points. The reader is referred to Jeffreys and Jeffreys (1988) and Press $et\ al.$ (1992) for further discussion.

While polynomial interpolation such as that illustrated by (9.19) or, in one dimension, (9.21), are common in many applications of particle-tracking methods, other approaches that are viable include the use of Hermitian interpolation, in which the polynomials use not only the value of the function at the grid points but also the value of the first derivative. Thus, considering again an example in one dimension for the sake of simplicity, if we have $n+1$ grid points $x_0, x_1, \ldots, x_n$ there are now a total of $2n+2$ pieces of information supplied to the interpolant, i.e. the $n + 1$ values of the function and $n + 1$ values of the derivative at the grid points $x_0, x_1, \ldots, x_n$. This in turn allows us to define a Hermitian polynomial of degree $2n + 1$, which incidentally shows that Hermitian polynomials are always of odd order.

It is convenient to express the interpolating polynomial in a form similar to (9.21),

$$f(x) \approx h_{2n+1}(x) = \sum_{j=0}^{n} f(x_j) H_{j,2n+1}(x) + \sum_{j=0}^{n} f'(x_j) G_{j,2n+1}(x), \quad (9.23)$$

where $H_{j,2n+1}$ and $G_{j,2n+1}$ represent the Hermite basis function associated with the function f and its first derivative f', both functions of degree $2n+1$.

The Hermite interpolants can be expressed in terms of the Lagrange interpolants (9.22) as

$$
\begin{aligned}
H_{j,2n+1}(x) &= L_{j,n}^2(x)\left[1 + 2L'_{j,n}(x_j)(x_j - x)\right] \\
G_{j,2n+1}(x) &= L_{j,n}^2(x)(x - x_j)
\end{aligned}
\quad (9.24)
$$

where $L_{j,n}(x)$ in (9.24) is given by (9.22). The error, $f(x) - h_{2n+1}(x)$, can be shown to decrease as $\mathcal{O}(\Delta^{2n+2})$ (e.g. see Hildebrand, 1956). If one considers as a specific example the use of third-order Hermitian interpolation for the function $f(x)$ at a point x_p lying between the points $[x_j, x_{j+1}]$, the formula (9.24) becomes

$$
\begin{aligned}
H_{j,3}(x) &= \zeta^2(3 - 2\zeta), \quad \zeta = \frac{x - x_{j-1}}{\Delta x}, \quad x_{j-1} \leq x \leq x_j \\
&= (1 - \zeta)^2(1 + 2\zeta), \quad \zeta = \frac{x - x_j}{\Delta x}, \quad x_j \leq x \leq x_{j+1}
\end{aligned}
\quad (9.25)
$$

$$
\begin{aligned}
G_{j,3}(x) &= \Delta\zeta^2(\zeta - 1), \quad \zeta = \frac{x - x_{j-1}}{\Delta x}, \quad x_{j-1} \leq x \leq x_j \\
&= \Delta\zeta(1 - \zeta)^2, \quad \zeta = \frac{x - x_j}{\Delta x}, \quad x_j \leq x \leq x_{j+1}.
\end{aligned}
\quad (9.26)
$$

Outside the range $[x_j, x_{j+1}]$, i.e. for $x_p < x_{j-1}$ and $x_{j+1} < x_p$, both $H_{j,3}$ and $G_{j,3}$ are zero.

The Hermite interpolation of the value $f(x_p)$ using two points $[x_j, x_{j+1}]$ and the polynomial (9.23) with the specific forms (9.25) and (9.26) yields a cubic polynomial with an error that decreases as $\mathcal{O}(\Delta^4)$. The asymptotic behavior of the error is superior to that resulting from the two-point Lagrange interpolation (9.21) and (9.22), which would yield a straight-line approximation with an error decreasing as $\mathcal{O}(\Delta^2)$. Achieving the same formal order in the error by Lagrange interpolation would require four points and, therefore, the use of information from other grid points.

One of the important aspects is the trade-off – a more rapid reduction of the error using the Hermitian interpolation (9.23) compared to the

Lagrange interpolation at the cost of additional operations in the formation of the interpolated value. This occurs not only in the representation
(9.23) but also in the formation of the basis functions (9.24). Additionally,
the Hermite interpolants require a knowledge of the derivative of the function being interpolated, in turn implying that this quantity is available. In
particle-tracking applications in which the fluid motion is described via solution of the Navier–Stokes equations, the first derivatives of the fluid velocity
are available for use in (9.23). Now, however, we must be aware that they
are themselves affected by truncation errors sensitive to the method used
to compute the numerical solution of the Navier–Stokes equations. Many
approaches used for solution of the Navier–Stokes equations employ discretizations of the spatial derivatives that possess truncation errors decreasing as $\mathcal{O}(\Delta^2)$. Consequently, the evaluation of the derivatives is sensitive
to the grid spacing and can be expected to influence the accuracy of the
interpolated result.

In practice, the numerical errors affecting the spatial derivatives of the
fluid velocity field and operations required may reduce the advantages of
schemes such as Hermite interpolation. In addition, depending upon the
algorithm used for solution of the Navier–Stokes equations, the spatial derivatives of the fluid velocity may not be conveniently available for use in an
interpolation formula. These aspects provide part of the motivation for the
application of spline interpolation. The most commonly used interpolants
are cubic splines, which are defined to provide an approximation of the first
derivative of the function by forcing the interpolant to have continuous second derivatives across each grid point.

It is possible to develop an approach to the formulas for cubic splines
using a modification of the relation (9.23):

$$f(x) \approx s_{j,3}(x) = \sum_{j=0}^{1} f(x_j) H_{j,3}(x) + \sum_{j=0}^{1} \delta f(x_j) G_{j,3}(x), \qquad (9.27)$$

where $s_{j,3}(x)$ is the cubic polynomial approximation to $f(x)$ in some interval j and $\delta f(x_j)$ is an approximation to the first derivative of $f(x_j)$.
Relationships for $\delta f(x_j)$ are obtained by enforcing continuity of the second
derivatives of $s_{j,3}(x)$. In one dimension, while considering the grid point x_j
bounding by the intervals j and $j+1$ as shown in Fig. 9.1, continuity of the
second derivative of the cubic polynomial across grid point x_j implies

$$\frac{d^2 s_{j,3}}{dx^2} = \frac{d^2 s_{j+1,3}}{dx^2}. \qquad (9.28)$$

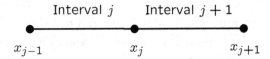

Fig. 9.1. Grid point and interval ordering for spline interpolation.

Evaluation of the above using (9.27) along with the relations (9.25) and (9.26) leads to an equation of the form,

$$\frac{\delta f(x_{j-1})}{\Delta x_j} + 2\delta f(x_j)\left(\frac{1}{\Delta x_j} + \frac{1}{\Delta x_{j+1}}\right) + \frac{\delta f(x_{j+1})}{\Delta x_{j+1}}$$
$$= -\frac{3f(x_{j-1})}{\Delta x_i^2} + 3f(x_j)\left(\frac{1}{\Delta x_j^2} - \frac{1}{\Delta x_{j+1}^2}\right) + \frac{3f(x_{j+1})}{\Delta x_{j+1}^2}. \qquad (9.29)$$

Applying (9.29) for each interval j leads to $N-1$ equations for the $N+1$ estimates of the first derivative $\delta f(x_j)$. Thus, two additional conditions are required to close the tridiagonal system of equations. Various conditions are possible, including requiring that the second derivatives be zero at the endpoints, a condition which yields the so-called natural cubic spline. Alternatively, estimates of δf at the endpoints can also be specified, e.g. the first derivative may be taken to vanish at each endpoint; this choice yields the so-called complete or clamped spline. Many more details and applications can be found in the excellent text by de Boor (1978).

Finally, while it is possible to differentiate between methods for interpolation based on their convergence properties with refinement in the mesh spacing in addition to other aspects such as their computational cost and the overhead introduced into a computer code, interpolation schemes are often applied to solutions possessing a range of time and length scales. In such cases the accuracy of an interpolation method is not uniform with respect to the wavelength of the field being interpolated. Intuitively, we would not expect for a given resolution (i.e. a given grid spacing) that the accuracy of an interpolated quantity with variations that occur on length scales close to the mesh spacing would be as accurately interpolated as one for which the length scale variations are resolved using several grid points. Application of particle-tracking methods in turbulent flows, for example, require interpolation of fluid properties that may possess significant local variation. Interpolation of variables that are well-resolved with respect to a given mesh spacing (such as the large length scales in a turbulent flow) can be accurately interpolated using simple interpolation procedures. For spatial variations occurring over shorter wavelengths, errors in the interpolated solution are generally more significant. The reader is referred to Balachandar and Maxey (1989) for details of an approach to assess the wavelength-dependent errors of some commonly used interpolation schemes.

9.3.2 Integration

In addition to the equation of motion used to determine the particle velocity, the displacement of each particle within the ensemble of N_p particles is evolved according to the kinematic relation (9.5). The differential equations (9.5) and (9.13) constitute initial value problems that describe the velocity and displacement of the particle which, in numerical simulations, require solution using appropriate schemes.

In general, both equations have the form,

$$\frac{dy}{dt} = y' = f(t, y), \qquad y(t = 0) = y_0, \tag{9.30}$$

for which numerical solutions are sought at discrete time intervals $m\Delta t$ where $m = 1, 2, \ldots$ and Δt is the time step, i.e. the interval between successive times at which the solutions are to be calculated. The solution methods for equations of this form commonly used in particle-tracking applications – referred to as time-marching methods – can be applied to linear or nonlinear ordinary differential equations. Such methods compute the numerical solutions as linear combinations of the dependent variable and its derivative y' at various time levels. Following Lomax *et al.* (2001), we denote the numerical solutions schematically as

$$y^{m+1} = g(\beta_1 y'^{m+1}, \beta_0 y'^m, \beta_{-1} y'^{m-1}, \ldots, \alpha_0 y^m, \alpha_{-1} y^{m-1}, \ldots), \tag{9.31}$$

where the coefficients $\beta_1, \beta_0, \beta_{-1}, \ldots$ and $\alpha_0, \alpha_{-1}, \alpha_{-2}, \ldots$ are specified in order to yield an approximation that matches a local Taylor series representation of the exact solution to a given order. The form (9.31) is useful to identify methods as belonging to either of two categories, explicit if $\beta_1 = 0$ and implicit otherwise. As an example, the simplest (and least useful in practice) time-marching scheme is the explicit Euler method,

$$y^{m+1} = y^m + \Delta t f(t^m, y^m), \qquad m = 0, 1, 2, \ldots \tag{9.32}$$

which yields a first-order approximation to the exact solution.

Time-marching methods for advancing the differential equations governing the particle velocity and displacement are dictated by considerations of accuracy and stability. As written above, the values of the coefficients for a particular method used to advance a set of ordinary differential equations can be obtained for a desired accuracy compared to a local Taylor series expansion of the exact solution. Given the local nature of such an estimate, an additional and useful distinguishing feature of a method is its global error, i.e. the error after its application for a finite time interval.

It is important to stress, however, that construction of a method on the basis of a Taylor series analysis to a certain formal order of accuracy does not yield information about the phase or amplitude characteristics of the error, nor does it shed light on its stability features. Amplitude and phase errors are important measures of the suitability of time-marching methods for many applications in particle-laden flows and the reader is referred to Lomax *et al.* (2001) for a detailed discussion.

Time-marching methods often employed in applications include those belonging to the class of linear multistep methods (LMM). For example, the explicit second-order Adams–Bashforth (AB2) method applied to (9.30) takes the form,

$$y^{m+1} = y^m + \frac{\Delta t}{2} \left[3f(t^m, y^m) - f(t^{m-1}, y^{m-1}) \right], \qquad m = 1, 2, \ldots \quad (9.33)$$

which yields estimates y^m possessing an error that is $\mathcal{O}(\Delta t^2)$. Another common LMM applied in practice is the implicit second-order Adams–Moulton (AM2) method, which takes the form,

$$y^{m+1} = y^m + \frac{\Delta t}{2} \left[f(t^{m+1}, y^{m+1}) + f(t^m, y^m) \right]. \qquad m = 0, 1, \ldots \quad (9.34)$$

Similar to AB2, the truncation error associated with (9.34) is also $\mathcal{O}(\Delta t^2)$. While AM2 is a linear multistep method, (9.34) is probably more commonly known as the trapezoid rule or the Crank–Nicholson method. Equation (9.33) shows that AB2 is not "self starting," because the method estimates the solution at level $m+1$ using information at level m and $m-1$; a different method must be applied to evaluate y_1.

Predictor–corrector methods are also common in application for particle tracking with the class of Runge–Kutta schemes widely used. For example, one of the Runge–Kutta schemes that possesses a truncation error that is $\mathcal{O}(\Delta t^2)$ is

$$y^* = y^m + \Delta t f(t^m, y^m)$$
$$y^{m+1} = y^m + \frac{\Delta t}{2} \left[f(t^{m+1}, y^*) + f(t^m, y^m) \right], \qquad m = 0, 1, \ldots \quad (9.35)$$

Important for the choice of a particular time-marching method for particle-tracking applications are its stability characteristics, a notion which in the present context refers to the maximum allowable time step for which the solution (e.g. the particle velocity) remains bounded. It is important to stress that a numerically stable (bounded) solution does not imply an accurate solution. Explicit methods such as (9.33) and (9.35) are subject to limits on the maximum allowable time step for stable computations, while implicit methods are almost always unconditionally stable, i.e. there is no limit on the time step for which the numerical solution will be bounded. We note that not all implicit methods are unconditionally stable, though the second-order Adams–Moulton method (9.34) is an example of an unconditionally stable method for ordinary differential equations. There are numerous references that provide great detail on the properties of schemes for the numerical integration of ordinary differential equations (e.g. see Lomax *et al.*, 2001 and references therein).

Finally, in addition to considerations of numerical accuracy and stability, a particular method dictated for use in a particle-tracking application will be somewhat dependent on the numerical approach used to resolve the Navier–Stokes equations for prediction of the fluid flow. A summary of a standard numerical approach for calculation of incompressible fluid flows is provided in the next section.

9.4 Numerical treatment of the fluid phase

Intrinsic to any particle-tracking application is a prediction of the underlying fluid flow. We have already seen in Chapter 2 a description of the projection or fractional step solution method for the Navier–Stokes equations and a second-order-accurate implementation of it. Here, we summarize a slightly different implementation having the same accuracy. As in our earlier discussion, the focus is on incompressible flows, first outlining in a general way the fractional-step method and then its application to the prediction of the simplest turbulent flow: homogeneous and isotropic turbulence.

There has been a large body of work on the development and analysis of fractional step methods for solution of the incompressible Navier–Stokes equations and, in addition to chapter 2, the reader is referred to Kim and Moin (1985), Dukowicz and Dvinsky (1992), Perot (1993), Burton and Eaton (2002) and references therein for more detailed discussions of many aspects connected to the development below.

In developing a numerical solution to the equations of conservation of mass and momentum for an incompressible fluid, it is useful to write (9.1) and (9.2) in the general form,

$$\mathsf{D}(\mathbf{u}) = 0 \,, \tag{9.36}$$

$$\frac{\partial \mathbf{u}}{\partial t} = \mathsf{N}(\mathbf{u}) + \nu \mathsf{L}(\mathbf{u}) - \mathsf{G}(p) \,, \tag{9.37}$$

where $N(u) = -\nabla \cdot (uu)$ represents the advection term in the momentum equations and D, L, and G are the divergence, Laplacian, and gradient operators, respectively. In this section, the variable p indicates the pressure divided by the fluid density. Numerical solutions of the above equations will introduce discrete representations of these operators, which will depend on the specifics of the numerical approach adopted.

The initial conditions for solution of the above system consist of a prescribed solenoidal velocity field. Boundary conditions on the velocity typically involve specification of the velocity components (e.g. no-slip) or the stresses. Boundary conditions on the pressure are dependent on the numerical representation: in some grid arrangements the pressure variable is not defined on the boundary and, consequently, explicit boundary conditions on the pressure are not required (see Section 2.4). Solution of (9.36) and (9.37) can appear subtle since the relation governing mass conservation imposes a constraint on the velocity field – that it be divergence free – and is not in a form similar to (9.37) which would permit one to use techniques for advancing in time systems of equations. As discussed in Chapter 2, this aspect of incompressible flow modeling motivates a widely applied treatment based on what is known as the fractional-step method.

9.4.1 A second-order fractional-step method for incompressible flows

The particular second-order-accurate implementation of the fractional-step method outline here parallels that presented in Dukowicz and Dvinsky (1992), Lund (1993), and Burton and Eaton (2002); additional variants can be found e.g. in Brown *et al.* (2001). We linearize the advection terms and consider an implicit discretization of (9.37) using the second-order Adams–Moulton method (9.34),

$$\frac{u^{n+1} - u^n}{\Delta t} = \frac{1}{2}\left[N(u)^{n+1} + N(u)^n\right]$$
$$+ \frac{\nu}{2}\left[L(u)^{n+1} + L(u)^n\right] - \frac{1}{2}\left[G(p)^{n+1} + G(p)^n\right], \qquad (9.38)$$

in which, as usual, Δt denotes the time step and the superscripts the time level. Upon rearranging this equation, we obtain

$$\underbrace{\left[\frac{I}{\Delta t} - \frac{1}{2}(N + \nu L)\right]}_{\mathcal{A}} u^{n+1} + \frac{1}{2}G(p^{n+1}) = \underbrace{\left[\frac{I}{\Delta t} + \frac{1}{2}(N + \nu L)\right]}_{\mathcal{B}} u^n - \frac{1}{2}G(p^n),$$

$$(9.39)$$

with I the identity matrix. By combining (9.39) with the continuity equation (9.36), we have

$$
\begin{bmatrix} \mathcal{A} & \frac{1}{2}G \\ D & 0 \end{bmatrix} \begin{bmatrix} u^{n+1} \\ p^{n+1} \end{bmatrix} = \begin{bmatrix} \mathcal{B} & -\frac{1}{2}G \\ 0 & 0 \end{bmatrix} \begin{bmatrix} u^n \\ p^n \end{bmatrix}. \tag{9.40}
$$

Solution of the above system is more efficient if, following Dukowicz and Dvinsky (1992) and Burton and Eaton (2002), we approximately factor it as

$$
\begin{bmatrix} \mathcal{A} & 0 \\ 0 & I \end{bmatrix} \begin{bmatrix} I & \frac{1}{2}\Delta tG \\ D & 0 \end{bmatrix} \begin{bmatrix} u^{n+1} \\ p^{n+1} \end{bmatrix}
$$
$$
= \begin{bmatrix} \mathcal{B} & 0 \\ 0 & I \end{bmatrix} \begin{bmatrix} I & -\frac{1}{2}\Delta tG \\ 0 & 0 \end{bmatrix} \begin{bmatrix} u^n \\ p^n \end{bmatrix} \tag{9.41}
$$
$$
+ \begin{bmatrix} \frac{1}{4}\Delta t\,(N+\nu L)G(p^{n+1}-p^n) \\ 0 \end{bmatrix}.
$$

The second term on the right-hand side of this equation is the factorization error and is proportional to Δt^2, since $p^{n+1}-p^n$ is $\mathcal{O}(\Delta t)$, and therefore formally of the same order as the truncation error associated with the time advance, which is also accurate to $\mathcal{O}(\Delta t^2)$.

The simplification introduced by neglecting the second term on the right-hand side of (9.41) presents an advantage for the numerical solution, as we now show. If we associate intermediate variables with the inner systems above, we can write

$$
\begin{bmatrix} I & \frac{1}{2}\Delta tG \\ D & 0 \end{bmatrix} \begin{bmatrix} u^{n+1} \\ p^{n+1} \end{bmatrix} = \begin{bmatrix} \widehat{u} \\ \widehat{p} \end{bmatrix}, \quad \begin{bmatrix} I & -\frac{1}{2}\Delta tG \\ 0 & 0 \end{bmatrix} \begin{bmatrix} u^n \\ p^n \end{bmatrix} = \begin{bmatrix} u^* \\ p^* \end{bmatrix}. \tag{9.42}
$$

With these new variables, (9.41) can now be written as

$$
\begin{bmatrix} \mathcal{A} & 0 \\ 0 & I \end{bmatrix} \begin{bmatrix} \widehat{u} \\ \widehat{p} \end{bmatrix} = \begin{bmatrix} \mathcal{B} & 0 \\ 0 & I \end{bmatrix} \begin{bmatrix} u^* \\ p^* \end{bmatrix}. \tag{9.43}
$$

Note that the second term on the right-hand side of (9.41) has now been neglected. The advantage of the approximate factorization now becomes more clear because (9.43) shows that the pressure has been decoupled from the equations governing the velocity, i.e.

$$
\mathcal{A}\widehat{u} = \mathcal{B}u^* \tag{9.44}
$$
$$
\widehat{p} = p^*. \tag{9.45}
$$

Further, (9.42) shows the sequence of steps,

$$\mathbf{u}^{n+1} + \frac{1}{2}\Delta t \mathsf{G}(p^{n+1}) = \widehat{\mathbf{u}} \tag{9.46}$$

$$\mathsf{D}(\mathbf{u}^{n+1}) = \widehat{p} \tag{9.47}$$

$$\mathbf{u}^n - \frac{1}{2}\Delta t \mathsf{G}(p^n) = \mathbf{u}^* \tag{9.48}$$

$$p^* = 0, \tag{9.49}$$

and that (9.45), (9.47), and (9.49) can be combined to recover the continuity constraint, $\mathsf{D}(\mathbf{u}^{n+1}) = 0$. Taking the divergence of (9.46) and utilizing the fact that the fluid velocity field at the next time level, $\mathbf{u}^{n+1}$, is divergence-free, yields a Poisson equation for the pressure,

$$\mathsf{DG}(p^{n+1}) = \frac{2}{\Delta t}\mathsf{D}(\widehat{\mathbf{u}}). \tag{9.50}$$

Equations (9.44), (9.46), (9.48), and (9.50) constitute the fractional-step algorithm which, in a logical order, can be rearranged as

$$\mathbf{u}^* = \mathbf{u}^n - \frac{1}{2}\Delta t\, \mathsf{G}(p^n) \tag{9.51}$$

$$\mathcal{A}\widehat{\mathbf{u}} = \mathcal{B}\mathbf{u}^* \tag{9.52}$$

$$\mathsf{DG}(p^{n+1}) = \frac{2}{\Delta t}\,\mathsf{D}(\widehat{\mathbf{u}}) \tag{9.53}$$

$$\mathbf{u}^{n+1} = \widehat{\mathbf{u}} - \frac{1}{2}\Delta t\, \mathsf{G}(p^{n+1}). \tag{9.54}$$

As written above, $\mathbf{u}^*$ computed in the first step represents an estimate of $\mathbf{u}^{n+1}$ in which half of the effect of the pressure gradient is taken into account. In the second step (9.52), the velocity $\widehat{\mathbf{u}}$ is a refined estimate that now includes the effects of convection and diffusion. Solution of the Poisson equation in (9.53) accounts for the remaining half of the pressure gradient and addition of the pressure gradient in (9.54) yields a fluid velocity field $\mathbf{u}^{n+1}$ which satisfies the continuity constraint.

9.4.2 Boundary conditions

The development beginning with (9.38) considered a discretization of the Navier–Stokes equations using time-advance schemes that are second-order accurate with respect to time and the use of an approximate factorization to simplify the system, eventually leading to the series of steps constituting the fractional step method. Not addressed in the preceding development is the imposition of boundary conditions. One distinction is whether the boundary conditions are applied to the system before or after it is approximately factored.

A logical approach that is discussed in greater detail in Dukowicz and Dvinsky (1992), Perot (1993), and Burton and Eaton (2002) is to apply the boundary conditions on the system before it is approximately factored. This in turn implies that the operators N, L, G, and D reflect not only the details of the approach used for the spatial discretization (e.g. finite difference, finite element, etc.), but also incorporate the effects of the boundary condition. Application of the boundary conditions to the system prior to it being approximately factored offers the strong advantage that the question of boundary conditions on the intermediate variables $\hat{\mathbf{u}}$ and $\mathbf{u}^*$ never arises. Further, if the grid points on which the pressure is computed are interior to the domain, then boundary conditions on the pressure are not explicitly required (but see the discussion on this delicate point in Chapter 2, Section 2.4). Such is the case using methods that resolve the Navier–Stokes equations on a network of grid points in which the dependent variables are staggered with respect to the center of a grid cell.

9.4.3 A Fourier series method for isotropic turbulence

The development of the fractional step method in Section 9.4.1 illustrated the basic outline of the approach and did not depend on a particular discretization of the spatial derivatives in the Navier–Stokes equations. As we have seen in earlier chapters, many approaches to the numerical solution of (9.36) and (9.37) are based on the discretization of the spatial derivatives using finite difference or finite volume approaches. The advantage of such approaches is the relative ease in computing the fluid flow within complicated geometries and in the presence of complex boundary conditions (e.g. with turbulent inflow and outflow).

While of great utility in many applications, finite difference approximations are less accurate than specialized methods which we now consider. This can be motivated by considering an arbitrary periodic function that is decomposed into its Fourier components, each component proportional to $e^{i\kappa x}$, where κ represents the wavenumber and $i = \sqrt{-1}$. Approximations to the spatial derivatives using finite differences introduce errors that are sensitive to the resolution of the function, i.e. its wavelength relative to the grid spacing. It is instructive to consider the consequences of such approximations by differentiating a representative component of our periodic function, $e^{i\kappa x}$.

It is sufficient to consider the first derivative, the exact value of which is

$$\frac{d}{dx}e^{i\kappa x} = i\kappa e^{i\kappa x}.$$ (9.55)

The standard centered approximation at the grid point x_j leads to

$$\frac{e^{i\kappa x_{j+1}} - e^{i\kappa x_{j-1}}}{2\Delta x} = \frac{(e^{i\kappa \Delta x} - e^{-i\kappa \Delta x})e^{i\kappa x_j}}{2\Delta x}$$

$$= i\frac{\sin \kappa \Delta x}{\Delta x}e^{i\kappa x_j}.$$ (9.56)

Comparison of (9.56) and (9.55) allows us to define what is known as the modified wavenumber associated with the centered approximation of the first derivative:

$$\kappa' = \frac{\sin \kappa \Delta x}{\Delta x},$$ (9.57)

(e.g. see Lomax *et al.* 2001). The modified wavenumber for the finite-difference expression (9.56) appears in place of the true wavenumber κ in the exact expression (9.55). The degree to which the modified wavenumber matches the exact wavenumber provides a measure of the accuracy of the finite-difference expression which provides useful information in addition to the truncation error, especially for differentiation of functions that possess variations over a range of wavelengths. The modified and exact wavenumbers are plotted in Fig. 9.2. The figure shows that the modified wavenumber follows closely the exact wavenumber for small $\kappa\Delta x$ and

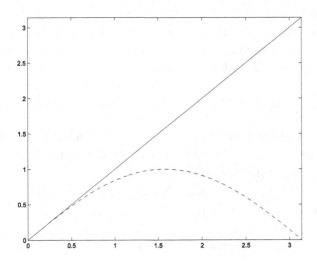

Fig. 9.2. The modified wavenumber for the central difference approximation of the first derivative.

then shows greater discrepancy for increasing $\kappa \Delta x$. Larger values of the wavenumber correspond to shorter wavelengths, and Fig. 9.2 indicates that the shorter wavelengths of the solution will not be treated as accurately as the longer wavelengths when the first derivative is approximated by centered finite differences.

Accuracy considerations over a range of wavelengths provide motivation for numerical approaches that incorporate a higher fidelity treatment of functions than can be achieved using finite differences. For some configurations, especially those in relatively simple geometries and for certain boundary conditions, it is possible to solve the Navier–Stokes equations using what are known as spectral numerical methods. In the following we consider one of the simplest examples – the solution of the Navier–Stokes equations within a cubic domain of side L with periodic boundary conditions applied in each of the three coordinate directions.

The periodic nature of the velocity and pressure fields solutions of this problem suggests expanding the dependent variables in a Fourier series, e.g. for the fluid velocity,

$$\mathbf{u}(\mathbf{x}, t) = \sum_{\kappa} \widehat{\mathbf{u}}(\boldsymbol{\kappa}, t)e^{i\boldsymbol{\kappa} \cdot \mathbf{x}}, \tag{9.58}$$

where $\widehat{\mathbf{u}}$ are the Fourier coefficients and each one of the components of the wavenumbers $\boldsymbol{\kappa}$ has the form

$$\kappa_m = \frac{2\pi m}{L}, \qquad m = -\frac{N}{2}, \dots 0, \dots \frac{N}{2} - 1. \tag{9.59}$$

The computation of the Fourier coefficients is numerically efficient because of the existence of the fast Fourier transform. Indeed, it is well known from the theory of this algorithm that the use of an equally spaced network of grid points (x_i, y_j, z_k), with

$$(x_i, y_j, z_k) = (i, j, k)\frac{L}{N}, \qquad i, j, k = 0, 1, \dots, N-1, \tag{9.60}$$

permits an efficient calculation of the Fourier coefficients. Conversely, knowledge of the latter permits the efficient reconstruction of the function at the points (x_i, y_j, z_k) of physical space.

Upon using the representation (9.58), spatial derivatives are obtained directly from the expansion, e.g.

$$\frac{\partial \mathbf{u}}{\partial x_j} = \sum_{\kappa} i\kappa_j \widehat{\mathbf{u}}(\boldsymbol{\kappa}, t)e^{i\boldsymbol{\kappa} \cdot \mathbf{x}}. \tag{9.61}$$

This equation shows that the Fourier coefficients of the derivative are $i\kappa_j \widehat{\mathbf{u}}$.

Thus, the use of the Fourier series transforms the operation of differentiation in physical space to multiplication by the wavenumber in Fourier space.

Applied to the Navier–Stokes equations, (9.58) for the velocity and the analogous expansion for the pressure result in a set of ordinary differential equations for the Fourier coefficients that we may write as

$$\frac{d\widehat{\mathbf{u}}}{dt} + \mathcal{F}\left(\mathbf{N}(\mathbf{u})\right) = -i\kappa\widehat{p} - \nu\kappa^2\widehat{\mathbf{u}} \tag{9.62}$$

where $(i\boldsymbol{\kappa}) \cdot (i\boldsymbol{\kappa}) = -\kappa^2$ denotes the square of the modulus of the wavenumber vector. The second term on the left-hand side of (9.62) denotes the Fourier transform of the nonlinear terms and the particular approach chosen for its calculation dictates the solution method for (9.62). Of importance in the computation of the nonlinear terms is the realization that formation of the product $\mathbf{u}\mathbf{u}$ in physical space (e.g. over a network of the N^3 grid points 9.60). The subsequent transformation to Fourier space results in components of the resulting Fourier series that reside at wavenumbers higher than those which can be supported on the grid. Because there is not sufficient resolution to support these contributions, these wavenumber components are "aliased" and appear as lower wavenumber contributions, an aspect that can lead to inaccurate solutions and possibly numerical instability (e.g. see Canuto *et al.*, 1988; Kravchenko and Moin, 1997).

Common approaches to the removal of aliasing errors include the use of such schemes as the "3/2 rule" in which Fourier transforms of length $3N/2$ are used to form the nonlinear terms. In this approach, the Fourier coefficients $\widehat{u}_i$, originally of length N in each dimension, are expanded to length $3N/2$ with zeros occupying the higher wavenumber components of the padded arrays. Subsequent transformation of the longer length arrays to physical space, formation of the velocity product over the grid, and then Fourier transformation still over the longer arrays accommodates the higher wavenumber contributions to the product. These contributions are then discarded so that the resulting Fourier transform of the product to physical space is of length N in each dimension. Spatial derivatives are then formed by multiplying the resulting set of Fourier coefficients by the wavenumber. The elimination of the aliasing error comes at the cost of computing longer transforms. The reader is referred to Canuto *et al.* (1988) for further discussion of these issues, other approaches to formation of the nonlinear terms, and a more in-depth development of spectral numerical methods in a wide range of applications.

An additional advantage of the use of Fourier series for our flow within a cubic domain can be illustrated by considering a very simple fractional-step method applied to (9.62), written as,

$$\frac{\widehat{\mathbf{u}}^* - \widehat{\mathbf{u}}^n}{\Delta t} = \left[-\mathcal{F}(\mathbf{N}(\mathbf{u})) - \nu \kappa^2 \widehat{\mathbf{u}} \right]^n, \tag{9.63}$$

where $\widehat{\mathbf{u}}^*$ represents the component of the solution advanced by the nonlinear and viscous terms as in Section 2.3 of Chapter 2. The Fourier coefficients at the new time level incorporate the effect of the pressure gradient:

$$\frac{\widehat{\mathbf{u}}^{n+1} - \widehat{\mathbf{u}}^*}{\Delta t} = -i\boldsymbol{\kappa}\widehat{p}, \tag{9.64}$$

and, as described in the previous section, we now determine the pressure by requiring the velocity at the new time level be divergence free. From the differentiation rule (9.61) it is clear that, in Fourier space, the continuity constraint dictates that the Fourier coefficients be orthogonal to the wavenumber vector:

$$\nabla \cdot \mathbf{u} = 0 \Leftrightarrow i\boldsymbol{\kappa} \cdot \widehat{\mathbf{u}} = 0. \tag{9.65}$$

Thus, by (9.63), multiplication of (9.64) by $i\boldsymbol{\kappa}$ leads to

$$\kappa^2 \widehat{p} = i\boldsymbol{\kappa} \cdot \frac{\widehat{\mathbf{u}}^*}{\Delta t} = i\boldsymbol{\kappa} \cdot \left[-\mathcal{F}(\mathbf{N}(\mathbf{u})) - \nu \kappa^2 \widehat{\mathbf{u}} \right]^n, \tag{9.66}$$

which is immediately solved for $\widehat{p}$. By substituting the Fourier coefficient of the pressure field obtained from this equation into (9.64) we find the Fourier coefficient of the divergenceless velocity field at the new time level. The method as written above is only first-order accurate in time and therefore insufficient for applications, though it shows one of the operational advantages, namely the ease of solution of the Poisson equation (9.66).

When generalized to temporal integration schemes of higher order accuracy, the fractional-step method using Fourier series as a means for achieving high accuracy over the entire range of resolved wavenumbers provides a useful tool for studying the most basic turbulent flow: homogeneous and isotropic turbulence. The initial conditions for a simulation of isotropic turbulence are nearly always specified in terms of the radial energy spectrum, e.g.

$$E(\kappa) = \frac{K}{\kappa_p} \left(\frac{\kappa}{\kappa_p} \right)^2 \exp\left(-2\frac{\kappa}{\kappa_p} \right), \tag{9.67}$$

where K is the kinetic energy of the fluid in the initial state and κ_p is the wavenumber corresponding to the peak in the radial spectrum $E(\kappa)$. The relation (9.67) is only one of many possible initial radial spectra and shows that the initial conditions require two parameters – a velocity and length scale – from which the amplitudes of the Fourier coefficients $\widehat{\mathbf{u}}$ may be generated. We note that initial conditions constructed from (9.67) allow determination of the initial amplitudes of the Fourier coefficients of the velocity, but do not constrain the phases of the Fourier coefficients. It should also be noted that, in the absence of an external mechanism to maintain its energy, isotropic turbulence will decay from its initial condition. This feature can complicate analysis and interpretation and in many studies an acceleration is added to the right-hand side of (9.62) which, at steady state, supplies energy to the flow that is balanced by viscous dissipation at the smallest scales. The reader is referred to Overholt and Pope (1998) for further discussion of these issues and various approaches to schemes used for forcing isotropic turbulence.

Regardless of the specific flow type, i.e. decaying or forced isotropic turbulence, in particle tracking applications it is important to maintain adequate resolution of the small-scale motions in the fluid. In turbulent flows the measure most commonly used is the value of $k_{\max}\eta$ where $k_{\max}$ is the highest resolved wavenumber and η is the Kolmogorov length scale. The highest resolved wavenumber $k_{\max}$ is dependent upon the de-aliasing scheme used. Balachandar and Maxey (1989) have shown that for low-order statistics, such as the particle kinetic energy, $k_{\max}\eta > 1$ is sufficient, with higher values of $k_{\max}\eta$ needed for the accurate interpolation of higher order quantities such as the velocity gradient.

9.5 Gas–solid flows

In this section we present a discussion of issues specific to point-particle methods when applied to gas–solid flows. The purpose is to highlight some of the additional considerations in this type of flows and then show recent results from two investigations on particle-laden turbulent flows.

9.5.1 The undisturbed fluid velocity in point-particle treatments

As explained in Section 9.2, for a flow containing N_p particles, the fluid velocity $\tilde{u}^n$ necessary to calculate the force on the n-th particle must be calculated taking into account the disturbances created by all other $N_p - 1$ particles in the flow but excluding that generated by the n-th particle itself. In principle, this in turn would require that a total of N_p flow fields are necessary to determine the motion of each particle. Only if the system is so dilute that the influence of the particles on the fluid flow can be neglected is $\tilde{u}^n$ the same for each particle within the system and identical to that calculated

from a solution of the single-phase fluid flow equations. In systems for which
the properties of the carrier flow are modified by momentum exchange with
the particles, and only in limiting regimes would it be possible to rigorously
construct $\tilde{\mathbf{u}}^n$. In general, approximations are required.

One approach that could be used to obtain the locally undisturbed fluid
velocities $\tilde{\mathbf{u}}^n$ for each particle would be to resolve the flow field $\mathbf{u}$ influenced
by the entire assembly of particles and then subtract the local perturbation
induced by the presence of particle n. Such an approach is feasible in the
linear limit in which it would be possible to precisely superpose the disturbances created by each particle with the surrounding fluid flow. As an
example, we consider very small particles such that the disturbance velocity
created in the fluid can be described using the linear solution developed by
Stokes. Due to the slow decay of velocity disturbances in these conditions,
this application further restricts the present analysis to dilute regimes, i.e.
to interparticle separations that are large compared to the particle diameter.

Under the point-force approximation, it is possible to show that $\mathbf{u}$ is the
sum of $\tilde{\mathbf{u}}^n$ and a local perturbation, the so-called Stokeslet (see, e.g. Happel
and Brenner, 1983; Saffman, 1973; Kim and Karrila, 1991):

$$\mathbf{u} = \tilde{\mathbf{u}}^n + \frac{3d}{4r}\left[\hat{\boldsymbol{I}} + \frac{\mathbf{rr}}{r^2}\right] \cdot \mathbf{v}^n \quad \text{with} \quad \mathbf{r} = \mathbf{x} - \mathbf{x}_p^n, \quad r = \mid \mathbf{x} - \mathbf{x}_p^n \mid, \quad (9.68)$$

where $\hat{\boldsymbol{I}}$ is the identity two-tensor. To obtain the locally undisturbed velocity
field for each particle, an iteration procedure could be developed using (9.68).
For a flow with a large number of particles, however, the computational cost
becomes prohibitive.

Equation (9.68) also sheds light on the representations of the disturbance
fields possible using the point-force approximation. For the Stokeslet given
by (9.68), the perturbation in the fluid due to the presence of a particle decays as the sum of two contributions, one proportional to $1/r$ ("long range")
and the other to $1/r^3$ ("short range"). For particles that are small relative
to the smallest relevant length scales of the flow, and for particles separated
by a distance L large compared to their diameter d, the most important
interactions are long range (see also Koch, 1990). The neglect of short-
range interactions is justifiable for particles with diameters smaller than the
Kolmogorov length scale of the flow field undisturbed by the presence of
the particle since in that case short-range perturbations are dissipated by
viscosity.

For practical purposes, with a large sample of particles N_p, and for intermediate particle Reynolds numbers, it is necessary to assume that, for each
particle, the locally undisturbed fluid velocity field $\tilde{\mathbf{u}}^n$ can be approximated

by $\mathbf{u}^n$. Therefore, the coupling force $\mathbf{f}_d^n$ may be expressed as

$$\mathbf{f}_d^n = \rho_p \frac{\mathbf{v}^n - \tilde{\mathbf{u}}^n}{\tau_p^n} \approx \rho_p \frac{\mathbf{v}^n - \mathbf{u}^n}{\tau_p^n} . \tag{9.69}$$

An estimate of the error made when using this approximation can be obtained from (9.68). If $\Delta \mathbf{u}^n$ denotes the error, (9.68) shows that

$$|\Delta \mathbf{u}^n| = |\tilde{\mathbf{u}}^n - \mathbf{u}| = \frac{3d}{4r} \left| \left(\hat{\boldsymbol{I}} + \frac{\mathbf{rr}}{r^2} \right) \cdot \mathbf{v}^n \right| . \tag{9.70}$$

In actual computations, the distance r between particle n and the neighboring grid nodes (where the locally undisturbed fluid velocity is approximated by the solution of the fluid-flow equations, $\mathbf{u}$) is of the order of the mesh size. Thus, on average, the relative error resulting from the use of (9.69) is $\mathcal{O}(d/\Delta x)$ and, in addition to the restriction that the particle size be small compared to the smallest relevant length scale of the carrier flow, the condition imposed by approximating the locally undisturbed velocity $\tilde{\mathbf{u}}^n$ by $\mathbf{u}^n$ requires that $d \ll \Delta x$.

9.5.2 Implementation of the coupling force

As described above, the system (9.1) and (9.4) governing the fluid flow using a point representation of the particulate phase and with the form of the coupling force written for gas–solid flows as shown in (9.69) implies several simplifications such as a negligible volumetric fraction of the dispersed phase and particle sizes that are smaller than the smallest relevant length scales in the carrier flow. In turbulent flows, the set (9.1), (9.4) together with an appropriate equation of motion such as (9.16) has been considered by several investigators (e.g. see Squires and Eaton, 1990; Elghobashi and Truesdell, 1993; Boivin *et al.*, 1998; Sundaram and Collins, 1999). In these studies the equations governing the fluid flow are solved without recourse to an explicit turbulence model at any scale of motion. As we have seen, the computations are not free from empiricism, however, since the force of the particle on the fluid is not obtained in an exact fashion by integrating the fluid stress over the particle surface, but is prescribed using a correlation. Furthermore, implicit in these approaches is a filtering of the solutions and an assumption that the local perturbations in the fluid flow due to the particle can be neglected, a reasonable assumption when the particle diameter is very small. While, due to these factors, it would be unjustified to refer to these calculations as "direct numerical simulations", this denomination is used in the literature to emphasize the fact that no turbulence model is used for the fluid flow.

These aspects can be clarified by considering the coupling term in (9.4),

$$\sum_{n=1}^{N_p} \mathbf{f}^n(\mathbf{x}_p^n, t)\delta(\mathbf{x} - \mathbf{x}_p^n), \tag{9.71}$$

where $\mathbf{f}^n(\mathbf{x}_p^n, t)$ is the force acting on particle n with center at position $\mathbf{x}_p^n$ at time t. In practice, numerical approximations are required to incorporate this relation into the integration of the equations for the fluid flow by "projecting" this coupling force from the particle position onto the surrounding grid nodes. This procedure is equivalent to a filtering and can be formally expressed as

$$\sum_{n=1}^{N_p} \int \mathbf{f}^n(\mathbf{x}_p^n, t)H_\Delta(\mathbf{x}_p^n - \mathbf{y})\, d\mathbf{y} \tag{9.72}$$

where H_Δ is a three-dimensional spatial low-pass filter with a characteristic width of the order of the mesh size Δx satisyfing

$$\int_V H_\Delta(\mathbf{x} - \mathbf{y})d\mathbf{y} = 1. \tag{9.73}$$

and the integration is taken over the entire computational volume V.

Some force coupling models also assume that groups of neighboring particles can be represented by a single test particle that is actually tracked in a computation. Such an approximation introduces further restrictions, in particular that the average spacing between test particles (i.e. computational particles actually tracked in the simulation) be much smaller than the integral length scale of the gas-phase fluid flow. In this case, the force in (9.72) is weighted by n_p, the number of actual particles represented by each computational particle p. For example, each computational particle represented $n_p = 100$ actual particles in the simulations performed by Elghobashi and Truesdell (1993).

The form of the filter function H_Δ is dictated by the method used for projection of the coupling force from the particle to the surrounding grid nodes. As an example, Elghobashi and Truesdell (1993) have used the top-hat filter,

$$H_\Delta(\mathbf{x} - \mathbf{y}) = \begin{cases} 1/V_\Delta & \text{if } |x_i - y_i| < \Delta x/2 \quad i = 1, 2, 3 \\ 0 & \text{otherwise} \end{cases} \tag{9.74}$$

where $V_\Delta = \Delta x^3$ is the volume of a computational cell. Alternatively, the filter H_Δ can be expressed in terms of a volume-weighted averaging, taking the form,

$$H_\Delta(\mathbf{x} - \mathbf{y}) = \begin{cases} (1-r_1)(1-r_2)(1-r_3) & \text{if } |x_i - y_i| < \Delta x \quad i = 1, 2, 3 \\ 0 & \text{otherwise} \end{cases}$$

$$(9.75)$$

where $r_i = |x_i - y_i|/\Delta x$. Using (9.75) along with (9.69) and (9.72) allows us to write the form of the coupling force under a point-particle treatment actually used in simulations:

$$\frac{1}{V_\Delta} \sum_{n_{p,c}} \alpha_{p,c} \left[\frac{m_p}{\tau_p^n} (\mathbf{v}^n - \mathbf{u}) \right], \tag{9.76}$$

where $n_{p,c}$ represents the number of particles that occupy the cells adjacent to the grid point under consideration, and $\alpha_{p,c}$ is a weighting factor that depends on the distance between particle n and each of the surrounding eight grid nodes.

9.6 Examples

We conclude this chapter by discussing two typical examples of studies employing the point-particle approach outlined in the previous sections. The first example concerns the transport of particles in homogeneous isotropic turbulence. The second example describes a study of a similar interaction in which, however, the particles are not homogeneously dispersed in the fluid but are injected in a "slab"-like region in the middle of a particle-free turbulent flow.

9.6.1 Particle-laden homogeneous and isotropic turbulence

The first example is an application of the techniques outlined above to calculate the particle motion in statistically stationary, homogeneous, and isotropic turbulence. The computations are from the work of Fevrier *et al.* (2005) and Fede (2004) in which the gas-phase turbulent carrier flow is computed using direct numerical simulation on a grid comprised of 128^3 mesh points. The Taylor-scale Reynolds number of the statistically stationary flow was $Re_\lambda = 52$. This Reynolds number provides a useful characterization of the flow and is defined by

$$Re_\lambda = \frac{u_{\mathrm{rms}}\lambda}{\nu}, \qquad u_{\mathrm{rms}} = \left[\frac{2}{3}q_f^2\right]^{1/2}, \qquad \lambda = \left[15\nu\frac{u_{\mathrm{rms}}^2}{\varepsilon}\right]^{1/2}, \tag{9.77}$$

where u_{rms} is the root-mean-square velocity fluctuation in the gas-phase turbulent flow, $\frac{1}{2}q_f^2$ is the fluid turbulence kinetic energy, and ε is the dissipation rate of the fluid turbulence. The length scale λ is known as the transverse Taylor microscale since it is related to the spatial correlation of the fluid velocities separated along a line that is normal to the velocity vector (see Pope, 2000 for additional details and discussion of two-point statistical measures). The small-scale resolution as characterized in terms of the parameter $k_{\max}\eta$ and discussed above was larger than unity.

Computation of a statistically stationary isotropic turbulent flow requires the addition of a stirring force that acts on the large scales of motion of the fluid. Use of a stirring force circumvents problems associated with the decay of isotropic turbulence – primarily the long-lasting influence of the initial conditions and the change in the length and time scales with time. For gas–solid flows this can complicate interpretation since the character of the interactions of the particles with the turbulent fluid flow is changing throughout the course of the simulation.

A widely applied method for achieving a statistically stationary flow and employed in the work of Fevrier *et al.* (2003) is that developed by Eswaran and Pope (1988). Their method can be expressed by augmenting the right-hand side of the Navier–Stokes equations with an additional acceleration. With this modification, equation (9.62) introduced above becomes

$$\frac{d\widehat{u}_{f,j}}{dt} + \mathcal{F}(N(u_{f,j})) = -i\kappa_j\widehat{p} - \nu\kappa^2\widehat{u}_{f,j} + \widehat{a}_{F,j}\,, \qquad (9.78)$$

where $a_{F,i}$ is the new forcing. In Eswaran and Pope (1988) the Fourier coefficients of this forcing are evolved in time according to an Uhlenbeck–Ornstein (UO) stochastic process. We note that artificially forcing the flow to achieve a statistically stationary condition introduces additional parameters into the simulations, including the forcing amplitude and time scale of the UO process. In addition, the range of wavenumbers over which the forcing is applied must also be considered. The reader is referred to Eswaran and Pope (1988), Yeung and Pope (1989), and Overholt and Pope (1998) for a detailed discussion of these and other issues.

The work reported in Fevrier *et al.* (2003) considers the interactions between the turbulent fluid flow and particles in the limit of one-way coupling, i.e. neglecting the momentum effect of the particles on the carrier fluid flow (equivalent to setting the coupling force in equation 9.76 to zero). In the limit of one-way coupling, the undisturbed fluid velocity needed for evaluation of the drag force in (9.69) is identical to the value computed from the solution of the Navier–Stokes equations and is available on the grid.

The interpolation scheme used to obtain the fluid velocity at the particle position can be assessed by comparison of energy spectra of the turbulent

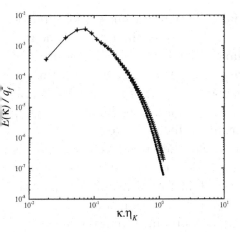
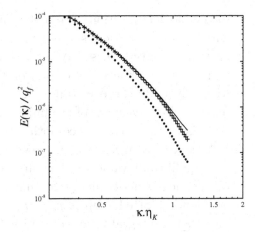

Fig. 9.3. Influence of interpolation on energy spectra: (a) energy spectra on the grid and following interpolation using cubic splines or linear interpolation; (b) zoomed view of the higher-wave-number range. Solid line: baseline on the grid; $+ + + ++$ after cubic spline interpolation; line with circles: after linear interpolation.

flow on the grid to that interpolated to an ordered array of positions located at the centers of each computational cell. The energy spectrum, $E(\kappa)$, yields information on the distribution of the kinetic energy of the flow with respect to wavenumber and its integral is the kinetic energy of the fluid turbulence,

$$\frac{1}{2}q_f^2 = \int E(\kappa)\,d\kappa. \tag{9.79}$$

The energy spectrum obtained following linear and cubic spline interpolation from the grid nodes to cell centers is compared to the reference spectrum on the grid in Fig. 9.3(a). At low wavenumbers (large scales of the fluid flow) there is adequate recovery of the energy spectrum by both interpolation schemes. The figure also shows that at smaller length scales (larger wavenumbers) there is a larger difference in the spectra obtained using linear interpolation or cubic splines, with the cubic spline interpolation yielding a spectrum in better agreement with that obtained on the grid. The figure illustrates the difficulty in accurately interpolating functions with variations that occur over relatively short length scales. This aspect is more clearly observed in Fig. 9.3(b) in which only the higher wavenumber range of Fig. 9.3(a) is shown. The figure shows that liner interpolation results in a larger underestimation of the energy spectrum compared to that obtained using cubic spline interpolation.

In the simulations, particles are introduced into the computational domain at random initial positions and with typical initial conditions on the particle velocity equal to zero or the value of the fluid velocity interpolated to the particle position. From the initial condition, a transient of approximately three particle response times is required for equilibration of the rms particle velocity (e.g. see Wang and Maxey, 1993). Statistical measures are then sampled for several (at least on the order of 10) "eddy turnover times," where the eddy turnover time can be defined in terms of the integral length scales and rms velocity of the fluid turbulence or using the kinetic energy q_f^2 and dissipation rate ε since both yield comparable estimates.

A statistical measure of interest in particle-laden isotropic turbulence is the equilibrium value of the particle-phase kinetic energy. The ratio of the particle kinetic energy to that of the fluid turbulence at the particle position is plotted against $\tau_{f@p}^t/\tau_{fp}^F$ in Fig. 9.4. The time scale $\tau_{f@p}^t$ is the Lagrangian integral time scale of the fluid flow obtained from the integral of the Lagrangian correlation of fluid element velocities following the particle. The time scale τ_{fp}^F is an averaged particle relaxation time,

$$\tau_{fp}^F = \frac{\rho_p}{\rho}\frac{d_p^2}{18\nu}\frac{1}{\left\langle \left(1 + 0.15 Re_p^{n^{0.687}}\right)\right\rangle}. \tag{9.80}$$

Figure 9.4 shows a reduction in the particulate-phase kinetic energy with increasing τ_{fp}^F, consistent with the increasingly sluggish response of a particle to the local fluid flow for larger particle inertia. Also shown in the figure is a theoretical estimate developed by Tchen (1947) and extended by Deutsch and Simonin (1991),

$$q_p^2 = \frac{\eta_r}{1 + \eta_r}q_{f@p}^2, \tag{9.81}$$

where $q_{f@p}^2$ is used to denote the fluid kinetic energy following the particle and $\eta_r = \tau_{f@p}^t/\tau_{fp}^F$ is the time scale ratio. As shown by Fig. 9.4, the dependence of q_p^2 on particle inertia is accurately predicted using (9.81) for larger-inertia particles. For smaller particle response times the figure shows that the theoretical prediction is below the simulation results. Equation (9.81) is developed by assuming that the fluid velocity correlation function along the particle path is a decaying exponential. As shown by Fede (2004), for the relatively low Reynolds numbers considered in the DNS the exponential form is a less accurate approximation for the smaller particle response times.

In addition to statistical measures of the particle-phase motion, structural interactions between particles and turbulence are of considerable interest. An example of these interactions is provided in Fig. 9.5. Shown in the figure are the instantaneous particle positions and fluid velocity vectors

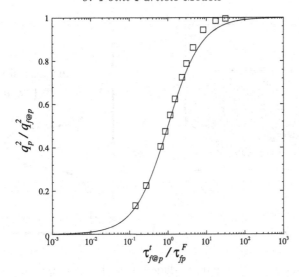

Fig. 9.4. Effect of particle inertia on the particulate-phase kinetic energy. □ results from DNS of isotropic turbulence and Lagrangian particle tracking by Fede (2004); ———— theoretical prediction using (9.81).

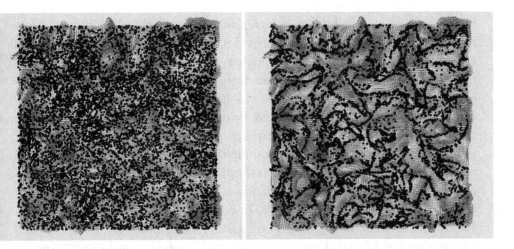

Fig. 9.5. Instantaneous particle positions in a single plane from a computation of particle-laden isotropic turbulence: (a) heavy particles; (b) light particles. See also http://www.cambridge.org/comp_mult_flow.

from a single plane of the DNS. The response time of the heavy particles made dimensionless using the Kolmogorov time scale of the fluid turbulence, $(\nu/\varepsilon)^{1/2}$, is 11.1. For the light particles the corresponding dimensionless response time is 0.96. Figure 9.5(a) shows that the heavy particles have an essentially random spatial distribution. The light particles in Fig. 9.5(b),

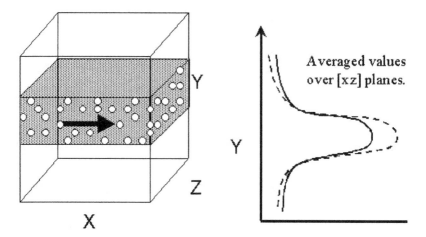

Fig. 9.6. Schematic of the particle-laden slab flow computed by Vermorel *et al.* (2003). Particles are injected within a "slab" occupying the central region of the computational domain. Statistics are accumulated over the homogeneous xz planes, leading to results that are a function of the y-coordinate.

on the other hand, exhibit a substantially different distribution and more organized structure. The figure provides an example of an inertial bias in particle motion, which results in higher particle concentrations in regions of low vorticity and/or high strain rate. Inertial bias leading to a preferential concentration of particles was first shown via analysis by Maxey (1987) and subsequently confirmed using DNS of the carrier phase flow along with particle tracking by Squires and Eaton (1991). Several other studies have provided detailed examinations of the effects of preferential concentration on particle concentration fields, settling velocity, interparticle collisions, and turbulence modulation (e.g. see Wang and Maxey, 1993; Sundaram and Collins, 1997, 1999; Boivin *et al.*, 1998; Wang *et al.*, 2000; Reade and Collins, 2000). The combination of DNS for the fluid flow and Lagrangian particle tracking have proven to be useful tools for such studies.

9.6.2 Particle-laden slab flow

The second example is a particle-laden slab flow reported by Vermorel *et al.* (2003) that highlights issues associated with computation of turbulence modulation by momentum exchange with heavy particles. The flow is shown schematically in Fig. 9.6 and consists of a particle "slab" with an initially large slip velocity between the carrier and the dispersed phases in an otherwise freely decaying isotropic and homogeneous turbulence. The

Table 9.1. Parameters characterizing the dispersed phase in the particle-laden slab flow of Vermorel *et al.* (2003). The "0" subscript refers to values at the instant of injection into the gas.

	d_p/η_0	Re_{p_0}	St_0	α_p	ϕ
Case 1	0.31	7.2	0.26	0.0014	1.0
Case 2	0.62	14.3	0.84	0.0112	5.0
Case 3	0.93	21.5	1.64	0.0380	27.0

flow with or without the dispersed phases evolves with time and therefore averaged quantities are accumulated in the statistically homogeneous "parallel" xz plane. With reference to Fig. 9.6, averaged quantities such as the mean velocity and velocity variance are then functions of the coordinate y and time.

In their work, the carrier phase flow of isotropic turbulence is initially generated using a Passot–Pouquet spectrum (e.g. see Hinze, 1975). The computations were performed within a cubic domain $L_{box}^3 = [-\pi; \pi]^3$ using a mesh comprised of 128^3 points. From its initial energy spectrum, the gas-phase turbulent flow is allowed to develop to a realistic state. The initial integration without particles in the domain is necessary to establish the transfer of energy from large to small scales. The Reynolds number of the gas-phase turbulence in the fluid velocity field at the time corresponding to injection of the particles is $Re_\lambda = 35$.

Following the development of the gas-phase turbulent flow, 2.68×10^5 particles are randomly distributed in a slab that occupies the central portion of the domain, $[-\pi; \pi]_x$, $[-0, 8; 0.8]_y$, $[-\pi; \pi]_z$. All the particles are injected with the same velocity u_{p_0} in the x-direction as shown in Fig. 9.6, with the initial particle velocity u_{p_0} larger than the initial root-mean-square turbulent velocity of the gas at the moment of injection, i.e. $u_{p_0}/u_{rms} = 7.8$. The regime considered by Vermorel *et al.* (2003) corresponds to large particle densities relative to the carrier gas-phase flow for which the particle equation of motion is given by (9.13), without the gravity term, and the drag force is prescribed using (9.8).

Studies of three different sets of particles are reported by Vermorel *et al.* (2003). Parameters characterizing the disperse phase at the time of their injection are summarized in Table 9.1, which also shows that the particle diameter is smaller than the Kolmogorov length scale of the background fluid flow, η_0 (the "0" subscript corresponds to properties of the gas-phase turbulent fluid flow at injection). The particle Reynolds number, averaged over the entire particle assembly, increases proportionally to the diameter. The particle Stokes number in the table, St_0, provides a dimensionless measure of the particle response time. In this work, the values in Table 9.1 are defined as $St_0 = \tau_p/\tau_{t0}$ where τ_{t0} is the so-called eddy turnover time and is defined as $\tau_{t0} = q_{f0}^2/\varepsilon_0$. Also given in Table 9.1 are the dispersed phase volume fraction, α_p, and mass loading, ϕ, which is the mass ratio of the particulate-to-gas phases. For each of the three particle sets shown in the table, the ratio of the particle-to-fluid densities was 711.

The mass loadings are sufficiently large that substantial modifications of the gas-phase carrier flow are possible. As in the previous example, in Vermorel et al. (2003) the effect the momentum transfer from the particles to the fluid was computed using DNS of the Navier–Stokes equations (i.e. without any explicit turbulence in the momentum equations for the gas) and using (9.76) for the form of the coupling force. Vermorel et al. (2003) investigated the dynamics of the slab flow, in addition to assessing closure models used in engineering calculations based on solution of the Reynolds-averaged Navier–Stokes equations.

Following injection, the initially high slip velocity between the two phases causes an abrupt acceleration of the fluid; the influence of the injection is apparent in the visualization of the results shown in Fig. 9.7. The figure shows four snapshots of the instantaneous fluid velocity in the x-direction from a single plane (note that the particles actually shown in the plane are those in the section $z_{\text{plane}} - \Delta z/2$ and $z_{\text{plane}} + \Delta z/2$ where Δz is the grid spacing in the z-direction and z_{plane} in the figure is the plane in the center of the domain). Following injection, the fluid velocity increases to a maximum value while the particles decelerate, with the details of these processes being sensitive to the parameters of each case. It is also apparent from Fig. 9.7 that the particle distribution is spreading in the cross-stream (y) direction with time and that the level of detail resolved in the calculations allows a detailed examination of fluid–particle interactions.

In Vermorel et al. (2003), as in all other similar calculations of two-way coupling using DNS of the carrier phase flow performed to date, the locally undisturbed fluid velocity required for the particle equation of motion is approximated by the value computed from the Navier–Stokes equations and available on the grid, i.e. the coupling force is approximated using (9.69). One of the errors associated with this approximation that can bias flow field statistics arises due to the possibly large relative velocity between phases.

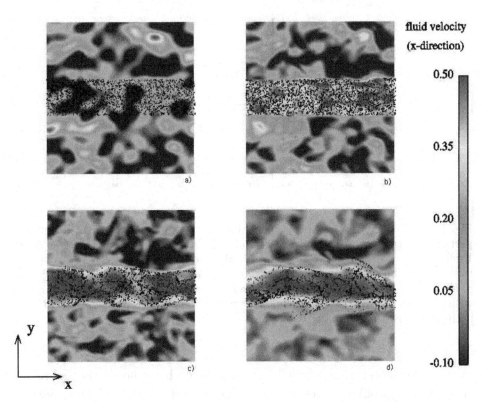

Fig. 9.7. Fluid velocity (*x*-component) normalized by the initial particle velocity v_0 and particle positions for Case 1 of Vermorel *et al.* (2003). Frames from (a) to (d) correspond to increasing evolution in time. See also http://www.cambridge.org/comp_mult_flow.

The projection of the force $(f_{d,i}^n)$ can influence the local instantaneous gas velocity at the particle position, resulting in a bias. Vermorel *et al.* (2003) assessed this error through computations using two populations of particles within the same simulation. Only the momentum of one population of particles is coupled to the fluid, while the second population represents "ghost" particles for which the fluid velocity interpolated to the particle position is the undisturbed value. If there is a negligible bias due to the influence of the force on the fluid velocity subsequently interpolated to the particle position, then the statistics for both particle populations should be identical, given the same parameters (i.e. mass loading, particle diameter, and mean relative velocity).

An example of the results of this comparison is shown Fig. 9.8 from the

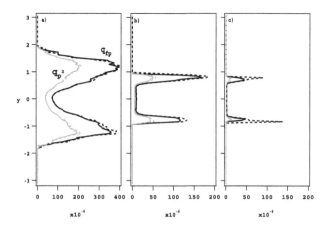

Fig. 9.8. Influence of force projection bias on the fluid–particle covariance and particle kinetic energy in the particle-laden slab flow of Vermorel *et al.* (2003). Profiles shown correspond to (a) Case 1, (b) Case 2, and (c) Case 3 at a representative time. The fluid–particle covariance, q_{fp}, is shown by the black curves, the particle kinetic energy q_p^2 by the gray curves. ———— shows profiles from particles with momentum coupling to the fluid; – – – – shows profiles from "ghost particles" without momentum coupling to the fluid.

work of Vermorel *et al.* (2003). Shown in each frame of the figure are cross-stream profiles of the particle kinetic energy, q_p^2, and fluid–particle covariance, $q_{fp} = \langle v'_{p,i} \tilde{u}'_{f,i} \rangle_p$ where $\langle \cdot \rangle_p$ reflects an average over the particles in a plane $y =$ const. at a given instant. As shown in the figure, for Case 1 there is good agreement between the statistics of the coupled and ghost particles, indicative of a negligible bias in the fluid velocity interpolated to the particle position due to the force projection. For Case 2 (Fig. 9.8b) relatively small differences are apparent in the two quantities, with the peaks in the profiles for the particle population with momentum feedback lower than for the other particle population. It is also apparent that in Case 3 (Fig. 9.8c) these differences are more significant and the error is obviously an important effect.

As a final point, we note that the filtering of the coupling force effected by the projection operation (9.76) implies that the fluid velocity **u** resolved on the grid represents a filtered velocity field, from which the small-scale perturbations induced by the particles (typically smaller than the mesh size) have been removed. It is important to remark that this is the case even in a DNS of the gas-phase flow and is a consequence of the point-particle treatment. Thus, the equations governing the evolution of the fluid velocity **u** (conservation of mass and momentum for an incompressible flow) should in principle be formally derived by an explicit filtering, analogous to the filtering operation employed in large eddy simulation. This would produce a "subgrid stress" accounting for the interactions of the scales captured on the grid and fluctuations in the velocity beneath the grid resolution, at scales comparable to the particle size.

In Vermorel *et al.* (2003), along with all other related studies using point-particle treatments to study modulation of turbulence in direct simulations, formal filtering is not applied and the "subgrid stress" resulting from interactions of the unresolved and resolved motions are not explicitly modeled. If the filtered motions are modeled by the introduction of an eddy viscosity, a comparison of the eddy viscosity to the molecular value yields estimates of the magnitude of the unresolved motions compared to the stresses explicitly computed and available on the grid. Implications of point-particle treatments and their implicit filtering on energy balances in gas–solid turbulent flows are discussed in Simonin and Squires (2003) and, from a slightly different viewpoint, by Crowe (2000).

10

Segregated methods for two-fluid models

The previous chapter, with its direct simulation of the fluid flow and a modeling approach to the particle phase, may be seen as a transition between the methods for a fully resolved simulation described in the first part of this book and those for a coarse-grained description based on the averaging approach described in Chapter 8. We now turn to the latter, which in practice are the only methods able to deal with the complex flows encountered in most situations of practical interest such as fluidized beds, pipelines, energy generation, sediment transport, and others. This chapter and the next one are devoted to numerical methods for so-called *two-fluid models* in which the phases are treated as interpenetrating continua describing, e.g. a liquid and a gas, or a fluid and a suspended solid phase. These models can be extended to deal with more than two continua and, then, the denomination *multifluid models* might be more appropriate. For example, the commercial code OLGA (Bendiksen *et al.*, 1991), widely used in the oil industry, recognizes three phases, all treated as interpenetrating continua: a continuous liquid, a gas, and a disperse liquid phase present as drops suspended in the gas phase. The more recent PeTra (*Petroleum Transport*, Larsen *et al.*, 1997) also describes three phases: gas, oil, and water. Recent approaches to the description of complex boiling flows recognize four interpenetrating phases: a liquid phase present both as a continuum and as a dispersion of droplets, and a gas/vapor phase also present as a continuum and a dispersion of bubbles. Methods for these multifluid models are based on those developed for the two-fluid model to which we limit ourselves.

In principle, one could simply take the model equations, discretize them, and solve them by a method suitable for nonlinear problems, e.g. Newton–Raphson iteration. In practice, the computational cost of such a frontal attack is nearly always prohibitive in terms of storage requirement and execution time. It is therefore necessary to devise different, less direct strategies.

Two principal classes of algorithms have been developed for this purpose. The first one, described in this chapter, consists of algorithms derived from the pressure-based schemes widely used in single-phase flow, such as SIMPLE and its variants (see, e.g. Patankar, 1980). In this approach, the model equations are solved sequentially and, therefore, these methods are often referred to as *segregated algorithms* to distinguish them from a second class of methods, the object of the next chapter, in which a coupled or semi-coupled time-marching solution strategy is adopted. Broadly speaking, the first class of methods is suitable for relatively slow transients, such as fluidized beds, or phenomena with a long duration, such as flow in pipelines. The methods in the second group have been designed to deal principally with fast transients, such as those hypothesized in nuclear reactor safety research. Since in segregated solvers the equations are solved one by one, it is possible to add equations to the mathematical model – to describe, e.g. turbulence – at a later stage after the development of the initial code without major modifications of the algorithm.

10.1 A general one-dimensional model

In order to describe the basic computational approach, it is not necessary to commit oneself to a specific model: it is sufficient to consider a general class of models suggested by the averaged equations derived in Chapter 8 and by the structure of the models which are encountered in actual practice.

We thus consider two phases, denoted by indices 1 and 2, with averaged continuity equations of the form

$$\frac{\partial}{\partial t}(\alpha_J \rho_J) + \frac{\partial}{\partial x}(\alpha_J \rho_J u_J) = 0, \qquad J = 1, 2. \tag{10.1}$$

Here and in the following we omit the explicit indication of averaging. Conservation of the total volume requires that

$$\alpha_1 + \alpha_2 = 1. \tag{10.2}$$

We take the momentum equations in the form

$$\frac{\partial}{\partial t}(\alpha_J \rho_J u_J) + \frac{\partial}{\partial x}(\alpha_J \rho_J u_J^2) = -\alpha_J \frac{\partial p}{\partial x} + F_J, \tag{10.3}$$

where

$$F_1 = -K_1 u_1 + H(u_2 - u_1) + C_1 \alpha_1 \alpha_2 \hat{\rho}(a_2 - a_1) + f_1 \tag{10.4}$$

$$F_2 = -K_2 u_2 + H(u_1 - u_2) + C_2 \alpha_1 \alpha_2 \hat{\rho}(a_1 - a_2) + f_2. \tag{10.5}$$

Here K_J is a parameter describing the drag force between the J-phase and the solid structure, e.g. the wall of a conduit, and H is an analogous interphase drag parameter. Writing these terms in the form shown by no means implies linearity: in general both K_J and H will be more or less complicated functions of the volume fractions, flow velocities, densities, and possibly other parameters. The added mass coefficients C_J in general are again functions of volume fractions and densities, $\hat{\rho}$ is a suitable density (e.g. the average density), and the accelerations a_J may be thought as given by

$$a_J = \frac{\partial u_J}{\partial t} + u_J \frac{\partial u_J}{\partial x} \qquad (10.6)$$

or by more complicated, but qualitatively similar, expressions such as those suggested by Drew *et al.* (1979). Finally, f_J indicates forces of other nature, such as the body force, viscous effects, a difference between the average and interfacial pressures, and possibly others; for example

$$f_J = \alpha_J \rho_J g + \frac{\partial}{\partial x}\left(\mu_J \frac{\partial u_J}{\partial x}\right) - (p - \overline{p}_J)\frac{\partial \alpha_J}{\partial x}. \qquad (10.7)$$

For the commercial CFX code[1], in one dimension, and with the neglect of phase change effects,

$$F_J = \frac{\partial}{\partial x}(\alpha_J \tau_J) + \alpha_J \rho_J \mathbf{g} + \sum_K c_{JK}(u_K - u_J) \qquad (10.8)$$

with $c_{JJ} = 0$ while $c_{JK} > 0$ represents the interphase drag coefficient when $K \neq J$. In addition to the viscous stress, the term τ_J also includes the turbulent stresses for which different models are available. In the multiphase literature, turbulence is often modelled as in single-phase flow, e.g. by means of a $k - \epsilon$ model with the same constants as the single-phase version. While use of such models may be justifiable in some situations, such as separated flows or very dilute disperse flows, in other cases it would be difficult to justify their adoption other than as a very crude approximation.

In general, one has to account for the compressibility of the phases, which requires the introduction of an equation of state relating pressure and density:

$$\rho_J = \rho_J(p). \qquad (10.9)$$

Although both ρ_J and p in reality are average quantities, this functional dependence is usually approximated by the same form valid for the microscopic unaveraged quantities. The momentum equations can also be written in

[1] See http://www-waterloo.ansys.com/cfx/products/cfx-5/.

non-conservation form by using the continuity equations in the standard way with the result

$$\frac{\partial u_J}{\partial t} + u_J \frac{\partial u_J}{\partial x} = -\frac{1}{\rho_J}\frac{\partial p}{\partial x} + \frac{F_J}{\alpha_J \rho_J}. \tag{10.10}$$

In a more complete model, temperature would also appear in the equation of state, which would have to be calculated from an energy equation for phase J. For simplicity, in this chapter we mostly disregard the energy equations.

In writing the momentum equations in the form (10.2) we have used a single pressure. Differences between the phase average pressures expressed in algebraic form as, for example, due to surface tension or hydrostatic effects, can be readily accommodated in the last term of equation (10.7). In general, however, the situation may get considerably more complex. Indeed, the validity of the equilibrium pressure assumption relies on the smallness of the ratio between the time scale for acoustic waves to propagate across the phases and the time scale of the problem. For very fast problems, such as the impact of a projectile on a target, this ratio may not be small and it may be necessary to resolve the individual pressure fields in the different materials. This step might require a multiscale approach based on the use of local velocity fields different from the average velocities entering the momentum equations (10.2). This is an active research subject for which standard procedures are still far from having been developed. Accordingly, we will not address problems of this type.

10.2 Discretization

For simplicity, we assume a uniform discretization of the computational domain of interest, $0 \leq x \leq L$ say, by means of nodes $x_1, x_2, \ldots, x_N$ spaced by the constant mesh size Δx. The use of a nonuniform mesh leads to somewhat more complicated equations without altering the substance of the methods.

In order to derive the discretized form of the model, we use the *finite-volume* approach, in which the equations are integrated over each interval or, in a multidimensional calculation, over each one of the control volumes (cells) into which the integration domain is subdivided. Over finite differences, the finite-volume method has the advantage that it can be used with the irregular discretizations of the computational domain which are sometimes necessary with complex geometries. (Of course, this aspect of the approach will not be apparent in the present one-dimensional context.) Furthermore, a finite-volume approach lends itself naturally to a direct application of the conservation laws and separates the two steps of domain discretization and approximation of the fluxes, thus affording somewhat greater flexibility.

It is well known from single-phase computational fluid dynamics that, when pressure and velocity are located at the same nodes, often spurious two-point oscillations arise. As mentioned earlier in Chapter 2, a powerful way to prevent this instability is to adopt a staggered-grid arrangement in which the scalar variables (volume fractions, pressure, densities) are located at the nodes x_i, while vectorial quantities (velocities, diffusive fluxes) are located at the midpoints between the successive nodes x_i and x_{i+1}, which we denote by $x_{i+1/2}$. We will adopt such a staggered grid arrangement for the moment. For computations on irregular domains it is often more convenient to use a co-located variable arrangement. In this case the pressure oscillations can be avoided by using the special interpolation procedures described later in Section 10.3.4.

10.2.1 Convection–diffusion equation

Before considering specifically the equations of the model described in the previous section, let us introduce the main ideas in the context of a typical convection–diffusion equation for some generic scalar variable ψ:

$$\frac{\partial \psi}{\partial t} + \frac{\partial}{\partial x}(u\psi) = \frac{\partial}{\partial x}\left(k\frac{\partial \psi}{\partial x}\right) + S, \tag{10.11}$$

in which k is the diffusivity of ψ and S a source term; in general $k = k(\psi)$ and $S = S(\psi)$. The continuity equations (10.1) have this form with $k = 0$ and $S = 0$. The structure of the momentum equations is significantly different due to the appearance of variables outside the differentiation operators, for example in the pressure and in the added-mass terms.

Suppose that a solution of (10.11) is available at time t^n. In order to advance it to time $t^{n+1} = t^n + \Delta t$, we integrate the equation over time between t^n and t^{n+1} and over x between $x_{i-1/2}$ and $x_{i+1/2}$, interchanging the order of the integrations as needed, and divide by $\Delta t \, \Delta x$ to find:

$$\frac{1}{\Delta x}\int_{x_{i-1/2}}^{x_{i+1/2}} dx \, \frac{\psi(x,t^{n+1}) - \psi(x,t^n)}{\Delta t} + \frac{1}{\Delta t}\int_{t^n}^{t^{n+1}} dt \, \frac{(u\psi)_{i+1/2} - (u\psi)_{i-1/2}}{\Delta x}$$

$$= \frac{1}{\Delta t}\int_{t^n}^{t^{n+1}} dt \, \frac{[k(\partial\psi/\partial x)]_{i+1/2} - [k(\partial\psi/\partial x)]_{i-1/2}}{\Delta x}$$

$$+ \frac{1}{\Delta t \Delta x}\int_{t^n}^{t^{n+1}} dt \int_{x_{i-1/2}}^{x_{i+1/2}} dx \, S. \tag{10.12}$$

The transformation of this (exact) expression into a usable discretized approximation requires suitable assumptions on the time- and space-dependence of the integrands. In order to execute the spatial integrations, we expand the integrands in a Taylor series around x_i to find

$$\frac{1}{\Delta x} \int_{x_{i-1/2}}^{x_{i+1/2}} dx \, \frac{\psi(x,t^{n+1}) - \psi(x,t^n)}{\Delta t} = \frac{\psi_i^{n+1} - \psi_i^n}{\Delta t} + O(\Delta x)^2, \quad (10.13)$$

where we use the usual notation $\psi_i^n = \psi(x_i, t^n)$, $\psi_i^{n+1} = \psi(x_i, t^{n+1})$.

In order to achieve first-order accuracy in time (i.e. an error which decreases linearly with the time step), it is sufficient to approximate the integrands in the time integration as constants. Here there are several possibilities. These constant values may be taken as the values at the final time t^{n+1}, to find

$$\frac{\psi_i^{n+1} - \psi_i^n}{\Delta t} + \frac{(u\psi)_{i+1/2}^{n+1} - (u\psi)_{i-1/2}^{n+1}}{\Delta x}$$
$$= \frac{1}{\Delta x} \left[\left(k\frac{\partial\psi}{\partial x}\right)_{i+1/2}^{n+1} - \left(k\frac{\partial\psi}{\partial x}\right)_{i-1/2}^{n+1} \right] + S_i^{n+1}. \quad (10.14)$$

In general, this *implicit discretization* gives rise to a nonlinear algebraic problem, e.g. because ψ and k may depend on u. The solution of problems of this type is usually difficult and time consuming. If, for example, a Newton–Raphson method is used, due to the nonlinearity, it is necessary to compute the Jacobian at each iteration. On the other hand, implicit methods are usually more stable (often unconditionally so) and enable one to take relatively large time steps. The difficulty due to the nonlinearity can be mitigated by linearizing equation (10.14). Two common procedures are the *Newton linearization*:

$$(u\psi)^{n+1} \simeq u^n\psi^{n+1} + u^{n+1}\psi^n - u^n\psi^n \quad (10.15)$$

and the *Picard linearization*[1]:

$$(u\psi)^{n+1} \simeq u^n\,\psi^{n+1}. \quad (10.16)$$

Here we prefer to write $u^n\,\psi^{n+1}$ in place of the seemingly equally possible $u^{n+1}\,\psi^n$ as we are solving an equation for ψ, rather than u. For the same reason, when applying this linearization to the momentum equation, we may

---

[1] The second method derives its name from Picard's iterative solution procedure of successive substitution, the first one from the Newton–Raphson solution method for nonlinear equations. For an elementary justification one may write $\psi^{n+1} = \psi^n + \Delta\psi$, with $\Delta\psi = \psi^{n+1} - \psi^n$, and similarly for u, and drop terms quadratic in the increments so that $(u^n + \Delta u)(\psi^n + \Delta\psi) \simeq u^n\psi^n + \Delta u\psi^n + \Delta\psi u^n$, which is (10.15). Since the increments can be expected to be of order Δt, the error incurred in dropping $\Delta\psi\,\Delta u$ is $O(\Delta t)^2$ and, therefore, compatible with the accuracy with which the time integrals have been approximated.

write $(\mathbf{u}^n \cdot \nabla)\mathbf{u}^{n+1}$. While both these procedures eliminate the nonlinearity, they still leave a system of coupled equations as, in (10.14), ψ_i^{n+1} depends on $\psi_{i\pm1}^{n+1}$.

The option leading to the simplest computational task is to evaluate the integrands *explicitly*, i.e. at the initial instant t^n, with the result

$$
\frac{\psi_i^{n+1} - \psi_i^n}{\Delta t} + \frac{(u\psi)_{i+1/2}^n - (u\psi)_{i-1/2}^n}{\Delta x}
$$

$$
= \frac{1}{\Delta x}\left[\left(k\frac{\partial\psi}{\partial x}\right)_{i+1/2}^n - \left(k\frac{\partial\psi}{\partial x}\right)_{i-1/2}^n\right] + S_i^n. \quad (10.17)
$$

Explicit schemes are less robust than implicit ones due to the requirement that the time step be sufficiently smaller than all the characteristic time scales of the problem. For example, the time scale for convection across a cell is of the order of $\Delta x/|u|$, which leads to the Courant–Friedrichs–Lewy, or CFL, stability condition (2.2.8)

$$
\frac{\Delta t}{\Delta x}\max_i\left|u_{i+1/2}\right| \leq 1. \quad (10.18)
$$

This limitation is particularly severe when compressibility effects are important or the phases are tightly coupled, and more will be said on this topic in Section 10.6 and in the next chapter. In spite of these shortcomings, explicit methods are appealing for their simplicity. Furthermore, the segregated solvers considered in this chapter usually require small time steps anyway.

In some problems, the source may depend on ψ in a linear or nonlinear way. Let us suppose that the major part of the nonlinearity of S can be captured by writing

$$
S = S^0 - S^1\psi, \quad (10.19)
$$

where S^0 and S^1 are not necessarily independent of ψ. Dimensionally, S^1 is an inverse time characterizing the time scale of the source and, if it is too large, the stability requirement $|S^1|\Delta t < 1$ that follows from an explicit treatment of S may be too stringent. In this case it is preferable to adopt a *linearization of the source term* by writing

$$
\frac{1}{\Delta t\Delta x}\int_{t^n}^{t^{n+1}} dt \int_{x_{i-1/2}}^{x_{i+1/2}} dx\, S \simeq (S^0)_i^n - (S^1)_i^n\psi_i^{n+1}, \quad (10.20)
$$

so that ψ_i^{n+1} in the left-hand side of (10.17) acquires the coefficient $(\Delta t)^{-1} + (S^1)_i^n$. If $(S^1)_i^n$ is negative, this procedure actually *impairs* the stability of the method by decreasing the coefficient of ψ_i^{n+1}. It is therefore important that S^1 be positive which, on physical grounds, is often the case anyway as, otherwise, the terms $\partial\psi/\partial t \simeq -S^1\psi$ would describe a process with a tendency toward instability (see, e.g. Patankar, 1980).

Whatever the choice made for the time integration, some work is still needed to put the equation into a final usable form. By expanding in a Taylor series around $x_{i+1/2}$ we have

$$\left(k\frac{\partial\psi}{\partial x}\right)_{i+1/2} \simeq k_{i+1/2}\frac{\psi_{i+1}-\psi_i}{\Delta x} + O(\Delta x)^2, \qquad (10.21)$$

with a similar relation for the corresponding term evaluated at $i-1/2$. The diffusivity k may depend on scalar quantities, such as ψ itself, and, therefore, it may be undefined at $i \pm 1/2$. Common options are[1]

$$k_{i+1/2} \simeq \frac{1}{2}\left[k(\psi_{i+1}) + k(\psi_i)\right] \qquad \text{or} \qquad k_{i+1/2} \simeq k\left(\frac{1}{2}(\psi_{i+1}+\psi_i)\right). \qquad (10.22)$$

The convection term $(u\psi)_{i\pm1/2} = \psi_{i\pm1/2}\,u_{i+1/2}$ offers a similar difficulty because, while the velocity is defined at the half-integer nodes, ψ is not. Thus, we need to develop an estimate of $\psi_{i\pm1/2}$ in terms of ψ_i and $\psi_{i\pm1}$. This is a more difficult task than it might appear at first sight.

10.2.2 Calculation of cell-surface quantities

The simplest way to estimate the value of ψ at $x_{1+1/2}$ is by means of a linear approximation which we may write in either one of two forms:

$$\psi = \psi_i + \frac{\psi_{i+1}-\psi_i}{\Delta x}(x-x_i) = \psi_{i+1} + \frac{\psi_{i+1}-\psi_i}{\Delta x}(x-x_{i+1}). \qquad (10.23)$$

Upon setting $x = x_{i+1/2}$ in either one of these equations, we find

$$\psi_{i+1/2} = \frac{1}{2}\left(\psi_{i+1}+\psi_i\right), \qquad (10.24)$$

which on a uniform grid is second-order accurate with a dispersive, or oscillatory, error. As we saw in Section 3.2, in convection-dominated flows, this error leads to spurious oscillations near regions of sharp gradients which corrupt the solution. This numerical artifact is of special concern in multiphase flow calculations as, for example, it may produce volume fractions outside the physical range [0,1], especially in the neighborhood of a transition between a two-phase and a single-phase region of the flow. A common

---

[1] On a nonuniform mesh, $k(\psi_i)$ and $k(\psi_{i+1})$ in the first expression would have weights $\Delta x_i/(\Delta x_i + \Delta x_{i+1})$ and $\Delta x_{i+1}/(\Delta x_i + \Delta x_{i+1})$ in place of $1/2$, with a similar modification for the average of ψ_i and ψ_{i+1} in the second expression.

way to overcome the problem is to use an *upwind* or *donor-cell* scheme, in
which the linear terms in (10.23) are omitted and the choice between one
form and the other is based on the direction of the flow:

$$\psi_{i+1/2} = \begin{cases} \psi_i & \text{for } u_{i+1/2} \geq 0 \\ \psi_{i+1} & \text{for } u_{i+1/2} < 0; \end{cases} \tag{10.25}$$

a multidimensional generalization is shown in Section 10.6. This scheme
is only first-order accurate and dissipative and usually leads to strongly
diffused solutions, particularly for long integration times or at low resolution;
an example was shown in Figure 3.2 on p. 46 and another one will be given
in Section 10.4.

The search for higher order methods free of spurious oscillations has generated a copious literature principally driven by the needs of gas dynamics.
We will return on this issue and introduce the concept of *Total Variation
Diminishing* (TVD) schemes in Section 11.3 of the next chapter. For the
time being, let us simply say that the general approach is to formulate a
rule for the calculation of $\psi_{i+1/2}$ which gives a result as close as possible to
the second-order central scheme (10.24) in smooth regions, degrading the
accuracy of the interpolation by the smallest possible amount sufficient to
avoid the appearance of spurious oscillations in regions of large gradients. In
the MUSCL (*M*onotone *U*pstream-centered *S*cheme for *C*onservation *L*aws)
method, pioneered by van Leer (1977), this objective is achieved by limiting
the slope of the linear interpolations used in (10.23).

It is convenient to adopt a notation introduced by Leonard and Mokhtari
(1990) which enables us to carry the two cases $u_{i+1/2} > 0$ and $u_{i+1/2} < 0$
at the same time. We denote by ψ_d the value of ψ at the node immediately
downstream of $x_{i+1/2}$, by ψ_c the value at the current node, immediately
upstream, and by ψ_u the value at the node immediately upstream of the
current node (Fig. 10.1); the convention is summarized in the following table:

	ψ_c	ψ_u	ψ_d
$u_{i+1/2} \geq 0$	ψ_i	ψ_{i-1}	ψ_{i+1}
$u_{i+1/2} < 0$	ψ_{i+1}	ψ_{i+2}	ψ_i

$$(10.26)$$

We write

$$\psi_{i+1/2} = \psi_c + \frac{1}{2}\ell_{i+1/2}(\psi_d - \psi_c). \tag{10.27}$$

in which $\ell_{i+1/2}$ is a suitably defined *slope limiter*. It is evident that the first-
order upwind scheme corresponds to taking $\ell_{i+1/2} = 0$, while the second-
order approximation (10.24) is obtained for $\ell_{i+1/2} = 1$. For more general
schemes, the magnitude of $\ell_{i+1/2}$ is decided on the basis of the direction of

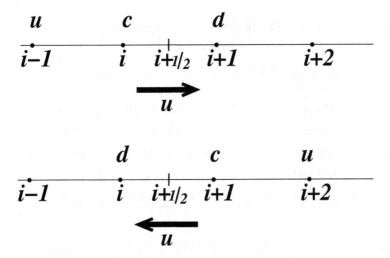

Fig. 10.1. Nodes used in the definition of the limiter.

the flow and an estimate of the magnitude of the local gradient, which is quantified by means of a *smoothness parameter*

$$\theta_{i+1/2} = \frac{\psi_c - \psi_u}{\psi_d - \psi_c} = \frac{1}{\psi_{i+1} - \psi_i} \begin{cases} \psi_i - \psi_{i-1} & \text{for } u_{i+1/2} \geq 0 \\ \psi_{i+2} - \psi_{i+1} & \text{for } u_{i+1/2} < 0. \end{cases} \quad (10.28)$$

In a smooth region of ψ, a Taylor series expansion shows that

$$\theta_{i+1/2} = 1 - \text{sgn}(u_{i+1/2})\Delta x \frac{\psi_c''}{\psi_c'} + O(\Delta x)^2. \quad (10.29)$$

Several considerations, which can be found, e.g. in Sweby (1984), Leonard and Mokhtari (1990), Thomas (1995), LeVeque (1992, 2002), and Syamlal (1998), constrain the dependence of the limiter ℓ_i on θ_i. If $(\psi_c-\psi_u)/(\psi_d-\psi_u)$ is less than 0 or greater than 1, there is a strong extremum between the nodes "d" and "u" and one takes $\ell_{i+1/2} = 0$, i.e. the upwind rule. Several choices have been proposed for the monotone case in which $0 \leq (\psi_c-\psi_u)/(\psi_d-\psi_u) \leq 1$. Table 10.1 shows some popular limiters as given in terms of the previous formulation (see also Syamlal, 1998). The Minmod limiter compares the two jumps $\psi_c - \psi_u$ and $\psi_d - \psi_c$ and, provided they have the same sign, it returns the smaller one of the two in magnitude. Thus, although it improves on the upwind rule, it is relatively close to it and is, accordingly, the most diffusive of the limiters shown. The Superbee limiter, on the other hand, tends to be overcompressive sharpening steep profiles into discontinuities. The other limiters have an intermediate behavior.

Table 10.1. Some popular limiters used to estimate cell-face values of the convected quantity ψ.

Discretization scheme	Limiter ℓ
van Leer	$(\theta + |\theta|)/(1 + |\theta|)$
Minmod	$\max[0,\min(1,\theta)]$
MUSCL	$\max[0,\min(2\theta,\frac{1}{2}(1+\theta),2)]$
UMIST	$\max[0,\min(2\theta,\frac{1}{4}(3+\theta),2)]$
SMART	$\max[0,\min(4\theta,\frac{1}{4}(3+\theta),2)]$
Superbee	$\max[0,\min(1,2\theta),\min(2,\theta)]$

For $u_{i+1/2} \geq 0$ and $u_{i+1/2} < 0$, the rule (10.27) gives, respectively,

and
$$\left.\begin{aligned}
\psi_{i+1/2} &= \left(1 - \tfrac{1}{2}\ell_{i+1/2}\right)\psi_i + \tfrac{1}{2}\ell_{i+1/2}\psi_{i+1}, \\
\psi_{i+1/2} &= \tfrac{1}{2}\ell_{i+1/2}\psi_i + \left(1 - \tfrac{1}{2}\ell_{i+1/2}\right)\psi_{i+1}.
\end{aligned}\right\} \quad (10.30)$$

In order to avoid having to continue to distinguish between the two cases, we set, with an obvious measuring of the symbols,

$$\psi_{i+1/2} = \xi_{i+1/2}\psi_{i+1} + \overline{\xi}_{i+1/2}\psi_i, \qquad \text{where} \qquad \xi_{i+1/2} + \overline{\xi}_{i+1/2} = 1, \quad (10.31)$$

and, for simplicity of writing, we define

$$[\![\psi]\!]_{i+1/2} = \xi_{i+1/2}\psi_{i+1} + \overline{\xi}_{i+1/2}\psi_i, \qquad [\![\psi]\!]_{i-1/2} = \xi_{i-1/2}\psi_i + \overline{\xi}_{i-1/2}\psi_{i-1}. \quad (10.32)$$

In what follows, we refer to quantities such as these, enclosed in double brackets, as *convected quantities*. When $0 \leq \xi, \overline{\xi} \leq 1$, which implies that $0 \leq \ell \leq 2$, $[\![\psi]\!]_{i\pm1/2}$ are weighted averages of ψ at the neighboring points. In this way, the convected quantities are bounded by the values of ψ at the neighboring cell centers, and spurious extrema cannot arise.

With (10.32), the convection term of (10.17) takes the form

$$\frac{(u\psi)_{i+1/2} - (u\psi)_{i-1/2}}{\Delta x} \simeq \frac{[\![\psi]\!]_{i+1/2}\, u_{i+1/2} - [\![\psi]\!]_{i-1/2}\, u_{i-1/2}}{\Delta x}. \quad (10.33)$$

Thanks to (10.32), this expression involves explicitly only the three nodes x_i and $x_{i\pm1}$, which permits the use of efficient algorithms such as the tri-diagonal Thomas algorithm. The parameters ξ and $\overline{\xi}$, however, also implicitly contain information about neighboring nodes through their dependence on θ.

10.2.3 Equations of the two-fluid model

Now that a few basic concepts have been introduced, we can proceed with the discretization of the model introduced in Section 10.1. The continuity equations have the form (10.11) considered before and a straightforward adaptation of (10.14) gives

$$\frac{(\alpha\rho)_i^{n+1} - (\alpha\rho)_i^n}{\Delta t} + \frac{[\![\alpha\rho]\!]_{i+1/2}^{n+1} u_{i+1/2}^{n+1} - [\![\alpha\rho]\!]_{i-1/2}^{n+1} u_{i-1/2}^{n+1}}{\Delta x} = 0, \quad (10.34)$$

where $[\![\alpha\rho]\!]_{i\pm1/2}$ are defined as in (10.32). A semi-explicit variant of (10.34) is

$$\frac{(\alpha\rho)_i^{n+1} - (\alpha\rho)_i^n}{\Delta t} + \frac{[\![\alpha\rho]\!]_{i+1/2}^n u_{i+1/2}^{n+1} - [\![\alpha\rho]\!]_{i-1/2}^n u_{i-1/2}^{n+1}}{\Delta x} = 0. \quad (10.35)$$

It is important to note that the velocities must be evaluated at the advanced time t^{n+1}. This is an essential feature of the algorithm as it enables us to determine the pressure, as we will see.

The momentum equations differ in a significant way from the model equation (10.11) in that they are not in conservation form, chiefly due to the volume fractions multiplying the pressure terms, and also possibly because of other differential terms such as the added-mass interaction. We start by considering the left-hand side of (10.2) and note that, since momentum is a vector quantity centered at half-integer nodes, the control volume (or interval) over which to carry out the integration is $x_i \leq x \leq x_{i+1}$. For the first term, proceeding as in (10.13), we find

$$\frac{1}{\Delta x} \int_{x_i}^{x_{i+1}} dx \, \frac{(\alpha\rho u)^{n+1} - (\alpha\rho u)^n}{\Delta t} \simeq \frac{(\alpha\rho u)_{i+1/2}^{n+1} - (\alpha\rho u)_{i+1/2}^n}{\Delta t}. \quad (10.36)$$

in which $\alpha_{i+1/2}$ and $\rho_{i+1/2}$ are evaluated by averaging, e.g.

$$\alpha_{i+1/2} \simeq \frac{1}{2}(\alpha_i + \alpha_{i+1}). \quad (10.37)$$

For the convective term we note that the expression $\alpha\rho u^2$ in equation (10.2) should really be understood as $u(\alpha\rho u)$, where the first u is the convection velocity and $\alpha\rho u$, the momentum density, is the quantity being convected and thus playing the role of ψ in equation (10.36) (cf. the comment after equation 10.16). Here we may use an explicit or implicit form; for the time being we do not commit ourselves to a specific choice by omitting the temporal superscript simply writing

$$\frac{1}{\Delta t} \int_{t^n}^{t^{n+1}} dt \, \frac{[(\alpha\rho u)u]_{i+1} - [(\alpha\rho u)u]_i}{\Delta x} \simeq \frac{[\![\alpha\rho u]\!]_{i+1} u_{i+1} - [\![\alpha\rho u]\!]_i u_i}{\Delta x} \quad (10.38)$$

where, on a regular grid,

$$u_i = \frac{1}{2}\left(u_{i+1/2} + u_{i-1/2}\right) \tag{10.39}$$

and, similarly to (10.32),

$$\begin{aligned}
[\![\alpha\rho u]\!]_{i+1} &= \xi_{i+1}(\alpha\rho u)_{i+3/2} + \overline{\xi}_{i+1}(\alpha\rho u)_{i+1/2}, \\
[\![\alpha\rho u]\!]_i &= \xi_i(\alpha\rho u)_{i+1/2} + \overline{\xi}_i(\alpha\rho u)_{i-1/2}.
\end{aligned} \tag{10.40}$$

The left-hand side of the momentum equation then becomes

$$\text{LHS} = \frac{(\alpha\rho u)^{n+1}_{i+1/2} - (\alpha\rho u)^{n}_{i+1/2}}{\Delta t} + \frac{[\![\alpha\rho u]\!]_{i+1}u_{i+1} - [\![\alpha\rho u]\!]_i u_i}{\Delta x}. \tag{10.41}$$

The advanced time $(\alpha\rho)^{n+1}$ in the time derivative may be eliminated by writing the discretized continuity equation (10.34) at position $i+1/2$ rather than i:

$$\frac{(\alpha\rho)^{n+1}_{i+1/2} - (\alpha\rho)^{n}_{i+1/2}}{\Delta t} + \frac{[\![\alpha\rho]\!]_{i+1}\,u_{i+1} - [\![\alpha\rho]\!]_i\,u_i}{\Delta x} = 0, \tag{10.42}$$

where again we leave the time level of the convected quantity $[\![\alpha\rho]\!]$ unspecified. Upon multiplying this equation by $u^{n+1}_{i+1/2}$ and subtracting from (10.41) we have

$$\begin{aligned}
\text{LHS} = (\alpha\rho)^{n}_{i+1/2}&\frac{u^{n+1}_{i+1/2} - u^{n}_{i+1/2}}{\Delta t} \\
&+ \frac{[\![\alpha\rho u]\!]_{i+1}u_{i+1} - [\![\alpha\rho u]\!]_i u_i}{\Delta x} - \frac{[\![\alpha\rho]\!]_{i+1}\,u_{i+1} - [\![\alpha\rho]\!]_i\,u_i}{\Delta x}\,u^{n+1}_{i+1/2}. \tag{10.43}
\end{aligned}$$

If all quantities are evaluated implicitly, this becomes

$$\begin{aligned}
\text{LHS} = (\alpha\rho)^{n}_{i+1/2}&\frac{u^{n+1}_{i+1/2} - u^{n}_{i+1/2}}{\Delta t} \\
&+ \frac{a^{n+1}_{i+1/2}u^{n+1}_{i+3/2} + b^{n+1}_{i+1/2}u^{n+1}_{i+1/2} + c^{n+1}_{i+1/2}u^{n+1}_{i-/2}}{\Delta x}, \tag{10.44}
\end{aligned}$$

where

$$a^{n+1}_{i+1/2} = \xi^{n+1}_{i+1}(\alpha\rho)^{n+1}_{i+3/2}u^{n+1}_{i+1} \tag{10.45}$$

$$b^{n+1}_{i+1/2} = \overline{\xi}^{n+1}_{i}(\alpha\rho)^{n+1}_{i-1/2}\,u^{n+1}_{i} - \xi^{n+1}_{i+1}(\alpha\rho)^{n+1}_{i+3/2}\,u^{n+1}_{i+1} \tag{10.46}$$

$$c^{n+1}_{i+1/2} = -\overline{\xi}^{n+1}_{i}(\alpha\rho)^{n+1}_{i-1/2}\,u^{n+1}_{i}. \tag{10.47}$$

As written, the coefficients a, b, and c are evaluated at the advanced time t^{n+1} which, in an iterative procedure, can be dealt with by evaluation at the previous iteration step.

If the convected quantities are evaluated explicitly, the resulting expression is

$$LHS = (\alpha\rho)^n_{i+1/2}\frac{u^{n+1}_{i+1/2} - u^n_{i+1/2}}{\Delta t} + \frac{C^n_{i+1/2} - d^n_{i+1/2}u^{n+1}_{i+1/2}}{\Delta x}, \qquad (10.48)$$

where

$$C^n_{i+1/2} = [\![\alpha\rho u]\!]^n_{i+1}u^n_{i+1} - [\![\alpha\rho u]\!]^n_i u^n_i, \qquad d^n_{i+1/2} = [\![\alpha\rho]\!]^n_{i+1}u^n_{i+1} - [\![\alpha\rho]\!]^n_i u^n_i. \qquad (10.49)$$

The expression (10.48) may be simplified further by approximating the factor $u^{n+1}_{i+1/2}$ multiplying the last term of (10.43) by $u^n_{i+1/2}$, which can be done without affecting the order of the discretization. Then one finds the same result as (10.44) with the coefficients a, b, and c and the velocities that they multiply evaluated at t^n rather than t^{n+1}. With these explicit choices, the complexity of the calculation is strongly reduced, but at the price of a CFL limitation on the time step.

A similar, although slightly less accurate, result can be derived starting from the momentum equation in the non-conservation form (10.10). With an explicit discretization one would find

$$\frac{1}{\Delta t \Delta x}\int_{t^n}^{t^{n+1}} dt \left(\frac{\partial u}{\partial t} + u\frac{\partial u}{\partial x}\right) \simeq \frac{u^{n+1}_{i+1/2} - u^n_{i+1/2}}{\Delta t} + \left\langle\frac{\delta u}{\delta x}\right\rangle^n_{i+1/2} u^n_{i+1/2} \qquad (10.50)$$

with

$$\left\langle\frac{\delta u}{\delta x}\right\rangle^n_{i+1/2} = \xi^n_{i+1/2}\frac{u^n_{i+3/2} - u^n_{i+1/2}}{\Delta x} + \bar{\xi}^n_{i+1/2}\frac{u^n_{i+1/2} - u^n_{i-1/2}}{\Delta x}. \qquad (10.51)$$

In the following we will assume the form (10.44) with the understanding that the coefficients a, b, and c will be zero with an explicit evaluation of the convective term, which would then appear as an explicit contribution to a suitably redefined momentum source in the right-hand side of the equation.

As noted before, the pressure term in the momentum equation is not in conservation form; we approximate it as

$$\frac{1}{\Delta t \Delta x}\int_{t^n}^{t^{n+1}} dt \int_{x_i}^{x_{i+1}} dx\, \alpha\frac{\partial p}{\partial x} \simeq \frac{1}{\Delta t \Delta x}\int_{t^n}^{t^{n+1}} dt\, \alpha_{i+1/2} \int_{x_i}^{x_{i+1}} dx\, \frac{\partial p}{\partial x}$$

$$\simeq \alpha^{n+1}_{i+1/2}\frac{p^{n+1}_{i+1} - p^{n+1}_i}{\Delta x}. \qquad (10.52)$$

Here we have adopted an implicit discretization for the volume fraction, although this choice is not critical. What is critical is the evaluation of the

pressure at the advanced time level, which is essential in order to update
the pressure field, as will soon be apparent.

We have seen earlier in connection with the simple convection–diffusion
equation that it may be necessary to treat part of the source term implicitly,
when the associated time scale is very short. In the present case of the momentum equation, this circumstance often arises with the drag terms when the
interphase or fluid-structure momentum transfer is strong. Thus, we write[1]

$$\frac{1}{\Delta t \Delta x} \int_{t^n}^{t^{n+1}} dt \int_{x_i}^{x_{i+1}} dx\, F_1 \simeq -\mathcal{A}_{i+1/2}^1 u_{1,i+1/2}^{n+1} + \mathcal{B}_{i+1/2}^1 u_{2,i+1/2}^{n+1} + \mathcal{C}_{i+1/2}^1$$

$$(10.53)$$

with an analogous expression for the integral of F_2. As an example, with
the expression (10.4) for F_1, we would have

$$\frac{1}{\Delta t \Delta x} \int_{t^n}^{t^{n+1}} dt \int_{x_i}^{x_{i+1}} dx\, F_1 \simeq -K_1 u_1^{n+1} + H(u_2^{n+1} - u_1^{n+1})$$

$$+ C_1 \alpha_1^n \alpha_2^n \rho^n \left(\frac{u_2^{n+1} - u_2^n}{\Delta t} + u_2^n \frac{\delta u_2}{\delta x} - \frac{u_1^{n+1} - u_1^n}{\Delta t} - u_1^n \frac{\delta u_1}{\delta x} \right) + f_1,$$

$$(10.54)$$

where the spatial index is $i + 1/2$; $(\delta u_J / \delta x)^n$ is an approximation to the
spatial derivative, and quantities carrying no superscript can be evaluated
at t^n or t^{n+1}. In this case then, explicitly, +

$$\mathcal{A}_{i+1/2}^1 = \left[K_1 + H + \frac{1}{\Delta t} C_1 \alpha_1 \alpha_2 \rho \right]_{i+1/2} \tag{10.55}$$

$$\mathcal{B}_{i+1/2}^1 = \left[H + \frac{1}{\Delta t} C_1 \alpha_1 \alpha_2 \rho \right]_{i+1/2} \tag{10.56}$$

$$\mathcal{C}_{i+1/2}^1 = \left[C_1 \alpha_1 \alpha_2 \rho \left(\frac{u_1 - u_2}{\Delta t} + u_2 \frac{\delta u_2}{\delta x} - u_1 \frac{\delta u_1}{\delta x} \right) + f_1 \right]_{i+1/2} \tag{10.57}$$

where, again, the time level can be t^n or t^{n+1}. In the latter case, a linearization as in (10.15) or (10.16) can be used, e.g. for the drag functions which
often depend nonlinearly on the velocities. The spatial velocity derivatives
could also be treated implicitly, and so on. What matters here is that all
these choices (which, in practice, will yield different degrees of accuracy,
robustness, and speed of convergence depending on the specific problem)
will lead to an expression having a structure similar to that exhibited in

---

[1] In a partially implicit approximation, one would evaluate u_1 at time level $n + 1$ and u_2 at
time level n, and conversely for the momentum equation for the other phase. This procedure
would obviate solving the system (10.58), (10.62) for the velocities but, of course, it would
affect the stability and robustness of the scheme.

(10.53), possibly with some coefficients carrying a superscript $n+1$, which can be dealt with by iteration.

Upon collecting the terms (10.44), (10.52), and (10.53), the momentum equation for phase 1 takes the form

$$\mathcal{M}^1_{i+1/2}(u_1)^{n+1}_{i+1/2} - \mathcal{N}^1_{i+1/2}(u_2)^{n+1}_{i+1/2}$$

$$= -\frac{\Delta t}{\Delta x}\,(\alpha_1)^{n+1}_{i+1/2}\left(p^{n+1}_{i+1} - p^{n+1}_i\right)$$

$$-\frac{\Delta t}{\Delta x}\left[(a_1)^n_{i+1/2}(u_1)^{n+1}_{i+3/2} + (c_1)^n_{i+1/2}(u_1)^{n+1}_{i-/2}\right] + \mathcal{P}^1_{i+1/2} \quad (10.58)$$

where

$$\mathcal{M}^1_{i+1/2} = (\alpha_1\rho_1)^n_{i+1/2} + \frac{\Delta t}{\Delta x}\,(b_1)^n_{i+1/2} + \mathcal{A}^1_{i+1/2}\Delta t, \qquad (10.59)$$

$$\mathcal{N}^1_{i+1/2} = \mathcal{B}^1_{i+1/2}\Delta t, \qquad (10.60)$$

$$\mathcal{P}^1_{i+1/2} = (\alpha_1\rho_1 u_1)^n_{i+1/2} + \mathcal{C}^1_{i+1/2}\Delta t. \qquad (10.61)$$

Similarly, the momentum equation for the other phase takes the form

$$\mathcal{M}^2_{i+1/2}(u_2)^{n+1}_{i+1/2} - \mathcal{N}^2_{i+1/2}(u_1)^{n+1}_{i+1/2}$$

$$= -\frac{\Delta t}{\Delta x}\,(\alpha_2)^{n+1}_{i+1/2}\left(p^{n+1}_{i+1} - p^{n+1}_i\right)$$

$$-\frac{\Delta t}{\Delta x}\left[(a_2)^n_{i+1/2}(u_2)^{n+1}_{i+3/2} + (c_2)^n_{i+1/2}(u_2)^{n+1}_{i-/2}\right] + \mathcal{P}^2_{i+1/2} \quad (10.62)$$

with the coefficients $\mathcal{M}^2_{i+1/2}$, $\mathcal{N}^2_{i+1/2}$, and $\mathcal{P}^2_{i+1/2}$ given by expressions similar to (10.59), (10.60), and (10.61).

When the interphase drag is strong, the drag parameter H will be large, which causes $\mathcal{B}$ and, therefore, $\mathcal{N}$, to be large as well. If the solution procedure of the problem involves an iteration where u_2 in equation (10.58) and u_1 in equation (10.62) were cast in the right-hand side and evaluated at the previous iteration step, convergence would be very slow or not occur at all. To circumvent this problem, equations (10.58) and (10.62) may be regarded as a linear system for the velocities at $i+\frac{1}{2}$ and solved to find, for $J = 1, 2$:

$$(u_J)^{n+1}_{i+1/2} = -\sum_{K=1}^{2}\left[r_{JK}(u_K)^{n+1}_{i-1/2} + s_{JK}(u_K)^{n+1}_{i+3/2}\right]$$

$$- \mathcal{V}^J_{i+1/2}\left(p^{n+1}_{i+1} - p^{n+1}_i\right) + \mathcal{U}^J_{i+1/2} \qquad (10.63)$$

where

$$\mathcal{U}^1_{i+1/2} = \frac{\mathcal{M}^2\mathcal{P}^1 + \mathcal{N}^1\mathcal{P}^2}{\mathcal{M}^1\mathcal{M}^2 - \mathcal{N}^1\mathcal{N}^2}, \qquad (10.64)$$

$$\mathcal{V}^1_{i+1/2} = \frac{\Delta t}{\Delta x} \frac{\alpha_1 \mathcal{M}^2 + \alpha_2 \mathcal{N}^1}{\mathcal{M}^1 \mathcal{M}^2 - \mathcal{N}^1 \mathcal{N}^2}, \tag{10.65}$$

$$r_{11} = \frac{\Delta t}{\Delta x} \frac{\mathcal{M}^2}{\mathcal{M}^1 \mathcal{M}^2 - \mathcal{N}^1 \mathcal{N}^2} c_1, \qquad r_{12} = \frac{\Delta t}{\Delta x} \frac{\mathcal{N}^1}{\mathcal{M}^1 \mathcal{M}^2 - \mathcal{N}^1 \mathcal{N}^2} c_2, \tag{10.66}$$

$$r_{21} = \frac{\Delta t}{\Delta x} \frac{\mathcal{N}^2}{\mathcal{M}^1 \mathcal{M}^2 - \mathcal{N}^1 \mathcal{N}^2} c_1, \qquad r_{22} = \frac{\Delta t}{\Delta x} \frac{\mathcal{M}^1}{\mathcal{M}^1 \mathcal{M}^2 - \mathcal{N}^1 \mathcal{N}^2} c_2; \tag{10.67}$$

here the spatial index is $i + 1/2$ and the temporal index the same as in equations (10.58) and (10.62). The quantities $\mathcal{U}^2_{i+1/2}$ and $\mathcal{V}^2_{i+1/2}$ are obtained by interchanging the indices 1 and 2 in the *numerators only* of (10.64) and (10.65), while the coefficients s_{JK} are obtained by replacing c_K by a_K. It is now easy to verify that $\mathcal{V}^J$ is finite in the limit $H \to \infty$, which avoids the problem mentioned before. What we have described is known as the *partial elimination algorithm*, or PEA, and is an extension of the source linearization idea described in connection with (10.20).

The PEA principle is readily adapted to the situation when the momentum equations couple more than two velocity fields, as in a higher number of space dimensions or in the presence of more than two phases. This extension is referred to as SINCE, for *SI*multaneous solution of *N*onlinearly *C*oupled *E*quations (see, e.g. Karema and Lo, 1999; Darwish *et al.*, 2001). Here the linear system analogous to equations (10.58) and (10.62) is first solved for the $(u_J)^{n+1}_{1+1/2}$ (keeping the neighboring-node velocities in the right-hand sides), and the provisional solution thus obtained substituted back into the right-hand sides of the original momentum equations, which are then treated in the normal way. A more strongly coupled approach is described by Kunz *et al.* (1998).

Equations (10.63) express the velocities in terms of the still unknown pressures. To determine these quantities, we adapt the procedure known in single-phase flow as SIMPLE (*Semi-Implicit Method for Pressure-Linked Equations*, see, e.g. Patankar, 1980; Ferziger and Perić, 2002) to the present two-phase flow situation.

10.3 Calculation of the pressure

There are essentially two ways to proceed, one based on the use of the global continuity equation, the other one on the conservation of volume relation $\alpha_1 + \alpha_2 = 1$. In both approaches, the basic idea is to use the error affecting mass or volume conservation to develop an iterative scheme in which successive pressure corrections are used to adjust densities and velocities until, at convergence, the required conservation relation is satisfied. In essence, in both approaches one derives a pressure Poisson equation, but in a slightly different way. For the first method, the continuity equations

$$\frac{\partial}{\partial t}(\alpha_J \rho_J) + \nabla \cdot (\alpha_J \rho_J \mathbf{u}_J) = 0, \qquad J = 1, 2, \tag{10.68}$$

are added to find

$$\frac{\partial}{\partial t}(\alpha_1 \rho_1 + \alpha_2 \rho_2) + \nabla \cdot (\alpha_1 \rho_1 \mathbf{u}_1 + \alpha_2 \rho_2 \mathbf{u}_2) = 0. \tag{10.69}$$

The momentum equations are discretized in time with a resulting expression of the form

$$(\alpha_J \rho_J \mathbf{u}_J)^{n+1} = -\alpha_J \nabla p^{n+1} + \mathbf{N}_J. \tag{10.70}$$

which, upon substitution into (10.69), gives the required Poisson equation for the pressure. A similar argument based on rewriting the continuity equations as

$$\frac{\partial \alpha_J}{\partial t} + \frac{1}{\rho_J}\left[\alpha_J \frac{\partial \rho_J}{\partial t} + \nabla \cdot (\alpha_J \rho_J \mathbf{u}_J)\right] = 0, \tag{10.71}$$

adding, and using $\alpha_1 + \alpha_2 = 1$, generates the other family of methods. A direct implicit implementation of these ideas faces the usual difficulties of large storage requirements and long computational times. Several ways to deal with these difficulties have been suggested in the literature.

The two approaches become very similar in the incompressible case and therefore, by way of introduction, we start with a simple treatment of this case based on a procedure described in Tomiyama *et al.* (1995; see also Tomiyama and Shimada, 2001). After this preliminary illustration, we will return to the complete problem and present versions of both methods.

10.3.1 A simple incompressible model

Let us assume the two phases to be incompressible and the right-hand side of the momentum equations to include drag only. We discretize the continuity equations semi-implicitly as in (10.35):

$$\frac{\alpha_i^{n+1} - \alpha_i^n}{\Delta t} + \frac{[\![\alpha]\!]_{i+1/2}^n u_{i+1/2}^{n+1} - [\![\alpha]\!]_{i-1/2}^n u_{i-1/2}^{n+1}}{\Delta x} = 0. \tag{10.72}$$

Upon adding the continuity equations for each phase, since $\alpha_1 + \alpha_2 = 1$, we find

$$([\![\alpha_1^n]\!] u_1^{n+1})_{i+1/2} - ([\![\alpha_1^n]\!] u_1^{n+1})_{i-1/2} + ([\![\alpha_2^n]\!] u_2^{n+1})_{i+1/2} - ([\![\alpha_2^n]\!] u_2^{n+1})_{i-1/2} = 0. \tag{10.73}$$

We discretize the momentum equations starting from the non-conservation form as in (10.50) to find, for phase 1,

$$\frac{(u_1)_{i+1/2}^{n+1} - (u_1)_{i+1/2}^n}{\Delta t} + (u_1)_{i+1/2}^n \left\langle \frac{\delta u_1}{\delta x} \right\rangle_{i+1/2}^n = -\frac{1}{\rho_1} \frac{p_{i+1}^{n+1} - p_i^{n+1}}{\Delta x}$$

$$-(\hat{K}_1)_{i+1/2}^n (u_1)_{i+1/2}^{n+1} + (\hat{H}_1)_{i+1/2}^n (u_2 - u_1)_{i+1/2}^{n+1}, \tag{10.74}$$

where $\hat{K}_1 = K_1/(\alpha_1 \rho_1)$, $\hat{H}_1 = H/(\alpha_1 \rho_1)$. Similarly, for phase 2,

$$\frac{(u_2)_{i+1/2}^{n+1} - (u_2)_{i+1/2}^n}{\Delta t} + (u_2)_{i+1/2}^n \left\langle \frac{\delta u_2}{\delta x} \right\rangle_{i+1/2}^n = -\frac{1}{\rho_2} \frac{p_{i+1}^{n+1} - p_i^{n+1}}{\Delta x}$$

$$-(\hat{K}_2)_{i+1/2}^n (u_2)_{i+1/2}^{n+1} + (\hat{H}_2)_{i+1/2}^n (u_1 - u_2)_{i+1/2}^{n+1}. \tag{10.75}$$

By applying the partial elimination algorithm we have

$$(u_J)_{i+1/2}^{n+1} = \mathcal{U}_{i+1/2}^J - \mathcal{V}_{i+1/2}^J \left(p_{i+1}^{n+1} - p_i^{n+1} \right) \tag{10.76}$$

where, for phase 1,

$$\mathcal{U}_{i+1/2}^1 =$$

$$\frac{[1 + \Delta t(\hat{K}_2 + \hat{H}_2)] \left[1 - \Delta t(\delta u_1/\delta x)^n\right] u_1^n + \Delta t \hat{H}_1 \left[1 - \Delta t(\delta u_2/\delta x)^n\right] u_2^n}{D_e} \tag{10.77}$$

$$\mathcal{V}_{i+1/2}^1 = \frac{1}{D_e} \frac{\Delta t}{\Delta x} \left[\frac{1 + (\hat{K}_2 + \hat{H}_2)\Delta t}{\rho_1} + \frac{\hat{H}_1 \Delta t}{\rho_2} \right] \tag{10.78}$$

$$D_e = (1 + \Delta t \hat{K}_1)(1 + \Delta t \hat{K}_2) + \Delta t \left[\hat{H}_1(1 + \Delta t \hat{K}_2) + \hat{H}_2(1 + \Delta t \hat{K}_1) \right] \tag{10.79}$$

and similarly for phase 2.

The algorithm for the solution of the problem exploits the relation (10.73) of conservation of the total volume by noting that it will be satisfied by the correct velocities which in turn, from (10.76), presuppose the correct pressures. Let us consider the left-hand side of (10.73) as a residue that needs to be driven to zero by determining the correct pressure field:

$$D_{i+1/2}^{\kappa+1}(p_{i+1}, p_i, p_{i-1}) \equiv (\llbracket \alpha_1^n \rrbracket u_1^{n+1})_{i+1/2} - (\llbracket \alpha_1^n \rrbracket u_1^{n+1})_{i-1/2}$$

$$+(\llbracket \alpha_2^n \rrbracket u_2^{n+1})_{i+1/2} - (\llbracket \alpha_2^n \rrbracket u_2^{n+1})_{i-1/2} \tag{10.80}$$

where the dependence of D on the pressure is manifest when the expressions (10.76) for u_J^{n+1} are substituted, and we have temporarily dropped the temporal superscript $n+1$ on the pressures. To find pressures such that

$D_{i+1/2}$ vanishes, we proceed iteratively applying the usual Newton–Raphson argument and looking for pressure increments δp such that

$$D^{\kappa}_{i+1/2}(p^{\kappa}_{i+1} + \delta p_{i+1}, p^{\kappa}_i + \delta p_i, p^{\kappa}_{i-1} + \delta p_{i-1}) = 0, \tag{10.81}$$

where κ is an iteration counter. In order to avoid solving a coupled system, in the original single-phase SIMPLE method, the neighboring-node increments are simply dropped, i.e. $p_{i\pm1}$ are kept frozen at their current value. Upon expanding in Taylor series we then have

$$D^{\kappa}_{i+1/2}(p^{\kappa}_{i+1}, p^{\kappa}_i, p^{\kappa}_{i-1}) + \frac{\partial D^{\kappa}_{i+1/2}}{\partial p^{\kappa}_i}\delta p_i \simeq 0, \tag{10.82}$$

which determines the pressure increment

$$\delta p_i = -\frac{D^{\kappa}_{i+1/2}}{\partial D^{\kappa}_{i+1/2}/\partial p_i}. \tag{10.83}$$

The derivative in the denominator may be calculated directly from (10.73) and (10.76):

$$\frac{\partial D_{i+1/2}}{\partial p_i} = \sum_{J=1}^{2}\left[(\llbracket \alpha^n_J \rrbracket \mathcal{V}^J)_{i+1/2} + (\llbracket \alpha^n_J \rrbracket \mathcal{V}^J)_{i-1/2}\right]. \tag{10.84}$$

While dropping the increments to $p_{i\pm1}$ saves some computation, it also slows down convergence. To restore an acceptable rate, over-relaxation proves useful. The increments given by (10.83) are multiplied by a factor ω for which a typical value of 1.7 is cited in the literature. Alternatively, one may retain all the pressure increments finding, in place of (10.82),

$$\frac{\partial D^{\kappa}_{i+1/2}}{\partial p_{i+1}}\delta p_{i+1}\frac{\partial D^{\kappa}_{i+1/2}}{\partial p_i}\delta p_i + \frac{\partial D^{\kappa}_{i+1/2}}{\partial p_{i-1}}\delta p_{i-1} \simeq -D^{\kappa}_{i+1/2}. \tag{10.85}$$

In one spatial dimension this is a tridiagonal system which can be solved very efficiently by the well-known Thomas algorithm. In two or three dimensions the matrix is still banded, but with a greater number of nonzero diagonals. In this case, it may be useful to apply (10.85) to each coordinate direction in turn, keeping the other values of p temporarily frozen.

Once the pressure increments have been found from (10.82) or (10.85), the pressures are updated as

$$p^{\kappa+1}_i = p^{\kappa}_i + \delta p_i. \tag{10.86}$$

The corresponding velocity increments are found from (10.76):

$$(\delta u_J)_{i+1/2} = -\mathcal{V}^J_{i+1/2}(\delta p_{i+1} - \delta p_i) \tag{10.87}$$

which permits us to update the velocities

$$(u_J)_{i+1/2}^{\kappa+1} = (u_J)_{i+1/2}^{\kappa} + (\delta u_J)_{i+1/2}. \tag{10.88}$$

Once this iterative process has converged, estimates of the new volume fractions are found from (10.72):

$$\tilde{\alpha}_i^{n+1} = \alpha_i^n + \frac{\Delta t}{\Delta x} \left([\![\alpha]\!]_{i+1/2}^n u_{i+1/2}^{n+1} - [\![\alpha]\!]_{i-1/2}^n u_{i-1/2}^{n+1} \right). \tag{10.89}$$

Due to numerical error, these estimates may fail to satisfy the constraint $\alpha_1 + \alpha_2 = 1$ to a sufficient accuracy. A simple way to avoid the accumulation of error due to this deficiency is to correct the result of (10.89) according to

$$\alpha_J^{n+1} = \frac{\tilde{\alpha}_J^{n+1}}{\tilde{\alpha}_1^{n+1} + \tilde{\alpha}_2^{n+1}}. \tag{10.90}$$

Another derivation of equations (10.82) or (10.85) which sometimes is encountered in the literature consists in replacing u^{n+1} in (10.73) by $u^{\kappa} + \delta u$, with δu given by (10.87). The equivalence between the two derivations is obvious given the linearity in p of (10.87).

10.3.2 Mass-conservation algorithms

In the simple case we have just discussed, implicitness was kept at an absolute minimum, which permitted a simple solution procedure. With the implicit discretizations introduced in Section 10.2, the procedure is more complex, although the main ideas still apply. Once again we replace the time index $n+1$ by an iteration counter κ and express the updated values of the variables at iteration step $\kappa + 1$ as those at the previous step plus an increment:

$$u_J^{\kappa+1} = u_J^{\kappa} + \delta u_J, \qquad p^{\kappa+1} = p^{\kappa} + \delta p. \tag{10.91}$$

To start the process, for $\kappa = 0$, we take $p^0 = p^n$, while u^0 is calculated from (10.63) with p^n in place of p^{n+1}. The velocity increment obtained by differentiating (10.63) is

$$(\delta u_J)_{i+\frac{1}{2}} = -\sum_{K=1}^{2} \left[r_{JK}(\delta u_K)_{i-\frac{1}{2}} + s_{JK}(\delta u_K)_{i+\frac{3}{2}} \right] - \mathcal{V}_{i+\frac{1}{2}}^J (\delta p_{i+1} - \delta p_i). \tag{10.92}$$

The coefficients of (10.63) depend in general also on other time-dependent quantities, which have been kept frozen (i.e. evaluated at the previous iteration step) in deriving (10.92). In principle, this additional dependence may be accounted for by including additional increments in (10.92), although the equations quickly become much more complicated. As written, (10.92) is not an explicit expression for the velocity increment $(\delta u_J)_{i+1/2}$ as the right-hand side includes the velocity increments of both phases at the neighboring nodes. As mentioned before, in the original single-phase SIMPLE method, the neighboring-node increments are dropped, which corresponds to using the velocities u_J^κ at the previous iteration in evaluating the contribution of the neighboring nodes to the convective term of the momentum equation. This step usually results in slow convergence. A faster convergence is obtained with the SIMPLEC method (which was, for example, the default algorithm in the earlier versions of the commercial code CFX) by the approximations $(\delta u)_{i-1/2} \simeq (\delta u)_{i+3/2} \simeq (\delta u)_{i+1/2}$. With these relations, and upon dropping the contribution of the other phase, (10.92) becomes

$$\left[1 + (r_{JJ} + s_{JJ})_{i+1/2}\right] (\delta u_J)_{i+1/2} = - \mathcal{V}_{i+1/2}^J (\delta p_{i+1} - \delta p_i). \tag{10.93}$$

Several other ways to deal with the velocity increments at the neighboring nodes have been devised, which give rise to the numerous versions of the SIMPLE algorithm to be found in the literature. Many of these variants can be adapted to two-fluid model equations as shown, e.g. in Darwish *et al.* (2001). For example, in the PISO algorithm, the velocity increments of the neighboring nodes are neglected in a first correction, just as in the original SIMPLE procedure, but then a second correction step is executed in which these velocity increments are estimated using the results of the first correction (see, e.g. Darwish *et al.* 2001; Ferziger and Perić 2002).

In order to calculate the pressure increments, we form a combined continuity equation by adding the two continuity equations (10.34) after dividing each one by a representative value of the corresponding density, $\rho_{J,\text{ref}}$, and suitably adjusting the iteration index:

$$\frac{1}{\rho_{1,\text{ref}}} \left[[\![(\alpha\rho)_1]\!]_{i+1/2}^\kappa (u_1)_{i+1/2}^{\kappa+1} - [\![(\alpha\rho)_1]\!]_{i-1/2}^\kappa (u_1)_{i-1/2}^{\kappa+1} \right]$$

$$+ \frac{1}{\rho_{2,\text{ref}}} \left[[\![(\alpha\rho)_2]\!]_{i+1/2}^\kappa (u_2)_{i+1/2}^{\kappa+1} - [\![(\alpha\rho)_2]\!]_{i-1/2}^\kappa (u_2)_{i-1/2}^{\kappa+1} \right]$$

$$+ \frac{\Delta x}{\Delta t} \left\{ \frac{[(\alpha_1^\kappa \rho_1^{\kappa+1})_i - [(\alpha\rho)_1]_i^n}{\rho_{1,\text{ref}}} + \frac{[(\alpha_2^\kappa \rho_2^{\kappa+1})_i - [(\alpha\rho)_2]_i^n}{\rho_{2,\text{ref}}} \right\} = 0. \tag{10.94}$$

Division by the reference densities is necessary to avoid the denser phase biasing the result. For an incompressible phase, the obvious choice is $\rho_{J,\text{ref}} = \rho_J$; for a compressible phase, one may take, for example, $\rho_{J,\text{ref}} = (\rho_J)_i^\kappa$. To account for compressibility, in the last term we write

$$(\rho_J)_i^{\kappa+1} \simeq (\rho_J)_i^{\kappa} + \left(\frac{\partial \rho_J}{\partial p}\right)_i^{\kappa} \delta p_i. \tag{10.95}$$

This procedure is only adequate in the presence of weak compressibility effects. For strong compressibility, a (nearly) fully coupled solution procedure is necessary as will be seen in the next chapter.

Equation (10.94) now plays the role of equation (10.73) in the simple example: by driving its residue to zero, we will determine the pressure. To this end, as before in the derivation of (10.85), we replace $u^{\kappa+1}$ by $u^{\kappa} + \delta u$, with δu given by (10.93), linearize, and find a tridiagonal system in the pressure increments. After solution of this system the pressures and velocities are updated from (10.91) with the velocity increments given by (10.93), with or without the correction term in the brackets in the left-hand side. The updated pressures permits one to update the densities $(\rho)_i^{\kappa+1}$ by (10.95) so that, in the phase continuity equations (10.34), only the volume fractions are unknown at this point. These can be updated directly from

$$(\alpha)_i^{\kappa+1}(\rho)_i^{\kappa+1} = (\alpha\rho)_i^n + \frac{\Delta t}{\Delta x} \left[[\![\alpha\rho]\!]_{i+1/2}^{\kappa} u_{i+1/2}^{\kappa+1} - [\![\alpha\rho]\!]_{i-1/2}^{\kappa} u_{i-1/2}^{\kappa+1} \right] \tag{10.96}$$

or, at the price of solving another tridiagonal system, from

$$(\alpha)_i^{\kappa+1}(\rho)_i^{\kappa+1} + \frac{\Delta t}{\Delta x} \left[[\![\alpha^{\kappa+1}\rho^{\kappa+1}]\!]_{i+1/2} u_{i+1/2}^{\kappa+1} - [\![\alpha^{\kappa+1}\rho^{\kappa+1}]\!]_{i-1/2} u_{i-1/2}^{\kappa+1} \right]$$
$$= (\alpha\rho)_i^n. \tag{10.97}$$

The renormalization (10.90) may be applied to limit the effect of departures from the conservation of volume relation (10.4).

Before the iterations have converged, it is possible that equations (10.96) or (10.97) generate volume fractions outside the physical range $0 \le \alpha \le 1$. Even though this problem would be corrected at convergence, it is possible that model relations using α (e.g. a constitutive relation for the drag including $\sqrt{\alpha}$) fail in these circumstances. A possible way to deal with this problem is simply to force the incorrect value back into the physical range after evaluation of (10.96) or (10.97). This procedure may, however, either slow down or destroy convergence. A less drastic remedy is to use underrelaxation calculating $\alpha_J^{\kappa+1}$ as the weighted average of the value given by (10.96) or (10.97) and the value at the previous iteration (Carver, 1982).

In summary, for each time step, the complete algorithm is the following:

(i) Generate a first estimate of the new velocities by solving implicitly the momentum equations (10.63) using the available pressure values

p^n; this can be considered a predictor step which is then corrected iteratively by the following steps;

(ii) Find the pressure increments by setting $u^{\kappa+1} = u^\kappa + \delta u$, with the δu's expressed in terms of pressure increments δp by (10.93), in the pressure correction equation (10.94);

(iii) Update the pressure, density and velocities by using (10.91) and (10.95);

(iv) Solve the phase continuity equations (10.96) or (10.97);

(v) Return to step 2 and continue until convergence.

As an example of results generated by a mass-conservation algorithm similar to the procedure just described, we reproduce in Fig. 10.2(a) (from Moukalled and Darwish, 2002) the convergence history to steady state for a problem in which solid particles are fluidized by an upward gas flow. Differences between the algorithm used here and the one described above are the adoption of the co-located variable arrangement described later in Section 10.3.4, and of under-relaxation. The quantity plotted is the mass residual, i.e. the normalized error with which the total mass conservation equation is satisfied, and the different lines correspond to different numbers of control volumes. Figure 10.2(b) (from the same reference) shows the convergence history for a PISO-like implementation of the algorithm. With both methods, the number of iterations necessary to reach a certain accuracy increases as the discretization is refined. The number of iterations required by the PISO method is slightly smaller, but it is found that the computation times are very similar.

10.3.3 The IPSA volume-conservation algorithm

The first pressure-correction strategy relying on conservation of volume, equation (10.4), was introduced by Spalding (1979, 1983) with the name IPSA, for *Inter-Phase Slip* Algorithm. Several variants of the idea have been developed in the intervening years (see, e.g. Issa and Oliveira, 1994; Darwish *et al.*, 2001; Miller and Miller, 2003; Moukalled and Darwish, 2004).

To explain the procedure, for simplicity, we consider the incompressible case. We introduce an iterative scheme in which volume fractions, velocity, and pressure are iteratively adjusted as

$$\alpha_J^{\kappa+1} = \alpha_J^\kappa + \delta\alpha_J, \qquad u_J^{\kappa+1} = u_J^\kappa + \delta u_J, \qquad p^{\kappa+1} = p^\kappa + \delta p. \quad (10.98)$$

After substitution into the incompressible version of the implicitly discretized

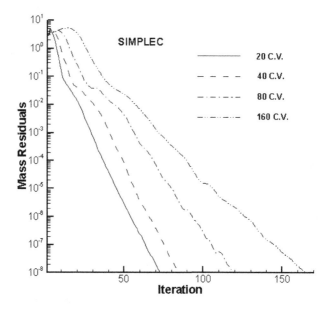

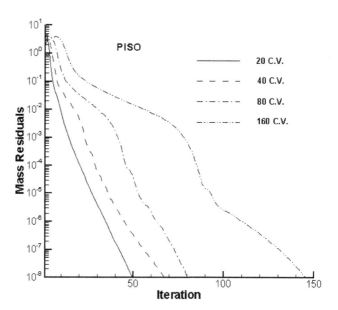

Fig. 10.2. Mass residual convergence history for the one-dimensional simulation of solid particles fluidized by a gas flow. The left diagram is for the SIMPLEC-like algorithm described in Section 10.3.3, the right diagram for a PISO-like method, both implemented with a co-located variables arrangement; the lines correspond to different discretizations (from Moukalled and Darwish 2002).

continuity equation (10.34)

$$\frac{(\alpha_J)_i^{n+1} - (\alpha_J)_i^n}{\Delta t} + \frac{[\![\alpha_J]\!]_{i+1/2}^{n+1}\, u_{J,i+1/2}^{n+1} - [\![\alpha_J]\!]_{i-1/2}^{n+1}\, u_{J,i-1/2}^{n+1}}{\Delta x} = 0, \quad (10.99)$$

and linearization, one finds

$$(\delta\alpha_J)_i + \frac{\Delta t}{\Delta x}\left([\![u_J^\kappa \delta\alpha_J]\!]_{i+1/2} - [\![u_J^\kappa \delta\alpha_J]\!]_{i-1/2}\right)$$

$$+ \frac{\Delta t}{\Delta x}\left([\![\alpha_J^\kappa \delta u_J]\!]_{i+1/2} - [\![\alpha_J^\kappa \delta u_J]\!]_{i-1/2}\right)$$

$$= -(\alpha_J^\kappa - \alpha_J^n) - \frac{\Delta t}{\Delta x}\left([\![u_J\alpha_J]\!]_{i+1/2}^\kappa - [\![u_J\alpha_J]\!]_{i-1/2}^\kappa\right). \quad (10.100)$$

This equation is simplified by dropping the coupling with the neighboring cells, after which the velocity increments are expressed in terms of pressure increments as in (10.93) to find an equation of the form

$$(\delta\alpha_J)_i = \frac{\Delta t}{\Delta x}\left[\tilde{\mathcal{V}}_{i+1/2}^J [\![\alpha_J]\!]_{i+1/2}^\kappa \delta p_{i+1}\right.$$

$$-\left(\tilde{\mathcal{V}}_{i+1/2}^J [\![\alpha_J]\!]_{i+1/2}^\kappa + \tilde{\mathcal{V}}_{i-1/2}^J [\![\alpha_J]\!]_{i-1/2}^\kappa\right)\delta p_i$$

$$\left. +\tilde{\mathcal{V}}_{i-1/2}^J [\![\alpha_J]\!]_{i-1/2}^\kappa \delta p_{i-1}\right] - (A_J)_i^\kappa \quad (10.101)$$

where

$$(A_J)_i^\kappa = (\alpha_J^\kappa - \alpha_J^n) + \frac{\Delta t}{\Delta x}\left([\![u_J\alpha_J]\!]_{i+1/2}^\kappa - [\![u_J\alpha_J]\!]_{i-1/2}^\kappa\right) \quad (10.102)$$

and

$$\tilde{\mathcal{V}}_{i+1/2}^J = \frac{\mathcal{V}_{i+1/2}^J}{1 + (r_{JJ} + s_{JJ})_{i+1/2}}; \quad (10.103)$$

the symbols $\mathcal{V}_{i+1/2}^J$, r_{JJ}, and s_{JJ} are defined in (10.78), (10.66), and (10.67), respectively. These expressions for $\delta\alpha_J$ are then substituted into the volume conservation relation (10.4):

$$(\alpha_1^\kappa + \delta\alpha_1)_i + (\alpha_2^\kappa + \delta\alpha_2)_i = 1 \quad (10.104)$$

to generate an equation for the pressure increments δp.

The algorithm is therefore the following:

(i) Generate a first estimate of the new volume fractions by solving implicitly the continuity equations (10.99) with the available velocities u_J^n

(ii) Generate a first estimate of the new velocities by solving implicitly the momentum equations (10.63) using the available pressure and volume fractions values

(iii) Find the pressure increments δp from the pressure correction equation (10.104).

(iv) Update the pressure, density and velocities by using (10.98).

(v) Return to step 3 and continue until convergence.

As in the previous case, several variants of this algorithm exist (see, e.g. Darwish *et al.*, 2001; Moukalled and Darwish, 2004). For example, rather than following the SIMPLEC approach in relating the velocity to the pressure increments, one could use the original SIMPLE method, which would amount to taking $r_{JJ} = s_{JJ} = 0$. Other possible variants include evaluating the volume fraction multiplying the pressure gradient in the momentum equations at the advanced time $n + 1$, which would lead to an additional term proportional to $\delta\alpha_J$ in the velocity correction equation (10.93). A procedure reminiscent of the SIMPLEC improvement over SIMPLE would be, in place of dropping the coupling terms $(\delta\alpha_J)_{i\pm1/2}$ in equation (10.100), to approximate them by $(\delta\alpha_J)_i$.

The case of compressible fluids is treated as before introducing a density increment expressed in terms of the pressure via an equation of state as in (10.95); see also Section 10.6.

Figure 10.3(a) (from Moukalled and Darwish, 2004) shows the convergence history to steady state for the same problem as in Fig. 10.2. Again a co-located variables arrangement and under-relaxation were used. The lines show the mass residual for different numbers of control volumes. With

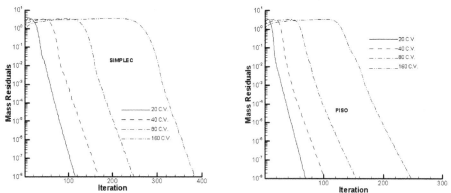

Fig. 10.3. Mass residual convergence history for the one-dimensional simulation of solid particles fluidized by a gas flow. The left diagram is for the SIMPLEC-like algorithm described in Section 10.3.3, the right diagram for a PISO-like method (from Moukalled and Darwish, 2004).

all discretizations, there is an initial phase during which the error remains nearly constant, followed by one in which it decreases rapidly. The first phase becomes longer as the discretization is made finer. The origin of this initial phase can be traced back to the terms neglected in going from (10.100) to (10.101) which seem to have a great influence on convergence at the beginning of the iterative process. Figure 10.3(b) (from the same reference) shows the convergence history for a PISO-like implementation of the algorithm. The number of iterations required is significantly smaller, but it is found that, due to the additional correction step of this algorithm, the difference in computational time is not large.

Each one of the two classes of methods described has its own strengths and weaknesses. The mass conservation approach appears better able to handle situations in which some regions of the flow become single-phase, e.g. due to the complete condensation of a vapor, as equation (10.94) readily degenerates to the single-phase form when one of the two volume fractions vanishes. In principle, the IPSA algorithm also exhibits the same feature as, when for example α_1 vanishes, $\delta\alpha_1$ also vanishes identically. In this case $\alpha_2 = 1$, $\delta\alpha_2 = 0$, and equation (10.101) becomes then essentially equal to the pressure correction equation of the single-phase SIMPLEC algorithm. Near these degenerate single-phase points, however, it is easy for the iteration method to produce volume fraction values outside the allowed range and, therefore, one is more likely to encounter numerical difficulties with the IPSA approach in these situations. On the other hand, the IPSA method appears to have some advantages when there is a large difference between the densities of the two phases (Darwish *et al.*, 2001).

If the energy equations are added to the model, the relation (10.95) between density and pressure would acquire a new term, e.g.

$$(\rho_J^{\kappa+1})_i \simeq (\rho_J)_i^\kappa + \left(\frac{\partial \rho_J}{\partial p}\right)_i^\kappa \delta p_i + \left(\frac{\partial \rho_J}{\partial T}\right)_i^\kappa \delta T_i. \tag{10.105}$$

To preserve the segregated nature of the algorithm, the temperature correction is neglected in the evaluation of the pressure correction. Velocities and volume fractions are updated as before, after which the energy equation is solved and, with the updated temperature, a new density is generated.

The segregated nature of these procedures permits the inclusion of additional equations, e.g. to model turbulence. The additional variables described by these equations would be kept frozen for the purposes of calculating the pressure correction, and would be adjusted at each iteration using the updated velocities.

10.3.4 Co-located variable arrangement

In place of the staggered grid described before, it is possible to use a co-located arrangement in which all the dependent variables are defined at the center of the control volume (or interval, in the present one-dimensional case)[1]. A crucial difference with the discretization adopted previously arises from the need to prevent the appearance of the well-known phenomenon of pressure oscillations (see e.g. Ferziger and Perić, 2002), which requires the introduction of a special interpolation of the fluxes, effectively amounting to the introduction of damping in the continuity equations (see e.g. Rhie and Chow, 1983; Miller and Schmidt, 1988; Ferziger and Perić, 2002).

The co-located arrangement is particularly advantageous with nonorthogonal coordinates because it permits the use of Cartesian velocity components which lead to simpler equations than contravariant or covariant components. With Cartesian components, the equations are in strict conservation form and there are no curvature terms in the momentum equations (see e.g. Ferziger and Perić, 2002). These terms involve derivatives of the scale factors and, therefore, their presence imposes a stringent constraint on the smoothness of the grid. Furthermore, the data structure on a nonstructured mesh is simplified. These properties make this approach particularly attractive with complex geometries and, in fact, most commercial CFD codes use the co-located arrangement with Cartesian velocity components. However, even with a regular grid, there may be some advantage as only one set of control volumes is required, which in particular means that the convective fluxes have the same form for all the variables (Perić *et al.*, 1988; Ghidaglia *et al.*, 2001).

We have seen that the mass conservation equation plays a crucial role in the determination of the pressure. When this equation is discretized implicitly in time and integrated over a control volume (interval), the convective term becomes

$$\frac{1}{\Delta x} \int_{x_{i-1/2}}^{x_{i+1/2}} \left(\frac{\partial(\alpha\rho u)}{\partial x} \right)^{n+1} dx = \frac{(\alpha\rho u)_{i+1/2}^{n+1} - (\alpha\rho u)_{i-1/2}^{n+1}}{\Delta x}. \tag{10.106}$$

Since the values of $(\alpha\rho u)_{i\pm1/2}$ are not available in a co-located arrangement, the natural thing to do is to use linear interpolation which, on a uniform mesh, expresses the values at $i + 1/2$ as the average of those at $i + 1$ and i

---

[1] The word "collocated" often encountered in the literature to denote this variable arrangement is incorrect and lexically indefensible. Since standard English does not seem to recognize the word "colocated," we will use the accepted hyphenated spelling "co-located."

so that

$$\frac{1}{\Delta x} \int_{x_{i-1/2}}^{x_{i+1/2}} \left(\frac{\partial(\alpha\rho u)}{\partial x}\right)^{n+1} dx \simeq \frac{(\alpha\rho u)_{i+1}^{n+1} - (\alpha\rho u)_{i-1}^{n+1}}{2\Delta x}. \tag{10.107}$$

In the normal procedure, $(\alpha\rho u)_i$ would be expressed from the momentum equation in terms of $\partial p/\partial x|_i \simeq (p_{i+1} - p_{i-1})/2\Delta x$ similarly to (10.74) which, when substituted into (10.106), would lead to a term proportional to $p_{i+2} - 2p_i + p_{i-2}$. In this way there is no coupling between pressure values at even and odd nodes, and it is this circumstance that causes pressure oscillations (see e.g. Wesseling, 2001; Ferziger and Perić, 2002).

A low-accuracy remedy for the problem (Wesseling, 2001) would be to use forward differences for $\alpha\rho u$ and backward differences for p (or vice versa) so that

$$\frac{1}{\Delta x} \int_{x_{i-1/2}}^{x_{i+1/2}} \left(\frac{\partial(\alpha\rho u)}{\partial x}\right)^{n+1} dx \simeq \frac{(\alpha\rho u)_{i+1}^{n+1} - (\alpha\rho u)_i^{n+1}}{\Delta x}, \tag{10.108}$$

$$\frac{1}{\Delta x} \int_{x_{i-1/2}}^{x_{i+1/2}} \alpha \frac{\partial p}{\partial x} dx \simeq \alpha_i \frac{p_i^{n+1} - p_{i-1}^{n+1}}{\Delta x}. \tag{10.109}$$

These approximations would be substituted into the continuity and momentum equations, respectively, after which the usual procedures can be applied.

A second-order accurate alternative was developed by Rhie and Chow (1983) and hinges on the special way in which the velocities on the faces of the control volume are to be interpolated. Again limiting oneself to a regular grid, it is easy to show that the relation

$$(\alpha\rho u)_{i+1/2} \simeq \frac{1}{2}\left[(\alpha\rho u)_i + (\alpha\rho u)_{i+1}\right] - \frac{\Delta x^2}{8} \frac{\partial^2}{\partial x^2}(\alpha\rho u)\bigg|_{i+1/2} \tag{10.110}$$

has an error of order Δx^4. The second derivative can be approximated as

$$\frac{\partial^2}{\partial x^2}(\alpha\rho u)\bigg|_{i+1/2} \simeq \frac{(\alpha\rho u)_{i+1} - 2(\alpha\rho u)_{i+1/2} + (\alpha\rho u)_i}{(\frac{1}{2}\Delta x)^2}. \tag{10.111}$$

The essence of the idea is to express the terms appearing here using the momentum equation, which introduces p_i and $p_{i\pm1}$ and therefore a coupling between even and odd nodes which is sufficient to remove the pressure oscillations. To show how the idea works, we write for brevity

$$(\alpha\rho u)_i^{n+1} = H_i - \alpha_i \Delta t \left. \frac{\partial p^{n+1}}{\partial x} \right|_i ,$$

$$(\alpha\rho u)_{i+1/2}^{n+1} = H_{i+1/2} - \frac{1}{2}(\alpha_i + \alpha_{i+1}) \Delta t \left. \frac{\partial p^{n+1}}{\partial x} \right|_{i+1/2} , \qquad (10.112)$$

where H is shorthand for all the terms other than the pressure gradient; then (10.111) becomes

$$\left. \frac{\partial^2}{\partial x^2}(\alpha\rho u) \right|_{i+1/2}^{n+1}$$

$$\simeq -\frac{\Delta t}{(\frac{1}{2}\Delta x)^2} \left[\alpha_{i+1} \left(\left. \frac{\partial p}{\partial x} \right|_{i+1}^{n+1} - \left. \frac{\partial p}{\partial x} \right|_{i+1/2}^{n+1} \right) - \alpha_i \left(\left. \frac{\partial p}{\partial x} \right|_{i+1/2}^{n+1} - \left. \frac{\partial p}{\partial x} \right|_{i}^{n+1} \right) \right]$$

$$+ \frac{H_{i+1} - 2H_{i+1/2} + H_i}{(\frac{1}{2}\Delta x)^2} . \qquad (10.113)$$

The values of the variables necessary to evaluate $H_{i+1/2}$ are obtained by interpolation between the available values at i and $i+1$. The differences between the pressure gradients are evaluated as, e.g.

$$\left. \frac{\partial p}{\partial x} \right|_{i+1/2} - \left. \frac{\partial p}{\partial x} \right|_{i} \simeq \frac{p_{i+1} - p_i}{\Delta x} - \frac{p_{i+1} - p_{i-1}}{2\Delta x} = \frac{p_{i+1} - 2p_i + p_{i-1}}{2\Delta x} . \qquad (10.114)$$

Near boundaries, it is necessary to maintain the second-order accuracy and, in this case, a one-sided second-order formula should be used.

   The pressure correction equation is now found similarly to (10.94), except that the momentum densities are evaluated as in (10.110) with the second derivative expressed as in (10.113), which justifies the denomination of *pressure-weighted interpolation*, or PWI, usually given to this procedure. In an intuitive sense, it may be said that this last term is sensitive to oscillations and smooths them out. When the time step is very small, since this term is multiplied by Δt, this smoothing effect weakens and sometimes oscillations are still encountered; ways to avoid this problem have been proposed in the literature (Shen *et al.*, 2001, 2003). The explicit presence of the time step in (10.113) has the consequence that the calculation of transient flows exhibits a weak dependence on Δt, which however disappears at steady state.

   As long as the pressure is smooth, the numerical error introduced by this procedure is consistent with the overall error and no problems are likely to arise. In the vicinity of large void fraction gradients, or of discontinuities of porosity, however, the pressure gradient can vary rapidly to balance other forces with an equally rapid variation. In these cases, it is advisable to extract from H_i, $H_{i+1/2}$ in (10.112) the rapidly varying term and to combine it with the pressure gradient in (10.113) as the combination will generally be smooth.

The co-located variable arrangement is widely used in the multiphase literature (e.g. Issa and Oliveira, 1994; Kunz *et al.*, 1998, 1999; Karema and Lo, 1999). In these applications, additional issues arise. For example, if the momentum equations contain a term proportional to $\partial \alpha / \partial x$, e.g. similar to the last one shown in (10.7), Kunz *et al.* (1998) find that its contribution should be included in (10.113) to avoid oscillations in the computed volume fractions.

10.4 Example: fluidized beds

In this section, we provide an illustration of the results that can be obtained by the methods described in the previous section for a realistic example of a two-dimensional bubbling gas–solid fluidized bed illustrated in Fig. 10.4. The situation is similar to one simulated by Guenther and Syamlal (2001). We focus in particular on the effect that different grid resolutions and orders of spatial discretization have on the shape of the bubble-like voids that form in the bed. The results illustrate the qualitative deficiency of the first-order upwinding scheme when used for this application and how it can be relieved by second-order spatial discretization. Another point is the problem of obtaining grid independence, which frequently plagues two-phase flow simulations as seen in Section 8.4.

The general form of the averaged equations of motion for the flow of multiphase mixtures has already been described in Chapter 8 and these equations can, of course, be solved only if they are supplemented by closure relations for the various terms appearing in them. Some examples of closures for the interphase interaction force were presented in Section 8.3 of Chapter 8, but we did not dwell on the subject of closures for the effective stresses in the different phases. Development of accurate closures has been, and still continues to be, a subject of much research. A discussion of these closures is beyond the scope of a book devoted, like this one, to numerical methods, so we will limit ourselves to some broad observations and cite references in the literature where the specific closure expressions used in the simulations have been discussed.

In general, in most gas–particle flows encountered in fluidized beds, the added mass, inertial lift, and history forces are dwarfed by the gravitational and drag forces. Hence, in these simulations, only the drag force contribution

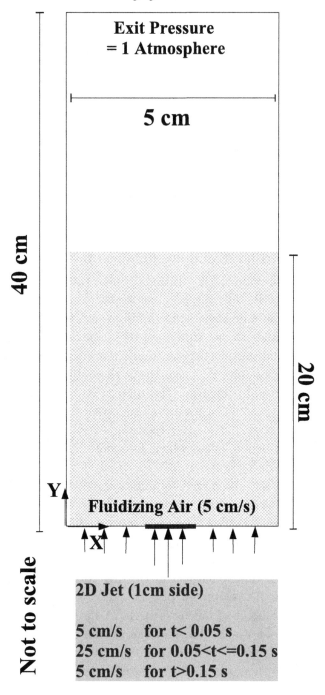

Fig. 10.4. Schematic of the two-dimensional jet bubbling bed.

to the interphase interaction force was taken into account. Specifically, the empirical model for the drag coefficient of Wen and Yu (1966), described in Section 8.3 of Chapter 8, was employed. A simple Newtonian form for the effective stress in the continuous phase is commonly assumed in simulations of fluid–particle flows. In gas–particle flows with appreciable mass loading of particles, the contribution of the deviatoric part in the effective fluid phase stress to the fluid and particle phase momentum balance equations is often negligibly small, and hence the accuracy of the closure for the effective viscosity of the fluid phase is often not an important consideration (e.g. Agrawal *et al.*, 2001). Here, we simply take the effective viscosity of the gas phase to be that of the pure gas. For the same reason, at sufficiently high mass loading of particles, the effect of fluid-phase turbulence is not important and, accordingly, it is disregarded here.

The effective stress in the particle phase can be partitioned into two parts: a streaming stress, resulting from the fluctuating motion of the particles; and a contact stress, transmitted through direct contact between particles. The former is simply the local average of $\langle \rho_S \left(\mathbf{u}_S - \langle \mathbf{u}_S \rangle \right) \left(\mathbf{u}_S - \langle \mathbf{u}_S \rangle \right) \rangle$, where the subscript "S" denotes the solid phase. The contact stress can arise as a result of (predominantly binary and essentially instantaneous) collisions between particles at low to modest particle concentrations, or through sustained contact between multiple particles which typically arises at high particle volume fractions and low shear rates. Bridging these two extremes to describe the rheological behavior in the transitional regime remains an active area of research. As a first approximation, the total particle phase stress is expressed as the sum of the rapid flow regime and quasistatic flow regime contributions.

The streaming stress and the part of the contact stress arising from the particle collisions in the rapid flow regime are usually considered together and have been closed either phenomenologically or by adapting the kinetic theory of dense gases; these closures are generally referred to as kinetic theory models.

Phenomenological closures (e.g. Anderson *et al.*, 1995) typically assume that the particle phase stress can be expressed in Newtonian form

$$\boldsymbol{\sigma}_S = -p_S \mathbf{I} + 2\mu_S \left(\frac{1}{2} \left[\boldsymbol{\nabla} \mathbf{u}_S + (\boldsymbol{\nabla} \mathbf{u}_S)^T \right] - \frac{1}{3} \mathbf{I}(\boldsymbol{\nabla} \cdot \mathbf{u}_S) \right). \qquad (10.115)$$

The particle phase pressure and viscosity are allowed to vary with particle volume fraction, $p_S = p_S(\alpha_S)$, $\mu_S = \mu_S(\alpha_S)$. In a fluid–solid suspension, both of these quantities should vanish as $\alpha_S \to 0$. Based on a speculation (e.g. Anderson *et al.*, 1995) that they should increase monotonically with

α_S, expressions such as

$$p_S = a_1 \alpha_S^3 \exp\left(\frac{a_2 \alpha_S}{\alpha_m - \alpha_S}\right), \qquad \mu_S = \frac{a_3 \alpha_S}{1 - (\alpha_S/\alpha_m)^{1/3}} \qquad (10.116)$$

where a_1, a_2, a_3 are empirical constants and α_m is the maximum particle volume fraction at random close packing, have been employed to study the dynamics of fluidized beds of particles. Clearly, one can choose a number of different functional forms for the pressure and viscosity which are consistent with the speculation that these quantities increase with α_S. Glasser *et al.* (1996, 1997) studied the hierarchy of bifurcation of nonuniform structures that emerge when a uniformly fluidized bed of particles becomes unstable, and demonstrated that the bifurcation hierarchy is robust and is independent of the specific choice made to model $p_S(\alpha_S)$ and $\mu(\alpha_S)$. The empirical closures have been useful in establishing that the gross dynamic features of the flow such as waves, bubbles, and clusters in fluidized suspensions are not associated with specific functional forms for $p_S(\alpha_S)$ and $\mu_S(\alpha_S)$.

Like phenomenological closures, kinetic-theory closures for the particle-phase stress recognize that the pressure and viscosity depend on the particle volume fraction, but they also add the effect of the intensity of the particle velocity fluctuations. While the phenomenological closures assume that the velocity fluctuations can be slaved to the particle volume fraction (so that the pressure and viscosity could be expressed as functions of α_S only), the kinetic theory approach allows them to remain as distinct dependent variables. In its simplest form, the kinetic theory assumes that the velocity fluctuations are nearly isotropic and introduces a quantity, commonly referred to as the granular temperature T_S, defined by

$$T_S = \frac{1}{3}\langle (\mathbf{u}_S - \langle \mathbf{u}_S \rangle) \cdot (\mathbf{u}_S - \langle \mathbf{u}_S \rangle) \rangle \qquad (10.117)$$

as an additional scalar variable to be tracked in the flow calculations. The granular (or fluctuation, or pseudothermal) energy balance equation (Lun *et al.*, 1984; Gidaspow, 1994, Koch and Sangani, 1999) takes the form

$$\frac{\partial}{\partial t}\left(\frac{3}{2}\rho_S \alpha_S T_S\right) + \nabla \cdot \left(\frac{3}{2}\rho_S \alpha_S T_S \mathbf{u}_S\right)$$
$$= -\nabla \cdot \mathbf{q}_S - \boldsymbol{\sigma}_S : \nabla\langle \mathbf{u}_S \rangle + \Gamma_{\text{slip}} - J_{\text{coll}} - J_{\text{vis}}. \qquad (10.118)$$

The first term on the right-hand side represents the diffusive transport of fluctuation energy, with $\mathbf{q}_S$ denoting the diffusive flux. The second and third terms represent rates of production of fluctuation energy by shear and gas–particle slip, respectively. The fourth and the fifth terms denote rates of dissipation of fluctuation energy through inelastic collisions and viscous damping, respectively. The early derivation of granular energy balance (e.g. Lun *et al.*, 1984) focused on granular flow in the absence of any appreciable effect due to the interstitial fluid and hence the third and the fifth terms on the right-hand side were not considered. Later developments (Savage, 1987; Koch, 1990; Ma and Ahmadi, 1988; Gidaspow, 1994; Balzer *et al.*, 1995; Boelle *et al.*, 1995; Koch and Sangani, 1999) have shed some light on these terms. The form used in the derivation of the results shown here is that of Agrawal *et al.* (2001).

Slow flow of dense assemblies of particles, commonly referred to as quasistatic flow, has been studied extensively in the soil mechanics community, and closure models for particle phase stress developed in the study of soil mechanics are used quite frequently in granular and two-phase flows. The quasistatic flow part used in the simulations shown here is based on a model proposed by Schaeffer (1987) and is used by Guenther and Syamlal (2001). In the quasistatic flow regime, the granular-phase shear stress has an order-zero dependence on the shear rate, and this brings in computational difficulties. Usually, in simulations of fluidized beds of fairly large particles, such as those considered here, the primary role played by the quasistatic stresses is to ensure that the particle volume fraction does not become unphysically large. In most of the bed, the particle-phase stress is determined only by the rapid flow part.

As often when average multifluid models are used, a word of caution about the closure relations is in order. In many fluid–particle flow problems, such as sedimentation and fluidization, the body force due to gravity, the fluid-phase pressure gradient and the drag force tend to be the most important terms in the right-hand side of the momentum balance equations. Thus, a few percent error in the estimate of the drag force can lead to appreciable quantitative changes in the predicted flow patterns. What we do not know at present is whether uncertainties in the drag force closure can lead to qualitatively incorrect predictions. For example, the drag-force closures mentioned above assume that the local drag force can be expressed in terms of local average velocities and the particle volume fraction. In any real flow problem, there will always be spatial gradients, and the potential impact of these gradients on the local drag force has not been thoroughly studied. Thus, it is prudent to view the results obtained by solving the averaged equations with a critical eye, performing sensitivity tests and whenever possible validating against experimental data. This procedure is indeed typical in the literature.

Boundary conditions at solid surfaces have been examined by several researchers. When the boundary is impervious to the particles, the normal velocity of the particle phase must obviously match that of the boundary.

The nature of the particle–wall collisions can be expected to influence the shear stress and fluctuation energy flux at the boundary. Hui *et al.* (1984) introduced an empirical parameter, known as the specularity coefficient, to quantify the nature of particle–wall collisions. A value of zero for the specularity coefficient implies specular collisions (which can be expected with smooth frictionless walls) and unity corresponds to diffuse collisions (which would occur on a very rough wall). Using the specularity coefficient and the coefficient of restitution for particle–wall collisions as model parameters, Johnson and Jackson (1987) constructed boundary conditions for the particle phase shear stress and fluctuation energy flux at bounding solid surfaces. Jenkins and Richman (1986) have derived boundary conditions that apply for inelastic collisions between smooth particles and a frictionless, but bumpy, boundary. Jenkins (1992) developed boundary conditions that apply for the interaction of frictional particles with flat frictional boundaries. These expressions do not involve the specularity coefficient, but require additional information on the nature of the bumpiness of the surface or the frictional properties. Virtually nothing is known about the possible influence of the interstitial fluid on the wall boundary conditions for the particle phase. Hence, it is common practice in the fluidization literature to consider different boundary conditions, all the way from free-slip to no-slip, and examine the sensitivity of the results to the choice of boundary conditions. In the simulations presented here as illustration, we simply consider no-slip boundary conditions for the particle phase.

Even though the fluid satisfies no-slip boundary conditions at the bounding surface, it is not necessarily correct to prescribe that the local *average* velocity of the fluid be equal to that of the surface. Unfortunately, our understanding of the boundary condition for the fluid phase is poor. In gas-fluidized bed problems, this is not a serious limitation, as the deviatoric stress in the gas phase plays a negligible role. In fact, many simulations have revealed that there is hardly any noticeable difference between the results obtained with free-slip and no-slip boundary conditions for the gas phase at bounding walls. With this in mind, in the simulations described here, no-slip boundary conditions were enforced at the walls.

There has been little systematic effort to develop and validate boundary conditions for two-phase flows at inlet and exit boundaries. Most simulators assume that the fluid-phase pressure is specified at the boundary through which fluid exits and impose a zero-normal-gradient condition at this boundary for all dependent variables (other than continuous phase pressure). All simulators typically assume that the dependent variables (other than continuous phase pressure) are fully specified at the inlet, while the pressure at the inlet is allowed to float. Because of the *a priori* specification of a boundary as either an inlet or an exit, most simulators in use today cannot handle mixed boundaries – for example, countercurrent flow of gas–particle mixture in a vertical pipe cannot be handled by commonly available simulators.

It is important to recognize that the closure expressions used in the simulations discussed below are simply examples. Our ability to solve numerically the two-fluid model equations is by no means limited to these closures. As new closures are developed, one can incorporate them in numerical simulations quite readily.

The simulations we now describe were performed using MFIX (*M*ulti-phase *F*low with *I*nterphase e*X*changes), which is a general-purpose code for describing the hydrodynamics, heat transfer, and chemical reactions in fluid–solid systems with high mass loading of particles. The code has been used in the literature to describe bubbling, spouted, and circulating fluidized beds. MFIX calculations give transient field data on the three-dimensional distribution of pressure, velocity, temperature, and species mass fractions.[1] The example considered below – an isothermal, nonreactive two-phase system without phase change – is one of the simplest applications of MFIX. MFIX can handle nonisothermal, reactive systems with phase change, containing multiple types of particle phases. Extensions of the concepts discussed here to these more general situation can be found in the MFIX documentation (Syamlal *et al.*, 1993; Syamlal, 1998).

MFIX can be used to perform one-, two-, and three-dimensional simulations of the averaged equations of motion. The code is based on a finite-volume method with staggered grids, and it employs the SIMPLE algorithm adapted to multiphase systems. The discretization of the governing equations through the finite-volume method is a fairly straightforward extension to more than one space dimension of the procedure described earlier in this chapter. In particular, we now consider an additional scalar variable, namely the granular temperature, and hence one needs to discretize the additional equation (10.118), but this circumstance poses no new conceptual challenges. The implicit backward-Euler discretization is used for time advancement, while several options are available for the spatial discretization: first-order upwind and second-order approximations with the possibility to choose from a slate of flux limiters such as those shown in Table 10.1 on p. 330 with the Superbee limiter being the preferred choice.

---

[1] The code has been parallelized for use on Beowulf clusters or massively parallel supercomputers and can be run in shared, distributed or hybrid memory modes. MFIX is an open-source multiphase flow code used by over 400 researchers from over 200 institutions all over the world. The documentation (Syamlal *et al.*, 1993; Syamlal, 1998), related technical reports, and the latest version of the source code are all available from http://www.mfix.org.

Table 10.2. Particle, flow, and geometric parameters for the
two-dimensional jet bubbling bed simulations of Section 10.4.

Particle diameter	$d_p = 200 \ \mu\text{m}$
Particle density	$\rho_s = 2500 \ \text{kg/m}^3$
Coefficient of restitution	0.95
Angle of internal friction	40°
Temperature of particles and gas	294 K
Average molecular weight of the fluid	29
Fluid viscosity	$1.8 \times 10^{-4} \ \text{g/cm s}$
Low resolution	$(\Delta x, \Delta y) = (2 \ \text{mm}, 2 \ \text{mm})$
High resolution	$(\Delta x, \Delta y) = (1 \ \text{mm}, 2 \ \text{mm})$

Figure 10.4 shows schematically a two-dimensional fluidized bed, where we assume at the outset that all the properties are uniform in the third direction. The bed is 0.05 m wide and 0.4 m tall, and it initially contains particles at a volume fraction of 0.55 filled to a height of 0.2 m. At time $t = 0$, fluidizing gas at a superficial velocity of 0.05 m/s is introduced uniformly over the entire 0.05 m width at the bottom of the bed. This flow is maintained at all subsequent times, with the exception that for a short time interval, $0.15 \ \text{s} < t < 0.25 \ \text{s}$, the gas velocity is increased to 0.25 m/s in the region $0.02 \ \text{m} < x < 0.03 \ \text{m}$. This transient gas jet leads to the formation of a bubble-like void with a considerably lower volume fraction of particles than the surrounding bed. This void travels up the bed through the action of buoyancy and the goal of the simulation is to track the evolution of the shape and location of this rising bubble. This is a simple test problem (Gidaspow, 1994) through which one can examine the issues associated with simulations of real fluidized beds containing many interacting bubbles.

The physical properties of the gas and particles used in these simulations are summarized in Table 10.2. The gas was assumed to behave like an ideal gas. The domain of interest was partitioned into a number of rectangular cells: 2 mm × 2 mm in the low-resolution simulations and 1 mm × 2 mm in a slightly higher resolution simulation. Simulations were run using first-order upwinding and Superbee schemes for each one of these two resolutions. Symmetry about the vertical midplane was not assumed. The time step was allowed to vary and was optimized by the code during the course of the simulation. Typically, the time steps were close to 10^{-4} s.

Figures 10.5 and 10.6 present snapshots of void fraction (i.e. gas volume fraction) at four different times obtained in the low-resolution simulations with the upwind and Superbee schemes, respectively. Both figures clearly show the formation and rise of the bubble-like void. Thus, both schemes are

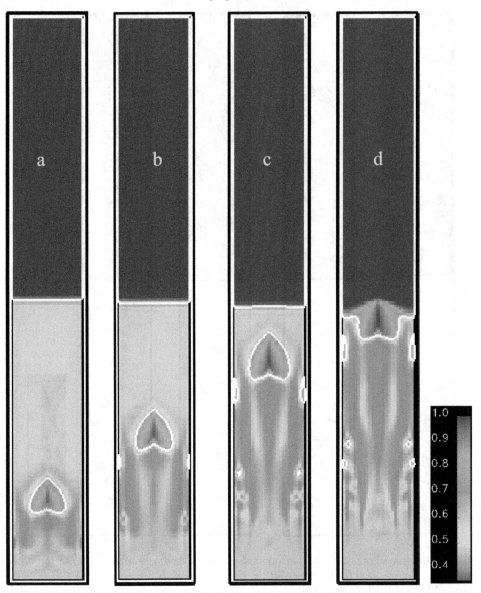

Fig. 10.5. Snapshots of void-fraction surface contours along with void-fraction contour line for $\alpha_G = 0.6$. Low-resolution simulation with first-order upwind. (a) $t = 0.28$ s, (b) 0.42 s, (c) 0.58 s, and (d) 0.67 s. Average bubble rise velocity between 0.42 and 0.58 s is 34.65 cm/s. See also http://www.cambridge.org/comp_mult_flow.

able to capture the existence of these structures. There are, however, clearly identifiable differences between the results shown in these two figures.

First, there is a noticeable difference between the shapes of the bubble nose. Figures 10.7 and 10.8 show an expanded view of the bubble-like void at

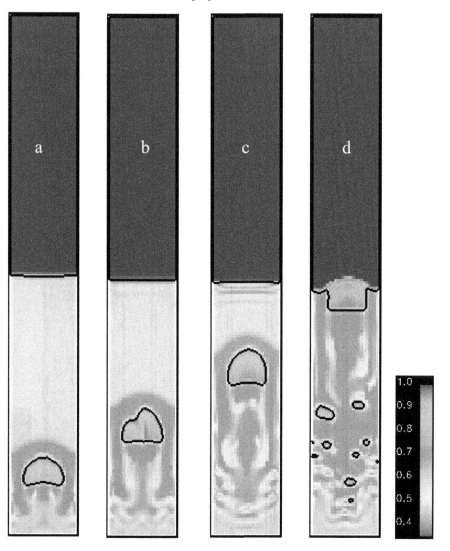

Fig. 10.6. Snapshots of void-fraction surface contours along with void-fraction contour line for $\alpha_G = 0.6$. Low-resolution simulation with Superbee. (a) $t = 0.28$ s, (b) 0.42 s, (c) 0.58 s, and (d) 0.8 s. Average bubble rise velocity between 0.42 and 0.58 s is 29.70 cm/s. See also http://www.cambridge.org/comp_mult_flow.

one particular time instant as obtained with the two schemes. Figures 10.7(a) and 10.8(a) show the distribution of void fraction in and around the bubbles, along with contours demarcating the bubbles. First-order upwinding (Fig. 10.7a) produces bubbles with a pointed nose, as noted by many authors (e.g. see Guenther and Syamlal, 2001), while the Superbee scheme (Fig. 10.8a) produces a more blunted bubble nose, which is closer to

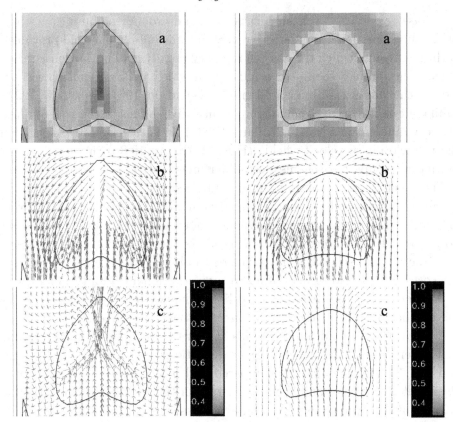

Fig. 10.7. (Left) Close-up view of the bubble at $t = 0.58$ s in Fig. 10.5(c). (a) void-fraction distribution along with void fraction contour line for $\alpha_G = 0.6$, (b) void-fraction contour line for $\alpha_G = 0.6$ along with solid velocity vectors, (c) void-fraction contour line for $\alpha_G = 0.6$ along with gas velocity vectors. See also http://www.cambridge.org/comp_mult_flow.

Fig. 10.8. (Right) Close-up view of the bubble at $t = 0.58$ s in Fig. 10.6(c). (a) void-fraction distribution along with void fraction contour line for $\alpha_G = 0.6$, (b) void-fraction contour line for $\alpha_G = 0.6$ along with solid velocity vectors, (c) void-fraction contour line for $\alpha_G = 0.6$ along with gas velocity vectors. See also http://www.cambridge.org/comp_mult_flow.

experimental observations. Guenther and Syamlal (2001) examined other limiters and found that higher-order schemes, in general, yield a blunted bubble nose. Thus it appears that the increased numerical dissipation of the upwind scheme is responsible for the pointed nose of the bubble. The bubble rise velocities obtained with the two schemes are also quite different (see the caption for Figs. 10.7 and 10.8).

Second, the quantitative details of the velocity fields are also very different between these two cases. Figures 10.7(c) and 10.8(c) present the gas-phase velocities at various locations in and around the bubbles in these two simulations. Both simulations reveal that the bubble-like void provides a pathway through which the gas can rise very rapidly, and in the dense region outside the bubbles there is a cloud within which the gas tends to recirculate. Such a gas circulation is well known (Davidson, 1961; Gidaspow, 1994). While in qualitative agreement, the quantitative details of the velocity field are very different in Figs. 10.7(c) and 10.8(c). It is then not surprising that the particle phase velocity fields (Figs. 10.7b and 10.8b) predicted by the upwind and Superbee schemes also differ quantitatively.

The effect of grid resolution on the bubble shape has been studied in the literature; for example, Guenther and Syamlal (2001) found that, with very fine grids, they could achieve grid independence in their simulations using both upwind and higher order schemes. As the resolution was increased, even the upwind scheme yielded rounded bubbles, which however came at the cost of nearly two orders of magnitude additional computational time.

We include here one example of a slightly higher grid resolution, in which the number of grids in the lateral direction has been increased by a factor of 2. Snapshots of void fraction fields at four different times for the upwind and Superbee schemes in these higher resolution runs are shown in Fig. 10.9 and Fig. 10.10. These can be compared with Fig. 10.5 and Fig. 10.6. It is clear that grid resolution affects the details of the satellite bubbles, the rise velocity of the primary bubble, and even its shape.

Further increase of the grid resolution may lead to grid-independent results, as found by Guenther and Syamlal (2001), but we need not go into this aspect in this illustration. Instead, let us note that 2 mm (the grid size in the low-resolution simulations) is only 10 particle diameters! Guenther and Syamlal (2001) observed grid independence when the grid size was down to about 3 particle diameters with higher order discretization while, with the upwind scheme, rounded bubbles were obtained when the grid size was ~ 1 particle diameter.

It is legitimate to wonder about the reason for the need to have such extraordinarily small grid sizes to get nearly converged solutions of the averaged equations of motion. One possibility is that the closures are deficient – for example, a larger granular-phase viscosity can be expected to give converged solutions with coarser grids. It is also possible that the need for fine grids stems from the sharp interfaces that accompany the bubble-like voids.

In real engineering and scientific applications, one invariably deals with flows over much larger domains, of the order of meters, and fluidized beds contain many bubbles. Grid sizes of the order of a few particle diameters

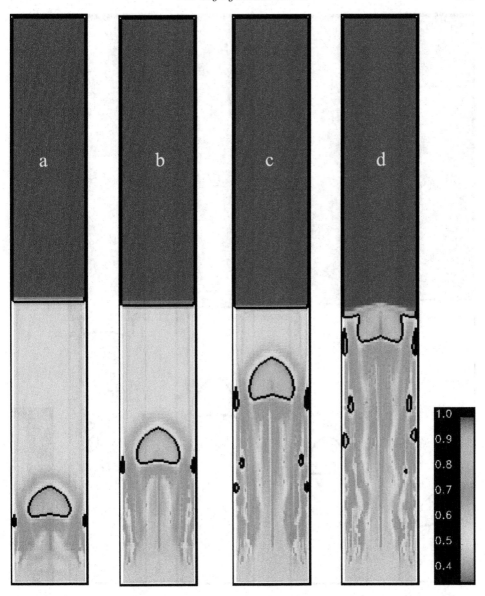

Fig. 10.9. Snapshots of void-fraction surface contours along with void-fraction contour line for $\alpha_G = 0.6$. High-resolution simulation with the upwind scheme. (a) $t = 0.28$ s, (b) 0.42 s, (c) 0.58 s, and (d) 0.72 s. Average bubble rise velocity between 0.42 s and 0.58 s is 0.2970 m/s. See also http://www.cambridge.org/comp_mult_flow.

are impractical for simulation of gas–particle flows in such large vessels and, in practice, one is forced to accept much coarser grids, of the order of tens to hundreds of particle diameters. Hence, simulations using the equations and closures employed in the illustrations presented here, but with coarse

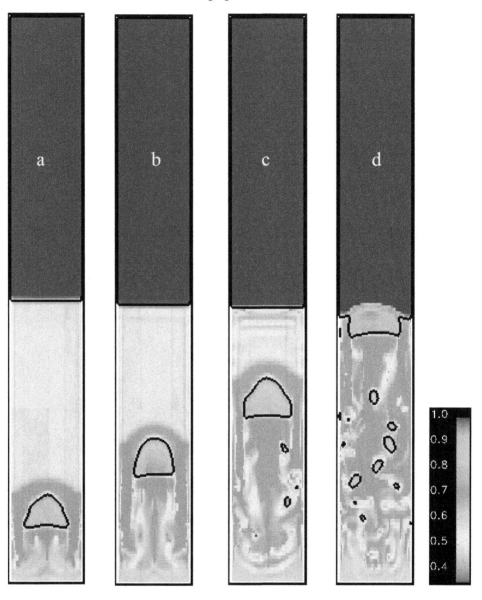

Fig. 10.10. Snapshots of void-fraction surface contours along with void-fraction contour line for $\alpha_G = 0.6$. High-resolution simulation with the Superbee scheme. (a) $t = 0.28$ s, (b) 0.42 s, (c) 0.58 s, and (d) 0.79 s. Average bubble rise velocity between 0.42 s and 0.58 s is 0.2696 m/s. See also http://www.cambridge.org/comp_mult_flow.

grids, should be viewed with caution, and a careful check of grid-size effect should always be performed.

Alternatively, one may attempt to derive averaged equations that are more appropriate for coarser grids – analogous to large-eddy simulation of turbulent flow – and develop closure relations which are appropriate for these coarsened equations (see, e.g. Sundaresan, 2000; Agrawal *et al.*, 2001; Zhang and VanderHeyden, 2002). It is worth noting that these coarsened equations have the same general form as the averaged equations which were solved in the example discussed here. Hence, all the numerical methods and issues that we have discussed apply to these coarsened equations as well.

10.5 An Eulerian–Lagrangian scheme

The derivation of a pressure equation by enforcing conservation of mass or volume that we have introduced in Section 10.3 is a procedure of broad applicability. We now describe how it is used in the context of a robust and versatile method of the Eulerian–Lagrangian type. The method as presented here is only first-order accurate in time. Its accuracy can be extended to second order, and its scope can be significantly broadened beyond the description given here. The approach originates in the work of Hirt *et al.* (1974), and has been widely applied to a variety of problems in single-phase flow, solid mechanics, fluid–structure interaction and others. The method as described here is at the basis of the code CartaBlanca (Zhang *et al.*, 2007).

The time advancement is decomposed into the succession of a *Lagrangian* and a *remapping* step. If the flow domain is partitioned into volumes, or cells, in the Lagrangian step each volume evolves in a Lagrangian way according to the velocity field on its boundary. At the end of this step, the material contained in these deformed volumes is redistributed among – or remapped onto – the volumes of the original underlying grid. For the multiphase case, the proper velocity of each phase is used in the Lagrangian step for that phase. In a more general version of this approach, the deformed volume at the end of the Lagrangian step may be remapped onto a volume with a shape different from the initial one. In this way one may control the deformation of the grid in situations where, for example, the computational domain itself evolves with time. We will not pursue this generalization here.

A useful feature of the two-step nature of this method is that it enables code developers to adopt an object-oriented structure which, among other advantages, makes the code easier to read and to maintain.

We will describe the procedure in the context of a multidimensional model consisting of N phases treated as interpenetrating continua. Each phase K is described by a continuity equation

$$\frac{\partial}{\partial t}(\alpha_K \rho_K) + \nabla \cdot (\alpha_K \rho_K \mathbf{u}_K) = \Gamma_K, \qquad K = 1, 2, \ldots, N \qquad (10.119)$$

in which Γ_K is the mass source for phase K, and by a momentum equation with the general form

$$\frac{\partial}{\partial t}(\alpha_K \rho_K \mathbf{u}_K) + \nabla \cdot (\alpha_K \rho_K \mathbf{u}_K \mathbf{u}_K)$$
$$= -\alpha_K \nabla \cdot p + \nabla \cdot [\alpha_K (\boldsymbol{\tau}_K + \boldsymbol{\sigma}_{K,Re})] + \alpha_K \rho_K \mathbf{g}_K + \mathbf{f}_K$$

(10.120)

where $\boldsymbol{\sigma}_{K,Re}$ is the Reynolds stress, $\mathbf{g}_K$ denotes additional forces including gravity, and $\mathbf{f}_K$ is the interphase interfacial force. A common form for this force similar to (10.4), (10.5) is

$$\mathbf{f}_K = \alpha_K \sum_{L=1}^{N} \alpha_L K_{KL}(\mathbf{u}_L - \mathbf{u}_K) + \alpha_K \sum_{L=1}^{N} \alpha_L C_{KL}(\mathbf{a}_L - \mathbf{a}_K),$$
(10.121)

where K_{KL} and C_{KL} are the drag and added-mass coefficients between phase K and phase L and $\mathbf{a}$ denotes the acceleration. For simplicity, we disregard the effects of phase change on the momentum equations. Energy equations may also be part of the model, but we do not need to write down a specific form.

10.5.1 Single-phase flow

In order to explain the procedure, let us start once again by considering a single-phase flow for which the general balance equation, in the terminology of Section 8.2 of Chapter 8, is written as

$$\frac{\partial}{\partial t}(\rho f) + \nabla \cdot (\rho f \mathbf{u}) = \nabla \cdot \boldsymbol{\phi} + \rho \theta.$$
(10.122)

The corresponding statement for a control volume $\mathcal{V}(t)$ the surface of which moves with a velocity $\mathbf{v}$ is

$$\frac{d}{dt}\int_{\mathcal{V}} F(\mathbf{x},t)\,dV + \int_S F(\mathbf{x},t)\,(\mathbf{u}-\mathbf{v})\cdot\mathbf{n}\,dS = \int_S \boldsymbol{\phi}(\mathbf{x},t)\cdot\mathbf{n}\,dS + \int_{\mathcal{V}} \rho\theta(\mathbf{x},t)\,dV,$$
(10.123)

in which, for convenience, we introduce the density per unit volume $F = \rho f$ of the generic conserved quantity f. In the Lagrangian step from t^n to t^{n+1}, we suppose that the surface S moves with the fluid so that $\mathbf{v} = \mathbf{u}$. Just before the time step is completed, however, the control volume "snaps back" to its initial position, which can be formally obtained by attributing to the points of S a velocity $\mathbf{v} = -\mathbf{d}\,\delta(t - t^{n+1-0})$, where $\mathbf{d}$ is the displacement of each point of the surface of the control volume during the interval $t^n \le t < t^{n+1}$ and the argument $(t - t^{n+1-0})$ of the δ-function indicates that the "snap-back" takes place just before the end of the time interval. Thus we write:

$$\mathbf{v} = \mathbf{u} - \mathbf{d}\delta(t - t^{n+1-0}),$$
(10.124)

Upon integrating (10.123) in time with this velocity field, we find

$$
\int_{\mathcal{V}^{n+1}} F(\mathbf{x}, t^{n+1})\, dV - \int_{\mathcal{V}^n} F(\mathbf{x}, t^n)\, dV
$$

$$
+ \int_{t^n}^{t^{n+1}} dt\, \delta(t - t^{n+1-0}) \int_S F(\mathbf{x}, t)\mathbf{d} \cdot \mathbf{n}\, dS
$$

$$
= \int_{t^n}^{t^{n+1}} dt \left(\int_S \boldsymbol{\phi} \cdot \mathbf{n}\, dS + \int_{\mathcal{V}} \rho\theta\, dV \right). \tag{10.125}
$$

But $\mathbf{d} \cdot \mathbf{n}\, dS$ is just the volume swept by the surface element dS during the time $\Delta t = t^{n+1} - t^n$, so that

$$
\int_{t^n}^{t^{n+1}} dt\, \delta(t - t^{n+1-0}) \int_S F(\mathbf{x}, t)\mathbf{d} \cdot \mathbf{n}\, dS = \int_{\mathcal{V}^{n+1} - \mathcal{V}^n} F(\mathbf{x}, t^{n+1})\, dV.
$$

$$
\tag{10.126}
$$

The integral in the right-hand side is over the volume comprised between the initial and final positions of the control volume. This volume is clearly equal to the material volumes convected in and out of the control volume.

Equation (10.125) shows that the total change of the quantity F within the control volume during the time interval $t^n \le t \le t^{n+1}$ can be calculated in two steps. At the end of the first one, the Lagrangian step, we have a provisional value

$$
\int_{\mathcal{V}(t^{n+1})} F^*(\mathbf{x}, t^{n+1})\, dV
$$

$$
= \int_{\mathcal{V}(t^n)} F(\mathbf{x}, t^n)\, dV + \int_{t^n}^{t^{n+1}} dt \left(\int_S \boldsymbol{\phi} \cdot \mathbf{n}\, dS + \int_{\mathcal{V}} \rho\theta\, dV \right). \tag{10.127}
$$

In this equation the integrands in the last term can be calculated ignoring the snap-back as they are bounded quantities which contribute nothing when integrated between t^{n+1-0} and t^{n+1}. The second step, remapping, accounts for the snap-back of the control volume: the control volume gains a certain inflow volume V_i and loses a certain outflow volume V_o and, therefore, it gains and loses the amount of F contained in these volumes. Since this jump takes place at the end of the time step, the quantity F contained in these volumes is F^* evaluated at the time t^{n+1}, so that

$$
\int_{\mathcal{V}^{n+1}} F(\mathbf{x}, t^{n+1})\, dV = \int_{\mathcal{V}^{n+1}} F^*(\mathbf{x}, t^{n+1})\, dV + \int_{V_i} F^*(\mathbf{x}, t^{n+1})\, dV
$$

$$
- \int_{V_o} F^*(\mathbf{x}, t^{n+1})\, dV. \tag{10.128}
$$

The procedure just described can be implemented equally well on a staggered or a co-located grid. The scalar fields are defined at the centers (or centroids) of their respective grid cells as volume averages:

$$F(\mathbf{x}, t) \simeq \frac{1}{\Delta V} \int_{\Delta V} F(\mathbf{x} + \boldsymbol{\xi}, t)\, d^3\xi. \tag{10.129}$$

The Lagrangian step is carried out first by writing

$$F^* = F^n + \frac{1}{\Delta V} \Gamma \Delta t \tag{10.130}$$

where F^n is the value of F at time level t^n and Γ denotes an approximation to the sources of F in the right-hand side of (10.125). Since in this step we follow the motion of the material, the advection term is automatically included and need not be considered explicitly. With an explicit time-stepping strategy, which is the one we consider here, Γ is evaluated at time t^n. If, however, $\Gamma = \Gamma(F)$ depends on F itself as, for example, the drag force in the momentum equation, writing in (10.130) $\Gamma = \Gamma(F^*)$ in place of $\Gamma = \Gamma(F^n)$ improves the stability of the calculation. In this case, it may be necessary to solve (10.130) by Picard or Newton–Raphson iteration. A more fully implicit time-stepping strategy can also be built on (10.125); we omit the details.

The surface integral in the right-hand side of (10.127) is approximated as

$$\int_{t^n}^{t^{n+1}} dt \int_S \boldsymbol{\phi} \cdot \mathbf{n}\, dS \simeq \Delta t \sum_{a=1}^{N_f} \boldsymbol{\phi}_\alpha \cdot \mathbf{n}_a\, \Delta A_a \tag{10.131}$$

where the summation extends to all the faces $a = 1, 2, \ldots, N_f$ of the cell, each one with area ΔA_a and outward normal $\mathbf{n}_a$.

After the Lagrangian step, the remapping step is performed by setting

$$F^{n+1} = F^* - \frac{1}{\Delta V} \sum_{a=1}^{N_f} F_a^* \Delta A_a \mathbf{u}_a^* \cdot \mathbf{n}_a \Delta t, \tag{10.132}$$

where $\Delta A_a |\mathbf{u}_a^* \cdot \mathbf{n}_a| \Delta t$ approximates the volume swept by the a-th face during Δt. This volume will be positive or negative depending on the direction of the normal velocity $\mathbf{u}_a^* \cdot \mathbf{n}_a$ on the surface. In a first-order scheme, one can use equivalently F^n or F^* in the summation, and similarly for the velocity. In either case, the right-hand side of (10.132) requires the value of F^n or F^* on the cell face. This value will in general be different from the cell-average (10.129). Within the context of this method, this is of course the same problem we dealt with in Section 10.2.2 and, as before, the numerical schemes used to approximate these cell-face values are essential to the success of the method; we will address this issue later in Section 10.5.4.

To calculate the momentum density at the end of the Lagrangian step, the momentum equation is integrated over the cell to find

$$\frac{\Delta V}{\Delta t}(\mathbf{u}^* - \mathbf{u}^n) = -\frac{1}{\rho^{n+1}}\sum_{a=1}^{N_f} p_a^{n+1}\mathbf{n}_a\,\Delta A_a + \frac{1}{\rho^{n+1}}\sum_{a=1}^{N_f}(\boldsymbol{\tau}_a^n + \boldsymbol{\sigma}_{Re}^n)\cdot\mathbf{n}_a\,\Delta A_a + \mathbf{g}.$$
$$(10.133)$$

Here we provisionally indicate a time level $n+1$ for the pressure. The way this point is handled will become clearer in Section 10.5.3.

In a staggered grid arrangement, the velocities computed from (10.133) are already the cell-face convection velocities needed to carry out the remapping step for the scalar variables. With a co-located grid arrangement, on the other hand, the convection velocities $\mathbf{u}_a \cdot \mathbf{n}_a$ necessary to calculate the convective fluxes on the a-th face of the generic cell need to be calculated by solving again the momentum equation. This second solution of the momentum equation is also needed with staggered grids in executing the remapping step for the velocities themselves as, for this purpose, the velocities on the faces of the momentum cells are required. In carrying out this step, only the normal velocity component is required. Thus, rather than integrating the momentum equation over the cell, it is simpler to use a finite difference discretization in the neighborhood of the center of the a-th face of the generic cell and project on the normal direction $\mathbf{n}_a$ to find:

$$\mathbf{u}_a^*\cdot\mathbf{n}_a = \mathbf{u}_a^n\cdot\mathbf{n}_a + \frac{\Delta t}{\rho_a^{n+1}}\left[-\nabla_a p^{n+1} + \nabla_a\cdot(\boldsymbol{\tau} + \boldsymbol{\sigma}_{Re})\right]\cdot\mathbf{n}_a + \mathbf{g}_a\cdot\mathbf{n}_a\Delta t. \quad (10.134)$$

As will be seen in Section 10.5.3 when we take up the multiphase flow situation, this equation will be used both to calculate the pressure and to effect the remapping step.

10.5.2 Multiphase flow

The adaptation of the Eulerian–Lagrangian method just described to the multiphase flow situation is relatively straightforward: in carrying out the Lagrangian step for each phase, use is made of the velocity of that phase. In applying this notion to the momentum equation, one must keep in mind the twofold role of velocity as momentum density and agent of convection in the remapping step (10.132).

As before, to calculate the momentum density at the end of the Lagrangian step, we integrate the momentum equation (10.120) over the cell to find

$$\frac{\Delta V}{\Delta t}\left(\mathbf{u}_K^* - \mathbf{u}_K^n\right) = -\frac{1}{(\alpha\rho)_K^{n+1}}\sum_{a=1}^{N_f}\alpha_{Ka}^n p_a^{n+1}\mathbf{n}_a\,\Delta A_a$$

$$+\frac{1}{\rho_K^{n+1}}\sum_{a=1}^{N_f}(\boldsymbol{\tau}_{Ka}^n+\boldsymbol{\sigma}_{Re}^n)\cdot\mathbf{n}_a\,\Delta A_a+\frac{1}{\rho_K^{n+1}}\sum_{L=1}^{N}\alpha_L^{n+1}K_{KL}(\mathbf{u}_L^*-\mathbf{u}_K^*)$$

$$+\frac{1}{\rho_K^{n+1}}\sum_{L=1}^{N}\alpha_L^{n+1}C_{KL}\left(\frac{\mathbf{u}_L^*-\mathbf{u}_L^n}{\Delta t}-\frac{\mathbf{u}_K^*-\mathbf{u}_K^n}{\Delta t}\right)+\mathbf{g}_K.$$

$$(10.135)$$

In the added-mass terms, the material acceleration for phase K is calculated as $(\mathbf{u}_K^* - \mathbf{u}_K^n)/\Delta t$ in keeping with the comment made earlier after equation (10.130). Note that here it is the Lagrangian accelerations of the other phases that enter rather than, for example, their acceleration relative to the motion of phase K. To understand this point, it is worth recalling that the added mass interaction is a volume source of momentum, in the sense that it arises from the interfaces contained within the finite volume used in the discretization and, therefore, it should be considered as part of the source term θ in (10.123) rather than of the flux term. It is true that the amount of the other phases L within the Lagrangian volume of phase K varies as the motion of this phase is followed in time. However, this change is of order Δt and, therefore, the error incurred by neglecting it is of order Δt^2, which is compatible with the other approximations introduced in (10.135). In equation (10.135), the coupling of the different phase velocities in the same cell can be dealt with similarly to the PEA or SINCE ideas described at the end of Section 10.2.

The velocities $\mathbf{u}_{Ka}\cdot\mathbf{n}_a$ on the cell faces necessary to calculate the convective fluxes of the transported quantities are calculated as before in equation (10.134) by a direct discretization of the momentum equation:

$$\mathbf{u}_{Ka}^*\cdot\mathbf{n}_a = \mathbf{u}_{Ka}^n\cdot\mathbf{n}_a - \frac{\Delta t}{\rho_{Ka}^{n+1}}\mathbf{n}_a\cdot\nabla_a p^{n+1}$$

$$+\frac{\Delta t}{(\alpha\rho)_{Ka}^{n+1}}\mathbf{n}_a\cdot\nabla_a\cdot[\alpha_K^n(\boldsymbol{\tau}_K^n+\boldsymbol{\sigma}_{K,Re}^n)]$$

$$+\frac{1}{\rho_{Ka}^{n+1}}\sum_{L=1}^{N}\alpha_{La}C_{KL}\left[(\mathbf{u}_{La}^*-\mathbf{u}_{La}^n)-(\mathbf{u}_{Ka}^*-\mathbf{u}_{Ka}^n)\right]\cdot\mathbf{n}_a$$

$$+\frac{\Delta t}{\rho_{Ka}^{n+1}}\sum_{L=1}^{N}\alpha_{La}K_{KL}(\mathbf{u}_{La}^*-\mathbf{u}_{Ka}^*)\cdot\mathbf{n}_a+\mathbf{g}_{Ka}\cdot\mathbf{n}_a.\qquad(10.136)$$

Again, the coupling of the different phase velocities can be dealt with by the partial elimination algorithm in one of its variants. The use of this formula requires that it be possible to express the normal component of the pressure gradient on the cell face in terms of the values at the surrounding cell centers. The simplest situation is when the normal is in the same direction as the line connecting the two nodal points on the two sides of the cell face, as in the case of a regular rectangular grid or a Voronoi-type grid. In this case, the projected value can be simply calculated as

$$\mathbf{n}_a \cdot \nabla_a p = \frac{p_r - p_l}{\Delta r} \qquad (10.137)$$

with an obvious meaning of the symbols. More generally, one can expand p in a two-term Taylor series and equate to the pressure values at the surrounding cell centers thus deriving a linear system for the components of the gradient. If more than three cells are used the system may be over-determined and can be solved in a least-squares sense. The method mentioned later in connection with equation (10.144) can also be used.

10.5.3 Pressure calculation

According to (10.130), as applied to the continuity equation (10.119) for phase K, the Lagrangian step gives

$$(\alpha\rho)_K^* = (\alpha\rho)_K^n + \frac{\Delta t}{\Delta V}\Gamma_K, \qquad (10.138)$$

where the time index on Γ_K can be n or $n+1$. The final macroscopic density $(\alpha\rho)_K^{n+1}$ must be obtained from the remapping step:

$$(\alpha\rho)_K^{n+1} = (\alpha\rho)_K^* - \frac{\Delta t}{\Delta V}\sum_{a=1}^{N_f}(\alpha\rho)_{Ka}^* \mathbf{u}_{Ka}^* \cdot \mathbf{n}_a A_a \qquad (10.139)$$

execution of which necessitates a knowledge of $\mathbf{u}_{Ka}^* \cdot \mathbf{n}_a$ which, in turn, requires that p^{n+1} be known. As before, we derive an equation for this quantity by imposing the conservation of volume

$$\sum_{K=1}^{N} \alpha_K^{n+1} = 1, \quad \text{or} \quad \sum_{K=1}^{N} \frac{(\alpha\rho)_K^{n+1}}{\rho_K(p^{n+1}, T_K^{n+1})} = 1, \qquad (10.140)$$

in which the dependence of ρ_K on pressure, and possibly temperature, is expressed by means of an equation of state. Upon using (10.139), this equation becomes

$$\sum_{K=1}^{N} \frac{1}{\rho_K^{n+1}}\left[(\alpha\rho)_K^* - \frac{1}{\Delta V}\sum_{a=1}^{N_f}(\alpha\rho)_{Ka}^* \mathbf{u}_{Ka}^* \cdot \mathbf{n}_a A_a \Delta t\right] = 1. \qquad (10.141)$$

When (10.136) is used to express $\mathbf{u}^*_{Ka} \cdot \mathbf{n}_a$, this becomes an equation for the pressure which can be solved locally for each cell without involving the neighboring ones similarly to the step leading from (10.81) to (10.82). For incompressible materials, it is the volume fractions rather than the densities that depend on the pressure and this approximation fails unless over-relaxation is used as in the procedure mentioned in Section 10.3.1. A faster convergence is achieved if, rather than proceeding cell by cell, the new pressure field is calculated by simultaneously solving all the equations (10.141) written for all the cells.

The calculational sequence can be arranged in various ways depending on the desired degree of implicitness and coupling. For example, if the temperature variable in the equation of state is approximated by its value at time level n, the procedure would be implemented as follows:

(i) The Lagrangian step for the macroscopic densities $(\alpha\rho)^*_K$ is executed according to (10.138) using t^n-time values to evaluate the source term Γ_K.

(ii) The face values $(\alpha\rho)^*_{Ka}$ are calculated by the upwinding rule or according to Section 10.5.4.

(iii) The normal velocities $\mathbf{u}^*_{Ka} \cdot \mathbf{n}_a$, expressed in terms of the new pressure from (10.136), are substituted into (10.141) and this equation is solved to find an improved estimate $p^{\kappa+1}$ of p^{n+1}.

(iv) This updated pressure is used to calculate new interfacial velocities from (10.136); whenever a velocity is found to change sign with respect to the velocity at the previous iteration step, new values of the convected quantities $(\alpha\rho)^*_{Ka}$ are calculated.

(v) The updated pressure is used to calculate the corresponding densities $\rho^{\kappa+1}$ from the equations of state keeping $T_K = T^n_K$, after which $(\alpha\rho)^{\kappa+1}_K$ is obtained from (10.139) and $\alpha^{\kappa+1}_K$ from $\alpha^{\kappa+1}_K = (\alpha\rho)^{\kappa+1}_K / \rho^{\kappa+1}_K$; if necessary, the mass source term Γ_K in (10.138) is updated and new $(\alpha\rho)^*_K$ calculated.

(vi) Steps (iii)–(v) are repeated until convergence.

(vii) At convergence, the Lagrangian step (10.135) for the velocities is executed with the final pressure value.

(viii) The final values for all the variables are calculated by executing the remapping step using $\mathbf{u}^*_{Ka} \cdot \mathbf{n}_a$ as given by (10.136) with the final pressure value.

10.5.4 Calculation of cell-surface quantities

As already remarked in connection with (10.132), execution of the remapping step requires a knowledge of F_a, the value of the convected quantity F on the cell surface a. One low-accuracy possibility is to simply use the upwind rule taking the value of F_a as the value at the center of the cell from which the convected volume comes. To achieve higher order accuracy, one needs to account for the effects of the gradient of the quantity F.

With a piecewise-linear representation, the value F at a point $\mathbf{x}$ near the center of the cell j can be written as

$$F = F_j + \mathbf{r} \cdot (\nabla F)_j, \qquad (10.142)$$

where $\mathbf{r} = \mathbf{x} - \mathbf{x}_j$, with $\mathbf{x}_j$ the centroid of the cell defined by

$$\mathbf{x}_j = \frac{1}{\Delta V} \int_{\Delta V} \mathbf{x} \, dV = \frac{1}{2 \Delta V} \int_S (\mathbf{x} \cdot \mathbf{x}) \, \mathbf{n} \, dS, \qquad (10.143)$$

in which the generalized divergence theorem has been used and S is the surface of the cell. Note that, as a consequence, F_j in (10.142) coincides with the cell average of F. The gradient $(\nabla F)_j$ can be calculated in several ways, two of which were described earlier in connection with the pressure gradient at the end of Section 10.5.2. An alternative method (see, e.g. Dukowicz and Kodis, 1987) is to construct a secondary cell $\Delta V'$ having as vertices the centers of the cells surrounding the cell j and calculate the integral

$$(\nabla F)_j = \frac{1}{\Delta V'} \int_{S'} F \mathbf{n} \, dS', \qquad (10.144)$$

e.g. by assuming a linear variation over the faces of the cell $\Delta V'$.

We already know from the one-dimensional case that, in general, use of (10.142) will not be satisfactory as it can contaminate the results with spurious oscillations. However, one can build on this expression by introducing a suitable limiter writing

$$F = F_j + \ell_j \mathbf{r} \cdot (\nabla F)_j, \qquad (10.145)$$

with the limiter ℓ_j chosen so that F as given by this formula does not lie outside the range established by F_{max} and F_{min}, the maximum and minimum values of F at the centers of the cells surrounding the cell j. Dukowicz and Kodis (1987; see also Swartz, 1999) extended van Leer's (1977) original approach to the multidimensional case in the following way. Let $\max\{F_v\}$ and $\min\{F_v\}$ be the maximum and minimum values of F at the vertices of the cell in question as calculated using (10.142) with the unlimited gradient.

Then choose ℓ_j according to the rule

$$\ell_j = \min[1, \ell_{\min}, \ell_{\max}] \tag{10.146}$$

where

$$\ell_{\max} = \max\left[0, \frac{F_{\max} - F_j}{\max\{F_v\} - F_j}\right], \quad \ell_{\min} = \max\left[0, \frac{F_{\min} - F_j}{\min\{F_v\} - F_j}\right]. \tag{10.147}$$

This algorithm is simple, but it has some shortcomings; Dukowicz and Kodis (1987) discuss other procedures.

A more satisfactory performance in some special cases, and in particular when the Courant number is close to 1, may be achieved by using (10.145) to evaluate F_a not at the center of the face a but, rather, at the center of the convected volume, i.e. a distance $\frac{1}{2}\mathbf{u}_a^* \cdot \mathbf{n}_a \Delta t$ upstream of the face a.

In may be expected that when, in the remapping step, the density ρ_{Ka} on the surface a is calculated using (10.145), the density will be positive everywhere provided that it is positive after the Lagrangian step and the Courant number is sufficiently small.

Another issue that arises in connection with the computation of cell-surface quantities is that of *compatibility*. Consider for example the enthalpy h. In principle, this quantity is calculated as the ratio $(\rho h)/\rho$ of the enthalpy per unit volume, ρh, which is the conserved variable in (one form of) the energy equation and the density. Since the spatial distributions of ρ and ρh are to some extent independent of each other, it is not obvious that the absence of spurious oscillations in both quantities will result in the absence of spurious oscillations in their ratio. In other words, the elimination of spurious oscillations from ρ and ρh is, in general, insufficient to guarantee that h be monotonic. A scheme which does ensure the smooth variation of the quotient is termed *compatible*. VanderHeyden and Kashiwa (1998) built on the van Leer scheme to derive a compatible scheme in the following way.

For ρh in the vicinity of node j, we have, with the same notation used in (10.142):

$$\rho(\mathbf{x})h(\mathbf{x}) = \rho_j h_j + \mathbf{r} \cdot \nabla(\rho h)_j = [\rho_j + \mathbf{r} \cdot (\nabla \rho)_j] h_j + \rho_j \mathbf{r} \cdot (\nabla h)_j. \tag{10.148}$$

The basic idea is to derive a limitation on $\nabla(h\rho)$ by enforcing monotonicity on h. This objective is attained by noting that, as a consequence of this relation, h is a nonlinear function of $\mathbf{r}$ in the cell:

$$h = h_j + \frac{\rho_j \mathbf{r} \cdot (\nabla h)_j}{\rho_j + \mathbf{r} \cdot (\nabla \rho)_j}. \tag{10.149}$$

To obtain a compatible flux on the cell surface, we use this relation to calculate the parameters $h_{\max}$, $h_{\min}$, and h_v in the definitions (10.147), with $(\nabla\rho)_j$ evaluated with the normal van Leer limiter (10.146) applied to ρ and $(\nabla h)_j$ unlimited and evaluated as in (10.137) using the cell center upstream of the cell of interest or (10.144). With the limiter for h, ℓ_h, obtained in this way, equation (10.148) is then rewritten as

$$\rho h \simeq \rho_j h_j + \mathbf{r} \cdot [\ell_\rho h \nabla\rho + \ell_h \rho \nabla h]_j \qquad (10.150)$$

and this expression is used to calculate ρh on the cell faces in the remapping step.

10.6 A partially coupled algorithm

In the segregated solution strategies described in the previous sections, the equations are solved in a sequential fashion using, at each iteration step, the currently available values. Depending on the set of nonlinear equations and on the specific problem, this procedure may not ensure robust or efficient convergence. As a matter of fact, the convergence properties of these schemes often demand small time steps in order to achieve a converged solution. When the problem of interest involves long integration times and large spatial regions, in order to prevent the accumulation of error, it is necessary to drive the iteration process to a very high degree of convergence, with a further increase in computation time. In such cases, it may be preferable to use a solution algorithm which more closely reflects the coupled structure of the problem at hand. A typical class of problems where this procedure is necessary arises in the oil industry, where it may be necessary to simulate gas–liquid flows in a pipeline with time durations of hours or even weeks. We describe here a two-dimensional simplified version of the method implemented in the commercial code OLGA which is widely used for this type of simulation (Bendiksen *et al.*, 1991; Moe and Bendiksen, 1993; Moe, 1993; Nordsveen and Moe, 1999). A key feature of this algorithm is that, by relying on relatively small time steps and on what amounts to a suitable correction added to the pressure equation, it is possible to avoid the iteration procedures of the methods described before. It should be pointed out that the development of coupled solution strategies for segregated methods is a topic of current research; for an example of more recent developments, see, e.g. the papers by Kunz *et al.* (1998, 1999, 2000).

Although the OLGA model recognizes three phases – continuous liquid, gas–vapor, and drops carried by the gas – in its application to stratified and intermittent flow the drop field plays no role and is disregarded here. Interphase mass transfer and the energy equation are also omitted for simplicity.

10.6.1 Mathematical model

We consider a model which is a straightforward extension to two spatial dimensions of the one-dimensional one of Section 10.1. In this application the two phases are gas and liquid and, accordingly, we use subscripts G and L. The continuity equations are given by

$$\frac{\partial m_J}{\partial t} + \nabla \cdot (m_J \mathbf{u}_J) = 0, \qquad J = G, L, \tag{10.151}$$

where, to simplify the equations that follow, we have introduced the macroscopic density

$$m_J = \alpha_J \rho_J. \tag{10.152}$$

The momentum equations have the form

$$\frac{\partial}{\partial t}(m_J \mathbf{u}_J) + \nabla \cdot (m_J \mathbf{u}_J \mathbf{u}_J) = -\alpha_J \nabla p + \nabla \cdot \boldsymbol{\tau}_J + \mathbf{F}_J, \tag{10.153}$$

with

$$\boldsymbol{\tau}_J = \alpha_J \mu_{J,\text{eff}} \left(\nabla \mathbf{u}_J + \nabla \mathbf{u}_J^T - \frac{2}{3}(\nabla \cdot \mathbf{u}_J)\mathbf{I} \right), \tag{10.154}$$

and

$$\mathbf{F}_G = H|\mathbf{u}_L - \mathbf{u}_G|(\mathbf{u}_L - \mathbf{u}_G) + \alpha_G \rho_G \mathbf{g}. \tag{10.155}$$

Here μ_{eff} is an effective viscosity given by the sum of the laminar viscosity and a turbulent contribution modelled by an extension of the single-phase mixing-length formula of Prandtl (Moe and Bendiksen, 1993), and H is the interphase drag parameter. The force $\mathbf{F}_L$ on the liquid phase is given by a similar expression with the indices G and L interchanged.

With a functional relation between pressure and density as in (10.9), the continuity equations in the form (10.71) may be rewritten as

$$\frac{\partial \alpha_J}{\partial t} + \frac{\alpha_J}{\rho_J c_J^2}\frac{\partial p}{\partial t} + \frac{1}{\rho_J}\nabla \cdot (m_J \mathbf{u}_J) = 0, \tag{10.156}$$

in which

$$c_J^2 = \frac{dp}{d\rho_J}, \tag{10.157}$$

is the speed of sound in phase J. By adding the two equations and using $\alpha_G + \alpha_L = 1$ we find

$$\left(\frac{\alpha_G}{\rho_G c_G^2} + \frac{\alpha_L}{\rho_L c_L^2}\right)\frac{\partial p}{\partial t} + \frac{1}{\rho_G}\boldsymbol{\nabla}\cdot(m_G \mathbf{u}_G) + \frac{1}{\rho_L}\boldsymbol{\nabla}\cdot(m_L \mathbf{u}_L) = 0, \quad (10.158)$$

which will be used to determine the pressure.

10.6.2 Discretization

We again consider a staggered grid arrangement in the x, y plane as in Chapter 2 (Figs. 2.1 and 2.2). For both phases the continuity equations are discretized implicitly by the upwind method as

$$\frac{m_{i,j}^{n+1} - m_{i,j}^n}{\Delta t} + \left\langle\frac{\delta(mu)}{\delta x}\right\rangle_{i,j}^{n+1} + \left\langle\frac{\delta(mv)}{\delta y}\right\rangle_{i,j}^{n+1} = 0 \quad (10.159)$$

where u and v are the velocity components in the x- and y-directions, respectively, and, in this section, the angle brackets denote upwind discretization; explicitly

$$\left\langle\frac{\delta(mu)}{\delta x}\right\rangle_{i,j} = \frac{\langle m\rangle_{i+1/2,j}\, u_{i+1/2,j} - \langle m\rangle_{i-1/2,j}\, u_{i-1/2,j}}{\Delta x} \quad (10.160)$$

$$\left\langle\frac{\delta(mv)}{\delta y}\right\rangle_{i,j} = \frac{\langle m\rangle_{i,j+1/2}\, v_{i,j+1/2} - \langle m\rangle_{i,j-1/2}\, v_{i,j-1/2}}{\Delta y} \quad (10.161)$$

with

$$\langle m\rangle_{i+1/2,j}\, u_{i+1/2,j} = \begin{cases} m_{i,j}\, u_{i+1/2,j} & \text{if } u_{i+1/2,j} \geq 0 \\ m_{i+1,j}\, u_{i+1/2,j} & \text{if } u_{i+1/2,j} < 0 \end{cases} \quad (10.162)$$

$$\langle m\rangle_{i,j+1/2}\, v_{i,j+1/2} = \begin{cases} m_{i,j}\, v_{i,j+1/2} & \text{if } v_{i,j+1/2} \geq 0 \\ m_{i,j+1}\, v_{i,j+1/2} & \text{if } v_{i,j+1/2} < 0. \end{cases} \quad (10.163)$$

The scalar variables with a half-integer index are evaluated by averaging the values at the two adjacent nodes as usual. In (10.159), the implicit evaluation of the velocity has the effect of eliminating a restrictive CFL time-step restriction based on the speed of pressure waves, and the implicit evaluation of the macroscopic densities ensures mass conservation at the end of the time step.

In order to maintain a desirable degree of coupling among the equations while removing the need for an iterative solution procedure, the momentum equations are discretized semi-implicitly with the Picard linearization

(10.16). The gas x-momentum equation is approximated as[1]

$$(m_G)_{i+1/2,j}^n \frac{(u_G)_{i+1/2,j}^{n+1} - (u_G)_{i+1/2,j}^n}{\Delta t} + \left\langle \frac{\delta(m_G u_G)^n (u_G)^{n+1}}{\delta x} \right\rangle_{i+1/2,j}$$

$$+ \left\langle \frac{\delta(m_G v_G)^n (u_G)^{n+1}}{\delta y} \right\rangle_{i+1/2,j}$$

$$= -\min\left((\alpha_G)_{i+1,j}^n, (\alpha_G)_{i,j}^n\right) \frac{p_{i+1,j}^{n+1} - p_{i,j}^{n+1}}{\Delta x} + \left(\frac{\delta(\tau_G)_{xx}}{\delta x}\right)_{i+1/2,j}$$

$$+ \left(\frac{\delta(\tau_G)_{xy}}{\delta y}\right)_{i+1/2,j} - (m_G)_{i+1/2,j}^n g \sin\phi$$

$$+ H\left[(u_L)_{i+1/2,j}^{n+1} - (u_G)_{i+1/2,j}^{n+1}\right]. \tag{10.164}$$

The angle ϕ is measured from the horizontal and is positive for upflow. The derivatives of the convective terms are evaluated by an upwinding rule based on the direction of the mass flux at time t^n. The x-derivative is taken as

$$\left\langle \frac{\delta(mu)^n u^{n+1}}{\delta x} \right\rangle_{i+1/2,j} = \frac{1}{\Delta x} \begin{cases} (mu)_{i+1/2,j}^n (u)_{i+1/2,j}^{n+1} - (mu)_{i-1/2,j}^n u_{i-1/2,j}^{n+1} \\ \quad \text{if } u_{i+1/2,j}^n \geq 0 \\ (mu)_{i+3/2,j}^n u_{i+3/2,j}^{n+1} - (mu)_{i+1/2,j}^n u_{i+1/2,j}^{n+1} \\ \quad \text{if } u_{i+1/2,j}^n < 0. \end{cases}$$

$$\tag{10.165}$$

Since the value of v is not available at $(i+\frac{1}{2}, j)$, the y-direction mass flux at this location is evaluated by averaging the four upwinded neighboring nodal values:

$$(\overline{mv})_{i+\frac{1}{2},j} = \frac{1}{4}\left[\langle mv \rangle_{i+1,j+\frac{1}{2}} + \langle mv \rangle_{i+1,j-\frac{1}{2}} + \langle mv \rangle_{i,j+\frac{1}{2}} + \langle mv \rangle_{i,j-\frac{1}{2}}\right], \tag{10.166}$$

and this estimate is used in the upwinding rule for the convective y-derivative:

$$\left\langle \frac{\delta(mv)^n u^{n+1}}{\delta x} \right\rangle_{i+\frac{1}{2},j} = \frac{1}{\Delta y} \begin{cases} (\overline{mv})_{i+\frac{1}{2},j}^n u_{i+\frac{1}{2},j}^{n+1} - (\overline{mv})_{i+\frac{1}{2},j-1}^n u_{i+\frac{1}{2},j-1}^{n+1} \\ \quad \text{if } (\overline{mv})_{i+\frac{1}{2},j}^n \geq 0 \\ (\overline{mv})_{i+\frac{1}{2},j+1}^n u_{i+\frac{1}{2},j+1}^{n+1} - (\overline{mv})_{i+\frac{1}{2},j}^n u_{i+\frac{1}{2},j}^{n+1} \\ \quad \text{if } (\overline{mv})_{i+\frac{1}{2},j}^n < 0 \end{cases}$$

$$\tag{10.167}$$

---

[1] As remarked after equation (10.16), since this is an evolution equation for u_G, the term $\delta(m_G u_G u_G)^{n+1}$ is linearized as $\delta(m_G u_G)^n u_G^{n+1}$. Picard's linearization is preferred here as, in more than one dimension, use of Newton's linearization (10.15) would couple different velocity components in the same equation (e.g $(uv)^{n+1} \simeq u^{n+1}v^n + u^n v^{n+1} - u^n v^n$), which would complicate the solution procedure.

The x-derivative of the stress is discretized as

$$\left(\frac{\delta\tau_{xx}}{\delta x}\right)_{i+1/2,j} = \frac{1}{\Delta x}\left[(\alpha\mu_{\text{eff}})_{i+1,j}^n\frac{u_{i+3/2,j}^{n+1} - u_{i+1/2,j}^{n+1}}{\Delta x}\right.$$
$$\left. -(\alpha\mu_{\text{eff}})_{i,j}^n\frac{u_{i+1/2,j}^{n+1} - u_{i-1/2,j}^{n+1}}{\Delta x}\right] \qquad (10.168)$$

and the y-derivative as

$$\left(\frac{\delta\tau_{xy}}{\delta y}\right)_{i+1/2,j} = \frac{1}{\Delta y}\left[(\overline{\alpha\mu}_{\text{eff}})_{i+1/2,j+1/2}^n\frac{u_{i+1/2,j+1}^{n+1} - u_{i+1/2,j}^{n+1}}{\Delta y}\right.$$
$$\left. -(\overline{\alpha\mu}_{\text{eff}})_{i+1/2,j-1/2}^n\frac{u_{i+1/2,j}^{n+1} - u_{i+1/2,j-1}^{n+1}}{\Delta y}\right] \qquad (10.169)$$

where $\overline{\alpha\mu}_{\text{eff}}$ is evaluated by averaging over the four adjacent nodes as in (10.166), but without the upwinding rule. In equations (10.165)–(10.169) all quantities refer to the gas phase but the subscript G has been dropped for simplicity of writing. The y-momentum equation for the gas phase is discretized in a similar manner:

$$(m_G)_{i+1/2,j}^n\frac{(v_G)_{i,j+1/2}^{n+1} - (v_G)_{i,j+1/2}^n}{\Delta t} + \left\langle\frac{\delta(m_G u_G)^n(v_G)^{n+1}}{\delta x}\right\rangle_{i,j+1/2}$$
$$+ \left\langle\frac{\delta(m_G v_G)^n(v_G)^{n+1}}{\delta y}\right\rangle_{i,j+1/2}$$
$$= -\min\left((\alpha_G)_{i,j}^n, (\alpha_G)_{i,j+1}^n\right)\frac{p_{i,j+1}^{n+1} - p_{i,j}^{n+1}}{\Delta y} + \left(\frac{\delta(\tau_G)_{yx}}{\delta x}\right)_{i,j+1/2}$$
$$+ \left(\frac{\delta(\tau_G)_{yy}}{\delta y}\right)_{i,j+1/2} - (m_G)_{i,j+1/2}^n g\cos\phi$$
$$+ H\left[(v_L)_{i,j+1/2}^{n+1} - (v_G)_{i,j+1/2}^{n+1}\right]. \qquad (10.170)$$

In particular, the y-derivative of the convective term is discretized analogously to (10.165) and the x-derivative analogously to (10.167) with $\overline{mu}$ evaluated by averaging similarly to (10.166). The liquid-phase momentum equations have the same form except that the terms $\min\left((\alpha_G)_{i+1,j}^n, (\alpha_G)_{i,j}^n\right)$, and $\min\left((\alpha_G)_{i,j}^n, (\alpha_G)_{i,j+1}^n\right)$ are replaced by $\max\left((\alpha_L)_{i+1,j}^n, (\alpha_L)_{i,j}^n\right)$ and $\max\left((\alpha_L)_{i,j}^n, (\alpha_L)_{i,j+1}^n\right)$, respectively. This particular treatment of the volume fractions in the pressure terms has a stabilizing effect as it reduces the effect of the common pressure gradient on the lighter phase.

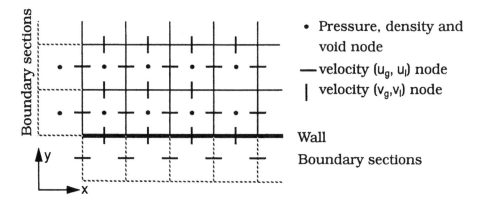

Fig. 10.11. Grid near boundaries showing the "ghost" cells used to impose boundary conditions.

The pressure equation (10.158) is also discretized semi-implicitly as

$$
\left(\frac{\alpha_G}{\rho_G c_G^2} + \frac{\alpha_L}{\rho_L c_L^2} \right)_{i,j}^n \frac{p_{i,j}^{n+1} - p_{i,j}^n}{\Delta t}
$$

$$
+ \frac{1}{(\rho_G)_{i,j}^n} \left[\left\langle \frac{\delta(m_G)^n (u_G)^{n+1}}{\delta x} \right\rangle_{i,j} + \left\langle \frac{\delta(m_G)^n (v_G)^{n+1}}{\delta y} \right\rangle_{i,j} \right]
$$

$$
+ \frac{1}{(\rho_L)_{i,j}^n} \left[\left\langle \frac{\delta(m_L)^n (u_L)^{n+1}}{\delta x} \right\rangle_{i,j} + \left\langle \frac{\delta(m_L)^n (v_L)^{n+1}}{\delta y} \right\rangle_{i,j} \right] = 0. \quad (10.171)
$$

10.6.3 Boundary conditions

Figure 10.11 shows the grid near a wall and an inlet/outlet boundary. It can be noticed that the computational domain is surrounded by a layer of "ghost" cells which facilitate the imposition of the boundary conditions as explained in Section 2.4. For a no-slip wall at $y = 0$, both velocity components vanish, which can be imposed by writing

$$
v(x,0) = 0, \qquad u(x, \Delta y/2) = -u(x, -\Delta y/2) . \qquad (10.172)
$$

For a free-slip wall at $y = 0$, such as a plane of symmetry, the proper conditions are

$$
v(x,0) = 0, \qquad u(x, \Delta y/2) = u(x, -\Delta y/2) , \qquad (10.173)
$$

in which the second one in effect has the consequence of imposing that the (x, y)-component of the viscous stress vanishes.

At the inlet, velocities and volume fractions of each phase are specified. At the outlet, a reference pressure and the volume fractions are specified.

10.6.4 Method of solution

It will be observed that, thanks to the particular choice of the time levels in the momentum equations, the four momentum and the pressure equations form a coupled set, but are uncoupled from the advanced time macroscopic densities in the continuity equations. This fact renders a separate solution possible. For each phase, the u-velocity couples to the upstream and downstream u-velocities for both phases, to the v-velocities at the same node and at the nodes above and below, and to the upstream and downstream pressures. Thus, although some of the coupling coefficients may vanish due to the upwinding, in principle there is a total of eight couplings. The equation for p_{ij}^{n+1} contains couplings to $u_{i\pm1/2,j}$ and $v_{i,j\pm1/2}$ for both phases and therefore it involves nine quantities. If the equations are ordered as shown in Fig. 10.12, a banded matrix results. With N_x and N_y nodes in the x- and y-directions, there is a total of $N_x(5N_y - 2)$ equations with a bandwidth of $10N_y - 3$. A difficulty which arises in the solution of the system is due to the large speeds of sound (in particular in the liquid), which causes the diagonal elements corresponding to $p_{i,j}^{n+1}$ to be small, which limits the convergence of iterative solvers. For this reason, initially the system was solved by a Gaussian band algorithm. More recently, with the use of suitable preconditioners, it has become possible to use faster methods (Nordsveen and Moe, 1999).

Once the updated velocities have been determined, the macroscopic densities are calculated from the continuity equations, which can be solved separately for each phase. This step clearly is a much smaller computational task. In the absence of phase change, the new densities can be obtained from the equations of state with the new pressures, and finally the volume fractions are found from (10.152):

$$\alpha^{n+1} = \frac{m^{n+1}}{\rho^{n+1}}. \tag{10.174}$$

While the partial coupling provided by this procedure avoids the need for the slowly converging iterations of the previous methods, the decoupling of the macroscopic densities from the velocities has the disadvantage that the mass fluxes $(mu)^{n+1}$ are not quite consistent, as $\mathbf{u}^{n+1}$ is calculated using the earlier-time densities m^n. Due to the way in which the continuity equations are discretized, this circumstance does not affect conservation of mass, but it does affect the volume fluxes αu. As a consequence, the end-of-time-step volume fractions do not sum up to 1 exactly. This error is remedied by the

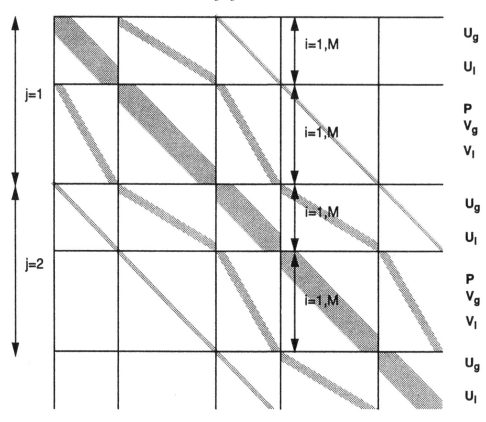

Fig. 10.12. Structure of the matrix for the method of Section 10.6. For each j-value (y-coordinate) the equations are arranged in the following order: x-momentum equations for gas and liquid (10.164), pressure equation (10.171), y-momentum equations for gas and liquid (10.170).

renormalization (10.90), but excessive reliance on this device compromises the accuracy of the calculation. In order to limit the magnitude of the error it is necessary to introduce a time-step control which is based on the pressure equation. If the problem were solved in a fully coupled way, (10.158) would be discretized as

$$
\left(\frac{\alpha_G}{\rho_G c_G^2} + \frac{\alpha_L}{\rho_L c_L^2}\right)^{n+1}_{i,j} \frac{p_{i,j}^{n+1} - p_{i,j}^n}{\Delta t}
$$

$$
+ \frac{1}{(\rho_G)_{i,j}^{n+1}} \left[\left\langle \frac{\delta(m_G u_G)^{n+1}}{\delta x}\right\rangle_{i,j} + \left\langle \frac{\delta(m_G v_G)^{n+1}}{\delta y}\right\rangle_{i,j}\right]
$$

$$
+ \frac{1}{(\rho_L)_{i,j}^{n+1}} \left[\left\langle \frac{\delta(m_L u_L)^{n+1}}{\delta x}\right\rangle_{i,j} + \left\langle \frac{\delta(m_L v_L)^{n+1}}{\delta y}\right\rangle_{i,j}\right] = 0. \quad (10.175)
$$

Because of the solution algorithm, however, the equation is solved in an approximate way, and it is this approximation which produces a difference between 1 and (the unrenormalized) $(\alpha_G + \alpha_L)^{n+1}$. By combining (10.159) and (10.175) written for both phases we find

$$
(\alpha_G + \alpha_L)^{n+1} - 1 \simeq \left(\frac{\Delta \alpha_G}{(\rho_G c_G^2)^n} + \frac{\Delta \alpha_L}{(\rho_L c_L^2)^n} \right)_{i,j} \left(p_{i,j}^{n+1} - p_{i,j}^n \right)
$$

$$
+ \frac{\Delta t}{(\rho_G)_{i,j}^n} \left[\left\langle \frac{\delta(\Delta m_G u_G^{n+1})}{\delta x} \right\rangle_{i,j} + \left\langle \frac{\delta(\Delta m_G v_G^{n+1})}{\delta y} \right\rangle_{i,j} \right]
$$

$$
+ \frac{\Delta t}{(\rho_L)_{i,j}^n} \left[\left\langle \frac{\delta(\Delta m_L u_L^{n+1})}{\delta x} \right\rangle_{i,j} + \left\langle \frac{\delta(\Delta m_L v_L^{n+1})}{\delta y} \right\rangle_{i,j} \right],
$$

$$
\text{(10.176)}
$$

where

$$
\Delta \alpha = \alpha^{n+1} - \alpha^n, \qquad \Delta m = m^{n+1} - m^n. \tag{10.177}
$$

Since the densities vary only slowly, t^n-levels have been used in place of updated t^{n+1} ones. By use of this equation, the volume error at the end of the time step can be estimated and if it is too large, the time step can be decreased until an acceptable level is attained. A crude estimate of the terms involving the spatial derivatives in this equation can be given by noting that, if the velocities are slowly varying, $\rho^{-1} | \langle \delta(mu)/\delta x \rangle | \Delta t < [\Delta(\alpha \rho)/\rho]|u|\Delta t/\Delta x < |u|\Delta t/\Delta x < 1$, which is just a CFL condition based on the convective velocities. This limitation on the time step is necessary anyway for reasons of stability and accuracy. Thus, the real effect of equation (10.176) is essentially to prevent an excessive change of the pressure in a single time step. With the value $|(\alpha_L + \alpha_G)^{n+1} - 1| \simeq 10^{-5}$ cited in the literature as acceptable (Moe and Bendiksen, 1993), and crude estimates $\Delta \alpha \simeq 1$, $\rho \simeq 1000$ kg/m^3, $c \simeq 1000$ m/s, one has a maximum pressure change during a single time step of about 10 kPa.

Figure 10.13 (from Moe, 1993) shows the emptying of a liquid-filled tube as calculated by the method just described. The upper figure shows the $\alpha = 0.5$ line, while in the lower one the $\alpha = 0, 0.5$, and 1 lines are shown. While the liquid–gas interface is tracked reasonably well by this method, the appreciable diffusion chiefly due to the use of upwinding is also evident.

Another example shown in Fig. 10.14 (from Moe and Bendiksen, 1993) is the evolution of a liquid-gas system in a square tank as computed by this method. A uniform 20×20 grid was used with a constant time step of 1 ms. The liquid density was 1000 kg/m^3, with a viscosity of 1 mPa s, and the

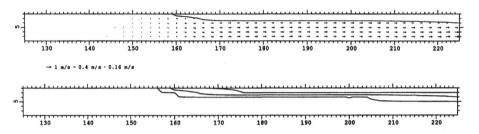

Fig. 10.13. Emptying of a horizontal liquid filled tube as computed by the method
of Section 10.6. The upper figure shows the $\alpha = 0.5$ line with some velocity vectors,
the lower one the isolines $\alpha = 0$, 0.5, and 1 (from Moe, 1993).

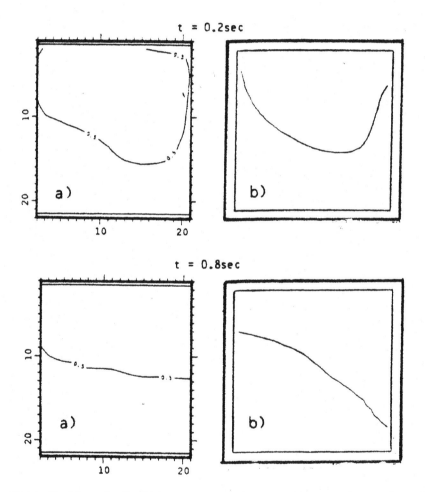

Fig. 10.14. Sloshing of a liquid in a square tank at two instants of time. Initially a
liquid and a gas are separated by a vertical plane with the liquid on the left. The
left panel in each figure is the result of the method of Section 10.6, the right panel
the result of the FLUENT commercial code for the same problem (from Moe and
Bendiksen, 1993).

gas density 1 kg/m^3. At the initial instant, the two fluids are separated by a vertical plane with the liquid on the left and the gas on the right. The right panel in each pair of figures shows the same problem as solved by the FLUENT commercial code, which uses a SIMPLE-like solution algorithm. In both calculations the interface is taken to correspond to the line $\alpha = 0.5$. The two predictions are in reasonable agreement at the earlier time $t = 0.2$ s (upper figures), but quite some differences arise at the later time $t = 0.8$ s. These differences are typical of the numerical predictions that result from the currently available two-phase flow codes and models.

11

Coupled methods for multifluid models

In the previous chapter we have presented segregated solution methods for multifluid models. When the interaction among the phases is very strong, or the processes to be simulated have short time scales, the methods that we now describe, in which the equations are more tightly coupled in the solution procedure, are preferable.

Work on methods of this type received a strong impulse with the development of nuclear reactor thermohydraulic safety codes in the 1970s and 1980s. This activity led to well-known codes such as RELAP[1], TRAC[2], SIMMER[3], and several others. The latest developments of these codes focus on the refinement of models, the inclusion of three-dimensional capabilities, better data structure, and vectorization, rather than fundamental changes in the basic algorithms[4]. By and large, the numerical methods they employ are an

---

[1] The code family RELAP (*R*eactor *E*xcursion and *L*eak *A*nalysis *P*rogram) was developed in the late 1960s starting from FLASH, a classified code for the design of naval nuclear reactors. Early versions of RELAP included very conservative assumptions. In the mid-1970s, the U.S. Nuclear Regulatory Commission prompted an orientation toward "best estimate" codes. Work on RELAP5, currently the most widely used program in the world for the analysis of the safety of nuclear power plants, was started in 1975 at EG&G Idaho Falls (now Idaho National Engineering and Environmental Laboratory) for the U.S. Nuclear Regulatory Commission. The first version of the code was released in 1982. The current version is RELAP5/MOD3.3 and was released in 2001. The code is limited to one spatial dimension allowing for a variable cross-section of the flow passage. More recently, INEEL has developed the code RELAP5-3D which has quasi-three-dimensional features (RELAP5-3D, 2003).

[2] TRAC (*T*ransient *R*eactor *A*nalysis *C*ode) was meant to include significantly more detail than RELAP and was to be used to check selected aspects of RELAP's results. Work on the code started at Los Alamos National Laboratories in 1974. The first version, TRAC-P1, was publicly released in late 1977. The current version, TRAC-PF1 MOD 2, carries the number 5.4 and was released in 1993 (Spore *et al.*, 1993). A boiling-water version, TRAC-BF1, was developed in parallel by EG&G (Borkowski and Wade, 1992).

[3] The SIMMER (S_n *I*mplicit *M*ultifield *M*ulticomponent *E*ulerian *R*ecriticality) codes were developed for the analysis of liquid-metal fast breeder nuclear reactors at Los Alamos National Laboratories in the 1970s (Bell *et al.*, 1977; Smith *et al.*, 1980). The next-generation version, SIMMER-III, is currently being developed by the Japan Nuclear Cycle Development Institute (JNC) jointly with several European organizations.

[4] An example is the code TRACE (*T*RAC/RELAP *A*dvanced *C*omputational *E*ngine) which is the result of the merger of RELAP with the pressurized-water and boiling-water versions

outgrowth of the ICE approach (*I*mplicit *C*ontinuous *E*ulerian) developed by Harlow and Amsden (1971) in the late 1960s. While very robust and stable, these methods, described in Section 11.2, are only first-order accurate in space and time and have other shortcomings. The more recent work, some of which is outlined in the second part of this chapter, is based on newer developments in computational fluid dynamics which are summarized in Section 11.3.

A tendency toward more strongly coupled solution methods is also evident in contemporary work springing from the segregated approach described in the previous chapter (see, e.g. Kunz *et al.*, 1998, 1999, 2000). These developments lead to a gradual blurring of the distinction between the two approaches.

11.1 Mathematical models

In order to gain a better appreciation of the computational task ahead, let us start by looking at some typical examples of Eulerian–Eulerian models; for the present purposes it is sufficient to consider the one-dimensional case.

Most of the nuclear two-fluid models for a liquid (index L) and a gas/vapor (index G) phase were designed to solve equations with the following general structure:

$$\frac{\partial}{\partial t}\left(\alpha_J \rho_J\right) + \frac{\partial}{\partial x}\left(\alpha_J \rho_J u_J\right) = \Gamma_J, \tag{11.1}$$

$$\frac{\partial}{\partial t}\left(\alpha_J \rho_J u_J\right) + \frac{\partial}{\partial x}\left(\alpha_J \rho_J u_J^2\right) + \alpha_J \frac{\partial p}{\partial x} = F_J, \tag{11.2}$$

$$\frac{\partial}{\partial t}\left(\alpha_J \rho_J e_J\right) + \frac{\partial}{\partial x}\left(\alpha_J \rho_J e_J u_J\right) + p\left[\frac{\partial \alpha_J}{\partial t} + \frac{\partial}{\partial x}(\alpha_J u_J)\right] = h_J \Gamma_J + Q_J. \tag{11.3}$$

Here the index $J = L, G$ denotes the component, α_J the volume fraction, ρ_J, u_J, e_J, h_J the average density, velocity, internal energy, and enthalpy[1], and p the common pressure. The terms in the right-hand sides express the coupling between the phases and with the solid structure: Γ_J represents the mass source per unit volume and time into phase J, F_J accounts for the forces, and Q_J for the heat transfer to the phase J other than that due to phase change. Conservation of mass and volume requires that

$$\Gamma_L + \Gamma_G = 0, \qquad \alpha_L + \alpha_G = 1. \tag{11.4}$$

of TRAC, including aspects of other codes such as PARCS (*P*urdue *A*dvanced *R*eactor *C*ode *S*imulator); see Mahaffy (2004).

[1] In TRAC, h_J is taken as the bulk enthalpy if phase J is decreasing (by evaporation or condensation), and as the saturation enthalpy if the amount of phase J is increasing.

The system must be closed with suitable equations of state relating ρ, e, h, T, and p for each phase, and with explicit closure relations expressing Γ_J, F_J, and Q_J in terms of the other variables. For simplicity of writing, here and in the following we drop the explicit indication of averaging.

In place of an equation for the internal energy, one could use one for the total energy from which the internal energy would be deduced by subtracting the mechanical energy. This procedure is followed in the complete OLGA model treated in a simplified form in the previous chapter (Bendiksen *et al.*, 1991), which is designed to simulate slow transients associated with mass transport in pipelines. This approach has the advantage that the total energy is accurately conserved. However, if the flow is such that the kinetic energy cannot be determined with reasonable accuracy due, e.g., to a complex flow regime, significant errors may be incurred in the calculation of the internal energy and, therefore, of the temperature. In these cases, as already remarked in Chapter 8, this approach may induce temperature fluctuations particularly in very low speed or supersonic flows (Harlow and Amsden, 1975). It may also be noted that different codes make different choices for the energy variables. For example, while both TRAC-PF1 and RELAP5 use the internal energy, TRAC-PD2 uses the temperature and COBRA-TF the product αh of the enthalpy and volume fraction. Temperature is convenient as physical properties are often given in terms of this variable, but it does not naturally appear in the conservation equations. The product αh is convenient in the evaluation of the Jacobian in implicit schemes. Each choice has positive and negative aspects and the best approach does not appear to have yet been sorted out (Frepoli *et al.*, 2003).

In the simplest case, the right-hand sides of equations (11.1)–(11.3) only contain algebraic terms. For example, in the nuclear thermohydraulic TRAC code, the force terms in the momentum equations are written as[1]

$$F_L^{\mathrm{T}} = c_i \left(u_G - u_L\right) \left|u_G - u_L\right| - c_{wL} u_L \left|u_L\right| + \Gamma_G u_L$$
$$- \Gamma^- \left(u_G - u_L\right) + \alpha_L \rho_L g_x \tag{11.5}$$

$$F_G^{\mathrm{T}} = -c_i \left(u_G - u_L\right) \left|u_G - u_L\right| - c_{wG} u_G \left|u_G\right| + \Gamma_G u_G$$
$$- \Gamma^+ \left(u_G - u_L\right) + \alpha_G \rho_G g_x \tag{11.6}$$

---

[1] The TRAC model, as actually solved in the code, is not in the form (11.1)–(11.3) as the equations are not written in conservation form and, in place of two separate energy equations, one total energy and one gas energy equation are solved. In this as in the other examples that follow we show the form that the TRAC equations would have if written in the form (11.1)–(11.3). The use of the non-conservative form permits the adoption of simpler numerical strategies, as shown in the next section. In the presence of rapid changes of the flow cross-section, however, the non-conservative form of the momentum equation can induce significant errors.

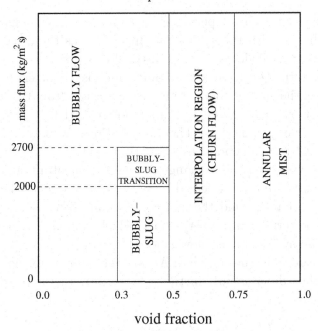

Fig. 11.1. The TRAC basic flow regime map for vertical flow.

Here g_x is the acceleration of gravity in the flow direction, c_i is the interphase drag parameter, and c_{wL}, c_{wG} are coefficients describing drag with the solid wall. These quantities depend on the flow regime, which is ascertained on the basis of simplified maps such as that shown in Fig. 11.1. In the terms expressing the liquid momentum source due to phase change, $\Gamma^- = \min(\Gamma_G, 0)$, so that

$$\Gamma_L u_L - \Gamma^- (u_G - u_L) = \begin{cases} \Gamma_L u_L = -\Gamma_G u_L & \text{for } \Gamma_G > 0 \\ \Gamma_L u_G = -\Gamma_G u_G & \text{for } \Gamma_G < 0. \end{cases} \qquad (11.7)$$

Thus, in the presence of evaporation ($\Gamma_G > 0$), the liquid loses momentum at the rate $\Gamma_G u_L$ while, with condensation, it gains it at the rate $\Gamma_L u_G$. The modelling is similar for the gas equation, in which $\Gamma^+ = \max(\Gamma_G, 0)$. The rate of phase change is expressed as

$$\Gamma_G = -\frac{q_G^i + q_L^i}{h_{\ell v}} \qquad (11.8)$$

where $h_{\ell v}$ is the enthalpy of vaporization and the q^i's are the interfacial heat fluxes expressed in terms of an interfacial heat transfer coefficient H_J multiplied by the interfacial area per unit volume a^i:

$$q_J^i = H_J a^i (T_{\text{sat}} - T_J), \qquad (11.9)$$

with T_{sat} the saturation temperature. A similar relation is used to model the heat exchange with the solid structures. Just as the drag coefficients, the heat transfer coefficients depend on the flow regime.

The source term Q_J in both the liquid and gas–vapor energy equations is expressed as the sum of conduction, heat transfers from the wall and the interface, augmented by an energy input not related to transfer processes (e.g., nuclear energy release). The contributions from the wall and the interface are modeled algebraically similarly to (11.9).

The terms expressing the exchange of mass, momentum, and energy between the phases are often referred to as *relaxation source terms* and tend to establish mechanical and thermal equilibrium. The characteristic time scales for these terms can be very short, which confers a numerically stiff character to many multiphase flow models.

The force models in the code RELAP are very similar to those of TRAC, from which they differ mainly in the presence of terms approximating the added-mass interaction[1]:

$$F_L^R = F_L^{\mathrm{T}} - C \alpha_L \alpha_G \rho_m \frac{\partial}{\partial t} (u_L - u_G) \tag{11.10}$$

$$F_G^R = F_G^{\mathrm{T}} + C \alpha_G \alpha_L \rho_m \frac{\partial}{\partial t} (u_L - u_G), \tag{11.11}$$

in which $F_{L,G}^{\mathrm{T}}$ are defined in (11.5) and (11.6), $\rho_m = \alpha_L \rho_L + \alpha_G \rho_G$ is the mixture density, and C is a flow-regime-dependent added-mass coefficient. The energy equations are also formally very similar to those of TRAC, except that the source terms Q_J include energy dissipation in the form $c_{wJ} \alpha_J \rho_J u_J^2$. As before, the heat transfer and phase change rates are given by algebraic expressions with coefficients dependent on the flow regime.

In the presence of phase change, the force term (11.9) of the commercial CFX code shown in Section 10.1 of the previous chapter is modified to the form

$$F_J = \frac{\partial}{\partial x} (\alpha_J \tau_J) + \alpha_J \rho_J \mathbf{g} + \sum_K c_{JK} (u_K - u_J) + \sum_K (\Gamma_{JK} u_K - \Gamma_{KJ} u_J) \tag{11.12}$$

in which Γ_{JK} is the mass source for phase J deriving from phase K and, therefore, the effect of phase change as written in (11.12) is equivalent to that of equations (11.5) and (11.6). Written in this form, it is evident that in the momentum equation mass transfer plays a role similar to drag. As mentioned before, in addition to the viscous stress, the term τ_J also includes turbulent stresses.

---

[1] Here we are only interested in the formal structure of the equations and we do not imply that the correlations giving the interphase drag, etc. are the same in the two codes.

More often than not, for several reasons, turbulence effects are not explicitly included in the models considered in this chapter. In the first place, as already noted in the previous chapter, the inclusion of new variables and new equations is simpler with segregated solvers, as there the equations are solved one by one rather than in a coupled manner as here. Secondly, for low-order schemes with upwinding, numerical diffusion is strong and would mask turbulent diffusion. As for the newer methods described in the second part of this chapter, they are still in relatively early stages of development which justifies their focus on a basic form of the model equations.

11.2 First-order methods

Until recently, first-order methods have been the most widespread tools for the solution of the two-fluid equations. With an explicit time discertization, pressure waves would impose a CFL stability condition of the form $|c \pm u| \Delta t / \Delta x \leq 1$, where c is the speed of sound, which would require prohibitively small time steps, particularly with fine spatial discretizations. The opposite approach of a fully implicit time discretization removes stability constraints on the time step, but the solution of the resulting large system of nonlinear equations is an onerous task. Therefore, substantial effort has been invested in the development of a compromise between implicitness and computational cost. A physical basis for such a hybrid *semi-implicit* approach is offered by the fact that, in general, the averaged equations for multiphase flow describe physically distinct processes which are often characterized by different time scales. Interphase exchanges of momentum and energy can be much faster than pressure wave propagation, which may in turn be much faster than convective transport. It may therefore be advantageous to pay the price of treating fast processes implicitly, while cheaper explicit methods can be used for the slow processes. In the last decade or so, with the progress in computational power, fully implicit schemes have become more practical, at least for one-dimensional calculations.

11.2.1 A semi-implicit method

We illustrate a classic semi-implicit method, based essentially on ideas of Liles and Reed (1978; see also Stewart and Wendroff, 1984; Mahaffy, 1993), with reference to equations (11.1), (11.2), and (11.3). This system is not hyperbolic and, therefore, it is necessary to use a dissipative numerical solution method. Upwind differencing on a staggered grid is a standard and effective way to dampen the instabilities. Of course there is no guarantee that, if not applied with great care, the method will not remove physical instabilities together with the unphysical ones. We will not elaborate on this difficulty as here we are mostly interested in the time discretization which can be based on similar considerations also in the case of better models.

The equations are discretized as

$$\frac{(\alpha\rho)_j^{n+1} - (\alpha\rho)_j^n}{\Delta t} + \frac{\langle(\alpha\rho)^n u^{n+1}\rangle_{j+1/2} - \langle(\alpha\rho)^n u^{n+1}\rangle_{j-1/2}}{\Delta x} = \Gamma_j^{n+1}, \quad (11.13)$$

where we omit the explicit indication of the phase, Δt and Δx are the time and space steps, and n and j the temporal and spatial indices. The angle brackets denote the mass fluxes calculated according to the upwinding rule:

$$\langle(\alpha\rho)^n u^{n+1}\rangle_{j+\frac{1}{2}} = u_{j+\frac{1}{2}}^{n+1} \begin{cases} (\alpha\rho)_j^n & \text{if } u_{j+\frac{1}{2}}^{n+1} \geq 0 \\ (\alpha\rho)_{j+1}^n & \text{if } u_{j+\frac{1}{2}}^{n+1} < 0. \end{cases} \quad (11.14)$$

The momentum equations are discretized in non-conservation form as

$$(\alpha\rho)_{j+\frac{1}{2}}^n \left[\frac{u_{j+\frac{1}{2}}^{n+1} - u_{j+\frac{1}{2}}^n}{\Delta t} + u_{j+\frac{1}{2}}^n \frac{\langle u_{j+\frac{1}{2}}^n - u_{j-\frac{1}{2}}^n \rangle}{\Delta x} \right] + \alpha_{j+\frac{1}{2}}^n \frac{p_{j+1}^{n+1} - p_j^{n+1}}{\Delta x}$$

$$= -C_{j+\frac{1}{2}}^n \Delta u_{j+\frac{1}{2}}^{n+1} + \hat{F}_{j+\frac{1}{2}}^{n+1}, \quad (11.15)$$

where we have written the force as $F_J = -C \Delta u_J + \hat{F}_J$ with C an interphase drag parameter and Δu_J the difference between the velocity of phase J and that of the other phase. All the terms in the right-hand side are assumed to depend algebraically in a known manner on the variables but not on their derivatives. Similarly to (11.14), the momentum flux is treated by upwinding:

$$u_{j+\frac{1}{2}}^n \langle u_{j+\frac{1}{2}}^n - u_{j-\frac{1}{2}}^n \rangle = u_{j+\frac{1}{2}}^n \begin{cases} (u_{j+\frac{1}{2}}^n - u_{j-\frac{1}{2}}^n) & \text{if } u_{j+\frac{1}{2}}^n \geq 0 \\ (u_{j+3/2}^n - u_{j+\frac{1}{2}}^n) & \text{if } u_{j+\frac{1}{2}}^n < 0 \end{cases} \quad (11.16)$$

and $(\alpha\rho)_{j+\frac{1}{2}}^n$ is evaluated as the simple average of $(\alpha\rho)_j^n$ and $(\alpha\rho)_{j+1}^n$. A disadvantage of the particular discretization of the convective terms of the momentum equation is that the conservative nature of the left-hand side of the original differential equation has been lost, which can lead to errors, e.g. in the presence of abrupt area changes[1], but the resulting finite difference equation has desirable robustness properties.

The time discretization adopted in equations (11.13) and (11.15) was based on the following considerations. In the model, the terms responsible for pressure wave propagation are combined with those responsible for all the other physical processes and cannot be cleanly separated. However,

---

[1] Corrections to mitigate this problem have been developed; see e.g. Spore et al., (1993).

in standard single-phase fluid mechanics, pressure waves rely on the time derivative and velocity terms of the continuity equation, and on the time derivative and pressure terms of the momentum equation, and it is these terms that have been evaluated at the new time level t^{n+1}. Phase change is often characterized by a short time scale, and, in addition, vapor production can change the compressibility of the system by orders of magnitude and strongly affect the fluid motion. Thus, Γ_J must be treated implicitly. Interphase drag can also be large, particularly in the case of disperse flow and, therefore, it requires an implicit discretization as well.

The energy equation is discretized similarly to the continuity equation as

$$\frac{(\alpha\rho e)_j^{n+1} - (\alpha\rho e)_j^n}{\Delta t} + \frac{\langle(\alpha\rho e)^n u^{n+1}\rangle_{j+1/2} - \langle(\alpha\rho e)^n u^{n+1}\rangle_{j-1/2}}{\Delta x}$$

$$+ p_j^{n+1}\left(\frac{\alpha_j^{n+1} - \alpha_j^n}{\Delta t} + \frac{\langle\alpha^n u^{n+1}\rangle_{j+1/2} - \langle\alpha^n u^{n+1}\rangle_{j-1/2}}{\Delta x}\right)$$

$$= h_J\Gamma_j^{n+1} + Q_j^{n+1}. \tag{11.17}$$

In some versions of the algorithm, the pressure term is evaluated at time level t^n. It proves useful to convect the internal energy per unit volume ρe, rather than e itself, as this procedure favors the coupling of the temperature and density fields (Harlow and Amsden, 1975).

In equations (11.15) and (11.17) the terms F and Q have been written indicating an implicit treatment. These terms may consist of several different contributions (e.g. wall friction and gravity) and this procedure may or may not be necessary for some or all of these contributions depending on the problem and the details of their dependence on the primary variables.

In order to explain the solution procedure, let us suppress temporarily the spatial index and write the momentum equations (11.15) for the two phases in a synthetic form as

$$u_L^{n+1} + \frac{\Delta t\, C}{\alpha_L\rho_L}\left(u_L^{n+1} - u_G^{n+1}\right) = -\frac{\Delta t}{\rho_L\Delta x}\left(p_{j+1}^{n+1} - p_j^{n+1}\right) + D_L \tag{11.18}$$

$$u_G^{n+1} + \frac{\Delta t\, C}{\alpha_G\rho_G}\left(u_G^{n+1} - u_L^{n+1}\right) = -\frac{\Delta t}{\rho_G\Delta x}\left(p_{j+1}^{n+1} - p_j^{n+1}\right) + D_G \tag{11.19}$$

where $D_{L,G}$ is short-hand for the terms that need not be shown explicitly, and omitted temporal superscripts imply evaluation at time n. An implicit evaluation of the force would give the terms $D_{L,G}$ a time level $n+1$, which can be treated by means of an iterative procedure or, in the case of a linear dependence on the velocity, can be incorporated in the left-hand side. By solving analytically the linear system (11.18) and (11.19), one can express

the velocity of each phase in terms of the pressure gradient with a result
having the structure

$$u_J^{n+1} = -a_J \left(p_{j+1}^{n+1} - p_j^{n+1} \right) + b_J. \tag{11.20}$$

It should be noted that an identical procedure would apply to a multidimensional situation, as well as to one with more than two components. In these cases the number of unknown velocities increases, but so does the number of momentum equations available.

Up to this point the solution strategy is very similar to those considered in Sections 10.2 and 10.3 of the previous chapter. A difference arises in the way in which the resulting equations are solved. The expressions (11.20) are substituted into the continuity and energy equations to give, in each cell, a nonlinear system involving α_J, ρ_J, e_J, and p, all evaluated at time level $n+1$. These are four equations for seven unknowns, but in addition one must consider the two equations of state $\rho = \rho(p, e)$ and the constraint $\alpha_1 + \alpha_2 = 1$. The system is therefore closed. By elimination – facilitated, where necessary, by linearization around the latest iterate values of the unknowns – a Helmholtz-type equation for the pressure (or pressure increment) is generated, from which all the other unknowns are then updated. The process is repeated to convergence. The procedure can be carried out cell-by-cell because of the discretization of the term $\partial(\alpha u)/\partial x$ of the energy equation as $\langle \alpha^n u^{n+1} \rangle_{j\pm 1/2}$. If this term were replaced by $\langle \alpha^{n+1} u^{n+1} \rangle_{j\pm 1/2}$, the volume fractions in adjacent cells would become coupled.

In principle, the reduction to a single unknown can be effected for a variable other than pressure. Density, however, is undesirable as it can undergo strong jumps across contact discontinuities (e.g. a liquid–vapor transition), while pressure would be continuous in such a situation. Furthermore, pressure is an extremely sensitive function of the liquid density, and use of this latter variable would strongly limit the convergence of the iterative process unless very small time steps are used. Finally, the density, energy, and volume fraction of a particular phase all have the disadvantage that they would cease to be well-defined if the phase in question were to disappear. Pressure therefore emerges as the most logical choice.

Many other variants of both discretization and solution procedure have been devised and a good review of the older work is provided by Stewart and Wendroff (1984). While in many cases differences between the various methods are limited in one space dimension, the generalization to higher dimensionality problems may prove difficult or impossible for some of these approaches. A case in point is the procedure developed by Trapp and Riemke

(1986; see also Barre *et al.*, 1993), in which the mass and energy equations are solved cell by cell to express the pressure in terms of the velocities, which are then found from the momentum equations. This procedure can only be carried out in one space dimension.

11.2.2 The SETS method

Since the convection of momentum is treated explicitly, the time step for the method just described is limited by a CFL condition based on the convective velocity, $\max_j (|u_L|, |u_G|) (\Delta t / \Delta x) < 1$. For high velocity, or when the coupling between the physical processes at work is very strong, the amount of implicitness introduced in this method may be insufficient. A strategy which combines some of the robustness of the fully implicit discretization with a less onerous solution procedure is the SETS (*S*tability *E*nhancing *T*wo-*S*tep) method (Mahaffy, 1982, 1993) implemented, e.g., in the nuclear safety code TRAC.

The first step consists in obtaining an estimate $\tilde{u}^{n+1}$ of the advanced-time velocities by solving a "stabilizer" momentum equation with the current pressures p^n:

$$(\alpha \rho)^n_{j+1/2} \left[\frac{\tilde{u}^{n+1}_{j+1/2} - u^n_{j+1/2}}{\Delta t} + u^n_{j+1/2} \frac{\langle \tilde{u}^{n+1}_{j+1/2} - \tilde{u}^{n+1}_{j-1/2} \rangle}{\Delta x} \right.$$
$$\left. + \beta \left(\tilde{u}^{n+1}_{j+1/2} - u^n_{j+1/2} \right) \frac{\langle \tilde{u}^n_{j+1/2} - \tilde{u}^n_{j-1/2} \rangle}{\Delta x} \right] + \alpha^n_{j+1/2} \frac{p^n_{j+1} - p^n_j}{\Delta x}$$
$$= -c^n_{j+1/2} |\Delta u^n_{j+1/2}| \left(2\Delta \tilde{u}^{n+1} - \Delta u^n \right)_{j+1/2} \qquad (11.21)$$

where we have omitted the term $\hat{F}$ for simplicity and have written the interphase drag $C\Delta u$ as $c|\Delta u| \Delta u$. The particular form of the discretization of this term can be understood by observing that, with $\Delta u^{n+1} = \Delta u^n + \delta$, $(\Delta u^{n+1})^2 \simeq (\Delta u^n) (\Delta u^n + 2\delta) = (\Delta u^n) (2\Delta u^{n+1} - \Delta u^n)$. A similar argument applied to the convective term gives $u^{n+1} \nabla u^{n+1} \simeq u^n \nabla u^{n+1} + (u^{n+1} - u^n) \nabla u^n$. In certain conditions, however, the second term has a destabilizing effect on the calculation. For this reason, in (11.21), this contribution is multiplied by a factor β defined as

$$\beta = \begin{cases} 0 & \text{if } \langle \tilde{u}^n_{j+1/2} - \tilde{u}^n_{j-1/2} \rangle < 0 \text{ or } u^n \tilde{u}^n < 0 \\ 1 & \text{if } \langle \tilde{u}^n_{j+1/2} - \tilde{u}^n_{j-1/2} \rangle \geq 0 \text{ and } u^n \tilde{u}^n \geq 0. \end{cases} \qquad (11.22)$$

In the present one-dimensional case, (11.21) is a simple linear system with a tridiagonal matrix.

With the new velocities estimated in this way, the "basic" momentum equations are rewritten as

$$
(\alpha\rho)^n_{j+1/2} \left[\frac{u^{n+1}_{j+1/2} - u^n_{j+1/2}}{\Delta t} + u^n_{j+1/2} \frac{\langle \tilde{u}^{n+1}_{j+1/2} - \tilde{u}^{n+1}_{j-1/2} \rangle}{\Delta x} \right.
$$
$$
\left. + \beta \left(u^{n+1}_{j+1/2} - u^n_{j+1/2} \right) \frac{\langle \tilde{u}^n_{j+1/2} - \tilde{u}^n_{j-1/2} \rangle}{\Delta x} \right] + \alpha^n_{j+1/2} \frac{\tilde{p}^{n+1}_{j+1} - \tilde{p}^{n+1}_j}{\Delta x}
$$
$$
= -c^n_{j+1/2} |\Delta u^n_{j+1/2}| \left(2\Delta u^{n+1} - \Delta u^n \right)_{j+1/2} . \tag{11.23}
$$

The "basic" mass and energy equations are written similarly to (11.13) and (11.17) for the semi-implicit method as

$$
\frac{(\tilde{\alpha}\tilde{\rho})^{n+1}_j - (\alpha\rho)^n_j}{\Delta t} + \frac{\langle (\alpha\rho)^n u^{n+1} \rangle_{j+1/2} - \langle (\alpha\rho)^n u^{n+1} \rangle_{j-1/2}}{\Delta x} = \tilde{\Gamma}^{n+1}_j \tag{11.24}
$$

$$
\frac{(\tilde{\alpha}\tilde{\rho}\tilde{e})^{n+1}_j - (\alpha\rho e)^n_j}{\Delta t} + \frac{\langle (\alpha\rho e)^n u^{n+1} \rangle_{j+1/2} - \langle (\alpha\rho e)^n u^{n+1} \rangle_{j-1/2}}{\Delta x}
$$
$$
+ \tilde{p}^{n+1}_j \left(\frac{\tilde{\alpha}^{n+1}_j - \alpha^n_j}{\Delta t} + \frac{\langle \alpha^n u^{n+1} \rangle_{j+1/2} - \langle \alpha^n u^{n+1} \rangle_{j-1/2}}{\Delta x} \right)
$$
$$
= \left(\tilde{h}\tilde{\Gamma} \right)^{n+1}_j + \tilde{Q}^{n+1}_j . \tag{11.25}
$$

Equations (11.23), (11.24), and (11.25) approximate a semi-implicit discretization of the equations in which the spatial coupling among adjacent cells has been removed from the momentum equation thanks to the preliminary "stabilizer" step[1]. These equations are now solved in a coupled way. To facilitate this step, the momentum equations (11.23) are first solved analytically to express the u^{n+1}_J in terms of the pressures $\tilde{p}$ as in (11.20). The result is substituted into (11.24) and (11.25) which, together with the equations of state expressing density and energy in terms of pressure and temperature, are solved by Newton iteration to find the new-time pressures and temperatures. After this step, the velocities u^{n+1}_J are calculated, and they can be used in the final calculation of pressure and temperature by rewriting the mass and energy equations in the "stabilizer" form

$$
\frac{(\alpha\rho)^{n+1}_j - (\alpha\rho)^n_j}{\Delta t} + \frac{\langle (\alpha\rho)^{n+1} u^{n+1} \rangle_{j+1/2} - \langle (\alpha\rho)^{n+1} u^{n+1} \rangle_{j-1/2}}{\Delta x} = \tilde{\Gamma}^{n+1}_j \tag{11.26}
$$

[1] In TRAC the "stabilizer" momentum equations are solved twice to obtain a better estimate of the interfacial drag force; see Mahaffy (1982).

$$\frac{(\alpha\rho e)_j^{n+1} - (\alpha\rho e)_j^n}{\Delta t} + \frac{\langle\!\langle(\alpha\rho e)^{n+1}u^{n+1}\rangle\!\rangle_{j+1/2} - \langle\!\langle(\alpha\rho e)^{n+1}u^{n+1}\rangle\!\rangle_{j-1/2}}{\Delta x}$$

$$+\tilde{p}_j^{n+1}\left(\frac{\tilde{\alpha}_j^{n+1} - \alpha_j^n}{\Delta t} + \frac{\langle\alpha^n u^{n+1}\rangle_{j+1/2} - \langle\alpha^n u^{n+1}\rangle_{j-1/2}}{\Delta x}\right)$$

$$= \left(\tilde{h}_J\tilde{\Gamma}\right)_j^{n+1} + \tilde{Q}_j^{n+1}. \tag{11.27}$$

It is stated that, for the full model including the energy equation, the solution of the stabilizer equations only requires about 20% more time than an explicit method, as compared with a factor of 5–6 for a fully implicit discretization (Mahaffy, 1982, 1993). A very useful consequence of the near-implicit treatment of the convection terms is that the scheme is linearly stable for any time step. Nonlinear effects can however cause instabilities when the time step is too large or the interfacial friction coefficient varies too rapidly (Barre *et al.*, 1993; Mahaffy, 1993; Frepoli *et al.*, 2003).

11.2.3 Implicit methods

With the progress in computational power, fully implicit schemes have become more practical, at least for one-dimensional calculations. Their advantage, of course, is the absence of stability limitations on the size of the time step. An example is the one-dimensional module of the French nuclear thermohydraulic code CATHARE which was developed at CEA-Grenoble to simulate a wide variety of accident scenarios in pressurized water reactors (Barre and Bernard, 1990; Bestion, 1990). The model used in this code is rendered hyperbolic by the use of added mass and interfacial pressure terms, but it is not necessary to show these terms explicitly for the present purposes.

The mass and energy equations are discretized similarly to (11.24) and (11.25) with the tildes removed and the convected quantities all evaluated at time level t^{n+1} rather than t^n. The momentum equations are written as

$$(\alpha\rho)_{j+1/2}^{n+1}\left[\frac{u_{j+1/2}^{n+1} - u_{j+1/2}^n}{\Delta t} + u_{j+1/2}^{n+1}\frac{\langle u_{j+1/2}^{n+1} - u_{j-1/2}^{n+1}\rangle}{\Delta x}\right]$$

$$+\alpha_{j+1/2}^{n+1}\frac{p_{j+1}^{n+1} - p_j^{n+1}}{\Delta x} = -c_{j+1/2}^{n+1}|\Delta u_{j+1/2}^{n+1}|\Delta u_{j+1/2}^{n+1}. \tag{11.28}$$

The set of coupled equations is solved by Newton iteration. In CATHARE, the derivatives necessary to evaluate the Jacobian are calculated analytically and the Jacobian itself is re-evaluated at each iteration. A difficulty that arises is the discontinuity or near-discontinuity of the derivatives at the boundaries between different flow regimes which, as can be seen from the example of Fig. 11.1, are often sharp and do not provide for a smooth transition. Furthermore, the exchanges of energy and momentum between the phases and the structure or between the phases themselves are very sensitive functions of the variables. As a consequence, convergence difficulties of the iteration procedure may force the use of a small time step in spite of the implicit discretization.

An alternative to the analytical evaluation of the derivatives is a numerical evaluation according to

$$\frac{\partial F_\alpha}{\partial X_\beta} \simeq \frac{F_\alpha(X_1, \ldots, X_\beta + \delta X_\beta, \ldots) - F_\alpha(X_1, \ldots, X_\beta, \ldots)}{\delta X_\beta} \qquad (11.29)$$

where F_α is the generic equation of the system and X_β one of the variables to be updated in the Newton iteration (Frepoli *et al.*, 2003). This procedure results in a greater flexibility of the code, in which the submodels and constitutive relations can be easily changed and not necessarily prescribed in analytical form. An important aspect is, however, how to choose the increments δX_β, which should be large enough to be above the round-off error of the numerical calculation but small enough that a good estimate of the derivative is obtained. Frepoli *et al.* (2003) find good results by following Dennis and Schnabel's (1983) suggestion according to which $\delta X_\beta = \epsilon X_\beta$, with ϵ of the order of the square root of the round-off error.

Recent work in this area focuses on the use of Jacobian-free Newton–Krylov methods with a physics-based preconditioner. Mousseau (2004) presents an interesting demonstration of the advantages of this approach.

11.2.4 Numerical diffusion and other difficulties

The stability and robustness of the methods described is acquired, at least in part, at the cost of the considerable numerical diffusion caused by upwinding, or donor-cell differencing. As already pointed out in Sections 3.2, this feature has several undesirable consequences. For example, while the two-fluid model could in principle deal with a gas–liquid interface, in practice numerical diffusion will quickly mix the two phases and the interface will lose its sharpness and disappear. As another example, in pressurized-water nuclear reactors, the core reactivity can be controlled by using boron dissolved in the cooling water of the primary loop. Undetected dilution of this neutron absorber is a safety concern which it is difficult to model numerically in the presence of spurious diffusion (Macian-Juan and Mahaffy, 1998). For the same reason, turbulent mixing is difficult to model and its effect to be identified in the presence of numerical mixing. A limited remedy is to use "weighted" donor-cell differencing in which, for example,

$$\rho_{j+1/2} = \frac{1}{2}\left(1 + \xi_{j+1/2}\right)\rho_j + \frac{1}{2}\left(1 - \xi_{j+1/2}\right)\rho_{j+1} \tag{11.30}$$

with ξ a weighting parameter. The choices $\xi = 0$ and $\xi = \mathrm{sgn}(u)$ result in central and full-upwind differencing, respectively. More generally, one may write

$$\xi_{j+1/2} = a_0 \frac{u_{j+1/2}\Delta t}{\Delta x} + b_0 \,\mathrm{sgn}(u_{j+1/2}), \tag{11.31}$$

with $0 \leq a_0,\, b_0 \leq 1$ to achieve a balance between the two extremes (Harlow and Amsden, 1975). When this procedure is adopted, it is important to ensure that the same proportions of donor-cell fluxes be used for mass and energy to avoid the tendency to develop inconsistencies in regions where the pressure should remain uniform but mass and energy densities have strong opposing gradients. This method mitigates somewhat the shortcomings of full upwinding, but real improvements can only by obtained by the different approaches described in Section 11.3.3 and Section 10.2.2 of the previous chapter.

While numerical diffusion is a difficulty also encountered in single-phase calculations, two other issues are specific to the multiphase case. The first one is the transition from two-phase to single-phase, e.g. due to complete condensation of the vapor phase or evaporation of the liquid phase. The equation set becomes singular in these cases and there is a strong potential for instabilities, especially when the liquid phase is modelled as incompressible. In most codes, such as RELAP5, COBRA-TF, CATHARE, and CFX, the problem is side-stepped by artificially preventing either phase from disappearing completely. This method has several shortcomings, particularly with the fully implicit scheme, as the void fraction is one of the variables updated in the procedure. The resetting of α to an artificial value at each iteration step may slow down convergence of even prevent it. In CATHARE these problems are mitigated by the use of artificial sources of mass and energy which cause the volume fractions to remain between the assigned bounds. These source terms also ensure saturation conditions for the disappearing phase and the use of a large drag coefficient forces mechanical equilibrium. An alternative approach to the problem is to switch from multiphase to single-phase equations when the volume fractions fall below preassigned limits, typically of the order of 10^{-6}, at either end of their range. A successful way to implement this idea is described by Frepoli *et al.* (2003).

Another problem is the so-called "water packing" (see e.g. Pryor *et al.*, 1978; Stewart and Wendroff, 1984; Frepoli *et al.*, 2003), the consequences

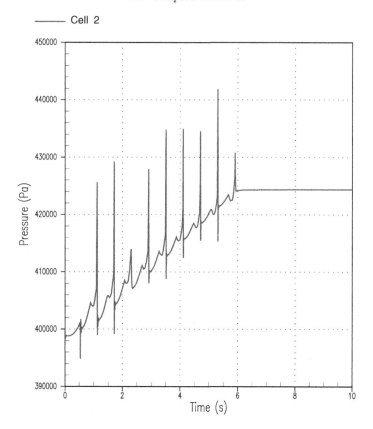

Fig. 11.2. Pressure at the bottom of a tube, initially filled with superheated steam, in which subcooled water is injected at a constant rate from below (from Frepoli *et al.*, 2003). The spikes are an artifact of the numerical method commonly called "water packing."

of which are illustrated in Fig. 11.2. The artificial pressure spikes promi- nent in this figure arise when a cell just fills with a nearly incompressible phase, the liquid in this case. This phenomenon is thought to be inherent in any Eulerian code, whether semi-implicit or fully implicit, but to be less severe with the latter discretization. It is remarkable that little progress seems to have been achieved in dealing with this difficulty in spite of several decades of effort (Stewart and Wendroff, 1984; Abe *et al.*, 1993; Aktas and Mahaffy, 1996).

11.3 Introduction to high-resolution schemes

The numerical schemes described in the previous sections are only first-order accurate in space and time. While this low accuracy results in efficiency and a pronounced stability, the methods are highly diffusive and tend to smear regions of rapid spatial variation of the solution. It is therefore of considerable interest to develop higher order schemes, which turns out to be a matter of perhaps unexpected complexity. Indeed, while it is relatively straightforward to formulate schemes with a formal accuracy of second or higher order, it turns out that they possess severe shortcomings which make them unsuitable for most cases of interest. This problem has generated a vast literature to which we can only present a very simplified and concise introduction. Excellent overviews of the subject are presented, e.g. in the books by LeVeque (1992) and Thomas (1995, vol. 2); see also Toro (1999).

It is useful to begin by looking again at first-order methods from a different viewpoint, more conducive to building higher accuracy algorithms.

11.3.1 Scalar conservation laws

Let us start from the initial value problem for a general scalar conservation equation of the form

$$\frac{\partial u}{\partial t} + \frac{\partial f(u)}{\partial x} = 0, \qquad u(x,0) = u^0(x), \tag{11.32}$$

where f is the flux of the conserved quantity u. In regions where the solution of (11.32) is smooth, it coincides with the solution of

$$\frac{\partial u}{\partial t} + f'(u)\frac{\partial u}{\partial x} = 0, \tag{11.33}$$

and it is readily verified by direct substitution that, in this case, the solution of (11.32) or (11.33) is given (implicitly) by

$$u(x,t) = u^0\left(x - tf'(u(x,t))\right). \tag{11.34}$$

For example, when $f = cu$ with c a constant, one has the well-known solution $u = u^0(x - ct)$.

It is well known that, depending on the initial data, the solutions of (11.32) may develop discontinuities, or *shocks*. In this case, (11.32) and (11.33) are no longer equivalent and use of the *conservation form* (11.32) is essential in order to guarantee the correct propagation speed of the discontinuity, which is given by

$$s(t) = \frac{f(u_R) - f(u_L)}{u_R - u_L}, \tag{11.35}$$

where the subscripts L and R denote limits taken as x approaches the discontinuity from the left and from the right, respectively. This is the jump condition, or *Rankine–Hugoniot relation*, for the special case of a scalar conservation law. Mathematically, discontinuous solutions belong to the class of *weak solutions* of the equation, a class which extends the notion of classical solution to functions possessing discontinuities and possibly other pathologies.

In the case of shock waves for the single-phase compressible Euler equations, in addition to the usual compression shocks, the equations formally admit expansion shocks. The unphysical nature of these spurious weak solutions can be proven by stability considerations, by considering the limit of the rapidly varying viscous solution as the viscosity parameter tends to zero, or by invoking the entropy inequality (see, e.g. Landau and Lifshitz, 1987; Whitham, 1974). The situation is similar in the case of equation (11.32): the unphysical expansion shocks can be identified and eliminated either by replacing the 0 in the right-hand side of (11.32) by a viscous-like term $\epsilon \partial^2 u / \partial x^2$ and taking the limit $\epsilon \to 0$, or by introducing a suitable entropy condition. It can be proven that the *vanishing viscosity solution* is the physically correct weak solution in the sense that it satisfies the proper *entropy condition*.

This is not the place to delve into this crucially important subject to any depth: the reader is referred to the abundant literature and its summaries, e.g. in Thomas (1995) or Toro (1999), for details. We simply mention that, when the flux f is convex – i.e. $f'' > 0$ – the entropy condition requires that

$$f'(u_L) > s(t) > f'(u_R). \tag{11.36}$$

This is essentially a stability condition: wavelets generated upstream, on the left of the shock, will catch up with the shock, while wavelets downstream will be overtaken by it. The shock acts as a "sink" for disturbances in its vicinity and is therefore stable[1]. Violation of (11.36) would correspond to the introduction of the analog of an expansion shock in place of the physically correct *expansion fan*, for which the solution is continuous, with the self-similar form

$$u = u(\eta), \qquad \eta = \frac{x - \overline{x}}{t - \overline{t}}, \qquad \eta_1 \le \eta \le \eta_2, \tag{11.37}$$

for suitable η_1, η_2, where $(\overline{x}, \overline{t})$ are the position and time where the fan forms.

---

[1] More generally, for nonconvex fluxes, it is necessary to require that, for all u between u_L and u_R,

$$\frac{f(u_L) - f(u)}{u_L - u} \ge \frac{f(u_R) - f(u)}{u_R - u}.$$

Another notion which arises in the following is that of a *contact discontinuity*. The idea is familiar from single-phase gas dynamics, in which these words designate a surface of discontinuity of density, and possibly temperature, across which pressure and velocity are continuous. In multiphase flow the typical analog is a gas–liquid interface and, in the context of equation (11.32), a surface across which u is continuous, but $\partial u/\partial x$ is not. Evidently, such a surface propagates with velocity $f'(u)$.

We have spent some time addressing the issues of discontinuities, uniqueness, and entropy in the context of the differential form of the equation as they figure prominently also in the construction of numerical methods suitable for its discretized solution.

11.3.1.1 Discretization

In order to derive a finite-difference scheme for equation (11.32), we divide the spatial integration domain along the x-axis into equal cells of width Δx and the temporal integration domain into intervals Δt. Upon integration over $x_j - \frac{1}{2}\Delta x \le x \le x_j + \frac{1}{2}\Delta x$ and $t^n \le t \le t^{n+1} = t^n + \Delta t$, we find

$$u_j^{n+1} = u_j^n - \frac{\Delta t}{\Delta x}\left(h_{j+1/2} - h_{j-1/2}\right) \qquad (11.38)$$

where

$$u_j = \frac{1}{\Delta x}\int_{x_j-\frac{1}{2}\Delta x}^{x_j+\frac{1}{2}\Delta x} u(x,t)\,dx\,, \qquad (11.39)$$

is the average of $u(x,t)$ over the cell centered at x_j and

$$h_{j\pm1/2} = \frac{1}{\Delta t}\int_{t^n}^{t^{n+1}} f\left(u(x_{j\pm1/2},t)\right)\,dt \qquad (11.40)$$

are the numerical fluxes at the cell edges $x_{j\pm1/2} = x_j \pm \frac{1}{2}\Delta x$. A discretization of the equation of the type (11.38) is called *conservative* to underline the fact that, upon carrying out the summation of equations (11.38) over all the cells, the numerical fluxes at interior cell boundaries cancel in pairs, leaving only the fluxes at the two ends of the computational domain. Similarly to the continuous case, the use of a conservative discretization is of special importance as a safeguard against an incorrect propagation velocity of discontinuous solutions[1].

In order to derive an actual scheme, it is necessary to approximate the

---

[1] The Lax–Wendroff theorem states essentially that, upon grid refinement, the finite difference solution generated by a conservative and consistent scheme converges to a weak solution of the original conservation law. Hou and LeFloch (1994) show explicitly that non-conservative schemes converge to a modified conservation law containing a source term concentrated at the discontinuities.

fluxes (11.40). *Two-level methods*, i.e. methods involving only time levels t^n and t^{n+1}, are particularly attractive as the use of more than two time levels requires more storage (a particular concern for systems of equations in several space dimensions) and special start-up procedures. Furthermore, computational efficiency is enhanced by the use of an explicit time discretization, in which $h_{j\pm1/2}$ is approximated using only values of $u(x, t^n)$. A general expression for $h^n_{j+1/2}$ possessing these requisites has the form

$$h^n_{j+1/2} = h_{j+1/2}(u^n_{j-p}, u^n_{j-p+1}, \ldots, u^n_{j+q}). \tag{11.41}$$

It can be shown that *consistency* of the discretization (i.e. the requirement that (11.38) reduce to (11.32) as $\Delta t, \Delta x \to 0$) requires that $h_{j+1/2}$ $(u, u, \ldots, u) = f(u)$.

A corresponding fully implicit scheme is obtained by replacing the time level n by $n+1$ in (11.41), while a partially implicit scheme would be found by using a weighted average of h^n and h^{n+1}. Most of the considerations that follow are also applicable to such discretizations which, accordingly, will not be treated explicitly. As usual, while implicitness mitigates or eliminates time-stepping limitations, this advantage is counter-balanced by the difficulties stemming from the need to solve coupled equations.

As applied to (11.32), a first-order upwind discretization might lead to

$$h^n_{j+1/2} = \begin{cases} f^n_j & \text{if } f'(u^n_j) \geq 0 \\ f^n_{j+1} & \text{if } f'(u^n_j) < 0 \end{cases} \tag{11.42}$$

or, alternatively, to

$$h^n_{j+1/2} = \begin{cases} f^n_j & \text{if } a^n_{j+1/2} \geq 0 \\ f^n_{j+1} & \text{if } a^n_{j+1/2} < 0 \end{cases} \tag{11.43}$$

where

$$a^n_{j+1/2} = \begin{cases} (f^n_{j+1} - f^n_j)/(u^n_{j+1} - u^n_j) & \text{if } u^n_{j+1} - u^n_j \neq 0 \\ f'(u^n_j) & \text{if } u^n_{j+1} - u^n_j = 0. \end{cases} \tag{11.44}$$

It will be recognized that $a^n_{j+1/2}$ is the exact speed of propagation (11.35) of discontinuous solutions of equation (11.32) (first line), or of continuous solutions with a discontinuous first derivative (second line). The difference between the two schemes is that the second one uses a finite difference approximation to f'. This apparently minor difference, however, causes the first scheme to be non-conservative, while the second one is. If $u^n_{j+1} \neq u^n_j$, a more compact expression of (11.43) is

$$h_{j+1/2}^n = \frac{1}{2}\left(f_j^n + f_{j+1}^n\right) - \frac{1}{2}\operatorname{sgn}[a_{j+1/2}^n]\left(f_{j+1}^n - f_j^n\right). \tag{11.45}$$

The second term represents the noncentered part of the numerical flux and contributes to the stability of the scheme. This point is apparent by considering, for example, a case with $\operatorname{sgn}[a_{j+1/2}^n] > 0$. Upon substitution of (11.45) into (11.38), we find

$$\frac{u_j^{n+1} - u_j^n}{\Delta t} + \frac{f_{j+1}^n - f_{j-1}^n}{2\Delta x} = \frac{\Delta x}{2}\frac{f_{j+1}^n - 2f_j^n + f_{j-1}^n}{\Delta x^2}. \tag{11.46}$$

The flux in the left-hand side arise from the first term of (11.45) and approximates the $\partial f/\partial x$ in the original equation. The terms in the right-hand side arise from the second term of (11.45) and approximate $D\,\partial^2 f/\partial x^2$ with $D = \frac{1}{2}\Delta x$. The diffusive nature of this term provides the stabilization of the method. When $f'(u_j^n) < 0 < f'(u_{j+1}^n)$, the solution produced by the upwind scheme is not the correct weak solution of the original equation, which will be shown below in equation (11.49). This problem obviously does not arise in the linear case $f \propto u$.

The diffusive nature of first-order schemes suggests that increasing the order of accuracy would be beneficial. A well-known example of a second-order scheme is the Lax–Wendroff scheme, one version of which is

$$h_{j+1/2}^n = \frac{1}{2}\left(f_{j+1}^n + f_j^n\right) - \frac{1}{2}\frac{\Delta t}{\Delta x}f_{j+1/2}'^n\left(f_{j+1}^n - f_j^n\right), \tag{11.47}$$

where $f_{j+1/2}'^n = f'\left(\frac{1}{2}(u_{j+1}^n + u_j^n)\right)$. As already noted in Sections 3.2 and 10.2, the problem with this and other higher order schemes is the appearance of spurious oscillations in the solution near discontinuities or strong gradients. Even aside from accuracy considerations, these oscillations present special difficulties in a multiphase flow context as, for example, they may cause the volume fractions to take on values outside the range $[0, 1]$, densities to become negative, or variables used in correlations to take values outside the domain of validity of the correlation itself. The issue is very significant as the oscillations are a feature inherent to any higher order explicit scheme as will be seen shortly.

Whether a scheme will generate oscillations or not can be ascertained by considering the time evolution of the *total variation* of the numerical solution. The total variation of a mesh function u_j is defined by

$$TV(\mathbf{u}) = \sum_{j=-\infty}^{\infty} |u_{j+1} - u_j|. \tag{11.48}$$

A method is said to be *total variation diminishing* (TVD) if $TV(u^{n+1}) \leq TV(u^n)$. It is evident that the appearance of a local maximum or minimum at time t^{n+1} in a mesh function which did not have it at time t^n corresponds to an increase in the total variation. Local maxima or minima also imply a loss of monotonicity and, therefore, TVD methods are also *monotonicity preserving* in the sense that, if u_j^n is monotonic, so is u_j^{n+1}. It can be proven that the exact solution of a scalar conservation relation (11.38) has the property that[1] $TV(u(\cdot, t_2)) \leq TV(u(\cdot, t_1))$ for $t_1 \leq t_2$.

The difficulty is that a monotonicity preserving two-level conservative method can only be at most first-order accurate in space. Furthermore, a *linear* TVD scheme is also necessarily only first-order accurate. A high-order TVD scheme, therefore, must be nonlinear even in the case of a linear equation[2]. This nonlinearity may be interpreted as an artificial viscosity which adapts itself point by point to the nature of the data. While this intuitive notion is appealing, in order to derive these methods it is better to impose the non-oscillatory requirement directly, as the resulting nonlinear viscosity coefficient may be complicated and not particularly intuitive.

11.3.1.2 Godunov's method

In the development of satisfactory numerical methods for systems of conservation laws, the re-interpretation of a relation such as (11.38) offered by Godunov plays a central role. Godunov observed that, instead of thinking of u_j^n as an approximation to $u(x, t^n)$, a scheme such as (11.38) is compatible with *replacing* $u(x, t^n)$ by a piecewise constant function $\overline{u}(x, t^n)$ in such a way that, in each cell $x_{j-1/2} < x < x_{j+1/2}$, $\overline{u}(x, t^n) = u(x_j, t^n)$. In this view, the piecewise function is considered as the initial condition for the next time step and evolved according to the original equation (11.32) up to $t = t^{n+1}$. The average of $\overline{u}(x, t^{n+1})$ over the intervals $x_{j-1/2} < x < x_{j+1/2}$ is then evaluated, used as the initial condition for the next time step, and so on. In practice, the evaluation of the average is not needed as the integrated form (11.38) of the conservation law (11.32) already gives the required average provided the fluxes are suitably approximated. This approach has

---

[1] For a function $v(x)$ the total variation is defined as

$$TV(v) = \lim_{\epsilon \to 0} \sup \frac{1}{\epsilon} \int_{-\infty}^{\infty} |v(x) - v(x - \epsilon)| \, dx$$

which, for a differentiable function, reduces to the integral of $|v'(x)|$.

[2] These negative features of TVD schemes have motivated the introduction of *essentially non-oscillatory schemes* (ENO) which satisfy the weaker requirement that $TV(u^{n+1}) \leq TV(u^n) + O(\Delta x^p)$ for some p.

the great virtue of permitting us to embed characteristic information into a conservative method, as will be clear from the following.

The solution of an equation such as (11.32) when the data have the form $u = u_L$ for $x < x_0$ and $u = u_R$ for $x > x_0$, with $u_{L,R}$ constants, is what is known as a *Riemann problem*. If the time step is such that the influence of each discontinuity remains confined within a cell, which in practice results in the CFL condition[1] $\Delta x < \frac{1}{2}|f'(u_k^n)|\Delta t$, Godunov's approach consists therefore in replacing the original discretized problem by a set of Riemann problems.

The solution of the Riemann problem $u = \overline{u}_j^n$ for $x < x_{j+1/2}$, $u = \overline{u}_{j+1}^n$ for $x > x_{j+1/2}$ is either a shock or an expansion fan but, in either case, $u(x_{j+1/2}, t)$ is a constant as long as information from the neighboring discontinuities does not reach $x_{j+1/2}$. The fundamental reason for this result is that the initial data of the Riemann problem have no intrinsic time or length scale, so that the solution can only depend on the similarity variable $(x - x_{j+1/2})/(t - t^n)$, introduced in equation (11.37), which equals 0 when $x = x_{j+1/2}$. This constant value is given by

$$\overline{u}(x_{j+1/2}, t) = \begin{cases} u_j^n & \text{for } a_{j+1/2}^n \geq 0 \\ u_{j+1}^n & \text{for } a_{j+1/2}^n < 0 \\ u_s^n & \text{if } f'(u_j^n) < 0 < f'(u_{j+1}^n) \end{cases} \qquad (11.49)$$

where $a_{j+1/2}^n$ was defined in (11.44) and u_s, the *sonic point*[2], is defined by $f'(u_s) = 0$. With this result, the numerical flux $h_{j+1/2}^n$ can be computed exactly from its definition (11.40) to find

$$h_{j+1/2}^n = f\left(\overline{u}(x_{j+1/2}, t)\right) = \begin{cases} f_j^n & \text{if } a_{j+1/2}^n \geq 0 \\ f_{j+1}^n & \text{if } a_{j+1/2}^n < 0 \\ f(u_s^n) & \text{if } f'(u_j^n) < 0 < f'(u_{j+1}^n). \end{cases} \qquad (11.50)$$

The first two lines in (11.49) and (11.50), which are just the upwind scheme (11.43), correspond to shocks or expansion fans that remain, respectively, to the right or to the left of the line $x = x_{j+1/2}$ for $t^n < t < t^{n+1}$. The last line corresponds to an expansion fan which straddles the line $x = x_{j+1/2}$ and embodies an important refinement of the upwind scheme in that it guarantees that, as the discretization is made finer, the solution of the discretized

---

[1] Here, as with all Godunov schemes, if the condition is relaxed to $\Delta x < |f'(u_k^n)|\Delta t$, the waves will interact with waves from neighboring Riemann problems, but the interaction will remain contained within the cell and the procedure still works.

[2] In general, a sonic point is a point where the eigenvalues of the Jacobian matrix of the fluxes (which in this case simply reduces to f') change sign. In gas dynamics, two of the eigenvalues are $u \pm c$ and, therefore, the sonic point occurs where the local flow velocity equals the speed of sound, which justifies the denomination.

equation approximates the physically meaningful entropy solution of the
original equation.

11.3.2 Systems of equations

To extend the previous considerations, we start from a linear system of the
form

$$\frac{\partial \mathbf{u}}{\partial t} + \mathsf{A}\frac{\partial \mathbf{u}}{\partial x} = 0, \qquad \mathbf{u}(x,0) = \mathbf{u}^0(x), \qquad (11.51)$$

where now $\mathbf{u}$ is an N-vector of unknown functions $u_K(x,t)$, $K = 1, 2, \ldots,$
N, and A a constant $N \times N$ matrix. We assume that A has real eigenvalues, so that the system is hyperbolic, and a full complement of linearly
independent eigenvectors $\mathbf{e}_1, \mathbf{e}_2, \ldots, \mathbf{e}_N$. In this case, the solution of the
Riemann problem consists of shocks and expansion fans equal in number to
the number of characteristics of the system (11.51).

Once again we wish to define numerical fluxes $\mathbf{h}^n_{j\pm1/2}$ in terms of which
the solution can be updated according to the analog of (11.38) as:

$$\mathbf{u}^{n+1}_j = \mathbf{u}^n_j - \frac{\Delta t}{\Delta x}\left(\mathbf{h}^n_{j+1/2} - \mathbf{h}^n_{j-1/2}\right). \qquad (11.52)$$

It is well known (and can readily be checked directly) that the matrix R
having as columns the eigenvectors:

$$\mathsf{R} = [\mathbf{e}_1 \ \mathbf{e}_2 \ \cdots \ \mathbf{e}_N] \qquad (11.53)$$

diagonalizes A:

$$\mathsf{R}^{-1}\mathsf{A}\mathsf{R} = \Lambda \qquad (11.54)$$

where $\Lambda = \mathrm{diag}(\lambda_1, \lambda_2, \ldots, \lambda_N)$ is a diagonal matrix having as elements the
eigenvalues of A. If the eigenvectors have unit length, as we assume, R is
orthogonal, i.e. R^{-1} equals the transpose R^{T} of R. Upon setting $\mathbf{u} = \mathsf{R}\mathbf{v}$ and
multiplying from the left by R^{-1}, the original linear system (11.51) becomes

$$\frac{\partial \mathbf{v}}{\partial t} + \Lambda\frac{\partial \mathbf{v}}{\partial x} = 0 \qquad (11.55)$$

which in reality consists of N uncoupled equations of the form

$$\frac{\partial v_K}{\partial t} + \lambda_K\frac{\partial v_K}{\partial x} = 0. \qquad (11.56)$$

Each one of these equations is a linear scalar conservation law the exact
solution of which, according to (11.34), is given by $v_K(x,t) = v^0_K(x - \lambda_K t)$,

where $v_K^0(x) = \left(\mathsf{R}^{-1}\mathbf{u}^0\right)_K$ is the initial condition. In the special case of a Riemann problem with $\mathbf{u}^0 = \mathbf{u}_L$ for $x < 0$ and $\mathbf{u}^0 = \mathbf{u}_R$ for $x > 0$, we may write

$$\mathbf{u}^0 = \mathbf{u}_L + H(x)\left(\mathbf{u}_R - \mathbf{u}_L\right) = \mathbf{u}_R - \left[1 - H(x)\right]\left(\mathbf{u}_R - \mathbf{u}_L\right), \qquad (11.57)$$

so that, if

$$\mathbf{u}_R - \mathbf{u}_L = \sum_{K=1}^{N} \beta_K \mathbf{e}_K, \qquad (11.58)$$

the solution is

$$\mathbf{u}(x,t) = \mathbf{u}_L + \sum_{K=1}^{N} \beta_K H(x - \lambda_K t)\mathbf{e}_K = \mathbf{u}_R - \sum_{K=1}^{N} \beta_K[1 - H(x - \lambda_K t)]\,\mathbf{e}_K.$$

$$(11.59)$$

Note that, due to the presence of the Heaviside functions H, the number of nonzero terms in the summations is usually less than N and, in fact, the combined number of nonzero terms in the two summations equals N. This feature becomes particularly clear upon evaluating this expression at $x = 0$, where we find

$$\mathbf{u}(0,t) = \mathbf{u}_L + \sum_{\lambda_K < 0} \beta_K\,\mathbf{e}_K = \mathbf{u}_R - \sum_{\lambda_K > 0} \beta_K\,\mathbf{e}_K. \qquad (11.60)$$

Similarly, for each one of the Riemann problems of Godunov's method, as long as waves from different discontinuities do not interact, for $t > t^n$ we may write

$$\mathbf{u}(x_{j+1/2}, t) = \mathbf{u}_j^n + \sum_{\lambda_K < 0} \beta_K\,\mathbf{e}_K = \mathbf{u}_{j+1}^n - \sum_{\lambda_K > 0} \beta_K\,\mathbf{e}_K \qquad (11.61)$$

where, as in (11.58), the β_K are given by

$$\mathbf{u}_{j+1}^n - \mathbf{u}_j^n = \sum_{K=1}^{N} \beta_K \mathbf{e}_K, \qquad (11.62)$$

or, more compactly, with $\boldsymbol{\beta}^{\mathrm{T}} = (\beta_1, \beta_2, \ldots, \beta_N)$, $\boldsymbol{\beta} = \mathsf{R}^{-1}\left(\mathbf{u}_{j+1}^n - \mathbf{u}_j^n\right)$. Note that, in this case, the CFL stability condition means $(\Delta t/\Delta x)\max|\lambda_K| \leq 1$.

On the basis of these results, we can construct the numerical fluxes to be substituted into (11.52). For this purpose, we decompose the diagonal matrix of eigenvalues as $\Lambda = \Lambda_+ + \Lambda_-$, where Λ_+ is a diagonal matrix formed with the nonnegative eigenvalues of A (in their proper position), and Λ_- the analogous matrix with the negative eigenvalues, and introduce the matrices

$$\mathsf{A}_\pm = \mathsf{R}\Lambda_\pm\mathsf{R}^{-1}. \qquad (11.63)$$

Then, upon substitution of the first form of (11.61) into the definition (11.40) of $h_{j+1/2}$ suitably generalized to the vector case, we have

$$h_{j+1/2}^n = A u_j^n + A \sum_{\lambda_K < 0} \beta_K e_K = A u_j^n + A_- \sum_{K=1}^N \beta_K e_K = A_+ u_j^n + A_- u_{j+1}^n,$$

(11.64)

where the last step follows from (11.62). In a similar fashion, starting from the second form of (11.61), we find

$$h_{j-1/2}^n = A_+ u_{j-1}^n + A_- u_j^n.$$

(11.65)

Written in this way, we see that the numerical flux has been split to separately account for the effect of left- and right-going waves. This *flux-vector splitting* (FVS) approach is very fruitful as will be seen later. An alternative form of the fluxes analogous to (11.45) can be found in terms of the matrix

$$|A| = A_+ - A_- = R|\Lambda|R^{-1},$$

(11.66)

namely

$$h_{j\pm1/2}^n = \frac{1}{2}A\left(u_{j\pm1}^n + u_j^n\right) \mp \frac{1}{2}|A|\left(u_{j\pm1}^n - u_j^n\right)$$

(11.67)

which may be seen as a combination of a centered term plus a stabilizing term as in (11.45). When (11.64) and (11.65) are substituted into (11.52), the solution at the new time level is found as

$$u_j^{n+1} = u_j^n - \frac{\Delta t}{\Delta x}\left[A_-(u_{j+1}^n - u_j^n) + A_+(u_j^n - u_{j-1}^n)\right].$$

(11.68)

Because of the linearity of the problem, this result is precisely the same one given by the Godunov scheme and by the upwind algorithm applied to each one of the scalar equations (11.56):

$$(h_K)_{j+1/2}^n = \begin{cases} \lambda_K(v_K)_j^n & \text{if } \lambda_K \geq 0 \\ \lambda_K(v_K)_{j+1}^n & \text{if } \lambda_K < 0. \end{cases}$$

(11.69)

Equation (11.68) may also be written as

$$u_j^{n+1} = u_j^n - \frac{\Delta t}{2\Delta x}A\left(u_{j+1}^n - u_{j-1}^n\right) + \frac{\Delta t}{2\Delta x}|A|\left(u_{j+1} - 2u_j + u_{j-1}\right).$$ (11.70)

The last term is a finite difference approximation to $\partial^2 \mathbf{u}/\partial x^2$, and explicitly shows the dissipative nature of the algorithm.

Upon comparison with (11.52), it is seen that, in equation (11.68), the flux difference $\mathbf{h}_{j+1/2} - \mathbf{h}_{j-1/2}$ has been split into right- and a left-going components. For this reason, methods of this type are called *flux-difference splitting* (FDS) methods. In the present linear context, both the FDS and FVS approaches give the same result, but this is not so in the presence of nonlinearities.

In the nonlinear case, the system to be solved has the form

$$\frac{\partial \mathbf{u}}{\partial t} + \frac{\partial \mathbf{f}(\mathbf{u})}{\partial x} = 0. \tag{11.71}$$

Just as in the scalar case, in regions of smoothness, the solution of this system coincides with that of

$$\frac{\partial \mathbf{u}}{\partial t} + \mathsf{A}(\mathbf{u})\frac{\partial \mathbf{u}}{\partial x} = 0 \tag{11.72}$$

where A is the Jacobian matrix with elements

$$\mathsf{A}_{ij} = \frac{\partial f_i}{\partial u_j}. \tag{11.73}$$

As we know, the system (11.71) is hyperbolic if this matrix possesses N distinct real eigenvalues.

An obvious approach to the solution of the nonlinear problem (11.72) is to adapt the previous method to a local linearization in which equation (11.72) is written as

$$\frac{\partial \mathbf{u}}{\partial t} + \mathsf{A}\left(\frac{1}{2}(\mathbf{u}_j^n + \mathbf{u}_{j+1}^n)\right)\frac{\partial \mathbf{u}}{\partial x} = 0. \tag{11.74}$$

With this step, the previous ideas carry over directly and so do the methods to achieve higher order spatial accuracy to be described later in Section 11.3.3. While simple, this approach is not free of drawbacks. In the first place, the repeated calculation of the Jacobian can incur a significant computational burden. Secondly, in a multiphase flow application, the flux $\mathbf{f}$ may be known only in terms of correlations with different domains of validity in neighboring parameter regions and the partial derivatives (11.73) may not exist. Finally, in many applications, discontinuous solutions are involved (e.g. the propagation of sharp mass fronts along a pipeline), and the form (11.72) would not guarantee the correct propagation velocity of these solutions. The flux-difference splitting method of Roe is an attempt to address these problems.

11.3.2.1 Roe's approximate Riemann solver

While for a scalar equation the application of Godunov's method is relatively straightforward as the problem of finding $\bar{u}(x_{j\pm1/2}, t)$ can be solved exactly, this is in general not possible for nonlinear systems.[1] To circumvent this difficulty, Roe observed that, since ultimately the details of the Riemann problem solution are lost as only the spatial averages of u^{n+1} are involved in the method, the use of a suitable approximate solution of the Riemann problem in place of the exact one might simplify the procedure without adversely affecting the accuracy of the final result.

On the basis of this remark, Roe's method relies on finding a local approximate solution of an equation of the form (11.72), but with a modified matrix $\hat{\mathsf{A}}$:

$$\frac{\partial \hat{u}}{\partial t} + \hat{\mathsf{A}} \frac{\partial \hat{u}}{\partial x} = 0. \tag{11.75}$$

Roe requires the matrix $\hat{\mathsf{A}}$ to satisfy the following conditions:

(i) $\hat{\mathsf{A}}(\mathbf{u}_L, \mathbf{u}_R) \to \mathsf{A}(\mathbf{u})$ (with A the Jacobian) smoothly as $\mathbf{u}_L, \mathbf{u}_R \to \mathbf{u}$;
(ii) $\hat{\mathsf{A}}$ is diagonalizable with real eigenvalues and linearly independent eigenvectors;
(iii)

$$\hat{\mathsf{A}}(\mathbf{u}_L, \mathbf{u}_R) (\mathbf{u}_R - \mathbf{u}_L) = \mathbf{f}(\mathbf{u}_R) - \mathbf{f}(\mathbf{u}_L), \tag{11.76}$$

where, as before, $\mathbf{u}_L = \mathbf{u}_j$ and $\mathbf{u}_R = \mathbf{u}_{j+1}$. Condition (i) is evidently necessary for consistency of the approximate system (11.75) with (11.72). Condition (ii) ensures that the approximate system is hyperbolic as the original one. Condition (iii) is crucial for several reasons. In the first place, it is easy to see that it ensures that the discretized form of (11.75) is conservative. Secondly, it is known from the theory of quasilinear hyperbolic systems that a shock or a contact discontinuity propagating with speed s satisfies the *Rankine–Hugoniot* condition analogous to (11.35):

$$\mathbf{f}(\mathbf{u}_R) - \mathbf{f}(\mathbf{u}_L) = s (\mathbf{u}_R - \mathbf{u}_L). \tag{11.77}$$

Condition (11.76) then implies that, in this case, $(\mathbf{u}_R - \mathbf{u}_L)$ is an eigenvector of $\hat{\mathsf{A}}$ with eigenvalue s and so the approximate solution $\hat{\mathbf{u}}$ of (11.75) also consists of this single jump propagating with the correct speed s and therefore is, in fact, exact. The Roe matrix may be considered as the result of the linearization of the exact Jacobian around some state intermediate

---

[1] In multiphase flow, explicit solutions of the Riemann problem are rare. Tang and Huang (1996) present analytical results for a liquid–gas system treated according to the homogeneous flow model.

between $\mathbf{u}_L$ and $\mathbf{u}_R$ and, for this reason, it is often referred to as Roe's average matrix.

For the special case of a single scalar equation, (11.75) is

$$\frac{\partial \hat{u}}{\partial t} + \hat{a}_{j+1/2}\frac{\partial \hat{u}}{\partial x} = 0 \tag{11.78}$$

and (11.76) gives

$$\hat{a}_{j+1/2} = \hat{a}(u_{j+1}, u_j) = \frac{f(u_{j+1}) - f(u_j)}{u_{j+1} - u_j}. \tag{11.79}$$

Equation (11.78) describes a simple non-dispersive wave propagating with the constant velocity $\hat{a}$ and, therefore, its solution is equivalent to the upwind scheme. Since the equation is linearized, however, possible sonic points have been lost and the case shown in the last line of (11.49) would be missed. Therefore, Roe's scheme needs an appropriate *entropy fix* near extrema of the flux, and the same fix is necessary when the method is applied to nonlinear systems (see, e.g. Engquist and Osher, 1980; LeVeque, 2002).

By following a development similar to that leading to (11.67) for the linear case, it can be shown that the numerical flux associated with the Roe-linearized equation (11.75) is given by

$$\mathbf{h}_{j+1/2} = \frac{1}{2}\left(\mathbf{f}_j^n + \mathbf{f}_{j+1}^n\right) - \frac{1}{2}|\hat{\mathbf{A}}|\left(\mathbf{u}_{j+1}^n - \mathbf{u}_j^n\right) \tag{11.80}$$

where $|\hat{\mathbf{A}}|$ is constructed in the same way as $|\mathbf{A}|$ in (11.66). The condition (11.76) plays an important role in the proof of this result. The expression (11.80) is a nonlinear generalization of (11.67) expressing the flux in the flux-difference splitting form.

The construction of the Roe matrix is in general nontrivial. Generally speaking, condition (11.76) may be regarded as a nonlinear algebraic system for the elements of $\hat{\mathbf{A}}$. However, since it consists of N equations for the determination of N^2 elements, its solution is not unique and it is unclear how to formulate universal criteria for the selection of a physically relevant A. Roe matrices have been explicitly constructed for several important cases such as gas dynamics and the shallow water equations. Entirely numerical procedures can also be devised (Romate, 1998; Fjelde and Karlsen, 2002), but they require extensive matrix operations.

One may look at equation (11.80) as the combination of a centered scheme and a stabilizing term just as (11.67), and one may therefore try to use different forms of the stabilizing term (see, e.g. Harten *et al.*, 1983). Obvious alternatives to the Roe average matrix are then

$$\hat{A}(\mathbf{u}_L, \mathbf{u}_R) = A\left(\frac{1}{2}(\mathbf{u}_L + \mathbf{u}_R)\right), \quad \text{or} \quad \hat{A}(\mathbf{u}_L, \mathbf{u}_R) = \frac{1}{2}[A(\mathbf{u}_j) + A(\mathbf{u}_{j+1})].$$

$$(11.81)$$

The first one is reminiscent of (11.74) and may be regarded as resulting from the linearization of the exact Jacobian (11.73) around $\frac{1}{2}(\mathbf{u}_L + \mathbf{u}_R)$ and is an example of the use of an *approximate state* Riemann solver (see, e.g. Toro, 1999). While such choices may result in good solutions in some cases, the matrices thus defined will satisfy (11.76) only in rare circumstances. Faille and Heintzé (1999) experimented with the second form in a drift-flux code describing flow in a pipeline and found that the dissipation induced by it was insufficient for stability. Another problem with approximate-state Riemann solvers in multiphase flow applications is that they may lead to a loss of positivity of the volume fractions (Romate, 2000).

11.3.2.2 Sources

So far we have dealt with homogeneous equations. However, a glance at the typical multiphase models shown in Section 11.1 is sufficient to see that in practice it is necessary to deal with source terms accounting, e.g., for drag and phase change. We are thus led to consider generalizations of (11.71) of the form

$$\frac{\partial \mathbf{u}}{\partial t} + \frac{\partial \mathbf{f}(\mathbf{u})}{\partial x} = \mathbf{s}(\mathbf{u}), \tag{11.82}$$

where $\mathbf{s}$ in general depends on $\mathbf{u}$ and also, possibly, on space and time. A standard way to deal with equations of this type is the use of *splitting*, also referred to as the method of *fractional time steps*. Let us consider the two problems

$$\frac{\partial \mathbf{u}}{\partial t} = \mathbf{s}(\mathbf{u}), \tag{11.83}$$

and

$$\frac{\partial \mathbf{u}}{\partial t} = -\frac{\partial \mathbf{f}(\mathbf{u})}{\partial x}. \tag{11.84}$$

If $T_1(\Delta t)$ denotes the solution operator for (11.83) which advances the solution of the equation over a time step Δt and $X_1(\Delta t)$ the analogous solution operator for (11.84), both having an error of order Δt^2, it is easy to see that

$$\mathbf{u}^{n+1} = X_1(\Delta t)\, T_1(\Delta t)\, \mathbf{u}^n + O(\Delta t)^2. \tag{11.85}$$

The validity of this relation is most easily appreciated in the linear case, where $\mathbf{f}(\mathbf{u}) = \mathbf{A} \cdot \mathbf{u}$, $\mathbf{s}(\mathbf{u}) = \mathbf{S} \cdot \mathbf{u}$. If, for example, one were to use an explicit method, then the solution of (11.82) would be

$$\mathbf{u}^{n+1} = \left(\hat{\mathbf{I}} - \frac{\Delta t}{\Delta x} \mathbf{A}\,\delta + \Delta t\,\mathbf{S} \right) \cdot \mathbf{u}^n + O(\Delta t)^2, \qquad (11.86)$$

where δ denotes a suitable finite difference operator. The solutions of (11.83) and (11.84) would be

$$\mathbf{u}^{n+1} = \left(\hat{\mathbf{I}} - \frac{\Delta t}{\Delta x} \mathbf{A}\,\delta \right) \cdot \mathbf{u}^n + O(\Delta t)^2 \quad \text{and} \quad \mathbf{u}^{n+1} = \left(\hat{\mathbf{I}} + \Delta t\,\mathbf{S} \right) \cdot \mathbf{u}^n + O(\Delta t)^2,$$
$$(11.87)$$

respectively, so that $\mathsf{X}_1 = \hat{\mathbf{I}} - (\Delta t/\Delta x)\mathbf{A}\,\delta$, $\mathsf{T}_1 = \hat{\mathbf{I}} + \Delta t\,\mathsf{S}$, and $\mathsf{X}_1\,\mathsf{T}_1\mathbf{u}^n$ reproduces (11.86) to order $(\Delta t)^2$.

Equation (11.85) shows that, to order Δt^2, the solution of (11.82) can be broken down into the sequence of steps

$$\mathbf{u}^{n+1/2} = \mathsf{T}_1(\Delta t)\,\mathbf{u}^n, \qquad \mathbf{u}^{n+1} = \mathsf{X}_1(\Delta t)\,\mathbf{u}^{n+1/2}. \qquad (11.88)$$

Thus, equation (11.83) is first solved over a time step Δt, and the result is used as initial condition for the solution of (11.84). Clearly, the final result is independent of the order in which the two steps are executed and in fact, in practice, it may be advisable to switch the order at every time step to avoid possible bias. Here we have followed the customary notation with fractional time exponents even though each step involves the full Δt.

This simple idea is used to great advantage, e.g., in the SIMMER code for the simulation of thermohydraulic processes in fast nuclear reactors. Typical scenarios relevant to these systems involve three phases of many different materials with an attendant great complexity of the models and a large number of equations. In order to limit the computational effort, the interphase mass and energy exchanges are solved implicitly in a first fractional step which treats each cell in isolation from the others. Then, in a second step, the interphase exchange rates are fixed and a propagation/convection problem is solved. It is evident that, depending on the problem, the lack of feedback between the effects of transport and sources may lead to inaccuracies unless a small time step is used (Stewart and Wendroff, 1984). A more recent application, in conjunction with the AUSM method described later in Section 11.5, is described by Niu (2001).

In general, if one were to follow the same procedure increasing the accuracy of T_1 and X_1 to second order, the resulting accuracy would not be $O(\Delta t)^2$ unless the two operators commute. Indeed, this procedure in general introduces a *splitting error*[1] of order $(\Delta t)^2$. Strang (1968; see also Toro 1999) showed that, if $T_2(\Delta t)$ and $X_2(\Delta t)$ are the solution operators for (11.83) and (11.84), both to second-order accuracy, then

$$\mathbf{u}^{n+1} = T_2\left(\frac{1}{2}\Delta t\right) X_2(\Delta t) T_2\left(\frac{1}{2}\Delta t\right) \mathbf{u}^n + o(\Delta t)^2$$

$$= X_2\left(\frac{1}{2}\Delta t\right) T_2(\Delta t) X_2\left(\frac{1}{2}\Delta t\right) \mathbf{u}^n + o(\Delta t)^2. \qquad (11.89)$$

In practice, to apply, for example, the first one of these relations, one would solve (11.83) over $t^n \le t \le t^n + \frac{1}{2}\Delta t$ using $\mathbf{u}^n$ as initial condition, then solve (11.84) over $t^n \le t \le t^{n+1}$ using as initial condition the result of the previous step, and then again solve (11.83) over $t^n + \frac{1}{2}\Delta t \le t \le t^{n+1}$ using the result of the second step as initial condition. It may be useful to adopt a similar approach also when $\mathbf{s}$ is not a true source term, but contains derivatives. For example, a viscous effect involving $\partial^2 u/\partial x^2$ can be treated by the second-order accurate Crank–Nicholson algorithm (see, e.g. LeVeque, 2002).

The advantage of these procedures is that the solution methods used for (11.83) and (11.84) can be optimized in dependence of the specific nature of these equations. For example, it is not unusual to encounter source terms with a much shorter intrinsic time scale than the flux terms. In this case, the solution of (11.83) can be advanced over the interval $\frac{1}{2}\Delta t$ by taking many time steps much smaller than Δt, possibly with an implicit discretization. One can take advantage of the same flexibility when (11.83) is stiff. A disadvantage, on the other hand, is that the method neglects the coupling between the two processes. For example, in approaching the steady solution of (11.82), the flux and the source terms are both larger than the time derivative and treating them independently of each other may lead to erroneous results (see, e.g. LeVeque, 2002).

11.3.2.3 Higher dimensionality

The schemes developed for one spatial dimension are usually extended to higher-dimensional spaces either by *dimensional splitting* on Cartesian grids or by finite-volume discretization on unstructured grids (see, e.g. Toro, 1999; LeVeque, 2002). The basis of the first approach is the fractional time-step

---

[1] It turns out that, often, the splitting error, while of formal order $(\Delta t)^2$, is actually small so that the same procedure (11.88) can be used with T_1, X_1 replaced by second-order accurate operators T_2, X_2; see LeVeque (2002).

method of the previous section. For example, the equation

$$\frac{\partial \mathbf{u}}{\partial t} + \frac{\partial \mathbf{f}}{\partial x} + \frac{\partial \mathbf{g}}{\partial y} = 0, \tag{11.90}$$

where the fluxes $\mathbf{f}$ and $\mathbf{g}$ depend on $\mathbf{u}$, can be solved to $O(\Delta t)$ by splitting as

$$\mathbf{u}^{n+1/2} = \mathsf{X}_1(\Delta t)\,\mathbf{u}^n, \qquad \mathbf{u}^{n+1} = \mathsf{Y}_1(\Delta t)\,\mathbf{u}^{n+1/2}, \tag{11.91}$$

where X_1 and Y_1 are first-order accurate solution operators for

$$\frac{\partial \mathbf{u}}{\partial t} + \frac{\partial \mathbf{f}}{\partial x} = 0 \qquad \frac{\partial \mathbf{u}}{\partial t} + \frac{\partial \mathbf{g}}{\partial y} = 0, \tag{11.92}$$

respectively. Each one of these two subproblems is effectively one dimensional and can be dealt with by the methods described before. Clearly, the order of the x- and y-integrations can be interchanged with an error of the same order. The formal order of the procedure can be increased by using Strang splitting.

With a finite-volume discretization, a corresponding effective reduction in dimensionality is obtained by considering the fluxes in the direction of the normal to each face of the control volume. A recent application of this technique to the modern reactor safety code ASTAR is described by Staedke *et al.* (2005).

While these methods are in widespread use, neither one is fully satisfactory as they are both equivalent to solving the Riemann problem in the direction of the grid, rather than the physical direction of the wave propagation. This circumstance may lead to the use of an incorrect local wave structure of the solution which, in some cases, may introduce an appreciable error. Better approaches would rely on the exact or approximate multidimensional solution of the Riemann problem. This is an area of current research; the reader is referred to the books by Toro (1999) and LeVeque (2002) as well as the current literature.

11.3.3 High-resolution schemes

All the methods described so far are first-order accurate in space and therefore affected by numerical diffusion. We have had a first look at this issue in Section 10.2.2 of the previous chapter and we have mentioned the fact that attempts to obtain schemes with a higher-order spatial accuracy are plagued by the appearance of spurious oscillations that cannot be removed if one insists on second- (or higher-)order accuracy everywhere. The best that can be hoped for is an increased accuracy where the solution is smooth, with some degradation near discontinuities. Thus, we will be concerned with schemes where the numerical flux function has the smoothing capability of a low-order scheme where necessary and the accuracy of a high-order scheme where possible. This is another topic with a vast literature, which we make no attempt to cover, limiting our considerations to two important approaches in common use.

11.3.3.1 Flux-limiter methods

Once again, it is useful to start with the case of a single scalar equation. We write the numerical flux in the form

$$h_{j+1/2} = h^L_{j+1/2} + \phi_j \left(h^H_{j+1/2} - h^L_{j+1/2} \right) \tag{11.93}$$

where h^L is a low-accuracy approximation to the flux (e.g. upwind) and h^H a high-accuracy expression. The *limiter* ϕ will be chosen so as to be close to 1 in smooth regions of the solution and close to 0 near steep gradients. Since the low-order flux is typically diffusive, the term $\phi(h^H - h^L)$ is sometimes referred to as the antidiffusive flux. It is evident that, to maintain this character, ϕ must be nonnegative. The formulation in terms of a numerical flux automatically ensures the conservative property of the discretization.

In a typical implementation, the low-order flux is taken as the upwind scheme (11.45) which we rewrite in the slightly modified form

$$h^{Ln}_{j+1/2} = \frac{1}{2} \left(f^n_{j+1} + f^n_j \right) - \frac{1}{2} |a^n_{j+1/2}| \left(u^n_{j+1} - u^n_j \right), \tag{11.94}$$

where $a^n_{j+1/2}$ is defined in (11.44), while the higher order flux would have the Lax–Wendroff form (11.47) which we rewrite as

$$h^{Hn}_{j+1/2} = \frac{1}{2} \left(f^n_{j+1} + f^n_j \right) - \frac{\Delta t}{2\Delta x} \left(a^n_{j+1/2} \right)^2 \left(u^n_{j+1} - u^n_j \right). \tag{11.95}$$

In order to adjust the limiter, we need a measure of the local smoothness of the solution, for which we use the same *smoothness parameter* introduced in the previous chapter and defined by

$$\theta^n_j = \frac{1}{u^n_{j+1} - u^n_j} \begin{cases} u^n_j - u^n_{j-1} & \text{when } a_{j+1/2} > 0 \\ u^n_{j+2} - u^n_{j+1} & \text{when } a_{j+1/2} < 0. \end{cases} \tag{11.96}$$

Clearly, when $a_{j+1/2} = 0$, the definition of ϕ is immaterial. The requirements stemming from the TVD property and other considerations impose constraints on the possible dependence of ϕ upon θ and several forms are in use in the literature. To the common examples already given in Section 10.2.2 of the previous chapter, we may add the Chakravarthy–Osher limiter:

$$\phi(\theta) = \max\{0, \min[\psi, \theta]\}, \tag{11.97}$$

where ψ is a parameter between 1 and 2. Evidently, for $\psi = 1$, this is identical to the Minmod limiter.

In extending the approach to a system of N equations, we set

$$\mathbf{h}^n_{j+1/2} = \left(\mathbf{h}^n_{j+1/2}\right)_{\text{Roe}} + \frac{1}{2}\sum_{K=1}^{N} |\lambda_K|^n_j \left(1 - \frac{\Delta t}{\Delta x}|\lambda_K|^n_j\right)(\phi_K)^n_j(\beta_K)^n_j(\mathbf{e}_K)^n_j \tag{11.98}$$

where $\left(\mathbf{h}^n_{j+1/2}\right)_{\text{Roe}}$ is the Roe flux defined in (11.80), possibly augmented by a suitable entropy fix. The other quantities λ_K, β_K, $\mathbf{e}_K$ are as defined in Section 11.3.2 with the matrix $\mathbf{A}$ taken as a Roe matrix satisfying the properties (i)–(iii) listed above in connection with equation (11.76).

11.3.3.2 Slope-limiter methods

The first-order Godunov scheme reconstructs the solution $u(x,t)$ in terms of a piecewise constant approximation. One may pursue the development of a higher order method by attempting the reconstruction of u in a piecewise linear form, i.e. by letting, in the cell centered at x_j,

$$u(x,t) \simeq u(x_j,t) + \sigma_j(t)\,(x - x_j), \tag{11.99}$$

where σ_j is the slope of the linear approximation (van Leer 1977). As already mentioned in Section 10.2.2 of the previous chapter, this approach is sometimes referred to as MUSCL, for *M*onotone *U*pstream-centered *S*cheme for *C*onservation *L*aws. Slope-limiter methods are similar to Godunov's method in that they use this linear approximation as an initial value with which to construct $u(x_j, t^{n+1})$ and σ_j^{n+1}. Of course, if one were to choose σ_j so as to enforce continuity of u at the cell edges, one would end up with a second-order method with the attendant problem of spurious oscillations. In order to avoid this problem, it is therefore necessary to limit the slopes.

The difficulty in developing this idea lies in the fact that, except for special cases, it is not possible to find an exact solution with data of the form (11.99). A possible way to sidestep the difficulty is to approximate the analytical flux $f(u)$, and then solve the resulting equation. The simplest option is to use a linear approximation, so that f' is a constant and the resulting equation solvable exactly. Again we omit details and simply show a common expression for the flux function for the scalar case:

$$h_{j+1/2}^n = h_{j+1/2}^{Ln} + \frac{1}{2}|a_{j+1/2}^n| \left(1 - \frac{\Delta t}{\Delta x}|a_{j+1/2}^n|\right) \Delta x\, \sigma_{j+\ell}^n \qquad (11.100)$$

where $h_{j+1/2}^{Ln}$ is the upwind flux (11.94) and $\ell = 0$ when $a_{j+1/2}^n > 0$, while $\ell = 1$ when $a_{j+1/2}^n < 0$. A common way to limit the slopes is to use the Minmod limiter already introduced in Section 10.2.2 of the previous chapter, which for the present purposes is conveniently expressed as

$$\sigma_j = \frac{1}{\Delta x} \text{Minmod}\,(u_{j+1} - u_j, u_j - u_{j-1}) \qquad (11.101)$$

where

$$\text{Minmod}\{a, b\} = \begin{cases} a & \text{if } |a| < |b| \text{ and } ab > 0 \\ b & \text{if } |b| < |a| \text{ and } ab > 0 \\ 0 & \text{if } ab < 0. \end{cases} \qquad (11.102)$$

If a and b have the same sign, this function selects the one that is smaller in modulus while, if a and b have different signs, the Minmod function returns zero. Indeed, in this case, u must have a local maximum or minimum and $\sigma_j = 0$ is necessary to maintain the TVD property.

In the vector case, in the notation of (11.98), the flux corresponding to (11.100) is

$$\mathbf{h}_{j+1/2}^n = \left(\mathbf{h}_{j+1/2}^n\right)_{\text{Roe}} + \frac{1}{2}\sum_{K=1}^{N} |\lambda_K|_j^n \left(1 - \frac{\Delta t}{\Delta x}|\lambda_K|_j^n\right) \Delta x(\sigma_K)_{j+\ell_K}^n (\mathbf{e}_K)_j^n$$

$$(11.103)$$

where $\ell_K = 0$ when $(\lambda_K)_j^n > 0$, and $\ell_K = 1$ when $(\lambda_K)_j^n < 0$, with

$$(\sigma_K)_j = \frac{1}{\Delta x} \text{Minmod}\{(\beta_K)_j, (\beta_K)_{j-1}\}. \qquad (11.104)$$

The extension of this technique to more than one space dimension can be obtained following the approach of Dukowicz and Kodis (1987) outlined in Section 10.5.4 of the previous chapter.

11.4 Application to multiphase flow models

The ideas described in the previous section have had a strong echo in the multiphase flow literature. Since all these approaches rely on the wave structure of Riemann problems, which require real eigenvalues of the Jacobian, much of this work has dealt with the homogeneous and drift flux models, which are hyperbolic, or with two-fluid models rendered hyperbolic by the addition of suitable terms. Even so, the analytic calculation of eigenvalues and eigenvectors often proves difficult in the multiphase flow case. Approximate methods based on the assumption of small Mach number have been developed for this purpose (Cortes *et al.*, 1998; Evje and Flåtten, 2003) and used in the development of numerical methods (Cortes, 2002).

A straightforward approach to the solution of a model based on the RELAP5 one shown in Section 11.1 was presented by Tiselj and Petelin (1997, 1998). They compare the results of the original RELAP model and numerics with the second-order solution of a model in which the momentum equations are augmented by an added-mass interaction of the form (Drew *et al.*, 1979)

$$CVM = c_{vm}\alpha(1-\alpha)\left(\frac{\partial u_G}{\partial t} + u_L\frac{\partial u_G}{\partial x} - \frac{\partial u_L}{\partial t} - u_L\frac{\partial u_L}{\partial x}\right) \qquad (11.105)$$

in which c_{vm} is an added-mass coefficient chosen so as to ensure a useful region of hyperbolicity. They write the governing equations in non-conservation form as

$$A\frac{\partial \mathbf{u}}{\partial t} + B\frac{\partial \mathbf{u}}{\partial x} = \mathbf{s} \qquad (11.106)$$

where $\mathbf{u}^T = [p, \alpha, u_L, u_G, e_L, e_G]$ and the matrices A, B and the source vector **s** can be written down from the equations given in Section 11.1.

Time integration is effected with the second-order Strang splitting algorithm (11.89) executing the following steps:

(i) integration of the sources over one half time step

$$\mathbf{u}_j^* = \mathbf{u}_j^n + \int_{t^n}^{t^n+\frac{1}{2}\Delta t} A^{-1}(\mathbf{u}_j^n)\,\mathbf{s}(\mathbf{u}_j(t))\,dt; \qquad (11.107)$$

(ii) convection substep

$$A\frac{\mathbf{u}^{**} - \mathbf{u}_j^*}{\Delta t} + B\frac{\partial \mathbf{u}^*}{\partial x} = 0; \qquad (11.108)$$

(iii) integration of the sources over the second half of the time step

$$\mathbf{u}_j^{n+1} = \mathbf{u}_j^{**} + \int_{t^{**}}^{t^{**}+\frac{1}{2}\Delta t} A^{-1}(\mathbf{u}_j)\,\mathbf{s}(\mathbf{u}_j(t))\,dt. \qquad (11.109)$$

The convection substep is executed by rewriting (11.108) in the form (11.74) as

$$\frac{\mathbf{u}^{**} - \mathbf{u}_j^*}{\Delta t} + C\frac{\partial \mathbf{u}^*}{\partial x} = 0 \qquad (11.110)$$

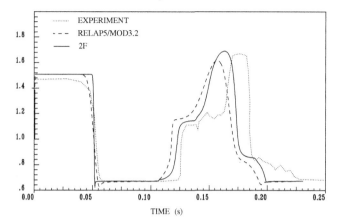

Fig. 11.3. Pressure vs time due to the abrupt closure at $t = 0$ of a valve at the end of a tube issuing from a tank with a free surface. At $t \simeq 0.05$ s the rarefaction wave caused by the reflection of the water hammer pressure at the free surface of the tank causes flashing to occur. The rarefaction wave is reflected as a compression wave and causes the collapse of the flashing region. The first-order RELAP5 results (thick dashes) show a greater dissipation than Tiselj and Petelin's (1998) second-order calculation (solid line). The dotted line is the experimental data.

with $\mathsf{C} = \mathsf{A}^{-1}\mathsf{B}$ and applying the flux-vector splitting approach described after equation (11.74) extended to second-order spatial accuracy with the Minmod flux limiter. The Jacobian C is evaluated at the point $x_{j+1/2}$ by averaging, $\mathsf{C}_{j+1/2} = \mathsf{C}\left(\frac{1}{2}(\mathbf{u}_j + \mathbf{u}_{j+1})\right)$. The source integrations (11.107) and (11.109) are effected with the second-order Euler method. When the mixture is close to equilibrium, the same time step as the convective one can be used for source terms. When the vapor–liquid system is far from equilibrium, however, the characteristic time scale for the sources is much shorter than the convection time scale and the source integration substeps may require a time step orders of magnitude shorter than Δt; a few hundred substeps may be required in such cases.

Figure 11.3 shows an example of the results of this work for the two-phase water hammer process caused by the sudden closure of a valve at the end of a tube issuing from a tank with a free surface. The figure shows the pressure near the valve. As the valve is closed at $t = 0$, the pressure jumps upward by about 5 atms and remains at that level until the expansion wave reflected from the free surface of the tank arrives at $t \simeq 0.05$ s and flashing occurs. In its turn, this expansion wave is reflected as a compression wave from the free-surface of the tank and causes the collapse of the flashing region, with a maximum pressure higher than the original pressurization due to the closure of the valve. A comparison of the RELAP5 results (thick dashes) with the results of this work (solid line) clearly shows the more dissipative nature of the RELAP5 algorithm. Both sets of results exhibit a slight unphysical undershoot around $t = 0.05$s, possibly due to the formulation in terms of non-conservative variables. The dotted line shows experimental results which, however, cannot be invoked to support one or the other calculation due to the rather drastic modeling assumptions.

Given the similarity between the two solutions and the fact that the RELAP5 model is not hyperbolic while the other one is, one may wonder as to the importance in the calculation of the hyperbolicizing term (11.105). In their 1997 paper, the authors compare the predictions obtained with two different hyperbolic models, one rendered hyperbolic with the added-mass term (11.105), and the other one with the interfacial pressure term used in the code CATHARE:

$$P_1 \frac{\partial \alpha}{\partial x}, \quad \text{where} \quad P_1 = \delta_i \frac{\alpha(1-\alpha)\rho_G \rho_L (u_G - u_L)^2}{\alpha \rho_L + (1-\alpha)\rho_G}. \tag{11.111}$$

The results for the pressure and gas velocity vs position in a shock-tube calculation are shown in Fig. 11.4 where the dashed lines show the results obtained with (11.111). The large differences appear surprising until it is realized that, in both these calculations, the source terms were set to zero. The source terms tend to establish mechanical and thermal equilibrium between the phases, so that the two-fluid model approaches a homogeneous flow model, the behavior of which is little affected by the differential terms which distinguish a hyperbolic from a nonhyperbolic formulation or one hyperbolic model from another one. This circumstance may lay at the root of the relative success of nonhyperbolic models.

A further comparison of the differences between a first- and a second-order calculation is afforded by Fig. 11.5, which shows the total energy distribution for a two-phase shock-tube calculation including added-mass effects, but no source terms (from Tiselj and Petelin, 1997). The case illustrated here is somewhat extreme as differences are not as marked at other times or for other variables, but it is nevertheless indicative of the dangers lurking in two-fluid models.

A famous experiment in the multiphase flow literature is the blow-down test of Edwards–O'Brien (1970) consisting in the measurement of pressure, volume fraction, etc. in a horizontal, 4 m long tube initially full of water at 7 MPa and 502 K, one end of which is suddenly opened to the atmosphere at the initial instant. A comparison of the pressure history as calculated by Tiselj and Petelin (1997) with the RELAP5 model (thick dashed line) and with the previous method with one first-order and two second-order

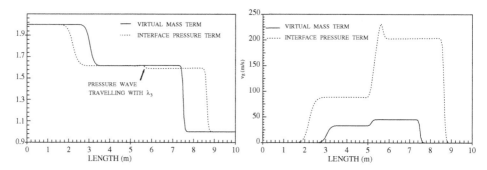

Fig. 11.4. The two-fluid model equations can be rendered hyperbolic in many ways. This figure shows the effect of two different choices, equation (11.105), which introduces added-mass effects (solid lines), and equation (11.111), which models interfacial pressure effects. The left figure shows the pressure and the right figure the gas velocity for a two-phase shock-tube problem (from Tiselj and Petelin, 1997).

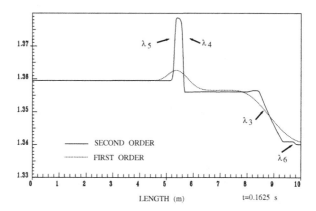

Fig. 11.5. Internal energy vs position for a two-phase shock-tube problem as calculated by a first-order (dotted) and a second-order method with a Superbee limiter (from Tiselj and Petelin, 1997). The model is hyperbolic with six equations and six characteristic speeds. This solution is the result of four waves; the remaining two are already outside the region shown.

formulations is shown in Fig. 11.6. The stronger numerical diffusion of the low-order methods is evident, as is also the relatively large effect of the particular flux limiters – Superbee and Minmod – used in the higher order method. The results confirm the diffusive tendency of the Minmod limiter and the over-compressive nature of the Superbee limiter. It should be noted that the difference with RELAP5 cannot be imputed entirely to the order of the calculation as the models used in the two codes are somewhat different.

   While relatively straightforward, the numerical approach used in these studies illustrates several of the issues encountered in trying to generate

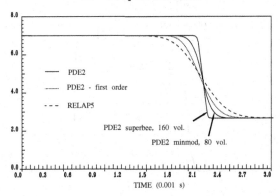

Fig. 11.6. Pressure history for the Edwards–O'Brien blow-down experiment as calculated with the RELAP5 code (thick dashed line), a first-order method, and two second-order methods (from Tiselj and Petelin, 1997).

high-resolution results for average-equations models. In the first place, the models are usually not in conservation form due to the presence of the volume fractions multiplying the pressure gradient in the momentum equations and, possibly, also to the structure of other terms such as the added-mass interaction. This fact hinders the direct application of the ideas described in the previous section. From this point of view, both the homogeneous and the drift flux models, which are or can be put in conservation form, present a clear advantage when they can be used. However, often these models have a complex structure due to the empirical submodels they contain. Sometimes no algebraic expression for the fluxes as functions of the conservative variables is available, which impedes straightforward analytic manipulations.

Secondly, in the model formulation, usually conservative variables, such as $(\alpha\rho)$, $(\alpha\rho u)$, etc. coexist with non-conservative, or primitive, variables, such as α, ρ, u. In particular, empirical or semi-empirical relations are most often expressed in terms of non-conservative variables. One is therefore presented by the dilemma of choosing between a numerical approach based on one or the other set. The task of expressing the primitive variables in terms of conservative ones often leads to nontrivial algebra, complicates the evaluation of the Jacobian, or may even be impossible except by numerical means when, for example, an empirical equation of state is used. The additional complexity and computational burden may be undesirable. The construction of a Roe, or Roe-like, matrix may be simpler in terms of primitive variables (Romate, 1998) but, on the other hand, the equations in primitive form may lead to an incorrect speed of propagation of discontinuities.

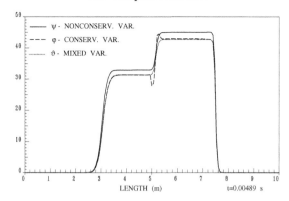

Fig. 11.7. Gas velocity for the shock-tube problem of Fig. 11.4 as computed with conservative (dashed), non-conservative (solid) and mixed variables (from Tiselj and Petelin, 1997). For the mixed variables calculation, the continuity and energy equations were written in terms of conservative variables, while the momentum equations were formulated in terms of the velocities.

Furthermore, the choice of which variables to use may have an impact on the appearance of oscillations near contact discontinuities such as the one shown in Fig. 11.7 (Tiselj and Petelin, 1997). This problem, which has been addressed by Karni (1994), Abgrall (1996), Saurel and Abgrall (1999), and Niu (2000), among others, is due to a failure of the algorithm to maintain pressure equilibrium across the discontinuity. It appears that conservative variables increase the possibility of such oscillations. The situation is complicated because while, generally speaking, the use of primitive variables may lead to good results for rapid transients, the use of conservative variables is essential, e.g., in problems encountered in the oil industry, where it is necessary to simulate flows in pipelines over long distances. In an attempt to combine the strengths of both approaches, in their drift flux calculation, Fjelde and Karlsen (2002) follow Toro (1998, 1999) in advocating the use of a primitive-variable formulation in smooth regions of the flow and of conservative variables in regions of steep gradients. They also demonstrate the use of the MUSCL algorithm of Section 11.3.3 for their drift-flux model.

Thirdly, the numerical method must be able to deal with the problem, already mentioned in Section 11.2.4, of the transition from two phases to a single phase or vice versa as, e.g. in condensation or boiling. Similarly, when the two velocities become equal, two of the eigenvalues of the system matrix coalesce and two of the eigenvectors become linearly dependent, so that diagonalization of the matrix C in (11.110) fails. In the work of Tiselj and Petelin (1997, 1998), as well as that of many others, these issues are simply resolved by artificially preventing either phase from disappearing completely or the two velocities becoming exactly equal, but more sophisticated ways to deal with these problems are clearly desirable. A related problem is the need to prevent oscillations around mechanical and thermal equilibrium states: the formulation of the source terms and the numerics must ensure that the sign of the relative velocity does not fluctuate, nor that oscillations between subcooled and superheated conditions occur.

A weak formulation of Roe's approximate Riemann solver suitable for systems which cannot be written in conservation form has been introduced in the multiphase flow literature by Toumi (1992; Toumi and Kumbaro, 1996) following work by Osher and collaborators (see, e.g. Engquist and Osher, 1981; Osher and Solomon, 1982). Let $\mathbf{v}(\mathbf{u}_L, \mathbf{u}_R; \sigma)$ be a path (i.e. a sequence of $\mathbf{u}$ states connecting the left and right states $\mathbf{u}_L$ and $\mathbf{u}_R$) in the space to which the solution vector $\mathbf{u}$ belongs, parameterized by a parameter σ and such that $\mathbf{v}(\mathbf{u}_L, \mathbf{u}_R; \sigma = 0) = \mathbf{u}_L$ and $\mathbf{v}(\mathbf{u}_L, \mathbf{u}_R; \sigma = 1) = \mathbf{u}_R$. If one considers the conservative system (11.72), it is obvious that

$$\int_0^1 \mathsf{A}(\mathbf{v}) \frac{\partial \mathbf{v}}{\partial \sigma} \, d\sigma = \mathbf{f}(\mathbf{u}_R) - \mathbf{f}(\mathbf{u}_L) \tag{11.112}$$

since $\mathsf{A}_{ij}(\mathbf{v}) \, (\partial v_j / \partial \sigma) = (\partial f_i / \partial v_j) \, (\partial v_j / \partial \sigma) = \partial f_i / \partial \sigma$. On the basis of this remark, condition (11.76) defining Roe's matrix may be rewritten as

$$\int_0^1 \mathsf{A}(\mathbf{v}) \frac{\partial \mathbf{v}}{\partial \sigma} \, d\sigma = \hat{\mathsf{A}} \left(\mathbf{u}_R - \mathbf{u}_L \right) \tag{11.113}$$

where the choice of the path is immaterial. The corresponding generalization of the Rankine–Hugoniot condition (11.77) is

$$\int_0^1 \left[\mathsf{A}(\mathbf{v}) - s\hat{\mathbf{I}} \right] \frac{\partial \mathbf{v}}{\partial \sigma} \, d\sigma = 0 \tag{11.114}$$

where, as in (11.77), s is the speed of the discontinuity and $\hat{\mathbf{I}}$ the identity matrix. We now observe that these definitions make sense also when the matrix A is not the Jacobian of a flux vector and, therefore, (11.113) represents a possible generalization of Roe's method to non-conservative systems. In this case, however, the choice of the path in $\mathbf{u}$-space does make a difference and this matter is, in general, complicated. Toumi (1992) treats by this method a vapor–liquid mixture described by the homogeneous model and bases the choice of the path on a heuristic argument which utilizes part of the saturation curve. In a later paper, Toumi *et al.* (2000) apply (11.114) to a drift-flux model by specifying a rectilinear integration path.

Toumi and Kumbaro (1996) consider a full four-equation model with an added-mass term similar to (11.105) and with the pressure gradients multiplied by the respective volume fractions in the phase momentum equations. They reduce the determination of $\hat{A}$ to the evaluation of suitable intermediate values of the volume fractions by writing

$$\int_{U_L}^{U_R} \alpha_k \frac{\partial p_i}{\partial s}\, ds = \overline{\alpha}(\alpha_k^L, \alpha_k^R)\, (p_i^R - p_i^L) \qquad (11.115)$$

which implicitly defines the path through the choice of the averaging function $\overline{\alpha}$. Toumi and Kumbaro (1996) take $\overline{\alpha} = 2\alpha^L \alpha^R/(\alpha^L + \alpha^R)$, while Evje and Flåtten (2003) choose the simple average $\overline{\alpha} = \frac{1}{2}(\alpha^L + \alpha^R)$.

In view of the difficulty in utilizing the recasting (11.113) of Roe's average matrix, and of the seemingly reasonable results obtained with its simplifications, other authors have used more straightforward approaches. For example, Adrianov $et\ al.$ (2003), who write the pressure terms in the momentum equations as $\partial(\alpha p)/\partial x) - p(\partial\alpha/\partial x)$, simply propose

$$\int_{x_{j-1/2}}^{x_{j+1/2}} p\frac{\partial \alpha}{\partial x}\, dx \simeq p_j \left(\alpha_{j+1/2}^* - \alpha_{j-1/2}^*\right) \qquad (11.116)$$

where α^* is a suitably defined intermediate value of α. The same approximation is used by Ghidaglia $et\ al.$ (2001) and its two-dimensional equivalent is essentially the choice also made by Cortes (2002; see also Dal Maso $et\ al.$, 1995).

The lack of a conservative form strikes at the very heart of multiphase flow modeling as it renders the convective terms nonunique. Indeed, if $C(u)(\partial u/\partial x)$ denotes the non-conservative part of the model, for smooth solutions, one has identically

$$\frac{\partial \mathbf{f}}{\partial x} + C(u)\frac{\partial \mathbf{u}}{\partial x} = \frac{\partial}{\partial x}(\mathbf{f} + \mathbf{g}) + \left(C - \frac{\partial \mathbf{g}}{\partial \mathbf{u}}\right)\frac{\partial \mathbf{u}}{\partial x} \qquad (11.117)$$

for any $\mathbf{g}(\mathbf{u})$ (here $\partial \mathbf{g}/\partial \mathbf{u}$ is the Jacobian of $\mathbf{g}$). The identity of the two sides of this relation fails, however, in the presence of discontinuities and, on a purely mathematical basis, it may not be easy to decide how to apportion the various terms of the model to the conservative and non-conservative components. This issue would be of little consequence, of course, were it not that the way in which non-conservative terms are treated has a strong impact on the accuracy with which discontinuities propagate (see, e.g. Adrianov $et\ al.$, 2003). Furthermore, the accuracy and stability of mass transport calculations which are of prime concern, for example, in the oil industry, are also affected. Clearly, finite-volume discretization of the equations is also ambiguous in this case.

11.5 Flux vector splitting

We have already mentioned in Section 11.3.2 the notion of flux vector splitting in the linear case in connection with equation (11.65). More generally, the procedure consists in splitting the flux vector $\mathbf{f}$ into a forward and a backward flux in the form

$$\mathbf{f} = \mathbf{f}^+ + \mathbf{f}^- \qquad (11.118)$$

in such a way that the Jacobian $\partial \mathbf{f}^+/\partial \mathbf{u}$ has only nonnegative eigenvalues, while $\partial \mathbf{f}^-/\partial \mathbf{u}$ has only nonpositive eigenvalues (see van Leer, 1982; Toro, 1999; Harten *et al.*, 1983). The simplest case is that in which $\mathbf{f}$ is a homogeneous function of $\mathbf{u}$:

$$\mathbf{f}(\mathbf{u}) = \mathsf{A}(\mathbf{u})\,\mathbf{u}, \qquad (11.119)$$

in which A is a matrix as, in this case, A is proportional to the Jacobian of $\mathbf{f}$. The decomposition (11.118) can then be effected simply by decomposing A as in (11.63) following the same procedure of Section 11.3.2 and setting $\mathbf{f}^\pm = \mathsf{A}_\pm \mathbf{u}$. The solution at the new time level is then given by (11.68).

An important instance is the Euler equations for the dynamics of a perfect gas, for which

$$\mathbf{u} = \begin{bmatrix} \rho \\ \rho u \\ \rho E \end{bmatrix}, \qquad \mathbf{f} = c \begin{bmatrix} M\rho \\ \left(M^2 + \frac{1}{\gamma}\right)\rho c \\ M\rho H \end{bmatrix} \qquad (11.120)$$

where $c = \sqrt{\gamma p/\rho}$ is the local speed of sound, $M = u/c$ is the Mach number, $E = e + \frac{1}{2}u^2 = H - p/\rho$, and $p = (\gamma - 1)\rho e$, with γ the ratio of specific heats. In this case, for $|M| \leq 1$, van Leer (1982; see also Toro 1999) finds

$$\mathbf{f}^\pm = \pm\frac{1}{4}\rho c \,(M \pm 1)^2 \begin{bmatrix} 1 \\ \frac{2c}{\gamma}\left(\frac{\gamma-1}{2}M \pm 1\right) \\ \frac{2c^2}{\gamma^2-1}\left(\frac{\gamma-1}{2}M \pm 1\right)^2 \end{bmatrix}. \qquad (11.121)$$

Note that $c(M \pm 1) = u \pm c$ are just the characteristics speeds associated with pressure wave propagation. The procedure can be generalized to fluxes which do not satisfy the homogeneity condition (11.119) as shown by van Leer (1982) and Harten *et al.* (1983).

A recent development in this area is a method termed AUSM (an acronym whimsically pronounced "awesome") for *A*dvection *U*pwind *S*plitting *M*ethod, which has been introduced by Liou and Steffen (1993) and improved to the form referred to as AUSM+ by Liou (1996), Edwards and Liou (1998) and others. This scheme attempts to combine the accuracy of Godunov-type methods with the efficiency of the original flux-vector splitting schemes. To explain the basic idea, let us consider again one-dimensional gas dynamics, noting that $p = \rho c^2/\gamma$. On the basis of the remark that convective transport and pressure wave propagation are distinct processes involving different physics and different time scales, the flux $\mathbf{f}$ in (11.120) is split into a convective part, $\mathbf{f}^c$, and a pressure part, $\mathbf{f}^p$:

$$\mathbf{f} = \mathbf{f}^c + \mathbf{f}^p, \qquad \mathbf{f}^c = u \begin{bmatrix} \rho \\ \rho u \\ \rho H \end{bmatrix} = Mc \begin{bmatrix} \rho \\ \rho u \\ \rho H \end{bmatrix}, \qquad \mathbf{f}^p = \begin{bmatrix} 0 \\ p \\ 0 \end{bmatrix}. \tag{11.122}$$

The convective part is now considered as a set of passive scalar quantities convected by a suitably defined velocity u or Mach number M. For the Riemann problem, $\mathbf{h}^c_{j+1/2}$, the discretized counterpart of $\mathbf{f}^c$, is defined in an upwind fashion as[1]

$$\mathbf{h}^c_{j+1/2} = M_{j+1/2}c_{j+1/2} \begin{bmatrix} \rho \\ \rho u \\ \rho H \end{bmatrix}_{\ell/r}, \tag{11.123}$$

where $[\ ]_{\ell/r} = [\ ]_j$ for $M_{j+1/2} \geq 0$ and $[\ ]_{\ell/r} = [\ ]_{j+1}$ for $M_{j+1/2} < 0$. The common speed of sound $c_{j+1/2}$ can be defined in several ways (Liou, 1996), the simplest (but not most accurate) one being the simple average $c_{j+1/2} = \frac{1}{2}(c_j + c_{j+1})$.

A difference with normal upwinding is the way in which $M_{j+1/2}$ is defined in terms of the neighboring states:

$$M_{j+1/2} = \mathcal{M}^+(M_j) + \mathcal{M}^-(M_{j+1}). \tag{11.124}$$

The split Mach numbers $\mathcal{M}^\pm$ can be defined in different ways. For example, as in (11.121) (Liou and Steffen, 1993),

$$\mathcal{M}^\pm = \begin{cases} \pm\frac{1}{4}(M \pm 1)^2 & \text{if } |M| \leq 1 \\ \frac{1}{2}(M + |M|) & \text{otherwise.} \end{cases} \tag{11.125}$$

The pressure term is similarly split as $p_{j+1/2} = \mathcal{P}^+(M_j)p_j + \mathcal{P}^-(M_{j+1})p_{j+1}$

---

[1] This is the definition in AUSM+; in the original AUSM $c_{j+1/2}$ is replaced by $c_{\ell/r}$.

with

$$\mathcal{P}^{\pm} = \begin{cases} \frac{1}{4}(M \pm 1)^2 (2 \mp M) & \text{if } |M| \leq 1 \\ \frac{1}{2}(M + |M|)/M & \text{otherwise.} \end{cases} \tag{11.126}$$

There is no uniqueness to these polynomial forms and, indeed, AUSM+ uses higher order polynomials for improved accuracy (Liou, 1996).

Since, in general, $\mathcal{P}^+(M_j) + \mathcal{P}^-(M_{j+1}) \neq 1$, $p_{j+\frac{1}{2}}$ is not necessarily bounded by p_j and p_{j+1}. It is interesting to note that the numerical flux can be rewritten as

$$\mathbf{h}_{j+\frac{1}{2}} = \frac{1}{2} M_{j+\frac{1}{2}} c_{j+\frac{1}{2}} \left(\begin{bmatrix} \rho \\ \rho u \\ \rho H \end{bmatrix}_{j+1} + \begin{bmatrix} \rho \\ \rho u \\ \rho H \end{bmatrix}_j \right)$$

$$-\frac{1}{2} |M_{j+\frac{1}{2}}| c_{j+\frac{1}{2}} \left(\begin{bmatrix} \rho \\ \rho u \\ \rho H \end{bmatrix}_{j+1} - \begin{bmatrix} \rho \\ \rho u \\ \rho H \end{bmatrix}_j \right) + \begin{bmatrix} 0 \\ p_{j+\frac{1}{2}} \\ 0 \end{bmatrix} \tag{11.127}$$

which is strongly reminiscent of (11.80). Unlike that expression, the first term is not a simple average of the two fluxes, but a Mach-number weighted average. The upstream bias of the convective terms uses a properly defined cell-interface velocity, while the pressure is dealt with using the direction of propagation of pressure waves. The second term is dissipative, but it has only a scalar coefficient $|M_{1/2}|$ rather than $|\mathbf{A}|$ as in (11.80), the calculation of which requires $O(N^2)$ operations.

AUSM+ exhibits a number of considerable advantages such as the exact resolution of one-dimensional contact and shock discontinuities, absence of spurious oscillations, simplicity and versatility. Furthermore, an extremely interesting property from the point of view of multiphase flow applications, is its positivity preserving, which makes it applicable to situations where one of the phases may disappear locally (Edwards *et al.*, 2000; Evje and Fjelde, 2003; Evje and Flåtten, 2003). Equally useful is the avoidance of an explicit calculation of the Jacobian, which may be difficult or impossible with many models. For these reasons, the emergence of AUSM schemes has generated active interest in the multiphase flow literature.

The first applications have been in the context of the homogeneous and drift-flux models, which are the closest to single-phase gas dynamics, for which the method was initially formulated, and where the issue of hyperbolicity does not arise. Edwards *et al.* (2000) describe homogeneous-model calculations of a two-phase liquid–vapor flow in a nozzle and over an obstacle, demonstrating good agreement with data and sharp discontinuities.

Paillere *et al.* (2001) also consider a homogeneous-flow model applied to a falling water jet and to a two-phase shock-tube problem.

Evje and Fjelde (2002, 2003) consider an isothermal drift-flux model consisting of one momentum and two continuity equations:

$$
\frac{\partial}{\partial t}
\begin{bmatrix}
\alpha_L \rho_L \\
\alpha_G \rho_G \\
\alpha_L \rho_L u_L + \alpha_G \rho_G u_G
\end{bmatrix}
+
\frac{\partial}{\partial x}
\begin{bmatrix}
\alpha_L \rho_L u_L \\
\alpha_G \rho_G u_G \\
\alpha_L \rho_L u_L^2 + \alpha_G \rho_G u_G^2 + p
\end{bmatrix}
=
\begin{bmatrix}
0 \\
0 \\
F
\end{bmatrix}
$$

$$(11.128)$$

in which F is an algebraic function. The relation between the phase velocities is assumed to have the form $(1 - K\alpha_G)u_G = K\alpha_L u_L + S(\alpha, \rho)$ with K a constant and S a function of the indicated variables. They split the pressure part from the flux term as in (11.122), but then further split the convective part into two terms, one for each phase:

$$
\begin{bmatrix}
\alpha_L \rho_L u_L \\
\alpha_G \rho_G u_G \\
\alpha_L \rho_L u_L^2 + \alpha_G \rho_G u_G^2 + p
\end{bmatrix}
=
\begin{bmatrix}
\alpha_L \rho_L u_L \\
0 \\
\alpha_L \rho_L u_L^2
\end{bmatrix}
+
\begin{bmatrix}
0 \\
\alpha_G \rho_G u_G \\
\alpha_G \rho_G u_G^2
\end{bmatrix}
+
\begin{bmatrix}
0 \\
0 \\
p
\end{bmatrix}
$$

$$
= \quad \mathbf{f}^L + \mathbf{f}^G + \mathbf{f}^P . \qquad (11.129)
$$

The discretized convective fluxes are then written as in (11.123):

$$
\mathbf{h}_{j+1/2}^J = (M_J)_{j+1/2}\, c_{j+1/2}
\begin{bmatrix}
J
\end{bmatrix}_{\ell/r}
, \qquad J = L, G \qquad (11.130)
$$

with an obvious meaning of the notation. The Mach number is calculated as $M_J = u_J/c$, with c a common speed of sound approximated as

$$
c =
\begin{cases}
c_L & \text{if } \alpha_G < \epsilon \\
p/[\rho_L \alpha_G \alpha_L] & \text{if } \epsilon \le \alpha_G \le 1 - \epsilon \\
c_G & \text{if } 1 - \epsilon < \alpha_G
\end{cases}
\qquad (11.131)
$$

where $c_{L,G}$ are the speeds of sound as given by the equations of state of the pure fluids and ϵ is a small parameter taken as 0.001. The pressure part of the flux is treated as in (11.126) but, here, the proper definition of the Mach number is less straightforward. The authors define it as u_m/c, where $u_m = \alpha_L u_L + \alpha_G u_G$ is the volumetric flux. The common speed of sound is taken as $c_{j+1/2} = \max(c_j, c_{j+1})$. Spatial accuracy is increased to second order using the MUSCL procedure and temporal accuracy by means of a two-stage Runge–Kutta scheme. The authors show that, with these choices, the scheme preserves positivity of pressure, densities, and volume fractions. An example of the results showing transition from two-phase to

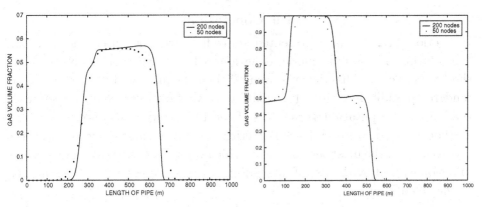

Fig. 11.8. Numerical results from Evje and Fjelde (2003) demonstrating the positivity preserving property of the AUSM+ scheme as applied to a drit flux model. Left: gas volume fraction vs. pipe length for a flow in which a two-phase mixture is pumped into a liquid-filled tube for a time, after which the gas flow rate is brought to 0. Right: gas volume fraction for a process in which the liquid flow rate vanishes for a time (in this latter simulation the two phases moved with the same velocity). The dotted lines shows the results with a grid four times as coarse as the solid lines.

pure liquid and from two-phase to pure gas is shown in Fig. 11.8, which also demonstrates convergence with grid refinement.

A necessary remark to be made in connection with (11.131) is that the two-phase sound speed shown here is only appropriate for a bubbly flow regime. The same form is used by Evje and Flåtten (2003), who apply the AUSM ideas to a full hyperbolic two-fluid model for which they calculate the characteristic speeds approximately assuming a small Mach number of the relative velocity. The need for more realistic models, which would distinguish different regimes, is illustrated by the results of Edwards *et al.* (2000) who, in one of their examples, are forced to consider a 10 m/s liquid jet falling in its vapor as a supersonic flow because they also use a bubbly liquid sound speed.

A variant of the AUSM method applied to a hyperbolic model with separate mass, momentum, and energy balance equations augmented by a turbulence model is described in Niu (2001). He limits the work to a disperse flow with a very small volume fraction of suspended particles or drops and demonstrates very sharp profiles for a shock-tube problem and a reasonable agreement with experiment for flow in a 90° bend. In order to deal effectively with the low Mach number regime, he uses the same rescaling of the sound speed based on the eigenvalues of the preconditioned equation system used in Edwards *et al.* (2000).

11.6 Conclusions

In Section 11.2 we have presented some first-order accurate methods for the solution of two-fluid model equations which have been widely used in the last 30 years. They have proven efficient, flexible, and robust and have rendered possible meaningful simulations of very complex systems even with the limited computational resources of the 1970s and 1980s. However, these methods embody a rather basic numerical technology and suffer from several shortcomings, the most serious one of which is probably the strong numerical diffusion caused by the use of upwinding, but also necessary for the stability of the calculation.

In the more recent past several attempts have been made to adapt to multiphase flow computations some of the major advances made in the simulation of single-phase flows. In discussing some of this work in the second part of this chapter, we have encountered recurring fundamental difficulties – uncertainties about the mathematical models, lack of hyperbolicity, non-conservation form – as well as practical ones, such as the strong non-linear coupling of the equations, the non-differentiability of constitutive relations across flow-regime transitions, stiffness of the source terms, multi-to-single phase transition and vice versa, positivity, and others. We have described some methods with the potential capability to deal with these practical aspects, but we have also seen that their application to two-fluid models is strongly hampered by the underlying fundamental problems. A case in point are the methods of the AUSM family, which offer several very attractive aspects but whose application encounters difficulties and ambiguities due to the very structure of the equations.

If the numerical methods currently available are not ideal, a point to keep in mind is that uncertainties in the correct formulation of the equations and the modeling of source terms may ultimately have a bigger impact on the result than the particular numerical method adopted. Thus, rather than focusing on the numerics alone, it makes sense to try to balance the numerical effort with the expected fidelity of the modeling. It seems justified to take a pragmatic view of the situation, using the available models and methods with open eyes and full awareness of their potential limitations. An exploration of the sensitivity of the results to the constitutive relations, parameter values, and numerics – always a good practice in computing – becomes a vital necessity in this field.

These considerations also suggest that, in setting up a new code, it may be advisable to adopt methods not too strongly reliant on a specific form of the mathematical model so that improvements can be easily incorporated and alternative options readily tested.

The formulation of a satisfactory set of averaged-equations models emerges as the single highest priority in the modeling of complex multiphase flows. While mathematics is and will remain a powerful guide for this endeavor, it is doubtful that much can be achieved without a solid physical foundation. Here we encounter a basic stumbling block which has stymied progress for decades, namely the inherent complexity of the phenomena and processes that we would like to model. The approach taken in the past has been to use relatively simple parameterizations, e.g. of the interphase exchange terms based on standard engineering practice, such as heat transfer correlations, drag coefficients, and the like. The usefulness of these concepts mostly rests on experience gained under steady conditions in nearly uniform flows, and it may therefore not be very surprising that their use in the time- and space-dependent context of averaged-equations models has proven problematic.

The complexity of the actual physical situation defies our intuition and limits our ability to develop accurate models. Two major tools can be relied on to improve the situation. The first one is – of course – experimentation. It must be recognized, though, that in spite of the recent explosion of experimental capabilities, multiphase flows offer some serious challenges, starting with the difficulty in gaining optical access to many situations of interest. The second major avenue toward gaining a better physical understanding is computation itself, especially the detailed type of simulations based on known and reliable models covered in the first part of this book. The information gained in this way is likely to supply the necessary guide to the development, e.g., of satisfactory closure relations, mechanisms for flow regime transitions, and others.

It may be safely concluded that numerical simulation is and will long remain an essential tool in multiphase flow research.

References

Y. Abe, H. Akimoto, H. Kamo, and Y. Murao, 1993. Elimination of numerical pressure spikes induced by two-fluid model. *J. Nucl. Sci. Technol.*, **30**:1214–1224.

R. Abgrall, 1996. How to prevent pressure oscillations in multicomponent calculations: a quasi-conservative approach. *J. Comput. Phys.*, **125**:150–160.

N. Adrianov, R. Saurel, and G. Warnecke, 2003. A simple method for compressible multiphase mixtures and interfaces. *Int. J. Num. Meth. Fluids*, **41**:109–131.

K. Agrawal, P. N. Loezos, M. Syamlal, and S. Sundaresan, 2001. The role of mesoscale structures in rapid gas–solid flows. *J. Fluid Mech.*, **445**:151–185.

G. Agresar, J. J. Linderman, G. Tryggvason, and K. G. Powell, 1998. An adaptive, cartesian, front-tracking method for the motion, deformation and adhesion of circulating cells. *J. Comput. Phys.*, **43**:346–380.

B. Aktas and J. H. Mahaffy, 1996. A two-phase level tracking method. *Nucl. Eng. Des.*, **162**:271–280.

I. Aleinov, E. G. Puckett, and M. Sussman, 1999. Formation of droplets in microscale jetting devices, fedsm99-7106. In *Proceedings of the 3rd ASME/JSME Joint Fluids Engineering Conference*, San Francisco, CA.

F. Alexander, S. Chen, and D. W. Grunau, 1993. Hydrodynamic spinodal decomposition: growth kinetics and scaling. *Phys. Rev. B*, **48**:634.

A. S. Almgren, J. B. Bell, P. Colella, and T. Marthaler, 1997. A cartesian grid projection method for the incompressible euler equations in complex geometries. *SIAM J. Sci. Comput.*, **18**:1289.

D. A. Anderson, J. C. Tannehill, and R. H. Pletcher, 1984. *Computational Fluid Mechanics and Heat Transfer*. New York, Hemisphere Publishing.

K. Anderson, S. Sundaresan, and R. Jackson, 1995. Instabilities and the formation of bubbles in fluidized beds. *J. Fluid Mech.*, **303**:327–366.

G. B. Arfken and H. J. Weber, 2000. *Mathematical Methods for Physicists*. New York, Academic Press.

E. S. Asmolov, 1999. The inertial lift on a spherical particle in a plane Poiseuille flow at large channel Reynolds number. *J. Fluid Mech.*, **381**:63–87.

E. S. Asmolov and J. B. McLaughlin, 1999. The inertial lift on an oscillating sphere in a linear shear flow. *Int. J. Multiphase Flow*, **25**:739–751.

E. S. Asmolov, 2002. The inertial lift on a small particle in a weak-shear parabolic flow. *Phys. Fluids*, **14**:15–28.

J. M. Augenbaum, 1989. An adaptive pseudospectral method for discontinuous problems. *Appl. Numer. Math.*, **5**:459–480.

T. R. Auton, 1987. The lift force on a spherical body in rotational flow. *J. Fluid Mech.*, **183**:199–218.

T. R. Auton, J. C. R. Hunt, and M. Prud'homme, 1988. The force exerted on a body in inviscid unsteady non-uniform rotational flow. *J. Fluid Mech.*, **197**:241–257.

S. Balachandar and M. R. Maxey, 1989. Methods for evaluating fluid velocities in spectral simulations of turbulence. *J. Comput. Phys.*, **83**:96–125.

P. Bagchi and S. Balachandar, 2002a. Effect of free rotation on the motion of a solid sphere in linear shear flow at moderate Re. *Phys. Fluids*, **14**:2719–2737.

P. Bagchi and S. Balachandar, 2002b. Shear versus vortex-induced lift force on a rigid sphere at moderate Re. *J. Fluid Mech.*, **473**:379–388.

P. Bagchi and S. Balachandar, 2002c. Steady planar straining flow past a rigid sphere at moderate Reynolds numbers. *J. Fluid Mech.*, **466**:365–407.

P. Bagchi and S. Balachandar, 2003a. Effect of turbulence on the drag and lift of a particle. *Phys. Fluids*, **15**:3496–3513.

P. Bagchi and S. Balachandar, 2003b. Inertial and viscous forces on a rigid sphere in straining flows at moderate Reynolds numbers. *J. Fluid Mech.*, **481**:105–148.

P. Bagchi and S. Balachandar, 2004. Response of a particle wake to isotropic turbulent flow. *J. Fluid Mech.*, **518**:95–123.

G. Balzer, A. Boelle, and O. Simonin, 1995. Eulerian gas–solid flow modelling of dense fluidized beds. In *Fluidization VIII*, pages 1125–1135.

F. Barre and M. Bernard, 1990. The CATHARE code strategy and assessment. *Nucl. Eng. Des.*, **124**:257–284.

F. Barre, M. Parent, and B. Brun, 1993. Advanced numerical methods for thermal-hydraulics. *Nucl. Eng. Des.*, **145**:147–158.

A. B. Basset, 1888. *Treatise on Hydrodynamics*. Deighton, Bell & Co., London, reprinted by Dover, 1961.

I. B. Bazhlekov, 2003. Non-singular boundary-integral method for deformable drops in viscous flows. PhD thesis, Technische Universiteit Eindhoven.

I. B. Bazhlekov, P. D. Anderson, and H. E. H. Meijer, 2004a. Boundary integral method for deformable interfaces in the presence of insoluble surfactants. *Lect. Notes Comput. Sci.*, **2907**:355–62.

I. B. Bazhlekov, P. D. Anderson, and H. E. H. Meijer, 2004b. Nonsingular boundary integral method for deformable drops in viscous flows. *Phys. Fluids*, **16**:1064–81.

R. C. Beach, 1991. *An Introduction to the Curves and Surfaces of Computer-Aided Design*. New York, Van Nostrand Reinhold.

M. Behr and T. E. Tezduyar, 1994. Finite element solution strategies for large-scale flow simulations. *Comput. Methods Appl. Mech. Engng.*, **112**:3–24.

C. R. Bell, P. B. Bleiweis, J. E. Boudreau, F. R. Parker, and L. L. Smith, 1977. SIMMER-I: An S_n, implicit multifield multicomponent Eulerian recriticality code. Technical Report LA-NUREG-6467-MS, Los Alamos Scientific Laboratory.

K. H. Bendiksen, D. Maines, R. Moe, and S. Nuland, May 1991. The dynamic two-fluid model OLGA: Theory and application. *SPE Prod. Eng.*, pages 171–180.

D. Bestion, 1990. The physical closure laws in CATHARE code. *Nucl. Eng. Des.*, **124**:229–245.

D. Bhaga and M. E. Weber, 1981. Bubbles in viscous liquids: Shapes, wakes and velocities. *J. Fluid Mech.*, **105**:61–85.

A. Biesheuvel and S. Spoelstra, 1989. The added mass coefficient of a dispersion of spherical gas bubbles in liquid. *Int. J. Multiphase Flow*, **15**:911–924.

R. B. Bird, R. C. Armstrong and O. Hassager, 1987. *Dynamics of Polymeric Liquids*, Vol. 1. New York: Wiley-Interscience.

T. D. Blake, J. Deconinck and U. Dortona, 1995. Models of wetting: immiscible lattice Boltzmann automata versus molecular kinetic theory. *Langmuir*, **11**: 4588–92.

J. J. Bluemink,, D. Lohse, A. Prosperetti and L. van Wijngaarden, 2008. A sphere in a uniformly rotating or shearing flow. *J. Fluid Mech.*, bf 600:201–233.

C. Bodart and M. J. Crochet, 1994. The time-dependent flow of a viscoelastic fluid around a sphere. *J. Non-Newtonian Fluid Mech.*, **54**:303–329.

A. Boelle, G. Balzer, and O. Simonin, 1995. Second-order prediction of the particle-phase stress tensor of inelastic spheres in simple shear dense suspensions. In D. E. Stock, M. W. Reeks, Y. Tsuji, E. E. Michaelides, and M. Gautam, editors, *Gas–Particle Flows, Proceedings of the ASME Fluids Engineering Division Summer Meeting*, pages 9–18, New York, ASME.

W. Böhm, G. Farin, and J. Kahmann, 1984. A survey of curve and surface methods in CAGD. *Comput. Aided Geom. Des.*, **1**:1–60.

M. Boivin, O. Simonin, and K. D. Squires, 1998. Direct numerical simulation of turbulence modulation by particles in isotropic turbulence. *J. Fluid Mech.*, **375**:235–263.

J. Boussinesq, 1885. *Application de Potentiels à l'Étude de l'Équilibre et de Mouvement des Solides Élastiques*. Gauthier-Villars, Paris, reprinted by Blanchard, Paris 1969.

M. Bouzidi, D. d'Humières, P. Lallemand, and L.-S. Luo, 2001. Lattice Boltzmann equation on a two-dimensional rectangular grid. *J. Comput. Phys.*, **172**:704–717.

M. Bouzidi, M. Firdaouss, and P. Lallemand, 2002. Momentum transfer of a lattice-Boltzmann fluid with boundaries. *Phys. Fluids*, **13**:3452–3459.

J. P. Boyd, 1989. *Chebyshev and Fourier Spectral Methods*. New York, Springer-Verlag.

J. U. Brackbill, D. B. Kothe, and C. Zemach, 1992. A continuum method for modeling surface tension. *J. Comput. Phys.*, **100**:335–354.

J. F. Brady and G. Bossis, 1988. Stokesian dynamics. *Ann. Rev. Fluid Mech.*, **20**:111–157.

J. F. Brady, 1993. Stokesian dynamics simulation of particulate flows, in *Particulate Two-Phase Flow* (M. C. Roco, ed.), pp. 971–998. London: Butterworth–Heinemann.

F. P. Bretherton, 1962. The motion of rigid particles in a shear flow at low Reynolds number. *J. Fluid Mech.*, **14**:284–304.

D. L. Brown, R. Cortez, and M. L. Minion, 2001. Accurate projection methods for the incompressible Navier–Stokes equations. *J. Comput. Phys.*, **168**:464–499.

A. Brucato, F. Grisafi, and G. Montante, 1998. Particle drag coefficients inturbulent fluids. *Chem. Eng. Sci.*, **53**:3295–3314.

J. Buckles, R. Hazlett, S. Chen, K. G. Eggert, D. W. Grunau, and W. E. Soll, 1994. Flow through porous media using lattice Boltzmann method. *Los Alamos Sci.*, **22**:112–121.

A. J. Bulpitt and N. D. Efford, 1996. An efficient 3D deformable model with a self-optimizing mesh. *Image Vision Comput.*, **14**:573–580.

B. Bunner, 2000. *Numerical Simulations of Gas–Liquid Bubbly Flows.* The University of Michigan.

B. Bunner and G. Tryggvason, 2002a. Dynamics of homogeneous bubbly flows. Part 1. Rise velocity and microstructure of the bubbles. *J. Fluid Mech.*, **466**:17–52.

B. Bunner and G. Tryggvason, 2002b. Dynamics of homogeneous bubbly flows. Part 2. Fluctuations of the bubbles and the liquid. *J. Fluid Mech*, **466**:53–84.

B. Bunner and G. Tryggvason, 2003. Effect of bubble deformation on the stability and properties of bubbly flows. *J. Fluid Mech.*, **495**:77–118.

T. M. Burton and J. K. Eaton, 2002. Analysis of a fractional-step method on overset grids. *J. Comput. Phys.*, **177**:336–364.

B. L. Buzbee, F. W. Dorr, J. A. George and G. H. Golub, 1971. The direct solution of the discrete Poisson equation on irregular regions. *SIAM J. Num. Anal.*, **8**:722–736.

J. W. Cahn and J. E. Hilliard, 1958. Free energy of a nonuniform system. I. Interfacial free energy. *J. Chem. Phys.*, **28**:258.

D. Calhoun and R. J. LeVeque, 2000. A cartesian grid finite-volume method for the advection-diffusion equation in irregular geometries. *J. Comput. Phys.*, **157**: 143–180.

C. Canuto, M. Y. Hussaini, A. Quarteroni, and T. A. Zang, 1988 (second revised printing 1990). *Spectral methods in fluid dynamics.* Berlin, Springer–Verlag.

M. Carlson, P. Mucha, and G. Turk, 2004. Rigid fluid: Animating the interplay between rigid bodies and fluid. *ACM Siggraph.*

M. B. Carver, 1982. A method of limiting intermediate values of volume fraction in iterative two-fluid computations. *J. Mech. Eng. Sci.*, **24**:221–224.

C. A. Catlin, 2003. The lift force on an arbitrarily shaped body in a steady incompressible inviscid linear shear flow with weak strain. *J. Fluid Mech.*, **484**:113–142.

E. J. Chang and M. R. Maxey, 1994. Unsteady flow about a sphere at low to moderate Reynolds number. Part 1. Oscillatory motion. *J. Fluid Mech.*, **277**:347–379.

Y. C. Chang, T. Y. Hou, B. Merriman, and S. Osher, 1996. Eulerian capturing methods based on a level set formulation for incompressible fluid interfaces. *J. Comput. Phys.*, **124**:449–464.

S. Chapman and T. G. Cowling, 1970. *The Mathematical Theory of Non-Uniform gases.* Cambridge Mathematical Library, third edition.

H. Chen, S. Chen, G. D. Doolen, Y. C. Lee, and H. A. Rose, 1989. Multithermodynamic phase lattice-gas automata incorporating interparticle potentials. *Phys. Rev. A*, **40**:2850.

H. Chen, S. Chen, and W. H. Matthaeus, 1992. Recovery of the Navier–Stokes equations using a lattice-gas Boltzmann method. *Phys. Rev. A.*, **45**:R5339.

H. D. Chen, B. M Boghosian, P. V. Coveney, and M. Nekovee, 2000. A ternary lattice Boltzmann model for amphiphilic fluids. *Proc. Royal Soc. of London,*

Series A, **456**:2043.

S. Chen, H. Chen, D. Martinez, and W. Matthaeus, 1991. Lattice Boltzmann model for simulation of magnetohydrodynamics. *Phys Rev. Lett.*, **67**:3776.

S. Chen and G. D. Doolen, 1998. Lattice Boltzmann method for fluid flows. *Annu. Rev. Fluid Mech.*, **30**:329–64.

H. Cheng, L. Greengard, and V. Rokhlin, 1999. A fast adaptive multipole algorithm in three dimensions. *J. Comput. Phys.*, **155**:468–98.

I.-L. Chern, J. Glimm, O. McBryan, B. Plohr, and S. Yaniv, 1986. Front tracking for gas dynamics. *J. Comput. Phys.*, **62**:83–110.

A. J. Chorin, 1968. Numerical solutions of the Navier–Stokes equations. *Math. Comput.*, **22**:745–762.

B. Cichocki, B. U. Felderhof, and R. Schmitz, 1988. Hydrodynamic interactions between two spherical particles. *PhysicoChem. Hyd.*, **10**:383–403.

B. Cichocki and B. U. Felderhof, 1989. Periodic fundamental solution of the linear Navier–Stokes equations. *Physica A*, **159**:19–27.

R. Clift, J. R. Grace, and M. E. Weber, 1978. *Bubbles, Drops, and Particles*. Academic Press.

P. Colella and P. R. Woodward, 1984. The piecewise parabolic method (PPM) for gas dynamical simulations. *J. Comput. Phys.*, **54**:174–201.

S. Corrsin and J. L. Lumley, 1956. On the equation of motion of a particle in a turbulent fluid. *Appl. Sci. Res.*, **A6**:114–116.

J. Cortes, A. Debusshce, and I. Toumi, 1998. A density perturbation method to study the eigenstructure of two-phase flow equation systems. *J. Comput. Phys.*, **147**:463–484.

J. Cortes, 2002. On the construction of upwind schemes for non-equilibrium transient two-phase flows. *Comput. Fluids*, **31**:159–182.

V. Cristini, J. BłAwzdziewicz, and M. Loewenberg, 1998. Drop breakup in three-dimensional viscous flows. *Phys. Fluids.*, **10**:1781–1784.

V. Cristini, J. BłAwzdziewicz, and M. Loewenberg, 2001. An adaptive mesh algorithm for evolving surfaces: Simulations of drop breakup and coalescence. *J. Comput. Phys.*, **168**:445–463.

V. Cristini, J. BłAwzdziewicz, M. Loewenberg, and L. R. Collins, 2003. Breakup in stochastic Stokes flows: Sub-Kolmogorov drops in isotropic turbulence. *J. Fluid Mech.*, **492**:231–50.

C. Crowe, M. Sommerfeld, and Y. Tsuji, 1998. *Multiphase Flows with Droplets and Particles*. Boca Raton FL, CRC Press.

G. Dal Maso, P. LeFloch, and P. Murat, 1995. Definition and weak stability of a non-conservative product. *J. Math. Pure Appl.*, **74**:483–548.

B. J. Daly, 1967. Numerical study of two fluid Rayleigh–Taylor instability. *Phys. Fluids*, **10**:297.

B. J. Daly, 1969a. Numerical study of the effect of surface tension on interface instability. *Phys. Fluids*, **12**:1340–1354.

B. J. Daly, 1969b. A technique for including surface tension effects in hydrodynamic calculations. *J. Comput. Phys.*, **4**:97–117.

D. S. Dandy and H. A. Dwyer, 1990. A sphere in shear flow at finite Reynolds number: effect of shear on particle lift, drag and heat transfer. *J. Fluid Mech.*, **216**:381–410.

M. Darwish, F. Moukalled, and B. Sekar, 2001. A unified formulation of the segregated class of algorithms for multifluid flow at all speeds. *Numer. Heat Transfer*, **B40**:99–137.

J. Davidson, 1961. Discussion. In *Symposium on Fluidization*, volume 39, pages 230–232. *Trans. Inst. Chem. Eng.*

C. de Boor, 1978. *A Practical Guide to Splines*. New York, Springer-Verlag.

J. E. Dennis Jr. and R. E. Schnabel, 1983. *Numerical Methods for Unconstrained Optimization and Nonlinear Equations*. Prentice Hall, Englewood Cliffs, Reprinted by SIAM, 1996.

M. Deserno and C. Holm, 1998. How to mesh up Ewald sums. I. A theoretical and numerical comparison of various particle mesh routines. *J. Chem. Phys.*, **109**:7678–93.

E. Deutsch and O. Simonin, 1991. Large eddy simulation applied to the modeling of particulate transport coefficients in turbulent two-phase flows. In *Proceedings of the 8th International Symposium on Turbulent Shear Flows, Volume 1*, page 1011. University of Munich.

V. Dhir, 2001. Numerical simulations of pool-boiling heat transfer. *A.I.Ch.E. Journal*, **47**:813–834.

D. d'Humières, 1992. Generalized lattice-Boltzmann equations. In B. D. Shizgal and D. P. Weaver, editors, *Rarefied Gas Dynamics: Theory and Simulations Progress in Astronautics and Aeronautics*, Volume 159 of *AIAA*, pages 450–458. Washington, DC.

D. d'Humières, M. Bouzidi, and P. Lallemand, 2001. Thirteen-velocity three-dimensional lattice Boltzmann model. *Phys. Rev. E*, **63**:066702.

D. d'Humières, I. Ginzburg, M. Krafczyk, P. Lallemand, and L.-S. Luo, 2002. Multiple-relaxation-time lattice Boltzmann models in three-dimensions. *Proc. R. Soc. Lond. A*, **360**:437–451.

A. Diaz, N. Pelekasis, and D. Barthès-Biesel, 2000. Transient response of a capsule subjected to varying flow conditions: Effect of internal fluid viscosity and membrane elasticity. *Phys. Fluids*, **12**:948–57.

A. Diaz, D. Barthès-Biesel, and Pelekasis N, 2001. Effect of membrane viscosity on the dynamic response of an axisymmetric capsule. *Phys. Fluids*, **13**:3835–8.

U. Dortona, 1995. Models of wetting: immiscible lattice Boltzmann automata versus molecular kinetic theory. *Langmuir*, **11**:4588.

D. A. Drew, L. Y. Cheng, and R. T. Jr. Lahey, 1979. The analysis of virtual mass effects in two-phase flow. *Int. J. Multiphase Flow*, **5**:233–242.

D. A. Drew, 1983. Mathematical modeling of two-phase flow. *Ann. Rev. Fluid Mech.*, **15**:261–291.

D. A. Drew and R. T. Jr. Lahey, 1993. Analytical modeling of multiphase flow. In Roco M. C., editor, *Particulate Two-Phase Flow*, pages 509–566. Boston. Butterworth-Heinemann.

J. K. Dukowicz and J. W. Kodis, 1987. Accurate conservative remapping (rezoning) for arbitary Lagrangian–Eulerian computations. *SIAM J. Sci. Stat. Comput.*, **8**:305–321.

J. K. Dukowicz and A. S. Dvinsky, 1992. Approximate factorization as a high-order splitting for the implicit incompressible flow equations. *J. Comput. Phys.*, **102**:336–347.

A. R. Edwards and F. P. O'Brien, 1970. Studies of phenomena connected with the depressurization of water reactors. *J. Brit. Nucl. Soc.*, **9**:125–135.

D. A. Edwards, H. Brenner, and D. T. Wasan, 1991. *Interfacial Transport Processs and rheology*. Boston, Butterworth-Heinemann.

J. R. Edwards and M.-S. Liou, 1998. Low-diffusion flux-splitting methods for flows at all speeds. *AIAA Journal*, **36**:1610–1617.

J. R. Edwards, R. K. Franklin, and M.-S. Liou, 2000. Low-diffusion flux-splitting methods for real fluid flows with phase transitions. *AIAA Journal*, **38**:1624–1633.

J. Eggers, 1995. Theory of drop formation. *Phys. Fluids*, **7**:941–953.

A. Elcrat, B. Fornberg, and K. Miller, 2001. Some steady axisymmetric vortex flows past a sphere. *J. Fluid Mech.*, **433**:315–328.

S. E. Elghobashi and G. C. Truesdell, 1993. On the two-way interaction between homogeneous turbulence and dispersed solid particles. I. Turbulence modification. *Phys. Fluids*, **5**:1790–1801.

B. Engquist and S. Osher, 1980. Stable and entropy-satisfying approximations for transonic flow calculations. *Math. Comp.*, **34**:45–75.

B. Engquist and S. Osher, 1981. One sided difference approximations for nonlinear conservation laws. *Math. Comp.*, **36**:321–351.

D. Enright, R. Fedkiw, J. Ferziger, I. Mitchell, 2002. A hybrid particle level set method for improved interface capturing. *J. Comput. Phys.*, **183**: 83–116.

A. Esmaeeli and G. Tryggvason, 1998. Direct numerical simulations of bubbly flows. Part I. Low Reynolds number arrays. *J. Fluid Mech.*, **377**:313–345.

A. Esmaeeli and G. Tryggvason, 1999. Direct numerical simulations of bubbly flows. Part II. Moderate Reynolds number arrays. *J. Fluid Mech.*, **385**:325–358.

A. Esmaeeli and G. Tryggvason, 2005. A direct numerical simulation study of the buoyant rise of bubbles at O(100) reynolds number. *Phys. Fluids*, **17**:093303.

V. Eswaran and S. B. Pope, 1988. An examination of forcing in direct numerical simulations of turbulence. *Comput. Fluids*, **16(3)**:257–278.

S. Evje and K. K. Fjelde, 2002. Hybrid flux-splitting schemes for a two-phase flow model. *J. Comput. Phys.*, **175**:674–701.

S. Evje and K. K. Fjelde, 2003. On a rough AUSM scheme for a one-dimensional two-phase flow model. *Comput. Fluids*, **32**:1497–1530.

S. Evje and T. Flåtten, 2003. Hybrid flux-splitting schemes for a common two-fluid model. *J. Comput. Phys.*, **192**:175–210.

E. A. Fadlun, R. Verzicco, P. Orlandi, and J. Mohd-Yusof, 2000. Combined immersed-boundary finite-difference methods for three-dimensional complex flow simulations. *J. Comp. Phys.*, **161**:35–60.

I. Faille and E. Heintze, 1999. A rough finite volume scheme for modeling two-phase flow in a pipeline. *Comput. Fluids*, **28**:213–241.

L. S. Fan and C. Zhu, 1998. *Principles of Gas–Solid Flows*. Cambridge: Cambridge University Press.

G. Farin, 1988. *Curves and Surfaces for Computer Aided Geometric Design. A Practical Guide*. New York, Academic Press.

R. Fedkiw, T. Aslam, B. Merriman, and S. Osher, 1999. A non-oscillatory eulerian approach to interfaces in multimaterial flows (the ghost fluid method). *J. Comput. Phys.*, **152**:457–492.

B. U. Felderhof and R. B. Jones, 1989. Displacement theorems for spherical solutions of the linear Navier–Stokes equations. *J. Math. Phys.*, **30**:339–42.

J. Feng, H. H. Hu and D. D. Joseph, 1994a. Direct simulation of initial value problems for the motion of solid bodies in a Newtonian fluid. Part 1: sedimentation. *J. Fluid Mech.*, **261**:95–134.

J. Feng, H. H. Hu and D. D. Joseph, 1994b. Direct simulation of initial value problems for the motion of solid bodies in a Newtonian fluid. Part 2: Couette and Poiseuille flows. *J. Fluid Mech.*, **277**:271–301.

J. Feng, P. Y. Huang and D. D. Joseph, 1995. Dynamic simulation of the motion of capsules in pipelines. *J. Fluid Mech.*, **282**:233–245.

J. Feng, P. Y. Huang and D. D. Joseph, 1996. Dynamic simulation of sedimentation of solid particles in an Oldroyd-B fluid. *J. Non-Newtonian Fluid Mech.*, **63**: 63–68.

B. Ferreol and D. H. Rothman, 1995. Lattice-Boltzmann simulations of flow-through fontainebleau sandstone. *Transp. Por. Med.*, **20**:3.

M. L. Ferrer, R. Duchowicz, B. Carrasco, J. G. García de la Torre, and A. U. Acuña, 2001. The conformation of serum albumin in solution: a combined phosphorescence depolarization-hydrodynamic modeling study. *Biophys. J.*, **80**:2422–30.

J. H. Ferziger and M. Perić, 2002. *Computational Methods for Fluid Dynamics*. New York, Springer, 3rd edn.

J. H. Ferziger, 2003. Interfacial transfer in Tryggvason's method. *Int J. Numerical Methods in Fluids*, **41**:551–560.

P. Fevrier, O. Simonin, and K. D. Squires, 2005. Partitioning of particle velocities in two-phase turbulent flows into a continuous field and a quasi-Brownian distribution: theoretical formalism and numerical study. *J. Fluid Mech.*, in press.

P. F. Fischer, G. K. Leaf, and J. M. Restrepo, 2002. Forces on particles in oscillatory boundary layers. *J. Fluid Mech.*, **468**:327–347.

K. K. Fjelde and K. H. Karlsen, 2002. High-resolution hybrid primitive-conservative upwind schemes for the drift flux model. *Comput. Fluids*, **31**:335–367.

R. Fletcher, 1976. Conjugate gradient methods for indefinite systems. In *Lecture Notes in Mathematics*, **506**. Berlin, Springer-Verlag.

B. Fornberg, 1988. Steady viscoud flow past a sphere at high Reynolds numbers. *J. Fluid Mech.*, **190**:471.

B. Fornberg, 1996. *A Practical Guide to Pseudospectral Methods*. Cambridge University Press.

L. P. Franca, S. L. Frey and T. J. R. Hughes, 1992a. Stabilized finite element methods: I. Application to the advective–diffusive model. *Comput. Methods Appl. Mech. Engng.*, **95**:253–276.

L. P. Franca and S. L. Frey, 1992b. Stabilized finite element methods: II. The incompressible Navier–Stokes equations. *Comput. Methods Appl. Mech. Engng.*, **99**:209–233.

D. Frenkel and B. Smit, 2002. *Understanding Molecular Simulation. From Algorithms to Simulations*. New York, Academic Press.

C. Frepoli, J. H. Mahaffy, and K. Ohkawa, 2003. Notes on the implementation of a fully-implicit numerical scheme for a two-phase three-field flow model. *Nucl. Eng. Des.*, **225**:191–217.

U. Frisch, B. Hasslacher, and Y. Pomeau, 1986. Lattice-gas automata for the Navier–Stokes equations. *Phys. Rev. Lett.*, **56**:1505.

U. Frisch, D. d'Humiéres, B. Hasslacher, P. Lallemand, Y. Pomeau, and J. P Rivet, 1987. Lattice gas hydrodynamics in two and three dimensions. *Complex Syst.*, 1:649.

M. J. Fritts, D. E. Fyre, and E. S. Oran, 1983. Numerical simulations of fuel droplet flows using a Langrangian triangular mesh. In *NASA CR-168263*.

I. J. Galea, 2004. *Permeability of foams with surfactant-covered interfaces.* PhD thesis, Yale University.

S. Ganapathy and J. Katz, 1995. Drag and lift forces on microscopic bubbles entrained by a vortex. *Phys. Fluids*, **7**:389–399.

P. R. Garabedian, 1998. *Partial Differential Equations.* American Mathematical Society, Providence RI, 2nd revised edition.

P. L. George, 1991. *Automatic Mesh Generation: Application to Finite Element Methods.* New York: John Wiley.

J. M. Ghidaglia, A. Kumbaro, and G. Le Coq, 2001. On the numerical solution to two fluid models via a cell centered finite volume method. *Eur. J. Mech. B-Fluids*, **20**:841–867.

D. Gidaspow, 1994. *Multiphase Flow and Fluidization.* Academic Press.

M. M. Giles, G. A. Jayne, S. Z. Rouhani, R. W. Shumway, G. L. Singer, D. D. Taylor, and W. L. Weaver, 1992. TRAC-BF1: An advanced best-estimate computer program for BWR accident analysis. Technical Report NUREG/CR-4356, U. S. Nuclear Regulatory Commission.

I. Ginzbourg and P. M. Alder, 1994. Boundary flow condition analysis for the three-dimensional lattice Boltzmann model. *J. Phys. II France*, **4**:191–214.

I. Ginzbourg and D. d'Humières, 1996. Local second-order boundary method for lattice Boltzmann models. *J. Stat. Phys.*, **84**:927–971.

I. Ginzbourg and D. d'Humières, 2003. Multi-reflection boundary conditions for lattice Boltzmann models. *Phys. Rev. E*, **68**:066614.

I. Ginzbourg, J.-P. Carlier, and C. Kao, 2004. Lattice Boltzmann approach to Richard's equation. pages 15–23.

I. Ginzbourg, 2005. Variable saturated flow with the anisotropic lattice Boltzmann methods.

B J. Glasser, I. G. Kevrekidis, and S. Sundaresan, 1996. One- and two-dimensional traveling wave solutions in gas-fluidized beds. *J. Fluid Mech.*, **306**:183–221.

B J. Glasser, I. G. Kevrekidis, and S. Sundaresan, 1997. Fully developed travelling wave solutions and bubble formation in fluidized beds. *J. Fluid Mech.*, **334**:157–188.

J. Glimm, 1982. Tracking of interfaces in fluid flow: Accurate methods for piecewise smooth problems, transonic shock and multidimensional flows. *Advances in Scientific Computing*, R. E. Meyer, ed., New York, Academic Press.

J. Glimm and O. McBryan, 1985. A computational model for interfaces. *Adv. Appl. Math.*, **6**:422–435.

J. Glimm, O. McBryan, R. Menikoff, and D. H. Sharp, 1986. Front tracking applied to Rayleigh–Taylor instability. *SIAM J. Sci. Stat. Comput.*, **7**:230.

R. Glowinski, T. Hesla, D. D. Joseph, T.-W. Pan and J. Periaux, 1997. Distributed Lagrange multiplier methods for particulate flows, in *Computational Science for the 21st Century* (M.-O. Bristeau, G. Etgen, W. Fitzgibbon, J. L. Lions, J. Periaux, M. F. Wheeler, eds.), pp. 270–279. Chichester: John Wiley.

R. Glowinski, T.-W. Pan, T. I. Hesla and D. D. Joseph, 1999. A distributed Lagrange multiplier/fictitious domain method for particulate flows. *Int. J. Multiphase Flow*, **25**:755–794.

T. G. Goktekin, A. W. Bargteil, and J. F. O'Brien, 2004. A method for animating viscoelastic fluids. *ACM Siggraph*.

D. Goldstein, R. Handler, and L. Sirovich, 1993. Modeling a no-slip flow boundary with an external force field. *J. Comput. Phys.*, **105**:354–366.

R. A. Gore and C. T. Crowe, 1990. Discussion of particle drag in a dilute turbulent two-phase suspension flow. *Int. J. Multiphase Flow*, **16**:359–361.

D. Gottlieb and S. A. Orszag, 1977. *Numerical Analysis of Spectral Methods: Theory and Applications*. SIAM.

W. G. Gray and P. C. Lee, 1977. On the theorem for local volume averaging for multiphase systems. *Int. J. Multiphase Flow*, **3**:333–340.

L. Greengard and V. Rokhlin, 1987. A fast algorithm for particle simulations. *J. Comput. Phys.*, **73**:325–48.

C. Guenther and M. Syamlal, 2001. The effect of numerical diffusion on isolated bubbles in a gas–solid fluidized bed. *Powder Tech.*, **116**:142–154.

A. K. Gunstensen, D. H. Rothman, S. Zaleski, and G. Zanetti, 1991. Lattice Boltzmann model of immiscible fluids. *Phys. Rev. A*, **43**:4320.

A. K. Gunstensen and D. H. Rothman, 1993. Lattice Boltzmann studies of immiscible two phase flow through porous media. *J. Geophys. Res.*, **98**:6431.

W. L. Haberman and R. M. Sayre, 1958. Motion of rigid and fluid spheres in stationary and moving liquids inside cylindrical tubes. David Taylor Model Basin Report No. 1143.

E. Hairer and G. Wanner, 1996. *Solving Ordinary Differential Equations II. Stiff and Differential-Algebraic Problems*. New York, Springer.

G. Hall and J. M. Watt, 1976. *Modern Numerical Methods for Ordinary Differential Equations*. Oxford, Clarendon Press.

B. Halle and M. Davidovic, 2003. Biomolecular hydration: From water dynamics to hydrodynamics. *Proc. Natl. Acad. Sci. USA*, **100**:12135–40.

P. Hansbo 1992. The characteristic streamline diffusion method for the time-dependent incompressible Navier–Stokes equations. *Comput. Methods Appl. Mech. Engng.*, **99**:171–186.

J. Happel and H. Brenner, 1983. *Low-Reynolds Number Hydrodynamics*. Noordhoff International Publishing, Leyden, The Netherlands.

S. E. Harding, 1995. On the hydrodynamic analysis of macromolecular conformation. *Biophys. Chem.*, **55**:69–93.

F. H. Harlow and J. E. Welch, 1965. Numerical calculation of time-dependent viscous incompressible flow of fluid with a free surface. *Phys. Fluid*, **8**:2182–2189.

F. H. Harlow and J. E. Welch, 1966. Numerical study of large-amplitude free-surface motions. *Phys. Fluid*, **9**:842–851.

F. H. Harlow and A. A. Amsden, 1971. Numerical fluid dynamics calculation method for all flow speeds. *J. Comput. Phys.*, **8**:197–231.

F. H. Harlow and A. A. Amsden, 1975. Numerical calculation of multiphase flow. *J. Comput. Phys.*, **17**:19–52.

S. Harris, 1971. *An Introduction to the Theory of the Boltzmann Equation.* New York, Dover.

A. Harten, P. D. Lax, and B. van Leer, 1983. On upstream differencing and Godunov-type schemes for hyperbolic conservation laws. *SIAM Review,* **25**:35–61.

H. Hasimoto, 1959. On the periodic fundamental solutions of the Stokes equation and their application to viscous flow past a cubic array of spheres. *J. Fluid Mech,* **5**:317–328.

X. He, L.-S. Luo, and M. Dembo, 1996. Some progress in lattice Boltzmann method. Part I. Nonuniform mesh grids. *J. Comput. Phys.,* **129**:357–363.

X. He and G. D. Doolen, 1997a. Lattice Boltzmann method on a curvilinear coordinate system: Vortex shedding behind a circular cylinder. *Phys. Rev. E,* **56**: 434–440.

X. He and G. D. Doolen, 1997b. Lattice Boltzmann method on curvilinear coordinates system: Flow around a circular cylinder. *J. Comput. Phys.,* **134**:306–315.

X. He and L.-S. Luo, 1997a. Lattice Boltzmann model for the incompressible Navier–Stokes equation. *J. Stat. Phys.,* **88**:927–944.

X. Y. He and L.-S. Luo, 1997b. *A priori* derivation of the lattice Boltzmann equation. *Phys. Rev. E,* **55**:R6333.

X. Y. He and L.-S. Luo, 1997c. Theory of the lattice Boltzmann method: From the Boltzmann equation to the lattice Boltzmann equation. *Phys. Rev. E,* **56**:6811.

X. He, L.-S. Luo, and M. Dembo, 1997a. Some progress in lattice Boltzmann method: Enhancement of reynolds number in simulations. *Physica A,* **239**: 276–285.

X. He, Q. Zou, L.-S. Luo, and M. Dembo, 1997b. Analytic solutions and analysis on non-slip boundary condition for the lattice Boltzmann BGK model. *J. Stat. Phys.,* **87**:115–136.

X. Y. He, X. W. Shan, and G. D. Doolen, 1998. Discrete Boltzmann equation model for non-ideal gases. *Phys. Rev. E.,* **57**:R13.

X. Y. He, S. Y. Chen, and R. Y. Zhang, 1999a. A lattice Boltzmann scheme for incompressible multiphase flow and its application in simulation of Rayleigh–Taylor instability. *J. Comput. Phys.,* **152**:642.

X. Y. He, R. Y. Zhang, S. Y. Chen, and G. D. Doolen, 1999b. On the three-dimensional Rayleigh–Taylor instability. *Phys. Fluids,* **11**:1143.

X. Y. He and G. D. Doolen, 2002. Thermodynamic foundation of kinetic theory and lattice Boltzmann models for multiphase flows. *J. Stat. Phys.,* **107**:309–328.

M. Herrmann, 2008. A balanced force refined level set grid method for two-phase flows on unstructured flow solver grids. *J. Comput. Phys.,* **227**: 2674–2706.

T. I. Hesla, A. Y. Huang, and D. D. Joseph, 1993. A note on the net force and moment on a drop due to surface forces. *J. Colloid Interface Sci.,* **158**:255–257.

T. Hibiki and M. Ishii, 2000. One-group interfacial area transport of bubbly flows in vertical round tubes. *Int. J. Heat Mass Transfer,* **43**:2711–2726.

T. Hibiki and M. Ishii, 2002. Distribution parameter and drift velocity of drift-flux model in bubbly flow. *Int. J. Heat Mass Transfer,* **45**:707–721.

T. Hibiki and M. Ishii, 2003. One-dimensional drift-flux model for two-phase flow in a large diameter pipe. *Int. J. Heat Mass Transfer,* **46**:1773–1790.

F. B. Hildebrand, 1956. *Introduction to Numerical Analysis*. New York, McGraw-Hill.

R. J. Hill, D. L. Koch, and A. J. C. Ladd, 2001a. The first effects of fluid inertia on flows in ordered and random arrays of spheres. *J. Fluid Mech.*, **448**:213–241.

R. J. Hill, D. L. Koch, and A. J. C. Ladd, 2001b. Moderate-Reynolds-number flows in ordered and random arrays of spheres. *J. Fluid Mech.*, **448**:243–278.

R. J. Hill and D. L. Koch, 2002. The transition from steady to weakly turbulent flow in a close-packed ordered array of spheres. *J. Fluid Mech.*, **465**:59–97.

M. Hilpert, J. F. McBride, and C. T. Miller, 2001. Investigation of the residual-funicular nonwetting-phase-saturation relation. *Adv. Water Res.*, **2**:157–177.

M. Hilpert and C. T. Miller, 2001. Pore-morphology-based simulation of drainage in totally wetting porous media. *Adv. Water Res.*, **24(3/4)**:243–255.

J. O. Hinze, 1975. *Turbulence*. New York, McGraw-Hill.

C. Hirsch, 1988. *Numerical Computations of Internal and External Flows*, Vol. I. New York, Wiley.

J. O. Hirschfelder, C. F. Curtiss, and R. B. Bird, 1954. *Molecular Theory of Gases and Liquids*. New York, Wiley.

C. W. Hirt, J. L. Cook, and T. D. Butler, 1970. A Lagrangian method for calculating the dynamics of an incompressible fluid with a free surface. *J. Comput. Phys.*, **5**:103–124.

C. W. Hirt, A. A. Amsden, and J. L. Cook, 1974. Arbitrary Lagrangian–Eulerian computing method for all flow speeds. *J. Comput. Phys.*, **14**:227–253.

C. W. Hirt and B. D. Nichols, 1981. Volume of fluid (VOF) method for the dynamics of free boundaries. *J. Comput. Phys.*, **39**:201–226.

D. J. Holdych, D. Rovas, J. G. Geogiadis, and R. O. Buckius, 1998. An improved hydrodynamics formulation for multiphase flow lattice Boltzmann methods. *Int. J. Mod. Phys. C*, **9**:1393.

S. L. Hou, X. W. Shan, Q. S. Zou *et al.*, 1997. Evaluation of two lattice Boltzmann models for multiphase flows. *J. Compu. Phys.*, **138**:695–713.

T. Y. Hou and P. LeFloch, 1994. Why non-conservative schemes converge to the wrong solutions: Error analysis. *Math. Comput.*, **62**:497–530.

H. H. Hu, D. D. Joseph and M. J. Crochet, 1992. Direct simulation of fluid particle motions. *Theor. Comput. Fluid Dynam.*, **3**:285–306.

H. H. Hu, A. Fortes and D. D. Joseph, 1993. Experiments and direct simulations of fluid particle motions. *Intern. Video J. Engng Res.*, **2**:17–24.

H. H. Hu, 1995. Motion of a circular cylinder in a viscous liquid between parallel plates. *Theor. Comput. Fluid Dynam.*, **7**:441–455.

H. H. Hu, 1996. Direct simulation of flows of solid–liquid mixtures. *Int. J. Multiphase Flow*, **22**:335–352.

H. H. Hu, N. Patankar and M. Zhu, 2001. Direct numerical simulations of fluid-solid systems using the Arbitrary-Lagrangian-Eulerian technique. *J. of Comp. Phys.*, **169**:427–462.

K. Huang, 2001. *Introduction to Statistical Physics*. Science.

P. Y. Huang, J. Feng and D. D. Joseph, 1994. The turning couples on an elliptic particle settling in a vertical channel. *J. Fluid Mech.*, **271**:1–16.

P. Y. Huang, J. Feng, H. H. Hu and D. D. Joseph, 1997. Direct simulation of the motion of solid particles in Couette and Poiseuille flows of viscoelastic fluids. *J. Fluid Mech.*, **343**:73–94.

P. Y. Huang, H. H. Hu and D. D. Joseph, 1998. Direct simulation of the sedimentation of elliptic particles in Oldroyd-B fluids. *J. Fluid Mech.*, **362**:297–325.

A. Huerta and W. K. Liu, 1988. Viscous flow with large free surface motion. *Comput. Methods Appl. Mech. Engng.*, **69**:277–324.

T. J. R. Hughes, W. K. Liu and T. K. Zimmerman, 1981. Lagrangian–Eulerian finite element formulation for incompressible viscous flows. *Comput. Methods Appl. Mech. Engng.*, **29**:329–349.

T. J. R. Hughes and G. M. Hulbert, 1988. Space-time finite element methods for elasto-dynamics: formulations and error estimates. *Comput. Methods Appl. Mech. Engng.*, **66**:339–363.

K. Hui, P. K. Haff, and R. Jackson, 1984. Boundary conditions for high-shear grain flows. *J. Fluid Mech.*, **145**:223–233.

G. J. Hwang and H. H. Shen, 1989. Modeling the solid phase stress in a fluid–solid mixture. *Int. J. Multiphase Flow*, **15**:257–268.

Y.-H. Hwang, 2003. Upwind scheme for non-hyperbolic systems. *J. Comput. Phys.*, **192**:643–676.

M. A. Hyman, 1952. Non-iterative numerical solution of boundary-value problems. *Appl. Sci. Res. Sec.*, **B2**:325–351.

T. Inamuro, T. Ogata, S. Tajime *et al.*, 2004. A lattice Boltzmann method for incompressible two-phase flows with large density differences. *J. Compu. Phys.*, **198**:628–644.

M. Ishii, 1975. *Thermo-Fluid Dynamic Theory of Two-Phase Flow*. Eyrolles, Paris.

M. Ishii and S. Kim, 2004. Development of one-group and two-group interfacial area transport equation. *Nucl. Sci. Eng.*, **146**:257–273.

M. Ishii and N. Zuber, 1979. Drag coefficient and relative velocity in bubbly, droplet or particulate flows. *A. I. Ch. E. J.*, **25**:843–855.

M. Ishii, T. C. Chawla, and N. Zuber, 1976. Constitutive equation for vapor drift velocity in two-phase annular flow. *AIChE J*, **22**:283–289.

M. Ishii, Sun X. D., and S. Kim, 2003. Modeling strategy of the source and sink terms in the two-group interfacial area transport equation. *Ann. Nucl. En.*, **30**:1309–1331.

M. Ishii, S. S. Paranjape, S. Kim, and J. Kelly, 2004. Interfacial structures and interfacial area transport in downward two-phase bubbly flow. *Int. J. Multiphase Flow*, **30**:779–801.

R. I. Issa and P. J. Oliveira, 1994. Numerical prediction of phase separation in two-phase flow through T-junctions. *Comput. Fluids*, **23**:347–372.

R. I. Issa and M. H. W. Kempf, 2003. Simulation of slug flow in horizontal and nearly horizontal pipes with the two-fluid model. *Int. J. Multiphase Flow*, **29**:69–95.

R. Jackson, 2000. *The Dynamics of Fluidized Particles*. Cambridge U.K., Cambridge University Press.

W. D. Jackson, 1999. *Classical Electrodynamics*. New York, Wiley.

D. Jacqmin, 1999. Calculation of two-phase Navier–Stokes flows using phase-field modeling. *J. Comput. Phys.*, **155**:96–127.

D. Jacqmin, 2000. Contact-line dynamics of a diffuse fluid interface. *J. Fluid Mech.*, **402**:57–88.

D. Jamet, O. Lebaigue, N. Coutris, and J. M. Delhaye, 2001. Feasibility of using the second gradient theory for the direct numerical simulations of liquid-vapor flows with phase-change. *J. Comput. Phys.*, **169**:624–651.

D. Jamet, D. Torres, and J. U. Brackbill, 2002. On the theory and computation of surface tension: The elimination of parasitic currents through energy conservation in the second-gradient method. *J. Comput. Phys.*, **182**:262–276.

H. Jeffreys and B. S. Jeffreys, 1988. *Methods of Mathematical Physics.* Cambridge University Press, Cambridge, England, 3rd edn.

J. T. Jenkins and M. W. Richman, 1986. Boundary conditions for plane flows of smooth, nearly elastic circular disks. *J. Fluid Mech.*, **171**:53–69.

J. T. Jenkins, 1992. Boundary conditions for rapid granular flow: Flat, frictional wall. *J. Appl. Mech.*, **114**:120–127.

J. Jimenez and P. Moin, 1991. The minimal flow unit in near-wall turbulence. *J. Fluid. Mech.*, **225**:213–240.

B. C. V. Johansson, 1993. Boundary conditions for open boundaries for the incompressible Navier–Stokes equation. *J. Comput. Phys.*, **105**:233–251.

A. Johnson and T. E. Tezduyar, 1994. Mesh update strategies in parallel finite element computations of flow problems with moving boundaries and interfaces. *Comput. Methods Appl. Mech. Engng.*, **119**: 73–94.

A. Johnson, 1995. Mesh update strategies in parallel finite element computations of flow problems with moving boundaries and interfaces. Ph. D. thesis, University of Minnesota.

A. Johnson and T. E. Tezduyar, 1996. Simulation of multiple spheres falling in a liquid-filled tube. *Comput. Methods Appl. Mech. Engng.*, **134**:351–373.

A. Johnson and T. E. Tezduyar, 1997. 3D simulation of fluid–particle interactions with the number of particles reaching 100. *Comput. Methods Appl. Mech. Engng.*, **145**:301–321.

A. Johnson and T. E. Tezduyar, 1999. Advanced mesh generation and update methods for 3D flow simulations. *Computational Mech.*, **23**:130–143.

P. C. Johnson and R. Jackson, 1987. Frictional-collisional constitutive relations for granular materials, with applications to plane shearing. *J. Fluid Mech.*, **130**: 187–202.

T. A. Johnson and V. C. Patel, 1999. Flow past a sphere up to a Reynolds number of 300. *J. Fluid Mech.*, **378**:19–70.

A. V. Jones and A. Prosperetti, 1985. On the suitability of first-order differential models for two-phase flow prediction. *Int. J. Multiphase Flow*, **11**:133–148.

A. V. Jones and A. Prosperetti, 1987. The linear stability of general two-phase flow models – II. *Int. J. Multiphase Flow*, **13**:161–171.

D. D. Joseph, 1996. Flow induced microstructure in Newtonian and viscoelastic fluids, in *Proc. 5th World Congress of Chem. Engng, Particle Technology Track*, San Diego, July 14–18. *AIChE*, **6**:3–16.

W. Kalthoff, S. Schwarzer, and H. J. Herrmann, 1997. Algorithm for the simulation of particle suspensions with inertia effects. *Phys. Rev.*, **E56**:2234–2242.

Q. H. Kang, D. X. Zhang, and S. Y. Chen, 2002a. Lattice Boltzmann simulation of chemical dissolution in porous media. *Phys. Rev. E*, **65**:Art. No. 036318.

Q. H. Kang, D. X. Zhang, and S. Y. Chen, 2002b. Unified lattice Boltzmann method for flow in multiscale porous media. *Phys. Rev. E*, **66**:Art. No. 056307.

H. Karema and S. Lo, 1999. Efficiency of interphase coupling algorithms in fluidized bed conditions. *Comput. Fluids*, **28**:323–360.

S. Karni, 1994. Multi-component flow calculations by a consistent primitive algorithm. *J. Comput. Phys.*, **112**:31–43.

G. E. Karniadakis, M. Israeli and S. A. Orszag, 1991. High-order splitting methods for the incompressible Navier–Stokes equations. *J. Comp. Phys.*, **97**:414–443.

G. E. Karniadakis and G. S. Triantafyllou, 1992. Three-dimensional dynamics and transition to turbulence in the wake of bluff objects. *J. Fluid Mech.*, **238**:1–30.

G. E. Karniadakis and S. J. Sherwin, 1999. *Spectral/HP Element Methods for CFD*. New York, Oxford University Press.

A. Karnis, H. L. Goldsmith and S. G. Mason, 1966. The flow of suspensions through tubes, V. Inertial effects. *Can. J. Chem. Engng.*, **44**:181–193.

H. Kato, M. Miyanaga, H. Yamaguchi, and M. M. Guin, 1995. Frictional drag reduction by injecting bubbly water into turbulent boundary layer and the effect of plate orientation. In *Proceedings of the 2nd International Conference on Multiphase Flow '95, ICMF95*, pages 31–38.

M. R. Kennedy, C. Pozrikidis, and R. Skalak, 1994. Motion and deformation of liquid drops and the rheology of dilute emulsions in simple shear flow. *Comput. Fluids*, **23**:251–278.

D. Kim and H. Choi, 2002. Laminar flow past a sphere rotating in the streamwise direction. *J. Fluid Mech.*, **461**:365–386.

I. Kim, S. Elghobashi, and W. A. Sirignano, 1993. 3-dimensional flow over 3-spheres placed side by side. *J. Fluid Mech.*, **246**:465–488.

I. Kim, S. Elghobashi, and W. A. Sirignano, 1998. On the equation for spherical-particle motion: effect of Reynolds and acceleration numbers. *J. Fluid Mech.*, **367**:3221–254.

J. Kim and P. Moin, 1985. Application of a fractional step method to incompressible Navier–Stokes equations. *J. Comput. Phys.*, **59**:308–323.

J. Kim, D. Kim, and H. Choi, 2001. An immersed-boundary finite-volume method for simulations of flow in complex geometries. *J. Comput. Phys.*, **171**:132–150.

S. Kim and S. J. Karrila, 1991. *Microhydrodynamics: Principles and Selected Applications*. London, Butterworth-Heinemann.

D. L. Koch, 1990. Kinetic theory for a monodisperse gas–solid suspension. *Phys. Fluids*, **A2**:1711–1723.

D. L. Koch and A. S. Sangani, 1999. Particle pressure and marginal stability limits for a homogeneous monodisperse gas fluidized bed: Kinetic theory and numerical simulations. *J. Fluid Mech.*, **400**:229–263.

Y. Kodama, A. Kakugawa, T. Takahashi, S. Nagaya, and K. Sugiyama, 2003. Microbubbles: Drag reduction and applicability to ships. In *Twenty-Fourth Symposium on Naval Hydrodynamics*, Naval Studies Board (NSB). Available at: http://books. nap. edu/books/NI000511/html/.

D. B. Kothe, R. C. Mjolsness, and M. D. Torrey, 1991. Ripple: A computer program for incompressible flows with free surfaces. Tech. Rep. LA-12007-MS, Los Alamos National Laboratory.

A. G. Kravchenko and P. Moin, 1997. On the effect of numerical errors in large eddy simulations of turbulent flows. *J. Comput. Phys.*, **131**:310–322.

A. M. Kraynik and Reinelt D. A., 1992. Extensional motions of spatially periodic lattices. *Int. J. Multiphase Flow*, **18**:1045–59.

R. F. Kunz, B. W. Siebert, W. K. Cope, N. F. Foster, S. P. Antal, and S. M. Ettorre, 1998. A coupled phasic exchange algorithm for three-dimensional multi-field analysis of heated flows with mass transfer. *Comput. Fluids*, **27**:741–768.

R. F. Kunz, W. K. Cope, and S. Venkateswaran, 1999. Development of an implicit method for multi-fluid flow simulations. *J. Comput. Phys.*, **152**:78–101.

R. F. Kunz, D. A. Boger, D. R. Stinebring, S. Chyczewski, J. W. Lindau, H. J. Gibeling, S. Venkateswaran, and T. R. Govindan, 2000. A preconditioned Navier–Stokes method for two-phase flows with application to cavitation prediction. *Comput. Fluids*, **29**:849–875.

R. Kurose and S. Komori, 1999. Drag and lift forces on a rotating sphere in a linear shear flow. *J. Fluid Mech.*, **384**:183–206.

S. Kwak and C. Pozrikidis, 1998. Adaptive triangulation of evolving, closed, or open surfaces by the advancing-front method. *J. Comput. Phys.*, **145**:61–88.

A. J. C. Ladd, 1994a. Numerical simulations of particulate suspensions via a discretized Boltzmann equation. Part 1. Theoretical foundation. *J. Fluid Mech.*, **271**:285–309.

A. J. C. Ladd, 1994b. Numerical simulations of particulate suspensions via a discretized Boltzmann equation. Part 2. Numerical results. *J. Fluid Mech.*, **271**:311–339.

B. Lafaurie, C. Nardone, R. Scardovelli, S. Zaleski, and G. Zanetti, 1994. Modelling merging and fragmentation in multiphase flows with surfer. *J. Comp. Phys.*, **113**:134–147.

M.-C. Lai and C. S. Peskin, 2000. An immersed boundary method with formal second-order accuracy and reduced numerical viscosity. *J. Comput. Phys.*, **160**:705–719.

P. Lallemand and L.-S. Luo, 2000. Theory of the lattice Boltzmann method: Dispersion, dissipation, isotropy, Galilean invariance, and stability. *Phys. Rev. E*, **61**:6546–6562.

P. Lallemand and L.-S. Luo, 2003. Lattice Boltzmann method for moving boundaries. *J. Computat. Phys.*, **184(2)**:406–421.

H. Lamb, 1932. *Hydrodynamics*. New York, Dover.

L. D. Landau and E. M. Lifshitz, 1987. *Fluid Mechanics*. Oxford, Pergamon.

J. Langford, 2000. Towards ideal large-eddy simulation. PhD. thesis, Department of Theoretical & Applied Mechanics, University of Illinois, Urbana, IL.

M. Larsen, E. Hustvedt, P. Hedne, and T. Straume, 1997. PeTra: A novel computer code for simulation of slug flow. Paper 3884 1, SPE Annual Technical Conference and Exhibition, San Antonio Texas, 5–8 October.

C. J. Lawrence and S. Weinbaum, 1988. The unsteady force on a body at low Reynolds number; the axisymmetric motion of a spheroid. *J. Fluid Mech.*, **189**:463–489.

C. J. Lawrence and R. Mei, 1995. Long-time behaviour of the drag on a body in impulsive motion. *J. Fluid Mech.*, **283**:307–327.

L. Lee and R. J. LeVeque, 2003. An immersed interface method for incompressible Navier–Stokes equations. *SIAM J. Sci. Comput.*, **25**:832–856.

T. Lee and C. Lin, 2005. A stable discretization of the lattice Boltzmann equation for simulation of incompressible two-phase flows at high density ratio. *J. Comput. Phys.*, **206**:16–47.

D. Legendre and J. Magnaudet, 1998. The lift force on a spherical bubble in a viscous linear shear flow. *J. Fluid Mech.*, **368**:81–126.

B. P. Leonard, 1979. A stable and accurate convective modelling procedure based on quadratic upstream interpolation. *Comput. Meth. Appl. Mech. Engrg.*, **19**:59–98.

B. P. Leonard and S. Mokhtari, 1990. Beyond 1st-order upwinding – The ultrasharp alternative for nonoscillatory steady-state simulation of convection. *Int. J. Numer. Meth. Eng.*, **30**:729–766.

R. J. LeVeque, 1992. *Numerical Methods for Conservation Laws*. Birkhäuser, Basel.

R. J. LeVeque and Z. Li, 1994. The immersed interface method for elliptic equations with discontinuous coefficients and singular sources. *SIAM J. Numer. Anal.*, **31**:1019.

R. J. LeVeque, 2002. *Finite Volume Methods for Hyperbolic Problems*. Cambridge U. P, Cambridge University Press.

J. Li and Kuipers J. A. M., 2003. Gas–particle interactions in dense gas-fluidized beds. *Chem. Eng. Sci.*, **58**:711–718.

X. L. Li, B. X. Jin, and J. Glimm, 1996. Numerical study for the three-dimensional Rayleigh–Taylor instability through the TVD/AC scheme and parallel computation. *J. Comput. Phys.*, **126**:1904.

D. R. Liles and W. H. Reed, 1978. A semi-implicit method for two-phase fluid dynamics. *J. Comput. Phys.*, **26**:390–407.

M.-S. Liou and C. Steffen, 1993. A new flux-splitting scheme. *J. Comput. Phys.*, **107**:23–39.

M.-S. Liou, 1996. A sequel to AUSM: AUSM+. *J. Comput. Phys.*, **129**:364–382.

L. Little and Y. Saad, 1999. Block LU preconditioners for symmetric and nonsysmmetric saddle point problem. University of Minnesota Supercomputing Institute Research report UMSI 99/104.

X. Liu, R. P. Fedkiw, and M. Kang, 2000. A boundary condition capturing method for Poisson's equation on irregular domains. *J. Comput. Phys.*, **160**:151–178.

M. Loewenberg and E. J. Hinch, 1996. Numerical simulations of a concentrated emulsion in shear flow. *J. Fluid. Mech.*, **321**:395–419.

M. Loewenberg and E. J. Hinch, 1997. Collision of two deformable drops in shear flow. *J. Fluid Mech.*, **338**:299–315.

H. Lomax, T. H. Pulliam, and D. W. Zingg, 2001. *Fundamentals of Computational Fluid Dynamics*. Springer-Verlag.

P. M. Lovalenti and J. F. Brady, 1993a. The force on a bubble, drop, or particle in arbitrary time-dependent motion at small Reynolds number. *Phys. Fluids*, **5**:2104–2116.

P. M. Lovalenti and J. F. Brady, 1993b. The force on a sphere in a uniform flow with small-amplitude oscillations at finite Reynolds number. *J. Fluid Mech.*, **256**:607–614.

P. M. Lovalenti and J. F. Brady, 1993c. The hydrodynamic force on a rigid particle undergoing arbitrary time-dependent motion at small Reynolds number. *J. Fluid Mech.*, **256**:561–605.

P. M. Lovalenti and J. F. Brady, 1995. The temporal behaviour of the hydrodynamic force on a body in response to an abrupt change in velocity at small but finite Reynolds number. *J. Fluid Mech.*, **293**:35–46.

J. Lu, A. Fernandez and G. Tryggvason, 2005. The effect of bubbles on the wall shear in a turbulent channel flow. *Phys. Fluids*, **17**:095102.

C. K. K. Lun, S. B. Savage, D. J. Jeffrey, and N. Chepurniy, 1984. Kinetic theories of granular flows: inelastic particles in Couette flow and slightly inelastic particles in a general flow field. *J. Fluid Mech.*, **140**:223–256.

T. S. Lund, 1993. Private communication.

L.-S. Luo, 1997. Analytic solutions of linearized lattice Boltzmann equation for simple flows. *J. Stat. Phys.*, **88**:913–926.

L. S. Luo, 1998. Unified theory of lattice Boltzmann models for nonideal gases. *Phys. Rev. Lett.*, **81**:1618.

L. S. Luo, 2000. Theory of the lattice Boltzmann method: lattice Boltzmann models for nonideal gases. *Phys. Rev. E.*, **62**:4982.

D. Ma and G. Ahmadi, 1988. A kinetic model for rapid granular flows of nearly elastic particles including interstitial fluid effects. *Powder Technol.*, **56**:191–207.

R. Macian-Juan and J. H. Mahaffy, 1998. Numerical diffusion and the tracking of solute fields in system codes. Part I. One-dimensional flows. *Nucl. Eng. Des.*, **179**:297–319.

J. Magnaudet, M. Rivero, and J. Fabre, 1995. Accelerated flows past a rigid sphere or a spherical bubble. 1. Steady straining flow. *J. Fluid Mech.*, **284**:97–135.

J. Magnaudet, 2003. Small inertial effects on a spherical bubble, drop or particle moving near a wall in a time-dependent linear flow. *J. Fluid Mech.*, **485**:115–142.

J. H. Mahaffy, 1982. A stability-enhanicng two-step method for fluid flow calculations. *J. Comput. Phys.*, **46**:329–341.

J. H. Mahaffy, 1993. Numerics of codes: Stability, diffusion, and convergence. *Nucl. Eng. Des.*, **145**:131–145.

J. H. Mahaffy, 2004. Purpose of simulation. Lecture Notes for Course NucE 470 "Power Plant Simulation", Pennsylvania State University, http://www. personal. psu. edu/faculty/j/h/jhm/470/lectures/1.html.

N. S. Martys and H. D. Chen, 1996. Simulation of multicomponent fluids in complex three-dimensional geometries by the lattice Boltzmann method. *Phys. Rev. E.*, **53**:743.

B. Maury, 1997. A many-body lubrication model. *C. R. Acad. Sci. Paris*, **325**, Serie I, 1053–1058.

B. Maury, 1999. Direct simulations of 2D fluid–particle flows in biperiodic domains. *J. Comp. Phys.*, submitted.

D. J. Mavriplis, 1997. Unstructured grid techniques. *Ann. Rev. Fluid. Mech*, **29**:473–514.

M. R. Maxey and J. J. Riley, 1983. Equation of motion for a small rigid sphere in a nonuniform flow. *Phys. Fluids*, **26**:883–889.

M. R. Maxey, 1987. The gravitational settling of aerosol particles in homogeneous turbulence and random flow fields. *J. Fluid Mech.*, **174**:441–465.

J. B. McLaughlin, 1991. Inertial migration of a small sphere in linear shear flows. *J. Fluid Mech.*, **224**:261–274.

J. B. McLaughlin, 1993. The lift on a small sphere in wall-bounded linear shear flows. *J. Fluid Mech.*, **246**:249–265.

J. B. McLaughlin, 1994. Numerical computation of particle–turbulent interaction. *Int. J. Multiphase Flow*, **20**:211–232.

G. McNamara and G. Zanetti, 1988. Use of the Boltzmann equation to simulate lattice-gas automata. *Phys. Rev. Lett.*, **61**:2332.

J. S. McNown, H. M. Lee, M. B. McPherson and S. M. Ebgez, 1948. *Proc. 7th Int. Congr. Appl. Mech.*, **2**:17.

D. M. McQueen and C. S. Peskin, 1989. A three-dimensional computational method for blood flow in the heart: (ii) contractile fibers. *Comput. Phys.*, **82**:289–297.

R. Mei and R. J. Adrian, 1992. Flow past a sphere with an oscillation in the free-stream and unsteady drag at finite Reynolds number. *J. Fluid Mech.*, **237**: 133–174.

R. Mei, 1994. Flow due to an oscillating sphere and an expression for unsteady drag on the sphere at finite Reynolds number. *J. Fluid Mech.*, **270**:133–174.

P. E. Merilees, 1973. The pseudospectral approximation applied to the shallow water equations on a sphere. *Atmosphere*, **11**:13–20.

C. L. Merkle and S. Deutsch, 1990. Drag reduction in liquid boundary layers by gas injection. *Prog. Astronaut. Aeronaut.*, **123**:351–412.

E. E. Michaelides, 1992. A novel way of computing the Basset term in unsteady multiphase flow computations. *Phys. Fluids*, **A4**:1579–1582.

E. Michaelides, 1997. The transient equation of motion for particles, bubbles, and droplets. *J. Fluids Eng.*, **119**:233–247.

T. F. Miller and F. W. Schmidt, 1988. Use of a pressure-weighted interpolation method for the solution of the incompressible Navier–Stokes equations on a non-staggered grid system. *Numer. Heat Transfer*, **14**:213–233.

T. F. Miller and D. J. Miller, 2003. A Fourier analysis of the IPSA/PEA algortihms applied to multiphase flows with mass transfer. *Comput. Fluids*, **32**:197–221.

R. Mittal and S. Balachandar, 1996. Direct numerical simulations of flow past elliptic cylinders. *J. Comput. Phys.*, **124**:351–367.

R. Mittal, 1999. A Fourier–Chebyshev spectral collocation method for simulating flow past spheres and spheroids. *Int. J. Numer. Meth. Fluids*, **30**:921–937.

R. Mittal and G. Iaccarino, 2005. Immersed boundary methods. *Ann. Rev. Fluid Mech.*, **37**:239–261.

R. Moe and K. H. Bendiksen, 1993. Transient simulation of 2D and 3D stratified and intermittent two-phase flows. Part I: Theory. *Int. J. Numer. Meth. Fluids*, **16**:461–487.

J. Mohd-Yusof, 1996. Interaction of massive particles with turbulence. PhD. thesis, Department of Mechanical Engineering, Cornell University.

J. Mohd-Yusof, 1997. Combined immersed boundaries b-spline methods for simulations of flows in complex geometries. CTR Annual Research Briefs, NASA Ames, Stanford University.

J. Montagnat, H. Delingette, and N. Ayache, 2001. A review of deformable surfaces: topology, geometry and deformation. *Image Vision Comput.*, **19**:1023–40.

F. Moukalled and M. Darwish, 2002. The performance of geometric conservation-based algorithms for incompressible multifluid flow. *Numer. Heat Transfer*, **B45**:343–368.

V. A. Mousseau, 2004. Implicitly balanced solution of the two-phase flow equations coupled to nonlinear heat conduction. *J. Comput. Phys.*, **200**:104–132.

W. Mulder, S. Osher, and J. A. Sethian, 1992. Computing interface motion in compressible gas dynamics. *J. Comput. Phys.*, **100**:449.

M. A. Naciri, 1992. Contribution á l'Étude des Forces Exercées par un Liquide sur un Bulle de Gaz. PhD thesis, École Centrale de Lyon.

R. Natarajan and A. Acrivos, 1993. The instability of the steady flow past spheres and disks. *J. Fluid Mech.*, **254**:323–344.

M. B. Nemer, X. Chen, D. H. Papadopoulos, J. Błazwzdziewicz, and M. Loewenberg, 2004. Hindered and enhanced coalescence of drops in Stokes flow. *Phys. Rev. Lett.*, **11**:4501-1–4.

R. I. Nigmatulin, 1979. Spatial averaging in the mechanics of heterogeneous and dispersed systems. *Int. J. Multiphase Flow*, **5**:353–385.

Y.-Y. Niu, 2000. Simple conservative flux splitting for multi-component flow calculations. *Num. Heat Transfer*, **B38**:203–222.

Y.-Y. Niu, 2001. Advection upwinding splitting method to solve a compressible two-fluid model. *Int. J. Numer. Meth. Fluids*, **36**:351–371.

W. F. Noh and P. Woodward, 1976. SLIC (simple line interface calculation). In A. I. van de Vooren and P. J. Zandbergen, editors, *Proceedings, Fifth International Conference on Fluid Dynamics*, Volume 59 of *Lecture Notes in Physics*, pages 330–340, Berlin, Springer.

T. Nomura and T. J. R. Hughes, 1992. An Arbitrary Lagrangian–Eulerian finite element method for interaction of fluid and a rigid body. *Comput. Methods Appl. Mech. Engng.*, **95**:115–138.

M. Nordsveen and R. Moe, 1999. Preconditioned Krylov subspace methods used in solving two-dimensional transient two-phase flows. *Int. J. Numer. Meth. Fluids*, **31**:1141–1156.

Nuclear Safety Analysis Division, Information Systems Laboratories, 2001. RELAP5/MOD3.3 Code Manual, Revision 1. Technical Report NUREG/CR-5535, U. S. Nuclear Regulatory Commission.

M. Ohta, S. Haranaka, Y. Yoshida, and M. Sussman, 2004. Three dimensional numerical simulations of the motion of a gas bubble rising in viscous liquids. *J. Chem. Eng. Japan*, **37**:968–975.

E. Olsson, G. Kreiss and S. Zahedi, 2007. A conservative level set method for two phase flow II. *J. Comput. Phys.*, **225**: 785–807.

S. A. Orszag, 1974. Fourier series on spheres. *Mon. Weather Rev.*, **102**:56–75.

W. R. Osborn, E. Orlandini, M. R. Swift, J. M. Yeomans, and J. R. Banavar, 1995. Lattice Boltzmann study of hydrodynamic spinodal decomposition. *Phys. Rev. Lett.*, **75**:4031.

C. W. Oseen, 1927. *Neuere Methoden und Ergebnisse in der Hydrodynamik.* Akademische Verlagsgesellschaft M. B. H., Leipzig.

S. Osher and F. Solomon, 1982. Upwind difference schemes for hyperbolic conservation laws. *Math. Comput.*, **38**:339–374.

S. Osher and R. Fedkiw, 2003. *Level Set Methods and Dynamic Implicit Surfaces.* Berlin, Springer Verlag.

S. Osher and J. A. Sethian, 1988. Fronts propagating with curvature-dependent speed: Algorithms based on Hamilton–Jacobi formulations. *J. Comput. Phys.*, **79(1)**:12–49.

M. R. Overholt and S. B. Pope, 1998. A deterministic forcing scheme for direct numerical simulations of turbulence. *Comput. Fluids*, **27**:11–28.

H. Paillere, A. Kumbaro, C. Viozat, S. Clerc, A. Broquet, and S. Corre, 2001. A comparison of Roe, VFFC, and AUSM+ schemes for two-phase water/steam flows. In E. F. Toro, editor, *Godunov Methods: Theory and Applications*, pages 677–683. Dordrecht, Kluwer.

B. J. Palmer and D. R. Rector, 2000. Lattice Boltzmann algorithm for simulating thermal two-phase flow. *Phys. Rev. E*, **61**:5295.

C. Pan, M. Hilpert, and C. T. Miller, 2001. Lattice-Boltzmann simulation of multiphase flow in water-wet porous media. In *In EOS Transactions No. 82(47), AGU Fall Meeting*, page F395. AGU, Washington, DC.

C. Pan, L.-S. Luo, and C. T. Miller, 2006. An evaluation of lattice Boltzmann methods for porous medium flow simulation. *Comput. Fluids* (to appear)

N. A. Patankar, 1997. Numerical simulation of particulate two-phase flow. Ph. D. thesis, University of Pennsylvania.

S. V. Patankar, 1980. *Numerical Heat Transfer and Fluid Flow. Comput. Fluids* (to appear), London, Taylor & Francis.

N. A. Patankar and H. H. Hu, 1996. Two-dimensional periodic mesh generation, in *Proc. 5th Int. Conf. on Numerical Grid Generation in Computational Field Simulations*, 1175–1184.

N. A. Patankar and H. H. Hu, 1997. Chain snapping in Newtonian and viscoelastic fluids. *Proc. 6th International Symposium on Liquid–Solid Flows*, ASME-FEDSM 97-3185, 1–7.

N. A. Patankar, P. Singh, D. D. Joseph, R. Glowinski and T.-W. Pan, 1999. A new formulation of the distributed Lagrange multiplier/fictitious domain method for particulate flows. *Int. J. Multiphase Flow*, submitted.

P. D. Patel, E. S. G. Shaqfeh, J. E. Butler, V. Cristini, J. Bławzdziewicz, and M. Loewenberg, 2003. Drop breakup in the flow through fixed fiber beds: An experimental and computational investigation. *Phys. Fluids*, **15**:1146–57.

R. B. Pember, J. Bell, P. Colella, W. Y. Crutchfield, and M. L. Welcome, 1995. An adaptive Cartesian grid method for unsteady compressible flow in irregular regions. *J. Comput. Phys.*, **120**:278.

M. Perić, R. Kessler, and G. Scheuerer, 1988. Comparison of finite-volume nuemrical methods with staggered and colocated grids. *Computers Fluids*, **16**:389–403.

J. B. Perot, 1993. An analysis of the fractional step method. *J. Comput. Phys.*, **108**:51–58.

C. S. Peskin, 1977. Numerical analysis of blood flow in the heart. *J. Comput. Phys.*, **25**:220.

L. Piegl and W. Tiller, 1997. *The NURBS Book*. New York, Springer.

J. E. Pilliod and E. G. Puckett, 1997. Second-order accurate volume-of-fluid algorithms for tracking material interfaces. Lawrence Berkeley National Laboratory Technical Report No. LBNL-40744 (submitted to *J. Comput. Phys.*).

H. Pokharna, M. Mori, and V. H. Ransom, 1997. Regularization of two-phase flow models: a comparison of numerical and differential approaches. *J. Comput. Phys.*, **134**:282–295.

C. Pozrikidis, 1992. *Boundary Integral and Singularity Methods for Linearized Viscous Flow*. Cambridge, Cambridge University Press.

C. Pozrikidis, 1997. *Introduction to Theoretical and Computational Fluid Dynamics.* New York, Oxford University Press.

W. H. Press, B. P. Flannery, S. A. Teukolsky, and W. T. Vetterling, 1992. *Numerical Recipes in FORTRAN: The Art of Scientific Computing.* Cambridge University Press, Cambridge, England, 2nd edn.

A. Prosperetti and A. V. Jones, 1984. Pressure forces in disperse two-phase flows. *Int. J. Multiphase Flow*, **10**:425–440.

A. Prosperetti and J. V. Satrape, 1990. Stability of two-phase flow models. In D. D. Joseph and D. G. Schaeffer, editors, *Two Phase Flows and Waves*, pages 98–117. New York, Springer.

A. Prosperetti and D. Z. Zhang, 1996. Disperse phase stress in two-phase flow. *Chem. Eng. Comm.*, **141–142**:387–398.

L. Proudman and J. R. A. Pearson, 1957. Expansions at small Reynolds numbers for the flow past a sphere and a circular cylinder. *J. Fluid Mech.*, **2**:237–262.

R. J. Pryor, D. R. Liles, and J. H. Mahaffy, 1978. Treatment of water packing effects. *Trans. Amer. Nucl. Soc.*, **30**:208–209.

Y. Qian, D. d'Humires, and P. Lallemand, 1992. Lattice BGK models for Navier–Stokes equation. *Europhys. Lett.*, **17**:479.

W. C. Reade and L. R. Collins, 2000. Effect of preferential concentration on turbulent collision rates. *Phys. Fluids*, **12**:2530–2540.

L. E. Reichl, 1998. *A Modern Course in Statistical Physics.* New York, Wiley.

RELAP5-3D Code Development Team, 2003. RELAP5-3D Code Manual, Revision 2.2. Technical Report INEEL-EXT-98-00834, INEEL.

M. Renardy and R. C. Rogers, 1993. *An Introduction to Partial Differential Equations.* New York, Springer, 2nd edition.

Y. Renardy and M. Renardy, 2002. PROST: A parabolic reconstruction of surface tension for the volume-of-fluid method. *J. Comput. Phys.*, **183**:400–421.

C. M. Rhie and W. L. Chow, 1983. Numerical study of the turbulent flow past an airfoil with trailing edge separation. *AIAA J.*, **21**:1525–1532.

J. F. Richardson and W. N. Zaki, 1954. Sedimentation and fluidization. I. Fundamental equations and wave propagation. *Trans. Inst. Chem. Eng.*, **32**:35–53.

W. J. Rider and D. B. Kothe, 1995. Stretching and tearing interface tracking methods. In *AIAA Paper 95-1717*.

A. M. Roma, C. S. Peskin, and M. J. Berger, 1999. An adaptive version of the immersed boundary method. *J. Comput Phys.*, **153**:509–534.

J. E. Romate, 1998. An approximate Riemann solver for a two-phase flow model with nuemrically given slip relation. *Comput. Fluids*, **27**:455–477.

J. E. Romate, 2000. Aspects of a numerical procedure for two-phase flow models. In H. Freistühler and G. Warnecke, editors, *Hyperbolic Problems: Theory, Numerics, Applications*, volume 2, pages 801–810, Basel, Birkhäuse.

D. Rothman and J. Keller, 1988. Immiscible cellular-automaton fluids. *J. Stat. Phys.*, **52**:1119.

D. Rothman and S. Zaleski, 2004. *Lattice-Gas Cellular Automata.* Cambridge University Press.

J. S. Rowlinson and B. Widom, 1982. *Molecular Theory of Capillarity.* Oxford University Press.

M. Rudman, 1997. Volume-tracking methods for interfacial flow calculations. *Int. J. Numer. Meth. Fluids*, **24**:671–691.

R. R. Rudolff and W. D. Bachalo, 1988. Measurement of droplet drag coefficients in polydispersed turbulent flow fields. *AIAA Paper*, 88-0235.

G. Ryskin and L. G. Leal, 1983. Orthogonal mapping. *J. Comput. Phys.*, **50**:71–100.

G. Ryskin and L. G. Leal, 1984. Numerical solution of free-boundary problems in fluid mechanics. Part 2. Buoyancy-driven motion of a gas bubble through a quiescent liquid. *J. Fluid Mech.*, **148**:19–35.

Y. Saad and M. H. Schultz, 1986. GMRES: A generalized minimal residual algorithm for solving nonsymmetric linear systems. *SIAM J. Sci. Stat. Comput.*, **7**:856–69.

Y. Saad, 1996. *Iterative Methods for Sparse Linear Systems*. PWS Publishing.

Y. Saad, 1996, *Iterative Methods for Sparse Linear Systems*. PWS Publishing.

P. G. Saffman, 1965. The lift on a small sphere in a slow shear flow. *J. Fluid Mech.*, **22**:385–400.

P. G. Saffman, 1968. Corrigendum to "The lift on a small sphere in a slow shear flow". *J. Fluid Mech.*, **31**:624.

P. G. Saffman, 1973. On the settling speed of free and fixed suspension. *Stud. Appl. Math.*, **52**:115–127.

E. M. Saiki and S. Biringen, 1994. Numerical simulation of particle effects on boundary layer flow, in Transition, Turbulence and Combustion, ed. M. Y. Hussaini, T. B. Gatski, and T. L. Jackson. Dordrecht, Kluwer Academic.

E. M. Saiki and S. Biringen, 1996. Numerical simulation of a cylinder in uniform flow: application of a virtual boundary method. *J. Comput. Phys.*, **123**:450–465.

A. S. Sangani and A. Acrivos, 1982. Slow flow through a periodic array of spheres. *Int. J. Multiphase Flow*, 8 (4):343–360.

A. S. Sangani, D. Z. Zhang, and A. Prosperetti, 1991. The added mass, Basset, and viscous drag coefficients in nondilute bubbly liquids undergoing small-amplitude oscillatory motion. *Phys. Fluids*, **A3**:2955–2970.

A. S. Sangani and A. K. Didwania, 1993. Dynamic simulations of flows of bubbly liquids at large Reynolds numbers. *J. Fluid Mech.*, **250**:307–337.

A. S. Sangani and A. Prosperetti, 1993. Numerical simulation of the motion of particles at large Reynolds numbers, in *Particulate Two-Phase Flow* (M. C. Roco, ed.), 971–998. London: Butterworth–Heinemann.

S. Sankagiri and G. A. Ruff, 1997. Measurement of sphere drag in high turbulent intensity flows. *Proc. ASME FED*, FED-Vol **244**:277–282.

K. Sankaranarayanan, X. Shan, I. G. Kevrekidis, and S. Sundaresan, 1999. Bubble flow simulations with the lattice Boltzmann method. *Chem. Eng. Sci.*, **54**:4817.

V. Sarin and A. Sameh, 1998. An efficient iterative method for the generalized Stokes problem. *SIAM J. Sci. Comput.*, **19**:206–226.

V. K. Saul'ev, 1963. On solving boundary-value problems on high-performance computers by fictitious-domain methods. *Sib. Math. J.*, **4**:912–925.

R. Saurel and R. Abgrall, 1999. A multiphase Godunov method for compressible multifluid and multiphase flow. *J. Comput. Phys.*, **150**:425–467.

S. B. Savage, 1987. Personal communication to S. Sundaresan.

R. Scardovelli and S. Zaleski, 1999. Direct numerical simulation of free-surface and interfacial flow. *Ann. Rev. Fluid Mech.*, **31**:567–603.

D. G. Schaeffer, 1987. Instability in the evolution equations describing incompressible granular flow. *J. Diff. Eq.*, **66**:19–50.

L. Schiller and Z. Naumann, 1935. A drag coefficient correlation. *Z. Ver. Deutsch. Ing.*, **77**:318.

J. A. Sethian, 1999. *Level Set Methods and Fast Marching Methods*. Cambridge University Press, 2nd edition.

X. W. Shan and H. D. Chen, 1993. Lattice Boltzmann model for simulating flows with multiple phases and components. *Phys. Rev. E*, **47**:1815.

X. W. Shan and H. D. Chen, 1994. Simulation of non ideal gases and liquid-gas phase transitions by the lattice Boltzmann equation. *Phys. Rev. E*, **49**:2941.

K. Shariff, 1993. Comment on "Coordinate singularities" by P. R. Spalart. Unpublished.

J. Shen and D. Yoon, 2004. A freeform shape optimization of complex structures represented by arbitrary polygonal or polyhedral meshes. *Int. J. Numer. Meth. Engng*, **60**:2441–66.

W. Z. Shen, J. A. Michelsen, and J. N. Sørensen, 2001. Improved Rhie–Chow interpolation for unsteady flow computations. *AIAA J.*, **39**:2406–2409.

W. Z. Shen, J. A. Michelsen, N. N. Sørensen, and J. N. Sørensen, 2003. An improved SIMPLEC method on collocated grids for steady and unsteady flow computations. *Numer. Heat Transfer*, **B43**:221–239.

S. Shin, S. I. Abdel-Khalik, V. Daru, and D. Juric, 2005. Accurate representation of surface tension using the level contour reconstruction method. *J. Comput. Phys.*, **203**:493–516.

S. Shirayama, 1992. Flow past a sphere: topological transitions of the vorticity field. *AIAA J.*, **30**:349–358.

C. W. Shu and S. Osher, 1989. Efficient implementation of essentially non-oscillatory shock capturing schemes, II. *J. Comput. Phys.*, **83**:32–78.

A. Sierou and J. F. Brady, 2001. Accelerated Stokesian dynamics simulations. *J. Fluid Mech.*, **448**:115–46.

O. Simonin, 1996. *Continuum Modeling of Dispersed Two-Phase Flows*. Combustion and Turbulence in Two-Phase Flows, VKI Lecture Series, Vol. 2. von Kármán Institute for Fluid Dynamics, Bruxelles.

P. Singh, D. D. Joseph, T. I. Hesla, R. Glowinski and T.-W. Pan, 1999. A distributed Lagrange multiplier/fictitious domain method for viscoelastic particulate flows. *J. Non-Newtonian Fluid Mech.*, submitted.

J. C. Slattery, 1967. Flow of viscoelastic fluids through porous media. *A. I. Ch. E. J.*, **13**:1066–1071.

P. Smereka. Level set methods for two-fluid flows. Lecture notes; INRIA short course.

L. L. Smith *et al.*, 1980. SIMMER-II: A Computer Program for LMFBR Disrupted Core Analysis. Technical Report LA-7515-M, NUREG/CR-0453, Rev. 1, Los Alamos National Laboratory.

Y. Sone, 2002. *Kinetic Theory and Fluid Dynamics*. Birkhäuser, Boston.

D. B. Spalding, 1979. Calculation of two-dimensional two-phase flows. In F. Durst et al., editor, *Heat and Mass Transfer in Chemical Process and Energy Engineering Systems*, Hemisphere.

D. B. Spalding, 1983. Developments in the IPSA procedure for numerical computation of multiphase flow phenomena. In T. M. Shi, editor, *Proceedings of the Second National Symposium on Numerical Methodologies in Heat Transfer*, pages 421–436. Hemisphere.

J. W. Spore, S. J. Jolly-Woodruff, T. K. Knight, R. A. Nelson, K. O. Pasamehmetoglu, R. G. Steinke, and C. Unal, 1993. TRAC-PF1/MOD2 code manual. Technical Report NUREG/CR-5673, U. S. Nuclear Regulatory Commission.

P. G. Squire and M. E. Himmel, 1979. Hydrodynamics and protein hydration. *Arch. Biochem. Biophys.*, **196**:165–77.

K. D. Squires and J. K. Eaton, 1990. Particle response and turbulence modification in isotropic turbulence. *Phys. Fluids*, **2**:1191–1203.

K. D. Squires and J. K. Eaton, 1991. Preferential concentration of particles by turbulence. *Phys. Fluids*, **3**:1169–1179.

H. Staedke, G. Francehllo, B. Worth, U. Graf, P. Romstedt, A. Kumbaro, J. García-Cascales, H. Paillère, H. Deconinck, M. Richhiuto, B. Smith, F. De Cachard, E. F. Toro, E. Romenski, and S. Mimouni, 2005. Advanced three-dimensional two-phase flow simulation tools for application to reactor safety (ASTAR). *Nucl. Eng. Des.*, **235**:379–400.

J. Sterling and S. Chen, 1996. Stability analysis of lattice Boltzmann methods. *J. Comp. Phys.*, **123**:196.

H. B. Stewart and B. Wendroff, 1984. Two-phase flow: models and methods. *J. Comput. Phys.*, **56**:363–409.

H. A. Stone, 1994. Dynamics of drop deformation and breakup in viscous fluids. *Annu. Rev. Fluid Mech.*, **26**:65–99.

G. Strang, 1968. On the construction and comparison of difference schemes. *SIAM J. Numer. Anal.*, **5**:506–517.

C. L. Streett and M. Macaraeg, 1989. Spectral multi-domain for large-scale fluid dynamics simulations. *Appl. Numer. Math.*, **6**:123–140.

S. Succi, 2001. *Lattice Boltzmann Equation for Fluid Dynamics and Beyond*. Oxford University Press.

S. Sundaram and L. R. Collins, 1997. Collision statistics in an isotropic particle-laden turbulent suspension. Part 1. Direct numerical simulations. *J. Fluid Mech.*, **335**:75–109.

S. Sundaram and L. R. Collins, 1999. A numerical study of the modulation of isotropic turbulence by suspended particles. *J. Fluid Mech*, **379**:105–143.

S. Sundaresan, 2000. Perspective: Modeling the hydrodynamics of multiphase flow reactors: Current status and challenges. *A. I. Ch. E. J.*, **46**:1102–1105.

M. Sussman, P. Smereka, and S. Osher, 1994. A level set approach for computing solutions to incompressible two-phase flows. *J. Comput. Phys.*, **114**:146–159.

M. Sussman, A. S. Almgren, J. B. Bell, P. Colella, L. H. Howell, and M. L. Welcome, 1999. An adaptive level set approach for incompressible two-phase flows. *J. Comput Phys.*, **148**:81–124.

M. Sussman and E. Fatemi, 1999. An efficient, interface preserving level set redistancing algorithm and its application to interfacial incompressible fluid flow. *SIAM J. Sci. Comput.*, 20(4):1165–1191.

M. Sussman, 2003. A second order coupled levelset and volume of fluid method for computing growth and collapse of vapor bubbles. *J. Comput. Phys.*, **187**:110–136.

M. Sussman and M. Y. Hussaini, 2003. A discontinuous spectral element method for the level set equation. *J. Sci. Comput.*, **19(1-3)**:479–500.

M. Sussman, 2005. A parallelized, adaptive algorithm for multiphase flows in general geometries. *Comput. Struct.*, **83**:435–444.

B. Swartz, 1999. Good neighborhoods for multidimensional Van Leer limiting. *J. Comput. Phys.*, **154**:237–241.

P. K. Sweby, 1984. High resolution schemes using flux limiters for hyperbolic conservation laws. *SIAM J. Numer. Anal.*, **21**:995–1011.

R. A. Sweet, 1974. A generalized cyclic reduction algorithm. *SIAM J. Numer. Anal.*, **11**:506–520.

M. R. Swift, W. R. Osborn, and J. M. Yeomans, 1995. Lattice Boltzmann simulation of nonideal fluids. *Phys. Rev. Lett.*, **75**:830.

M. R. Swift, E. Orlandini, W. R. Osborn, and J. M. Yeomans, 1996. Lattice Boltzmann simulation of liquid–gas and binary fluid system. *Phys. Rev. E.*, **54**:5041.

M. Syamlal, W. Rogers, and T. J. O'Brien, 1993. *MFIX* Documentation: Theory Guide. Technical Report DOE/METC-94/1004 (DE9400087), Morgantown Energy Technology Center, Morgantown WV, Available from http://www. mfix. org.

M. Syamlal, 1998. *MFIX* Documentation: Numerical Technique. Technical Report DOE/MC31346-5824 (DE98002029), Morgantown Energy Technology Center, Morgantown WV, Available from http://www. mfix. org.

S. Takagi, H. N. Ogz, Z. Zhang, and A. Prosperetti, 2003. Physalis: A new method for particle simulation. Part II. Two-dimensional Navier–Stokes flow around cylinders. *J. Comput. Phys.*, **187**:371–390.

H. Takewaki, A. Nishiguchi, and T. Yabe, 1985. Cubic interpolated pseudo-particle method (CIP) for solving hyperbolic-type equations. *J. Comput. Phys.*, **61**:261–268.

H. Takewaki and T. Yabe, 1987. The cubic-interpolated pseudo particle (CIP) method: application to nonlinear and multi-dimensional hyperbolic equations. *J. Comput. Phys.*, **70**:355–372.

H. S. Tang and D. Huang, 1996. A second-order accurate capturing scheme for 1-D inviscid flows of gas and water with vacuum zones. *J. Comput. Phys.*, **128**:301–318.

C. M. Tchen, 1947. Mean values and correlation problems connected with the motion of small particles suspended in a turbulent fluid. PhD thesis, Technical University of Delft.

A. Ten Cate and S. Sundaresan, 2005. Analysis of unsteady forces in ordered arrays. *J. Fluid Mech.*, submitted.

S. Teng, Y. Chen, and H. Ohashi, 2000. Lattice Boltzmann simulation of multiphase fluid flows through the total variation diminishing with artificial compression scheme. *Int. J. Heat and Fluid Flow*, **21**:112–121.

T. E. Tezduyar, J. Liou, and M. Behr, 1992a. A new strategy for finite element computations involving moving boundaries and interfaces – the DSD/ST procedure: I. The concept and the preliminary numerical tests. *Comput. Meths. Appl. Mech. Engng.*, **94**:339–351.

T. E. Tezduyar, J. Liou, M. Behr and S. Mittal, 1992b. A new strategy for finite element computations involving moving boundaries and interfaces – the DSD/ST procedure: II. Computation of free-surface flows, two-liquid flows, and flows with drafting cylinders. *Comput. Meths. Appl. Mech. Engng.*, **94**:353–371.

J. W. Thomas, 1995. *Numerical Partial Differential Equations.* Berlin, Springer.

P. J. Thomas, 1992. On the influence of the Basset history force on the motion of a particle through a fluid. *Phys. Fluids*, **A4**:2090–2093.

I. Tiselj and S. Petelin, 1997. Modelling of two-phase flow with second-order accurate scheme. *J. Comput. Phys.*, **136**:503–521.

I. Tiselj and S. Petelin, 1998. First and second-order accurate schemes for two-fluid models. *J. Fluids Engng.*, **120**:363–368.

A. G. Tomboulides and S. A. Orszag, 2000. Numerical investigation of transitional and weak turbulent flow past a sphere. *J. Fluid Mech.*, **416**:45–73.

A. Tomiyama, T. Matsuoka, T. Fukuda, and T. Sakaguchi, 1995. A simple numerical method for solving an incompressible two-fluid model in general curvilinear coordinate system. In A. Serizawa, T. Fukano, and J. Bataille, editors, *Multiphase Flow '95*, pages NU-23–NU-30, Kyoto.

A. Tomiyama and N. Shimada, 2001. A nuemrical method for bubbly flow simulation based on a multi-fluid model. *J. Press. Vessel. Technol.*, **123**:510–516.

E. F. Toro, 1998. Primitive, conservative and adaptive schemes for hyperbolic conservation laws. In E. F. Toro and J. F. Clarke, editors, *Numerical Methods for Wave Propagation*, pages 323–385. Dordrecht, Kluwer.

E. F Toro, 1999. *Riemann Solvers and Numerical Methods for Fluid Dynamics.* Berlin, Springer, 2nd edition.

I. Toumi, 1992. A weak formulation of Roe's approximate Riemann solver. *J. Comput. Phys.*, **102**:360–373.

I. Toumi, 1996. An upwind numerical method for two-fluid two-phase flow models. *Nucl. Sci. Eng.*, **123**:147–168.

I. Toumi and A. Kumbaro, 1996. An approximate linearized Riemann solver for a two-fluid model. *J. Comput. Phys.*, **124**:286–300.

I. Toumi, A. Bergeron, D. Gallo, E. Royer, and D. Caruge, 2000. Flica-**4**: a three-dimensional two-phase flow computer code with advanced numerical methods for nuclear applications. *Nucl. Eng. Des.*, 200:139–155.

L. B. Tran and H. S. Udaykumar, 2004. A particle-level set-based sharp interface Cartesian grid method for impact, penetration, and void collapse. *J. Comput. Phys.*, **193**:469–510.

J. A. Trapp and R. A. Riemke, 1986. A nearly implicit hydrodynamic numerical scheme for two-phase flows. *J. Comput. Phys.*, **66**:62–82.

G. Tryggvason, 1988. Numerical simulations of the Rayleigh–Taylor instability. *J. Comput. Phys.*, **75**:253.

G. Tryggvason and S. O. Unverdi, 1990. Computations of three-dimensional Rayleigh–Taylor instability. *Phys. Fluids A*, **2**:656.

G. Tryggvason, B. Bunner, A. Esmaeeli, D. Juric, N. Al-Rawahi, W. Tauber, J. Han, S. Nas, and Y.-J. Jan, 2001. A front tracking method for the computations of multiphase flow. *J. Comput. Phys.*, **169**:708–759.

H. S. Udaykumar, H. C. Kan, W. Shyy, and R. Tran-Son-Tay, 1997. Multiphase dynamics in arbitrary geometries on fixed Cartesian grids. *J. Comput. Phys.*, **137**:366–405.

H. S. Udaykumar, R. Mittal, and W. Shyy, 1999. Computation of solid–liquid phase fronts in the sharp interface limit on fixed grids. *J. Comput. Phys.*, **153**:535–574.

H. S. Udaykumar, R. Mittal, P. Rampunggoon, and A. Khanna, 2001. A sharp interface Cartesian grid method for simulating flows with complex moving boundaries. *J. Comput. Phys.*, **174**:345–380.

P. H. T. Uhlherr and C. G. Sinclair, 1970. The effect of freestream turbulence on the drag coefficients of spheres. *Proc. Chemca.*, **1**:1–12.

S. O. Unverdi and G. Tryggvason, 1992. A front-tracking method for viscous, incompressible, multi-fluid flows. *J. Comput Phys.*, **100**:25–37.

B. van Leer, 1977. Toward the ultimate conservative difference scheme. IV. A new approach to numerical convection. *J. Comput. Phys.*, **23**:276–299.

B. van Leer, 1982. Flux vector splitting for the Euler equations. In H. Araki, J. Ehlers, K. Hepp, R. Kippenhahn, H. A. Weidenmüller, and J. Zittarz, editors, *8th International Conference on Numerical Methods in Fluid Dynamics. Lecture Notes in Physics*, Vol. 170, pages 507–512, Berlin, Springer.

W. B. VanderHeyden and B. A. Kashiwa, 1998. Compatible fluxes for van Leer advection. *J. Comput. Phys.*, **146**:1–28.

O. Vermorel, B. Bédat, O. Simonin, and T. Poinsot, 2003. Numerical study and modelling of turbulence modulation in a particle-laden slab flow. *J. Turbulence*, **4**: Paper No. 025, http://jot. iop. org.

M. Verschueren, F. N. van de Vosse, and H. E. H. Meijer, 2001. Diffuse-interface modelling of thermo-capillary flow instabilities in a hele-shaw cell. *J. Fluid Mech.*, **434**:153–166.

P. Vlahovska, J. Bławzdziewicz, and M. Loewenberg, 2005. Deformation of surfactant-covered drops in linear flows. In review.

D. J. Vojir and E. E. Michaelides, 1994. Effect of the history term on the motion of rigid spheres in a viscous fluid. *Int. J. Multiphase Flow*, **20**:547–556.

G. W. Wallis, 1969. *One-Dimensional Two-Phase Flow*. New York, McGraw-Hill.

L. P. Wang, A. S. Wexler, and Y. Zhou, 2000. Statistical mechanical description and modelling of turbulent collision of inertial particles. *J. Fluid Mech.*, **415**:117–153.

L. P. Wang and M. R. Maxey, 1993. Settling velocity and concentration distribution of heavy particles in homogeneous isotropic turbulence. *J. Fluid Mech.*, **256**:27–68.

W. D. Warnica, M. Renksizbulut, and A. B. Strong, 1994. Drag coefficient of spherical liquid droplets. *Exp. Fluids*, **18**:265–270.

C. Y. Wen and Y. H. Yu, 1966. Mechanics of fluidization. *Chem. Engng. Prog. Symp. Ser.*, **62**:100–111.

P. Wesseling, 2001. *Principles of Computational Fluid Dynamics*. New York, Springer.

P. Wesseling, 2004. *An Introduction to Multigrid Methods—Corrected Reprint*. R. T. Edwards, Philadelphia.

S. Whitaker, 1969. Advances in the theory of fluid motion in porous media. *Ind. Engng. Chem.*, **61**:14–28.

G. B. Whitham, 1974. *Linear and Nonlinear Waves*. New York, Wiley.

L. van Wijngaarden, 1976. Hydrodynamic interaction between gas bubbles in liquid. *J. Fluid Mech.*, **77**:27–44.

R. C. Wilson and E. R. Hancock, 2000. Bias-variance analysis for controlling adaptive surface meshes. *Comput. Vision Image Understanding*, **77**:25–47.

J.-S. Wu and G. M. Faeth, 1994a. Effect of ambient turbulence intensity on sphere wakes at intermediate Reynolds numbers. *AIAA J.*, **33**:171–173.

J.-S. Wu and G. M. Faeth, 1994b. Sphere wakes at moderate Reynolds numbers in aturbulent environment. *AIAA J.*, **32**:535–541.

Q. Wu, S. Kim, M. Ishii, and S. G. Beus, 1999. One-group interfacial area transport in vertical bubbly flow. *Int. J. Heat Mass Transfer*, **41**:1103–1112.

T. Yabe, F. Xiao, and T. Utsumi, 2001. The constrained interpolation profile (CIP) method for multi-phase analysis. *J. Comput. Phys.*, **169**:556–593.

N. N. Yanenko, 1971. *The Method of Fractional Steps for Solving Multi-Dimensional Problems of Mathematical Physics in Several Variables*. Berlin, Springer-Verlag.

A. Yang, C. T. Miller, and L. D. Turcoliver, 1996. Simulation of correlated and uncorrelated packing of random size spheres. *Phys. Rev. E*, **53**:1516–1524.

B. Yang, A. Prosperetti, and S. Takagi, 2003. The transient rise of a bubble subject to shape or volume changes. *Phys. Fluids*, **15**:2640–2648.

T. Ye, R. Mittal, H. S. Udaykumar, and W. Shyy, 1999. An accurate Cartesian grid method for viscous incompressible flows with complex immersed boundaries. *J. Comput. Phys.*, **156**:209–240.

S. Y. K. Yee, 1981. Solution of Poisson's equation on a sphere by double Fourier series,. *Mont. Weather Rev.*, **109**:501–505.

P. K. Yeung and S. B. Pope, 1989. Lagrangian statistics from direct numerical simulations of isotropic turbulence. *J. Fluid Mech.*, **207**:531–586.

S. Yon and C. Pozrikidis, 1998. A finite-volume/boundary-element method for flow past interfaces in the presence of surfactants, with application to shear flow past a viscous drop. *Comput. Fluids*, **27**:879–902.

D. L. Youngs, 1982. Time dependent multimaterial flow with large fluid distortion. In K. M. Morton and M. J. Baines, editors, *Numerical Methods for Fluid dynamics*, pages 27–39, New York, Academic Press. Institute of Mathematics and its Applications.

D. L. Youngs, 1984. Numerical simulation of turbulent mixing by Rayleigh–Taylor instability. *Physica D*, **12**:32.

D. Yu, R. Mei, W. Shyy, and L.-S. Luo, 2002. Force evaluation in the lattice Boltzmann method involving curved geometry. *Phys. Rev. E*, **65**:041203.

H. Yu, L.-S. Luo, and S. S. Girimaji, 2005. MRT lattice Boltzmann method for LES of turbulence: Computation of square jet flow. *Comput. Fluids* (to appear).

S. T. Zalesak, 1979. Fully multidimensional flux-corrected transport algorithms for fluids. *J. Comput. Phys.*, **31**:335–362.

N. A. Zarin and J. A. Nicholls, 1971. Sphere drag in solid rockets – non-continuum and turbulence effects. *Comb. Sci. Tech.*, **3**:273–280.

E. Zauderer, 1989. *Partial Differential Equations of Applied Mathematics*. New York, Wiley, 2nd edition.

D. Zhang, R. Zhang, and S. Chen, 2000. Pore scale study of flow in porous media: Scale dependency, REV, and statistical REV. *Geophys. Res. Lett.*, **27**:1195.

D. Z. Zhang and W. B. VanderHeyden, 2002. The effects of mesoscale structures on the macroscopic momentum equations for two-phase flows. *Int. J. Multiphase Flow*, **28**:805–822.

D. Z. Zhang, W. B. VanderHeyden, Q. Zou, X. Ma and P. T. Giguere, 2007. CartaBlanca Theory Manual: Multiphase Flow Equations and Numerical Methods. Technical Report LA–UR–07–3621. Los Alamos National Laboratory, http://www.lanl.gov/projects/CartaBlanca/webdocs/theory.pdf

D. Z. Zhang and A. Prosperetti, 1994. Averaged equations for inviscid disperse two-phase flow. *J. Fluid Mech.*, **267**:185–219.

Q. Zhang and M. J. Graham, 1998. A numerical study of Richtmyer–Meshkov instability driven by cylindrical shocks. *Phys. Fluids*, **10**:974.

Y. Zhang and J. M. Reese, 2003. The drag force in two-fluid models of gas–solid flows. *Chem. Engng. Sci.*, **58**:1641–1644.

Z. Zhang and A. Prosperetti, 2003. A method for particle simulations. *J. Appl. Mech.*, **70**:64–74.

Z. Zhang and A. Prosperetti, 2005. A second-order method for three-dimensional particle simulation. *J. Comput. Phys.*, To appear.

H. Zhao and A. J. Pearlstein, 2002. Stokes-flow computation of the diffusion coefficient and rotational diffusion tensor of lysozyme, a globular protein. *Phys. Fluids*, **14**:2376–87.

X. Zheng, J. Lowengrub, A. Anderson, and V. Cristini, 2005. Adaptive unstructured volume remeshing. II. Application to two- and three-dimensional level-set simulations of multiphase flow. *J. Comput. Phys.*, **208**:626–650.

H.-X. Zhou, 2001. A unified picture of protein hydration: prediction of hydrodynamic properties from known structures. *Biophys. Chem.*, **93**:171–9.

M.-Y. Zhu, 1999. Direct Numerical simulation of particulate flow of Newtonian and viscoelastic fluids. Ph. D. thesis, University of Pennsylvania.

A. Z. Zinchenko and R. H. Davis, 2000. An efficient algorithm for hydrodynamical interaction of many deformable drops. *J. Comput. Phys*, **157**:539–87.

A. Z. Zinchenko and R. H. Davis, 2002. Shear flow of highly concentrated emulsions of deformable drops by numerical simulations. *J. Fluid Mech.*, **455**:21–62.

A. Z. Zinchenko, M. A. Rother, and R. H. Davis, 1997. A novel boundary-integral algorithm for viscous interaction of deformable drops. *Phys. Fluids*, **9**:1493–1511.

A. Z. Zinchenko, M. A. Rother, and R. H. Davis, 1999. Cusping, capture, and breakup of interacting drops by a curvatureless boundary-integral algorithm. *J. Fluid Mech.*, **391**:249–92.

N. Zuber, 1964. On the dispersed two-phase flow in the laminar flow regime. *Chem. Eng. Sci.*, **19**:897–917.

Index

Printed in the United States
by Baker & Taylor Publisher Services